Climate Changes Mitigation and Sustainable Bioenergy Harvest Through Animal Waste

Muhammad Arshad
Editor

Climate Changes Mitigation and Sustainable Bioenergy Harvest Through Animal Waste

Sustainable Environmental Implications of Animal Waste

Springer

Editor
Muhammad Arshad
Department of Basic Sciences
University of Veterinary and Animal
Sciences (Jhang-campus)
Lahore, Pakistan

ISBN 978-3-031-26223-4 ISBN 978-3-031-26224-1 (eBook)
https://doi.org/10.1007/978-3-031-26224-1

This Springer imprint is published by the registered company Springer Nature Switzerland AG
The registered company address is: Gewerbestrasse 11, 6330 Cham, Switzerland

The present book is dedicated to
MY DAUGHTERS
Tayyba Arshad and Tooba Arshad
A life's greatest blessing

Preface

Wish to get infinite renewability is need of the hour and can be achieved when the resources are managed sustainably. Animal waste has been proved to be the best substrate for sustainable bioenergy harvest and is also ecofriendly. To produce bio-renewables, slaughterhouse waste is a blessing. Major objective of this publication is to provide awareness about the fact that animal waste is environment friendly and sustainable resource if it is carried out properly. Because it can be a heavy source of pollution which leads to add up big problems for water quality and even human health if disposed off improperly. Moreover, novel and low capital cost technologies for valorization of animal waste for sustainable bioenergy have been discussed in this book.

The intention behind this book, is to provide satisfactory solutions of animal waste management such as bioenergy production, to reclaim salt degraded lands and use through most functional approaches. Competent researchers from academia and industry described it in an efficient manner. Proficient and experienced authors were invited to explain the mitigation of climate changes through sustainable use of animal waste. I am very grateful to all the authors and reviewers for their admirable work. They must be proud of such great achievement.

Jhang, Pakistan
November 2022

Muhammad Arshad

Acknowledgements I would like to say thanks to every individual, I have had the opportunity to lead or to be led by their leadership, and for being the inspiration and foundations in completing this book. Particularly, thanks to series editor. I especially want to send my appreciation and deepest gratitude to all authors for their support in compiling this book.

Also, I am very thankful to my family, friends and students especially Aqsa Mumtaz and Sadia Batool for their support in any way.

Contents

Chapter 1
Animal Waste: An Environmentally Sustainable Management Approach

Justus Amuche Nweze, Shruti Gupta, Joseph Akor, Charles O. Nwuche, Julius Eyiuche Nweze, and Victor U. Unah

Abstract Some localities still use non-sustainable management techniques to deal with animal waste. Animal waste has the potential to be profitable if properly managed, but it can also pose severe risks to human health. The quality of the land, water and air may be threatened by improper animal waste handling. Animal manure can be adequately handled using a variety of approaches, ranging from simple, low-cost procedures to complex strategies. Microorganisms play a significant role in the multifaceted approach to sustainable animal waste management that benefits farmers, the general population and the environment. It is possible to efficiently revive contaminated areas by utilizing the unique characteristics of microorganisms. Microorganisms can be used as "miracle cures" for biodegradation and the remediation of contaminated sites. At different levels, rules and policies have been put in place in many countries to support sustainable animal manure treatment. Proper animal manure management not only reduces the amount of synthetic fertilizer required on fields, but it also contributes to lower net greenhouse gas emissions from livestock waste and has an impact on climate change. This chapter delves into the properties of various forms of animal waste and shows how microorganisms can be employed effectively for waste management and sustainability.

J. A. Nweze · J. E. Nweze (✉)
Department of Ecosystem Biology, Faculty of Science, University of South Bohemia, Ceske Budejovice, Czech Republic
e-mail: julius.nweze@bc.cas.cz

J. A. Nweze · J. Akor
Department of Science Laboratory Technology, Faculty of Physical Sciences, University of Nigeria, Nsukka, Nigeria

J. A. Nweze · S. Gupta · J. E. Nweze
Institute of Soil Biology and Biogeochemistry, Biology Centre, Czech Academy of Sciences, Ceske Budejovice, Czech Republic

C. O. Nwuche · V. U. Unah
Department of Microbiology, Faculty of Biological Sciences, University of Nigeria, Nsukka, Nigeria

M. Arshad (ed.), *Climate Changes Mitigation and Sustainable Bioenergy Harvest Through Animal Waste*, https://doi.org/10.1007/978-3-031-26224-1_1

Graphical Abstract

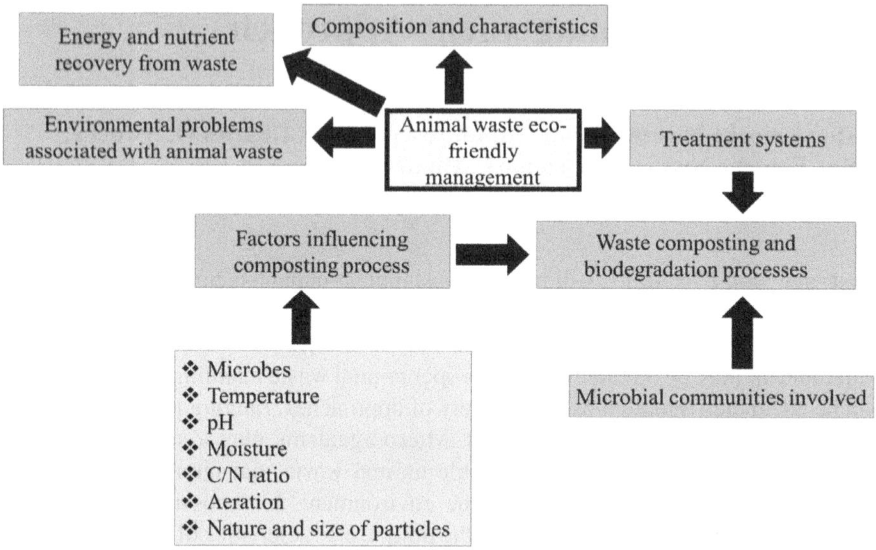

Keywords Sustainability · Organic matter · Composting · Biogas · Fertilizer · Nutrient management · Renewable energy

1.1 Introduction

Animal waste is defined as waste generated from livestock and meat production. When we think of animal waste, we usually think of the excreta of living animals. However, waste may also include wood crisps, hay, straw or other organic material, depending on the production process. The number of animal farms is growing yearly due to the increasing demand from the growing human population. In Europe, America, Australia, Africa and Asia, hectares of land are used to raise numerous herds of cattle, poultry, sheep and pigs for meat, milk, eggs and hides. Even though technological advances are mainly overtaking it, the ever-expanding agricultural industry continues to be an essential part of the global economy. Animal waste is generated in large quantities around the world every year, and if not properly collected, stored and treated, it can pollute soil, water and air. We can not only clean our environment but also save money on fertilizers if we adequately manage these animal wastes. A proper plan should be made to find a long-term solution to animal waste management (Malomo et al. 2018; Martinez et al. 2009a, b; Girotto and Cossu 2017; Arshad 2017).

Animal waste has serious adverse effects on human health and the environment, and it also raises greenhouse gas emissions and lowers water and air quality. The spread of pathogens from livestock waste into water supplies can occur through direct leakage of waste in buildings or warehouses into sewage systems or indirectly through the spread of waste onto land if not adequately treated (Penakalapati et al. 2017).

In addition, infections can enter the water phase from faeces deposited when animals graze on grasslands or dead animals. The pathway from soil to the water stream varies depending on soil type and conditions. Bacteria, protozoa and viruses are undoubtedly present in both aerosols and wastes, but how long they remain viable depends primarily on environmental conditions. Also, many livestock production practices often use antibiotics, which can enter the environment through waste and contribute to developing antibiotic-resistant microorganisms. Improper management of such waste can affect animal health, impacting disease transmission between animal production facilities and from animals to humans. The structure of the microbial population and the microbes participating in the degradation are both affected by antibiotic residues (Epps and Blaney 2016; Tasho and Cho 2016).

Composting is often an environmentally sound method of converting all animal waste into high-quality organic fertilizers for agriculture. However, the specific chemical composition of animal waste and its effects on the physicochemical and microbiological properties of compost are poorly understood. However, it is generally accepted that the type of animal waste affects microbial activity, metabolism and abundance, all of which depend on the physicochemical properties of the waste. The activity of a vast range of microorganisms that play a key role in the breakdown of organic material is required for microbial waste degradation. Bacteria and fungus are the most active and abundant microorganisms in waste degradation. Bacteria play key roles in the majority of the heat generated in compost and its breakdown, while fungi have the ability to decompose complicated polymers. While microbial populations evolving overtime during the various stages of composting has been thoroughly studied, there is little understanding of how the composition of the initial raw materials metataxonomy affects waste decomposition (Wan et al. 2021; Fernandez-Bayo et al. 2020; Akari and Uchida 2021).

The metataxonomic composition of the waste at the initial time is vital because it influences the mesophilic microbiota proliferation, which is responsible for the quick rise in decomposition temperature and the establishment of a favourable environment for successive or secondary microbes throughout the decomposition process (Akari and Uchida 2021; Sun et al. 2020). Temperature, moisture content and *C/N* ratio are among the physicochemical properties that affect microbial degradation of animal waste. However, depending on the type of raw material in animal waste, it is still difficult to determine how these different properties affect the diversity, composition and structure of microbial communities (Sun et al. 2020).

In this chapter, we have discussed (Fig. 1.1) the factors affecting the decomposition of animal waste, the characteristics of different types of animal waste, microbial composition, microbial succession and how these microorganisms can be successfully used for waste management and sustainability.

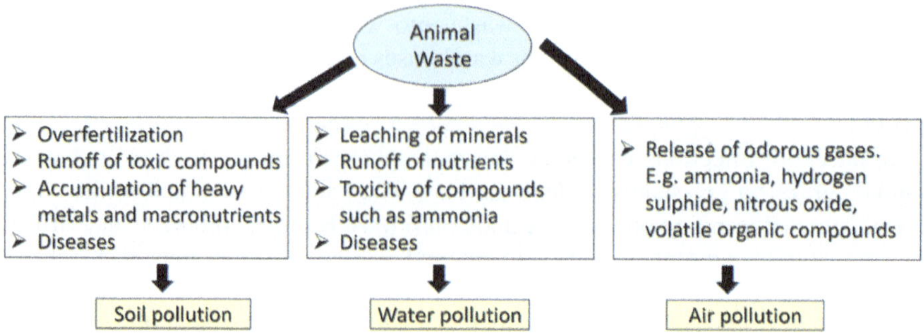

Fig. 1.1 Some environmental problems linked to improper management of animal manure

1.2 Animal Waste

1.2.1 What Are Animal Wastes?

Animal wastes are wastes generated during the production, processing, transportation and marketing of animals. They are used as a source for biomass-based conversion processes, especially in the production of biofertilizers and bioenergy. Feed waste or residues, effluents, wastes from hatcheries, slaughterhouses and manure are some possible sources of wastes generated during animal production. The most common sources of waste include effluents from dairy barns, which consist of urine, wash water, manure, feed residues and milk residues; poultry litter, which is a mixture of spilled feed, water, manure, litter material and feathers, and dairy manure, cleaning products and other wastes from animal finishing (Girotto and Cossu 2017; Girotto and Cossu 2017).

1.2.2 Characteristics and Composition of Animal Waste

Numerous factors, including the environment, the age or growth stages of the animals, the type of animal, the digestibility of the ration or feed, productivity, the content of fibre and protein, waste collection and handling methods and the amount of water in the waste, affect the production and characteristics of the billions of metric tons of waste that are produced annually by the animal production industry. The wastes are categorized as solids, slurry (liquid) and wastewater (effluent), depending on the type of stock and their physical form (Table 1.1) (Martín-Marroquín and Hidalgo 2014). Solid waste (20–25% solids), which may include livestock manure, animal carcasses or the remains of the slaughter process in abattoirs, is primarily collected by dry mucking out the waste and stacking and picking it up with a forklift, and drying or composting it. Most liquid or slurry waste comes from the animals' urine, excreta or wastewater, as well as from residues generated by washing the stalls, cages and the

Table 1.1 Forms and composition of animal waste

Waste form	Examples	Composition
Solid	Dung	Organic matter (20%), moisture (77%), nitrogen (0.32%), phosphorous (0.14%), calcium (0.4%), potassium (0.3%)
	Wasted feeding material	It includes food that is discarded or lost or uneaten
	Soiled bedding material	It includes wood shavings, straw, saw dust, paper-based bedding materials, etc.
Liquid	Urine	3–40 ml/kg bwt/Day
	Washed water	25–70 L/Animal/Day

animals themselves with water. Converting the liquid waste to solid waste requires draining the liquids either by wet-mucking or dry-mucking, followed by drying or bedding. Stable waste treatment characteristics vary depending on the solids present. Wastes with a solids content of 4–10% can usually be disposed of as a liquid, although special pumping may be required. Wastes with a solids content of 0–4% are treated as a liquid with an irrigation or flushing consistency (Martín-Marroquín and Hidalgo 2014; Eliot 2015).

In addition to what has already been said, animal species, feeding management, production capacity, nutrient intake, digestion, absorption, feed wastage (especially in pigs and poultry), disposal systems, nutrient content, other additives and environmental factors all influence the composition of animal waste. The amount and type of structural carbohydrates, proteins, nitrogen and other indigestible materials (silica) in the rations of animals whose wastes are collected have a major influence. Some wastes contain about 70% total solids, of which 95% are volatile solids, and in wastes containing slightly more than 2% organic nitrogen, the crude protein value is nearly 13%. The amino acids contained in this material account for slightly less than 40% of the crude protein value, the remainder being accounted for by the other nitrogenous non-protein compounds. Some articles have convincingly discussed the composition of animal waste on the basis of animal species (Patton and Turner 2008; Müller 1980; Nimmi n.d.).

1.3 Present-Day Environmental Problems

Livestock farming constitutes one of the major drivers of environmental pollution in recent times. The discharge of animal wastes (manures), agrochemicals, toxic and odorous gases and dissemination of different populations of microorganisms in the form of aerosols are linked to the degradation of the quality of soils, air as well as surface and ground waters in many locations around the world. Interests created as a result of the contribution of anthropogenic activities to climate change have brought livestock farming into focus due to increase in the generation of several greenhouse gases (GHG). GHG gases are known to be significantly connected to the steady decline in the global climatic and environmental conditions (Fig. 1.1). Presently, government guidelines are in force in many countries aimed at regulating livestock operations in order to be able to control the rate of emissions of environmental contaminants as a mitigation strategy. Equally, multiple streams of research are presently ongoing with efforts to characterize the emissions as well as identify their impacts on the health of man and animals in order to fully understand their chemistry and pathologies (Arshad 2017; Arshad et al. 2022).

1.3.1 Soil Pollution

Manures were traditionally disposed of on agricultural fields without recourse to proper management plan pertaining to the amounts admissible in a space. This practice led to over-application in many areas leading to overfertilization of soils, run-off of toxic constituents, leaching of contaminants and accumulation of heavy metals and macronutrients, principally phosphorus (P) and nitrogen (N). Heavy metals [e.g. zinc (Zn) and copper (Cu)] pose a significant health hazard to soil animals because they are passed through the food chain during grazing and also contribute directly to the causation of autosomal recessive diseases and to impaired metabolism and liver function (Giola et al. 2012; Maillard and Angers 2013). Endocrine disrupting compounds (EDC), e.g. steroids are another group of compounds found in animal wastes as residues. They originate from drugs administered animal feeds and have the tendency to trigger critical hormonal responses by mimicking normal androgenic and estrogenic signalling in man and animals. This condition could result in health and birth defects in the animals and their off springs (Combalbert et al. 2012). One major apprehension in the application of this class of drugs is that they retain the capacity to display activity even at very minute (i.e. parts-per-trillion or nanogram-per-litre) concentrations.

1.3.2 Water Pollution

Water pollution caused by animal waste results from the leaching of minerals and runoff of nutrients from soils due to overfertilization with manure. It may also result from the direct discharge of animal wastewater into municipal waters. Stormwater run-off is the gateway through which leached nutrients, especially nitrogen and phosphorus, move from manure-saturated lands to surface waters, where they cause pollution. In the aquatic matrix, free ammonia (NH_3) has been reported to be capable of causing higher levels of toxicity to much marine life than the salt compound ammonium (NH_4); for example, as little as 5 mg/L of ammonia is known to cause detectable levels of lethality in salmon (Martinez et al. 2009a, b).

Surface waters are highly susceptible to contamination from manure and are potential sources of infection because they are contaminated with microorganisms found in livestock effluent. Microorganisms make their way through a combination of sorption and suspension (Edwards and Daniel 1992). There are three main pathways by which potential contaminants in manure enter surface waters from farmed areas. They can be bound or adsorbed to soil particles, transported in suspensions/solutions or carried in particulate form (Martín-Marroquín and Hidalgo 2014). Ammonia and phosphorus are bound to soil particles and can be transported by erosion, while carbon (C), phosphorus (P) and nitrogen (N) are transferred in particulate form. Pollutants that are transported in solutions include soluble forms of carbon, phosphorus, ammonium, nitrates and uric acid (Martín-Marroquín and Hidalgo 2014). Domestic use of such water without proper treatment could, therefore, inevitably lead to severe morbidity and mortality (Gilchrist et al. 2007). There is, thus, a need to develop better management practices for manure application to protect streams and rivers from the reckless and unwarranted dumping of animal wastewater. The cooperation of all stakeholders, i.e. farmers, community leaders, opinion leaders, and policymakers, is critical in enforcing the appropriate legislation and ensuring compliance.

1.3.3 Air Pollution

Emission of ammonia

Livestock farming contributes significantly to the generation and emission of odorous pollutants such as ammonia, methane and carbon dioxide. A significant proportion of gases that emerge from livestock farms are ammonia. It is a secondary particulate precursor that reacts with other compounds in the atmosphere, such as nitric and sulphate acids, to form ammonium salts, a deadly form of particulate matter. Hence, livestock farming is one most prominent contributors to ammonia outflows in the ecosystem. The gases are produced by microbial fermentation in stored mixtures of animal faeces and urine (Vanotti et al. 2009). Under this condition, urea in urine is broken down by urease to liberate ammonia. However, the rate of emission varies from one facility to another because the conversion of liquid ammonium (NH_4) to the gaseous phase (i.e. ammonia (NH_3) is governed by a set of factors which include temperature, pH and wind speed (Martín-Marroquín and Hidalgo 2014). Presently, manure on land is considered a nuisance, particularly in densely populated areas, due to the discomfort linked to the malodorous discharges from fertilized lawns and animal shelters. Other components in airborne emissions from manure and animal farm settings include hydrogen sulphide, volatile organic compounds, endotoxins and particulates. Many studies indicate that several pulmonary conditions are linked to the prolonged exposure of these compounds to individuals, particularly farm workers. They include bronchitis, mucus membrane irritation and asthma (May et al. 2012). In one report, endotoxins and organic aerosols were found to be behind the onset of respiratory disorders among swine workers and neighbourhood residents (Leytem et al. 2011).

Methane and nitrous oxide emissions and climate change

Methane and carbon dioxide are greenhouse gases associated with the phenomenon of global warming. The increase in the concentration of these gases in the atmosphere is the cause of global warming. These gases inevitably cause global temperatures to rise, leading to economic and environmental disasters in some countries. Both gases originate from the metabolism of anaerobic microorganisms, whose activities dominate the digestive processes in ruminant diets. In addition, soils previously treated with livestock manure release nitrous oxide (N_2O), another greenhouse gas. Methane and nitrous oxide are essential for regulating ozone concentrations in the atmosphere.

Dusts, volatile organic compounds and particles

Particulate emissions can occur through the aerosol-assisted movement of livestock manure during ammonia emission (Cambra-López et al. 2010). They may also be propagated by the shaking up of litter or other materials during the movement of animals or equipment or through the exhaust system of the ventilation set installed in the animal shelters. Volatile organic compounds (VOC) are released mainly from fermented feeds and fresh faecal wastes of livestock (Martín-Marroquín and Hidalgo 2014). VOCs contribute to the photochemical activities that lead to the production of ozone. In the presence of sunlight, VOCs drive the oxidation of NO to NO_2, eventually culminating with the synthesis of ozone (O_3) (Ling and Guo 2014). Livestock rearing engenders the release of volatile odoriferous compounds due to the microbial modification of materials in their feed and excreta. However, the odorants released are not necessarily correlated to the presence or amounts of pathogens or indicator organisms but mainly promote the sensory recognition of the presence of the volatile organic compounds in any material or environment. VOCs can accumulate in confined spaces to pose health risks to animals and farm staff, particularly in settings where the stocking density of the livestock is relatively high (Schiffman et al. 2000).

Source of disease causing agents

Many types of pathogenic organisms including bacteria, viruses and parasites are present in livestock wastes and could potentially constitute hygiene risks during collection, packing and subsequent dispersal on agricultural fields. Although biological agents such as obligate parasites do not pose significant dangers outside of their hosts, bacteria and viruses are capable of surviving for extended periods in the fields. Some confirmed cases indicate that while the risk of zoonotic infections is low, it is the transmission of infection to other livestock that accounts by far the most significant numbers of confirmed cases of disease outbreaks (Burton 2009) often spreading to nearby farms.

Food crops, particularly those eaten raw, often expose consumers to the risks of infections because they harbour pathogens related to the application of livestock manures in soils during farming (Bezanson et al. 2014; Blaiotta et al. 2016). Common enteric pathogens such as campylobacter and salmonella are often implicated, although reported cases of food poisoning by the pathogens are comparatively low. However, efforts to protect the retailers and the public from a potential full outbreak are driving the enforcement of certain regulations on the use of animal wastes as manure in several countries. The health of farm workers is equally exposed due to frequent direct contact with the wastes as well as the dusts and gaseous emissions spewing out from the confined spaces of the animal enclosures (Burton 2009). Equally, the threat of transmission of zoonotic diseases is potent because run-offs usually carry materials from manure-dosed farms into surface waters which serve the domestic needs of many people in the locality. Zoonoses may also be transmitted when manure comes into contact with food or contaminate water used for irrigation or washing crops like leafy vegetables which are consumed raw (Cliver 2009).

1.4 Brief Background of Animal Waste Treatment Systems

Animal waste treatment system is a process used to reduce biomass, manage pathogens, concentrate nutrients and generate by-products like fertilizer or energy (Sobsey et al., 2006). Before disposal, a robust waste gathering and storing system is unavoidable. In most European and North American countries, these processes have simplified by using mixing and separation methods, which reduces clogging problems and ease carriage. These approaches can help to reduce environmental effect in some circumstances by resulting in a more consistent application of nutrients (Vanotti et al. 2009; Arshad et al. 2021).

Unlike direct application to land, animal waste treatment currently uses technology to alter its chemical or physical properties. This can be achieved by biological, chemical, physical or mechanical methods or their combinations. The major options of animal waste treatment include solid–liquid separation, nutrient partitioning, composting and digestion (missing citation). While sieving works well for cattle slurry containing 30–40% solids, centrifugation works better for poultry and pig slurry containing finer particles. Gravity sedimentation in large, shallow bins results in sludge with a dry solid concentration of 5–10%. With the right C/N ratio, moisture content, aeration and time, composting can produce an environmentally stable by-products using oxygen-consuming bacteria and fungi (missing citation). Digestion methods (aerobic/anaerobic) can be used for the removal of nitrogen and organic load from animal waste (Vanotti et al. 2009). In aerobic treatment, for example, aerobic microorganisms oxidize bioavailable oxygen-consuming compounds such as nitrogenous and organic compounds, which is a means of reducing odour and ammonia emissions. Nitrogen removal is accomplished through the processes of nitrification and denitrification. Microbial activity is expected to break down organic material and reduce the biomass load, producing carbon dioxide and water. However, this method has rarely been used for the treatment of slurry or manure, mainly because of the costs associated with the operation of the machine required to supply sufficient oxygen to the aerobic microorganisms. Under anaerobic conditions, acetic acid is formed, which is then utilized by methanogens to produce energy, mainly biogas (CH_4), the yield of which varies according to animal waste. Digestates from animal waste can be a valuable fertilizer, but this may require additional technology and cost due to its high moisture content. Anaerobic digestion of animal wastes is mostly popular in Europe because it benefits biogas production, which is used to generate heat and electricity (Arshad et al. 2021).

Concentration, separation and exportation are other strategies for removing ineradicable components of animal manure, such as heavy metals and phosphorus. If there are defined, accepted levels for these constituents, this process may be the efficient method to remove surplus nutrients, including nitrogenous and organic compounds. These methods yield dry solid products that can be utilized, blended with other products, or composted to create valuable natural ingredients that can occasionally be sold (Vanotti et al. 2009). Precipitation of some animal waste components can be done with some chemicals such as flocculants or lime, but their use alone is not often sufficient or sustainable (Vanotti et al. 2009).

1.5 Microorganisms in Animal Waste Recycle

1.5.1 Microbes Found in Animal Waste

Animal manure contains a variety of bacteria that change organic materials through various chemical processes in addition to disease microbes. The physical factors surrounding microorganisms, particularly the humidity, temperature and oxygen content, impact the chemical reactions. The metabolic activity of bacteria alters these conditions (Wan et al. 2021). The most prevalent and active microbes participating in the process are bacteria and fungi. Bacteria carry out the majority of decomposition and heat production, but fungi are also capable of degrading complex polymers (Akdeniz 2019). For instance, Proteobacteria are more prevalent in cattle manure, but Firmicutes frequently predominate in pig manure. Basidiomycota and Ascomycota are the main phyla of fungus found in chicken and cow manures. Such variations may result from the decomposition process. Still, they may also result from initial variations in the waste microbiota composition, which is influenced by the nutrition and microbiota of the animal's gut (Teira-Esmatges and Flotats 2003).

Animal wastes frequently carry high levels of disease-causing microbes from humans, spilled feed, bedding material, fur, process-generated wastewater, undigested feed leftovers, faeces, as well as urine. These microorganisms are also involved in the degradation processes. The amounts and kinds of disease-causing agents seen in animal wastes differ depending on the animal species, state of health, animals' age, physical/chemical features of the dung generated and the manure's storage installations (Burkholder et al. 2007; Hutchison et al. 2005).

1.5.2 Nitrogen Cycle and Microorganisms

Microbes (bacteria and fungi) play crucial functions in the nitrogen cycle in the natural environment. Nitrogen is the nutrient most susceptible to changes that increase the likelihood of wasteful losses. Mineralization to ammonium, immobilization, oxidation (nitrification) and denitrification are among the changes. It has commonly been recorded that total nitrogen is typically conserved amid the process of anaerobic digestion (Schievano et al. 2011). On the other hand, researchers compared biogas digesters' nutritional inputs plus outputs and discovered gross nitrogen depletion of 18% (Möller 2015). Net nitrogen losses of 5–10% were also recorded by Schievano and co researchers (Schievano et al. 2011). The biogas stream's nitrogen content of the ammonium ion flux, which comprises methane, carbon (iv) oxide, water vapour, minimum amounts of ammonium ion, hydrogen sulphide, as well as other elements, accounted for only about 10% of the depletions; other factors like incomplete sedimentation of organic or inorganic matter, formation of struvite, precipitation, as well as final reservation in the digesters, are ascribed to the remainder (Massé et al. 2007; Möller and Müller 2012). Moreover, following anaerobic digestion, animal manures slurry that has been digested rarely forms a "natural" top crust by suspended fibre particles in manure depots, as it does in undigested slurry reserves. Ammonia losses from the slurry that has been digested were comparable to those from the slurry that hasn't been treated over the winter, according to Clemens and other scholars (Clemens and Huschka 2001).

Nitrification has been observed in bacteria, archaea and fungi (autotrophs) (Laughlin et al. 2008; Leininger et al. 2006). Heterotrophic nitrification, on the other hand, occurs when NH_3 is directly oxidized or organic materials decompose to nitrate by heterotrophic bacteria. This procedure is known to occur in a variety of bacteria, and some, like *Paracoccus denitrificans* and *Pseudomonas putida*, have amoA sequences that differ from autotrophic nitrifiers (Maeda et al. 2011). Normally, heterotrophic denitrifiers convert NO_2^- or NO_3^- produced by nitrifiers into nitrous oxide, dinitrogen, or just nitrogen gas before releasing it into the environment. Despite the fact that nitrous oxide depletion is thermodynamically advantageous and nitrous oxide is a good electron acceptor, certain denitrifiers create nitrous oxide as an end product. This could be due to the fact that nitrous oxide is nontoxic to some microbes but may be poisonous to some bacterial cells (Schneider and Einsle 2016).

1.5.3 Systems of the Manure Recycle and Treatment

Indiscriminate disposal of manure may result in pollution of surface and groundwater. In the light of the above, various manure recycling and treatment systems, including activated sludge systems, lagoons, compaction, composting and other methods, are required for the recycling and treatment of animal manure before its use in the soil (Malomo et al. 2018).

An activated sludge system is one of the most predominant approaches to waste treatment. In the activated sludge, microbes absorb and assimilate nitrogen, phosphorus compounds as well as other nutrients in the wastewater. They as well nitrify as well as denitrify nitrogen compounds to nitrogen gas. The H_2O can be recycled for agricultural or domestic usage, whereas sludge deposited in the system which is a biomass of microbial cells is recycled as the fertilizer (Waki et al. 2018). Physical techniques such as pelletizing and baling can help to improve the storage as well as management of heap solid manures. These approaches are aimed at delivering manure's nutrients in a more cost-effective and dust-free manner, and dung conditioning before bio-energy transformation. Compacting a loose material into pellets like poultry litter enhances its consistency dramatically (McMullen et al. 2005).

The widely accepted standard practice for recycling waste is composting. It eliminates raw waste from areas where it could contaminate streams and groundwater. Pathogens are eliminated, and a safe soil amendment is produced by effective composting (Teira-Esmatges and Flotats 2003). When a heap of garbage is created, composting begins. Microbes begin to decompose by consuming oxygen as well as transforming it to CO_2, water vapours, plus heat (Sorathiya et al. 2014). It is possible to have an open or closed composting system, and compost can be piled or stacked in rows or deposited in a closed reactor or container (Haug 2018). Because of its technical complexity, the open system is seldom employed in low-income nations. The waste should ideally be piled and left for the same period of time in a four-pole fence that is surrounded by boards or chicken wire. This creates a rich compost that can be applied as a fertilizer for fields as well as gardens (Akdeniz 2019). Sanitation, odour elimination and safe storage are among the benefits of composting animal dungs over manure that has not been processed applied directly to the soil. On the other hand, the cost of installation and management, as well as the need for vast storage and operation spaces, is potential downsides of composting (Narula et al. 2011).

1.5.4 Microbial Flora of Animal Faeces After Excretion

The microbial flora of fresh faeces from animals has been extensively studied. For example, the most common faecal bacteria in pigs include Bacteroidaceae, Peptococcaceae, Eubacteria, Lactobacilli, Bifidobacteria, Spirillaceae and Enterobacteriaceae (Cox et al. 2005; Dowd et al. 2008; Lim et al. 2018). In the rectum of chickens, Bacteroidaceae, Lactobacilli, Enterobacteriaceae and Streptococci predominate (Nodar et al. 1990). Similarly, Bacteroidaceae, Spirillaceae, Enterobacteriaceae and Streptococci are the most common microbial groupings found in bovine faeces (Dowd et al. 2008). Despite the fact that the microbial flora in faeces has been extensively studied, little research has been done on how the microbial flora alters after expulsion. Animal excretion microbial flora may affect the performance of microbiological treatment systems for animal wastes (Hagey et al. 2019).

1.5.5 Microorganisms and Their Function in the Animal Waste Lagoon

In animal waste lagoon that operate normally, acid formers as well as methane formers, are two noticeable types of bacteria. Biodegradable organic matter is transformed to volatile acids by acid formers while methane and carbon dioxide are produced by converting these volatile acids by methane formers. Moreover, a level of equilibrium of biological responses by the two types of bacteria is attained under optimum conditions. This balance will be disturbed, and overweening odour and sludge accumulation are produced due to environmental variations (e.g. temperature fluctuations), indecorous design and lousy management. Consequently, anaerobic digestion is employed to stabilize manure, diminish pathogens plus emanations of odour and as well generate energy via production of biogas. The primary mechanism for natural animal waste treatment is anaerobic digestion, in open anaerobic digesters as well as anaerobic lagoons (MacSAFLEY et al. 1992; Nakai 2001). Anaerobic lagoons are typically designed for a storage period of twenty to one hundred and fifty days and for the treatment of wastewater. They're normally eight to fifteen feet deep and function similarly to septic tanks. The effluent from an anaerobic lagoon will need to be treated further (Leffert et al. 2008).

The lagoon system uses a combination of physical, biological as well as chemical approaches to treat waste. Although a few approaches employ aeration devices to provide O_2 to the wastewater, the majority of the treatment is done organically. Aeration enhances treatment effectiveness and reduces the amount of ground area required. Soil type, size of available land, as well as weather have an impact on the layout of the system. Waste from a lagoon may require additional treatment or "polishing" to remove pathogens or nutrients before it is released into the environment (Deviney et al. 2020).

1.5.6 Microorganisms in the Composting Process

An aerobic method of converting organic waste into a humus-like substance through microbial activity is termed composting. Composting is as well an approach to produce soil conditioner or fertilizer. In a normal composting process, bacteria as well as fungi exist and function (Jusoh et al., 2013). Researches beforehand have unveiled that mesophilic organic acid-generating bacteria like *Lactobacillus* species as well as *Acetobacter* species are the considerable groups of bacteria in the baseline of the composting process (Pan et al. 2011). Thereafter, in the thermophilic stage, bacterial species (e.g. Bacillus species and Actinobacteria) predominate. However, it has been suggested that the most effective composting process is achieved by mixed communities of bacteria and fungi (Malińska and Zabochnicka-Świtek 2013; Zhang et al. 2010).

Furthermore, composting is a three-phase process that involves microbes (e.g. bacteria and fungi), as well as mesophiles such as *Streptomyces rectus* and thermophiles such as *Actinobifida chromogena* (*Thermomonospora fusca*), etc., ultimately, transforming organic waste into humus. The substrate is depleted amid the first phase because of sugar as well as protein degradation by the mesophilic microbes' activity, as well as a rise in carbon dioxide levels in tandem with a rise in temperature (Novinscak et al. 2008; Zeng et al. 2011). In the second phase, the temperature of the compost heap increases from 45 to 70 °C, and thermophilic microorganisms replace mesophilic microbes. Several harmful individuals are minimized at this moment. The third process begins with the lowering of the temperature of the compost heap (Schloss et al. 2003).

1.6 The Microbial Community Profiles of Different Animal Waste

Animal wastes are home to a diverse spectrum of microbial communities, including both beneficial and pathogenic microorganisms (Mawdsley et al., 1995). Despite its extensive usage in agriculture, there is a variation in microbial diversity as a result of various treatment processes, which also vary based on the waste source (Table 1.2). For example, at various handling stages, manure from a dairy farm in the California Central Valley was sampled for 16S rRNA study of composition and diversity of microbial communities. The study revealed that there are variations in microbial population between the solid and liquid waste. For example, the bacterial genus *Thermos* was only present in the solid samples, while *Sulfuriomonas* was only observed in liquid samples. The genus *Clostridium* was abundant in both liquid and solid samples (Pandey et al. 2018).

1.6.1 Cow Waste

Cow waste, particularly dung, contains a diverse group of bacteria including *Kluyvera sp., Bacillus sp., Klebsiella pneumoniae, Lactobacillus sp., Corynebacterium sp.,*

Table 1.2 Some dominant microorganisms found in different animal wastes

Waste source	Type of microbes	Microorganisms	References
Cow	Bacteria	*Bacillus* sp., *Lactobacillus* sp., *Corynebacterium* sp., *Bacteroides, Paludibacter, Alistipes, Anaerovorax, Ruminococcus, Turicibacter, Lysinibacillus, Stenotrophomonas*	Randhawa and Kullar (2011), Girija et al. (2013), Mao et al. (2012)
	Fungi	*Candida, Saccharomyces cerevisiae, Aspergillus, Thermomyces, Myriococcum, Fusarium oxysporum, Alternaria, Ascobolus* sp.	Randhawa and Kullar (2011), Jiang et al. (2020), Thilagam et al. (2015), Tan and Cao (2013)
	Archaea	*Methanobrevibacter, Methanocorpusculum, Methanosphaera*	Cendron et al. (2020)
Poultry	Bacteria	*Bacillus, Lactobacillaceae, Brachybacterium, Azomonas agilis, Streptococcus sp., Proteus vulgaris, Aeromonas hydrophila, Proteus vulgaris, Echerichia coli, Sarcina maxima, Lactobacillus sp., Staphylococcus aureus*	Lovanh et al. (2007)
	Fungi	*Candida sp., Mucor sp., Cladosporium spp., Aspergillus* sp., *Penicillium* sp., *Saccharomycopsis, Sporendonema sp., Kloeckera sp., Zygosaccharomyces sp.*	Adegunloye and Adejumo (2014), Emmanuel-Akerele and Adamolekun (2021)
Swine	Bacteria	*Clostridium, Bacillus, Lactobacillus, Novibacillus, Planifilum, Corynebacterium, Virgibacillus, Terrisporobacter petrolearius*	Lim et al. (2018), Chen et al. (2017), Kumar et al. (2020)
	Fungi	*Aspergillus, Melanocarpus, Debaryomyces hansenii, Geotrichum sp., Acremonium strictum, Fusarium, Geotrichum sp, Mucorales, Wallemia*	Wan et al. (2021), Kumar et al. (2020), Kristiansen et al. (2012), Kim (2009)
	Archaea	*Methanobrevibacter, Methanosarcina, Methanobacterium, Methanothermobacter, Methanocorpusculum, Methanofollis*	Tuan et al. (2014), Qin et al. (2013)
Sheep	Bacteria	*Lysinibacillus, Clostridium, Enterococcus, Escherichia, Streptococcus, Bifidobacterium, Anaerocolumn, Tissierella, Anaerocolumna, Muricomes*	Shabana et al. (2020)
	Fungi	*Ascobolus, Preussia, Mortierella*	Tan and Cao (2013)
Goat	Bacteria	*Escherichia, Anaerotignum, Ruminococcus, Prevotella, Butyrivibrio*	Shabana et al. (2020)
	Fungi	*Neocallimastix, Caecomyces, Piromyces*	Peng et al. (2021)
	Archaea	*Methanobrevibacter sp., Methanosphaera stadtmanae*	Peng et al., 2021

Pseudomonas sp., *Citrobacter koseri*, *Providencia stuartii*, *Staphylococcus sp.*, *Klebsiella oxytoca*, *Morgarella morganii*, *Enterobacter aerogenes*, *Providencia alcaligenes*, *Pasteurella sp.*, and *Escherichia coli* (Sawant et al. 2007; Randhawa and Kullar 2011; Gupta and Rana 2016). About 60 bacterial species are found in dung, generally dominated by *Bacillus sp.*, *Lactobacillus sp.*, and *Corynebacterium sp.* A culture independent 16S rDNA techniques identified dominant genera in cow dung as *Bacteroides*, *Paludibacter*, *Alistipes* (Bacteroidetes), *Bacillus*, *Clostridium*, *Anaerovorax*, *Ruminococcus* (Firmicutes), *Pseudomonas*, *Acinetobacter*, *Rheinheimera*, *Rhodobacter*, *Stenotrophomonas* (alpha- and beta-Proteobacteria), and *Akkermansia* (Verrucomicrobia). About 87.5% of Firmicutes and 83.3% of Bacteroidetes constituted the unculturable bacteria (Girija et al. 2013). Mao et al. (2012) reported the abundance of Firmicutes, Proteobacteria, Actinobacteria, Bacteroidetes and Tenericutes in cow faecal bacterial community. The most dominant groups are T*uricibacter*, *Lysinibacillus*, *Stenotrophomonas*, *Solibacillus silvestris* and the family Lachnospiraceae.

It also contains other microorganisms, such as yeast (*Candida* and *Saccharomyces cerevisiae*), about 100 species of protozoa (Randhawa and Kullar 2011), fungi (*Trichoderma*, *Actinomycetes* and *Aspergillus*) (Munshi et al. 2019) and archaea (Cendron et al. 2020). Jiang et al. (2020) identified *Aspergillus*, *Thermomyces*, *Myriococcum*, *Mycothermus*, *Cladosporium*, *Scedosporium* and unclassified Microascaceae as fungal communities in cow manure using high-throughput sequencing. Of all the 25 fungal species belonging to 20 genera recorded in dung samples from the Lawspet area of Puducherry Union Territory of India, *Aspergillus fumigatus* was the dominant species by *Fusarium oxysporum* and *Alternaria alternata*. Other species include Aspergillus *clavatus*, *Penicillum sp.*, *Cladosporium cladosporioides*, *Scopulariopsis sp.*, *Arthrinium sp.*, *Acremonium sp.*, *Arthrobotrys sp.*, *Cephaliophora sp.*, *Myrothecium sp.*, *Trichoderma sp.*, *Fusarium oxysporum*, *Drechslera sp.*, *Pithomyces sp.*, *Nigrospora oryzae*, *Paecilomyces sp.*, *Phialophora sp.* and *Oidiodendron sp* (Thilagam et al. 2015). Tan and Cao (2013) also reported that the fungal diversity in cow faeces is dominated by the phylum Ascomycota (*Ascobolus* sp. and *Candida*) followed by Basidiomycota, and Chytridiomycota.

Various groups of archaea and methanogens belonging to the Methanomicrobiaceae have been detected in cattle manure (Kim et al. 2014). Cendron et al. (2020) reported archaeal phylum Euryarchaeota which includes five genera, *Methanobrevibacter*, *Methanocorpusculum*, *Methanosphaera*, unclassified Methanobacteriaceae and uncultured Methanomethylophilaceae.

1.6.2 Poultry Waste

Poultry waste contains a wide range of intestinal microbiota, primarily Proteobacteria-derived species, which may contain pathogens that pose a health risk. A taxonomic analysis of the 16S rRNA sequences showed that it is dominated by Firmicutes followed by Proteobacteria and Bacteroidetes (Zhang et al. 2018). The 16S rRNA sequencing identified *Bacillus, Lactobacillaceae, Brachybacterium sp., Arthrobacter sp., Corynebacterium sp., Enterococcaceae, Brevibacterium sp., Staphylococcus, Corynebacteriaceae, Aerococcaceae* and *Actinomycetes* (Lu et al. 2003; Lovanh et al. 2007). *Bacillus cereus, Azomonas agilis, Streptococcus sp., Proteus vulgaris, Escherichia coli, Staphylococcus aureus, Sarcina maxima, Thiocapsa lumicola, Xanthomonas fragariae* and *Enterococcus sp.* were isolated from turkey faeces. Other species found in duck samples included *Bacillus cereus, Aeromonas hydrophila, Proteus vulgaris, Echerichia coli, Sarcina maxima, Lactobacillus sp., Streptococcus sp., Streptobacillus moniliformis, Enterococcus sp.* and *Staphylococcus aureus.* Fungal species present in both samples were *Candida sp., Mucor sp., Cladosporium spp., Aspergillus fumigatus, Penicillium sp., Aspergillus flavus, Alternaria sp., Fusarium sp.* and *Varicosporium elodea* (Adegunloye and Adejumo 2014). Characterization and identification of bacteria from poultry droppings showed the presence of *Pseudomonas picketti, Streptococcus pluranimalium, Micrococcus holobium, Cellobiococcus sciuri, Enterobacter agglomerans, Bacillus pumilus, Staphylococcus aureus, Staphylococcus alrettae, Salmonella enteritidis* and *Staphylococcus saprophyticus.* The identified fungal species were *Saccharomyces sp., Candida tropicalis, Aspergillus fumigatus, Aspergillus niger, Saccharomycopsis, Sporendonema sp., Kloeckera sp., Fusarium oxysporum, Candida sp., Zygosaccharomyces sp.* (Emmanuel-Akerele and Adamolekun 2021). Nauanova et al. (2020) also reported cellulose-degrading bacteria from poultry manure, such as *Bacillus megaterium, Lentzea chajnantorensis, Burkholderia xenovorans, Enterobacter hormaechei* and *Sphingomonas trueperi.*

1.6.3 Swine Waste

Pig waste contains a diverse group of microorganisms that play an important role in the waste decomposition, including *Clostridium, Bacillus* and *Lactobacillus.* Firmicutes (*Clostridium, Lactobacillus, Streptococcus and Turicibacter*) and Actinobacteria (*Corynebacterium*) have been found to be the most abundant phyla in swine manure at different temperatures and storage times (Lim et al. 2018; Chen et al. 2017). According to most studies (Wan et al. 2021; Kumar et al. 2020), Firmicutes are commonly the most abundant phylum in pig and chicken waste. The bacterial genera profile of pig manure showed the presence of *Bacillus, Novibacillus* and *Planifilum.* In the same samples, the fungal group was dominated by *Aspergillus* and *Melanocarpus.* In another pig manure, *Bacillus, Corynebacterium,* Virgibacillus, *Actinobacteria, Pseudomonas, Pediococcus and Lactobacillus* were the predominant genera (Chen et al. 2017). A compositional analysis of swine slurry at different times using 16S rRNA metagenomic sequencing approach identified *Clostridium saudience, Clostridium leptum, Terrisporobacter petrolearius, Butyrivibrio hungatei and Lactobacillus ultunensis* as the most significantly abundant bacteria (Kumar et al. 2020).

Similarly, a liquid swine manure studied using DGGE/PCR of 16S rDNA identified *Clostridium disporicum, Clostridium butyricum*, two *Rhodanobacter* sp., a *Pedobacter* sp., a spirochete and seven uncultured eubacteria (Leung and Topp 2001). The microbial composition analysis of the pig particulate matter (faeces, hair, bedding particles, feedstuff, and animal skin) also showed that *Clostridium* was the most predominant followed by *Bacillus* and *Terrisporobacter. Other abundant species were Lactobacillus, Turicibacter, Prevotella, Curvibacter, Staphylococcus, Blautia, Weissella, Roseburia* and *Sediminibacterium* (Hong et al. 2021).

The archaeal community commonly found in swine wastes are members of the genera Methanobrevibacter, *Methanosarcina*, M*ethanobacterium* and *Methanothermobacter* (Tuan et al. 2014). Qin et al. (2013) reported the detection of *Methanocorpusculum, Methanofollis, Methanogenium, Methanoculleus, Methanocorpusculum labreanum* Z, *Methanosaeta concilii, Methanosarcina siciliae* and *Methanofollis ethanolicus* in swine manure.

Fungi are also part of the microbial communities in swine waste and a diverse species have been identified, including *Debaryomyces hansenii, Geotrichum sp., Acremonium strictum, Fusarium sporotrichioides, Fusarium sporotrichioides, Monographella nivalis, Cladosporium sphaerospermum, Acremonium alternatum, Pleurotus eryngii, Malassezia globosa, Myriangium durosai, Rhodotorula glutinis* and *Malassezia restricta. Geotrichum sp. (Saccharomycetes)* was the most abundant species, followed by *Acremonium strictum, Monographella nivalis* and *Pleurotus eryngii* (Kim 2009). Kristiansen et al. (2012) also identified *Mucorales, Wallemia and Russulales* as the most abundant fungal.

1.6.4 Sheep Waste

Firmicutes and Bacteroidetes have been found to be the most prevalent bacterial phyla in sheep faecal matter, accounting for 80% of the total population (Mamun et al. 2019). From a taxonomic standpoint, the sheep faecal bacteria appear to be comparable to that of other ruminants, with Firmicutes as the dominant phylum (Tanca et al. 2017). Instead of Bacteroidetes, Shabana et al. (2020) found Proteobacteria to be the second most abundant core bacterial phylum in sheep 6 months after birth, with *Lysinibacillus* being the most abundant genus, followed by *Clostridium, Enterococcus, Escherichia, Streptococcus, Bifidobacterium, Anaerocolumn, Tissierella, Anaerocolumna* and *Muricomes.* In ITS, 28S and 18S study of fungal community composition of sheep faeces, Tan and Cao (2013) reported that Ascomycota, Basidiomycota and Chytridiomycota are the most abundant phyla. The most detected genera were *Ascobolus* (ITS, 28S and 18S), *Preussia* (ITS and 28S) and *Mortierella* (ITS and 18S).

1.6.5 Goat Waste

The phylum Proteobacteria was discovered to be the most dominant community in goat faeces 6 months after birth, with *Escherichia* and *Anaerotignum* being highly prevalent. The goat faeces share the same core bacteria genera with sheep. At one year of age, goats had significantly higher abundance of the phylum Firmicutes than sheep, but sheep had higher abundance of the phylum Proteobacteria than goats. (Shabana et al. 2020). A metagenomic analysis of goat faecal microbial communities revealed that about 33.3% of the constructed metagenome-assembled-genomes (MAGs) were Firmicutes and Bacteroidetes, with more than half belonging to the Ruminococcaceae and Rikenellaceae families. *Ruminococcus, Prevotella* and *Butyrivibrio* are among the most abundant genera. The archaeal MAGs recovered were dominated by the genus *Methanobrevibacter sp.* as well as the class *Thermoplasmata* and the species *Methanosphaera stadtmanae.* The fungal MAGs studied in this research are members of the subphylum Neocallimastigomycota, with the majority belonging to the genus *Neocallimastix.* Other MAGs recovered from only the first generation of enrichment cultures are from the *Caecomyces* and *Piromyces* genera (Peng et al. 2021).

1.7 Composting Process

Composting simply enhances the process of decomposition by creating an ideal environment (nutrients, warm temperatures, moisture and sufficient oxygen) for bacteria, fungi and other decomposers (such as worms, nematodes, and sow bugs) (Bernal et al. 2009). Mesophilic bacteria (*Bacillus sp., Streptococcus, Pseudomonas, Streptosporangium sp, Proteus, Seratia, Streptomyces, Actinomyces, Methylomonas sp* and some faecal coliforms) and fungi (*Rhizopus* and *Trichothecium sp*) that flourish in temperatures of 20–45 °C begin physical breakdown of biodegradable materials a few days after composting begins (Chinakwe et al. 2019). These mesophiles are supersede by thermophilic bacteria (*Bacillus sp, Seratia sp, Methylomonas sp, Streptosporangium sp*) and fungi (*Aspergillus fumigatus*) after a few days and can last for some days or even several months (Taiwo and Oso 2004). At this point, temperatures have dropped sufficiently for mesophiles to reclaim dominance of the compost pile and complete the breakdown of the remnant organic materials into useful humus (Neher et al. 2013; Mingyan 2011).

1.7.1 Types of Composting

Depending on the nature of decomposition process, composting can be divided into two types.

Aerobic composting

Aerobic composting occurs when there is adequate oxygen in the system. Aerobic microorganisms decompose organic matter, producing ammonia, carbon dioxide, water, heat and humus in the process (Kim et al., 2015). These microorganisms continue to break down intermediate compounds such as organic acids, despite the fact that aerobic composting produces them (Cai et al. 2018). The intermediate products are relatively unstable, and the compost is completely safe. The heat generated in the process accelerates the degradation of complex carbohydrates (cellulose and hemicellulose), proteins and lipids. As a result, processing time is shorter and many pathogenic microorganisms that may infect humans and plants are killed, as they are not adapted to these environmental conditions (Millner et al., 2014). The heat also aids the growth of beneficial bacterial species such as mesophiles, psychrophiles and thermophiles. Although aerobic composting leads to more nutrient loss from the waste, it is considered more efficient and beneficial for agricultural production than anaerobic composting (Cai et al. 2019; Mehta and Sirari 2018).

Anaerobic composting

In anaerobic composting, decomposition occurs in the absence of or with a limited supply of oxygen. Anaerobic microorganisms thrive and take control of the community in this situation, resulting in the production of chemical intermediates such as carboxylic acids (–COOH), CH_4, H_2S and other toxic pollutants. In the absence of oxygen, these compounds build up and are not digested. Most of these compounds have a foul odour, and some of them may be harmful to animals and plants. Because anaerobic composting is a low-temperature process, organic materials and pathogens do not decompose. Moreover, the procedure usually takes longer than aerobic composting. These drawbacks typically overshadow the method's advantages (Eze and Okonkwo 2013; Mehta and Sirari 2018).

1.7.2 Factors Affecting the Animal Waste Composting Process

The composting of animal waste is influenced by a number of parameters, each of which has the potential to significantly affect the process. Such parameters (Fig. 1.2) include the size of the feedstock, pH, temperature, C/N ratio, moisture, the interaction of oxygen and aeration and other parameters (Bernal et al., 2009; Guo et al. 2012; Ameen et al., 2016; Chen et al. 2020). Controlling these elements helps speed up the natural composting process.

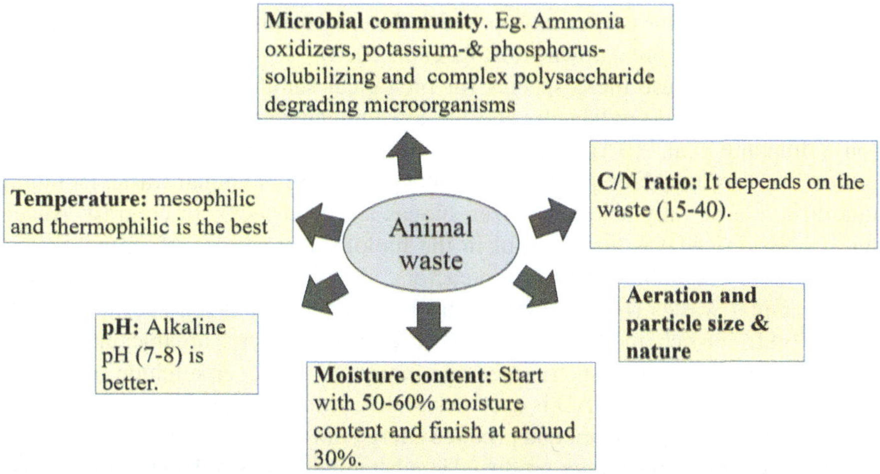

Fig. 1.2 Some parameters that affect animal waste composing process

1.8 Animal Waste Biodegradation

1.8.1 Bacterial Degradation Potential

Animal wastes include organic materials, decomposing animal body parts, urea (in the case of mammals), uric acid (in the case of birds), faeces and waste feed (Dinh Tuan et al. 2006). These components of organic wastes can be broken down by anaerobic and or aerobic bacteria such as *Escherichia coli* and *Salmonella spp.*, and there are three ways to introduce microorganisms to the waste for degradation processes (Jang et al. 2017). The first method involves spreading an isolated bacterium over the accumulated animal excreta. Animal wastes will be combined with soil and broken down by soil microorganisms if there is no microbial isolate available for this purpose (Briški and Domanovac 2017). This soil is best obtained from a moist, shaded area, such as beneath trees, as moist soil has more microorganisms than dried dirt (Hoitink and Boehm 1999). To save time, money and effort, the wastes should be stacked adjacent to the organic matter source, such as a field or a harvesting area (Dobermann et al. 2000). The final method uses bacteria linked to animal faeces. Animal faeces are a source of varying-quality organic nutrients, which puts microbial communities in a resource-contest and changes the structure and makeup of the soil microbiome. This is achieved by the excretion of specialized enzymes that convert complex polysaccharides, proteins and fats, such as cellulose, into simple nutrients that are ingested, such as sugars, amino acids and fatty acids. The heat produced by the biological process also aids in the stability and biodegradation of animal manure as the temperature rises. The other elements of animal manure are also taken into consideration. For example, the enzyme urease catalyzes urea hydrolysis, which is typically finished within one day of urine output, as opposed to the enzyme uricase, which takes longer to catalyze uric acid breakdown (Rastogi et al. 2020).

It is feasible to use these species for biological stabilization and treatment to produce valuable end products by altering the ambient and physio-chemical conditions (Brandelli et al. 2015). Anaerobic bacteria have been found to be useful in the production of value-added products from the processing of animal waste, including minerals, volatile fatty acids, fertilizer, biogas and feedstocks. In contrast, Anaerobic Digestion (AD) is also helpful in the biological treatment of animal wastes using outdoor anaerobic ponds or bioreactors, allowing for the sustainable utilization of animal wastes (Li et al. 2021). In the absence of oxygen and in the presence of nutrient-rich medium, anaerobic digestion converts organic matter into volatile organic molecules, such as methane, ammonia, hydrogen sulphide and carbon dioxide. Around the world, AD is used to stabilize animal wastes like manure, lower pathogen and odour emissions and produce energy through biogas (Durán-Lara et al. 2020). The energy in the biogas produced is substantially greater than what is needed, even though additional heat may be needed to maintain the proper temperatures (Chibuike 2013). The biogas synthesized from animal waste can be processed and utilized as fuel, injected into the transmission lines or used to produce heat and/or power.

1.8.2 Degradation by Plant and Animal Feed-Associated Bacteria

The environmentally responsible treatment of animal manure has benefited from plants. Animal faeces, which are a source of nutrients for plants, really promote plant growth far more than synthetic fertilizer does. Through their roots, plants primarily take up nutrients from animal waste and transform the soluble chemical components into plant tissues. Considering this, it is feasible to use plants to treat animal waste, which would have the twofold advantages of accelerating development and biochemical transformation while also recycling necessary animal wastes into plant feed, berries or dry materials. Based on the plant's species, growth stage, root length and dispersion, soil moisture, temperature and a variety of other parameters, the proportion of total absorbed by the roots varies (Ramachandra et al. 2018). In contrast, decomposers, which are often found in animal feed, absorb simple sugars and easily digested carbon compounds. They also bind soluble chemicals like nitrogen in their cell membranes, which helps with the organic recycling of carbon (Jambon et al. 2018).

1.8.3 Microfungal and Mycorrhizal Degradation

Roots of vascular plants and fungi have a symbiotic interaction known as mycorrhiza (Al-Maliki and AL-Masoudi 2018). The plant gives the fungi glucose, and the mycorrhizal fungi increase the roots surface area, enabling plant roots to absorb more water and nutrients from the soil and boosting the plant's resistance to disease (Jacoby et al. 2017). Free-living saprophytes and ectomycorrhizal or arbuscular mycorrhizal (AM) fungi are both well known for their substantial degradative abilities and effects that promote plant growth, making them appealing candidates for use in organic matter degradation (Jansa et al. 2013). AM fungus assists in the decomposition of animal dung by enhancing the activity of bacteria. Based on their capacity to stimulate the creation of bio-catalysts like pectinases, cellulases and hemicellulases, which are in charge of the breakdown process, mycorrhizal fungal species can decompose animal waste (Toljander et al. 2008). Because AM fungi lack saprotrophic abilities and depend on saprotrophic microorganisms to digest organic materials like animal wastes, this demonstrates that the breakdown of organic wastes by AM fungus is not considered direct (Etesami et al. 2021). As a result, there is a greater amount of organic nitrogen available for AM fungi to absorb (Wilkes 2021).

Furthermore, the AM fungus may indirectly affect the decomposition process by producing significant amounts of bio-compounds that aid the soil's microbial community in degrading organic waste. The bacterial population in soil is increased by low molecular weight carbohydrates and organic acids released by AM fungi, according to prior research (Jdruchniewicz 2018). However, Filion et al. found that some soil microbes were stimulated while others were inhibited by the hyphal exudates of AM fungi (Batstone et al. 2002). This suggests that instead of seeding microorganisms, earth could be used as the source of microorganisms for treating animal waste with microfungi.

1.8.4 Degradation by Algae

Algae and other aquatic plants' photosynthetic ability has proved successful in recycling carbon and other nutrients from animal wastes, as well as in environmental bio-remediation. According to this theory, diluted nutrients from animal excreta are converted into higher and lower plants by photosynthetic processes (Fernández et al. 2018). Animal waste can be effectively reused or converted into usable products and energy through the process of algal degradation. Algae could help people economically by using the energy and chemicals in animal waste (Puyol et al. 2017). Despite decades of algal seeding on animal waste effluent, just a few projects have reached commercial scale (Shah et al. 2014). Methane biosynthesis is the end product of several biological breakdown processes, including hydrolysis, acidogenesis, acetogenesis and methanogenesis. The products from previous phases are transformed into methane and carbon dioxide through the hydrogenotrophic and acetotrophic pathways (Phillips et al. 2017). Algae's capacity to break down animal waste can result in the production of significant biofertilizers in addition to methane. Additionally, a pilot plant that can transform pig poo into single cell protein has been created. It consists of a group of bacteria and algae. A high-temperature strain of *Chlorella vulgaris* was utilized the generating organism. According to research, 30–35% of the nitrogen waste from animal waste can be converted into single-cell protein. Algae biomass production, on the other hand, is more difficult than bacterial biomass production. This is due to the fact that algae biomass production necessitates control over culture depths, retention time and the amount of nutrients in the solution (Ozi et al. 2022).

1.9 Recovery of Nutrients and Energy from Animal Waste

There are many strategies or new technologies to recover high-value products and low-value by-products from animal waste in terms of environmental sustainability, which can also be incorporated into the value chain (Table 1.3). The anaerobic digestion can be used to convert waste into biogas for energy production and into a nutrient-rich digestate for use as fertilizer. Microbial technologies have the ability to convert waste into animal feed. Currently, processes such as anaerobic digestion, composting, worm culture and lime stabilization are used to process waste as well as to recycle and recover nutrients from the waste, and these can establish supply chains that are part of the bioeconomy. The use of treated digestate or organic waste as fertilizer and the CH_4 produced by anaerobic digestion for energy is considered waste recycling methods.

Table 1.3 Beneficial high- and low-value by-products of animal wastes

By products	Description
Biogas	It is produced through anaerobic digestion of animal waste. It has various applications in cooking, drying, cooling, heating, electricity generation, etc.
Digestate	It is a nutrient-rich material left at the end of anaerobic digestion and can be used as a fertiliser
Animal feeds	Animal wastes can be used as source of feed nutrients for, aquaculture, livestock, pig and poultry. To maintain the nutrient composition and increase the palatability and feeding values of the waste, dehydration, ensiling, chemical and physical treatments can be used. With proper treatment, animal waste is rich in nutrients and the worms can be the source of proteins

1.9.1 Recovery of Energy from Animal Waste

Animal waste with a high organic content is degraded by microbes in both natural and artificial environments. These methods stabilize the waste, reduce pathogens and odours and convert waste streams into biogas, which are rich in methane. In the context of the circular economy, the biogas produced can be used to generate heat or electricity (Arshad et al. 2022). The economic worth of biogas is mainly determined by its ability to produce heat. The biogas produced by a biodigester can be combusted on-site in a low-treatment processing plant to offset the plant's heat and energy needs. According to Fredheim et al., biogas produced at an animal processing plant can offset on-site heating needs (20–50%). This equates to an 83% reduction in carbon emissions and savings in energy costs (Fredheim 2017). A modelled scenario for producing biogas from swine wastewater resulted in a comparable reduction in fossil fuel use (25%) (Wiedemann et al. 2016). The uses of biogas are not limited to on-site energy production, as we have described in other chapters. Biogas can be upgraded to increase CH_4 content and injected into flexible and easily storable fuel as biomethane, or it can be stored in cylinders as an alternative to liquefied natural gas (Wiedemann et al. 2016; Malomo et al. 2018; Girotto and Cossu 2017). In addition, biochar, a valuable soil conditioner, can be produced as a by-product of waste treatment processes such as pyrolysis and liquefaction (Maroušek et al. 2019).

1.9.2 Animal Waste as a Source of Animal Feeds

As mentioned earlier, anaerobic digestion of waste is not only an effective treatment method and fertilizer source, but it can also be used to propagate microorganisms to produce nutrient-rich biomass. Slaughterhouse waste, including blood, can also be used as animal feed. The market for microbial-based proteins is growing due to the protein-rich biomass from microbes, which can be used as high-protein livestock feed or as high-performance feed additives for livestock (Ramirez et al. 2021). Microbes are chosen based on their lipid and protein contents, ability to thrive, productivity and efficiency in extracting nutrients from wastes. In addition to microbes such as *Chlorella* sp., *Lemna minor, Cladophora* sp., *Scenedesmus* sp., *Rhizoclonium* sp., *Rhodopseudomonas* sp., *Rhodobacter* sp. and *Ulothrix* sp., rumen microorganisms from rumen wastes rich in lignocellulosic plant fibres could be used in solid-state fermentation as a substrate to produce a product rich in protein that can be used directly in animal feed (Ramirez et al. 2021; Vadiveloo et al. 2019; Nwoba et al. 2017). According to studies, microbial biomass products are superior to both animal and plant feeds in terms of probiotic potential, protein content, essential vitamin or amino acids content, conversion efficiency and land footprint in different climates. The use of animal waste as feed for livestock and aquaculture has been shown to be cost-effective and straightforward and can meet the nutritional needs of animals (Delamare-Deboutteville et al. 2019; Ramirez et al. 2021; Matassa et al. 2015). Enzymatic digestion of solid animal waste such as cattle hair or wool residues has been shown to provide amino acids or peptides that can be used as feed supplement. In the animal production cycle, fermentation of waste materials to produce the required amino acids is an important method of recovering nutrients from waste (Navone and Speight 2018; Ramirez et al. 2021).

1.10 Conclusion

Scientists are focusing much of their attention on the current challenge of environmental sustainability. It is recognized as necessary at the highest levels and requires urgent attention at the global level because it is essential for progress. Animal waste management is a critical issue that requires the most incredible attention if sustainability is to be ensured. As society struggles to find a sustainable way to remediate polluted environments and wastes, interest in using various microorganisms has recently increased and gained importance. The potential of microbes for specific applications has attracted more attention and speculation with the development of biotechnology. The nature of microorganisms is unusual and even unpredictable. Diverse microorganisms can effectively solve numerous environmental problems. They can help decompose animal manure and return nutrients to the soil. The primary nutrients and vital components for plant health—nitrogen, potassium and phosphorus—are released through nutrient recycling. The successful development and application of microbiological waste management techniques are essential for environmental remediation and value creation.

References

Adegunloye DV, Adejumo FA (2014) Microbial Assessment of Turkey (*Meleagris ocellata* L.) and Duck (*Anas platyrhynchos* L.) Faeces (Droppings) in Akure metropolis. Adv Microbiol 04:774–779. https://doi.org/10.4236/aim.2014.412085

Akari M, Uchida Y (2021) Survival rates of microbial communities from livestock waste to soils: a comparison between compost and digestate. Appl Environ Soil Sci 2021:1–15. https://doi.org/10.1155/2021/6645203

Akdeniz N (2019) A systematic review of biochar use in animal waste composting. Waste Manage 88:291–300. https://doi.org/10.1016/j.wasman.2019.03.054

Al-Maliki S, AL-Masoudi M (2018) Interactions between mycorrhizal fungi tea wastes, and algal biomass affecting the microbial community, soil structure, and alleviating of salinity stress in corn yield (Zea mays L.). Plants 7:63. https://doi.org/10.3390/plants7030063

Ameen A, Ahmad J, Raza S (2016) Effect of pH and moisture content on composting of municipal solid waste. Int J Sci Res Publ 6. http://www.ijsrp.org/research-paper-0516/ijsrp-p5310.pdf

Arshad M, Ansari AR, Qadir R, Tahir H, Nadeem A, Mehmood T, Alhumade H, Khan N (2022) Green electricity generation from biogas of cattle manure: an assessment of potential and feasibility in Pakistan. Front Energy Res 2022:1256

Arshad M, Javed S, Ansari AR, Fatima A, Shahzad MI (2021) Biogas: a promising clean energy technology. Bioenergy Resour Technol 1:91–120

Arshad M (2017) Clean and sustainable energy technologies. In: Clean energy for sustainable development. Academic Press, pp 73–89

Batstone DJ, Keller J, Angelidaki I et al (2002) The IWA anaerobic digestion model no 1 (ADM1). Water Sci Technol 45:65–73. https://doi.org/10.2166/wst.2002.0292

Bernal MP, Alburquerque JA, Moral R (2009) Composting of animal manures and chemical criteria for compost maturity assessment. A Review. Bioresour Technol 100:5444–5453. https://doi.org/10.1016/j.biortech.2008.11.027

Bezanson GS, Ells TC, Prange RK (2014) Effect of composting on microbial contamination and quality of fresh fruits and vegetables—a mini-review. Acta Horticulturae, 631–638. https://doi.org/10.17660/actahortic.2014.1018.70

Blaiotta G, Cerbo AD, Murru N et al (2016) Persistence of bacterial indicators and zoonotic pathogens in contaminated cattle wastes. BMC Microbiol 16:1–11. https://doi.org/10.1186/s12866-016-0705-8

Brandelli A, Sala L, Kalil SJ (2015) Microbial enzymes for bioconversion of poultry waste into added-value products. Food Res Int 73:3–12. https://doi.org/10.1016/j.foodres.2015.01.015

Briški F, Domanovac MV (2017) Environmental microbiology. Phys Sci Rev 2. https://doi.org/10.1515/psr-2016-0118

Burkholder JA, Libra B, Weyer P et al (2007) Impacts of waste from concentrated animal feeding operations on water quality. Environ Health Perspect 115:308–312. https://doi.org/10.1289/ehp.8839

Burton CH (2009) Reconciling the new demands for food protection with environmental needs in the management of livestock wastes. Bioresour Technol 100:5399–5405

Cai L, Gong X, Sun X et al (2018) Comparison of chemical and microbiological changes during the aerobic composting and vermicomposting of green waste. PLoS ONE 13:e0207494. https://doi.org/10.1371/journal.pone.0207494

Cai X, Huang X, Ji H et al. (2019) A study on aerobic composting of organic waste. Author(s)

Cambra-López M, Aarnink AJA, Zhao Y et al (2010) Airborne particulate matter from livestock production systems: A review of an air pollution problem. Environ Pollut 158:1–17. https://doi.org/10.1016/j.envpol.2009.07.011

Cendron F, Niero G, Carlino G et al (2020) Characterizing the fecal bacteria and archaea community of heifers and lactating cows through 16S rRNA next-generation sequencing. J Appl Genet 61:593–605. https://doi.org/10.1007/s13353-020-00575-3

Chen Q, Liu B, Wang J et al (2017) Diversity and dynamics of the bacterial community involved in pig manure biodegradation in a microbial fermentation bed system. Ann Microbiol 67:491–500. https://doi.org/10.1007/s13213-017-1278-y

Chen Y, Li X, Li S, Xu Y (2020) Effect of C/N ration on disposal of pig carcass by co-composting with swine manure: experiment at laboratory scale. Environ Technol, 1–11. https://doi.org/10.1080/09593330.2020.1760358

Chibuike GU (2013) Use of mycorrhiza in soil remediation: a review. Sci Res Essays 8:679–1687. https://doi.org/10.5897/sre2013.5605

Chinakwe EC, Ibekwe VI, Ofoh MC et al (2019) Effect of temperature changes on the bacterial and fungal succession patterns during composting of some organic wastes in greenhouse. J Adv Microbiol 15:1–10. https://doi.org/10.9734/jamb/2019/v15i130075

Clemens J, Huschka A (2001) The effect of biological oxygen demand of cattle slurry and soil moisture on nitrous oxide emissions. Nutr Cycl Agroecosyst 59:193–198. https://doi.org/10.1023/a:1017562603343

Cliver DO (2009) Disinfection of animal manures food safety and policy. Biores Technol 100:5392–5394. https://doi.org/10.1016/j.biortech.2009.04.038

Combalbert S, Bellet V, Dabert P et al (2012) Fate of steroid hormones and endocrine activities in swine manure disposal and treatment facilities. Water Res 46:895–906. https://doi.org/10.1016/j.watres.2011.11.074

Cox P, Griffith M, Angles M et al (2005) Concentrations of pathogens and indicators in animal feces in the Sydney watershed. Appl Environ Microbiol 71:5929–5934. https://doi.org/10.1128/aem.71.10.5929-5934.2005

Delamare-Deboutteville J, Batstone DJ, Kawasaki M et al (2019) Mixed culture purple phototrophic bacteria is an effective fishmeal replacement in aquaculture. Water Res X 4:100031

Deviney A, Classen J, Bruce J, Sharara M (2020) Sustainable swine manure management: a tale of two agreements. Sustainability 13:15. https://doi.org/10.3390/su13010015

Dinh Tuan V, Porphyre V, Farinet J-L, Duc Toan T (2006) Composition of animal manure and co-products. In: Porphyre V, Coi NQ (eds) Pig production development, animal-waste management and environment protection: a case study in Thai Binh Province, Northern Vietnam. CIRAD-PRISE publications, pp 127–143

Dobermann A, Dawe D, Roetter RP, Cassman KG (2000) Reversal of rice yield decline in a long-term continuous cropping experiment. Agron J 92:633–643. https://doi.org/10.2134/agronj2000.924633x

Dowd SE, Callaway TR, Wolcott RD et al (2008) Evaluation of the bacterial diversity in the feces of cattle using 16S rDNA bacterial tag-encoded FLX amplicon pyrosequencing (bTEFAP). BMC Microbiol 8:125. https://doi.org/10.1186/1471-2180-8-125

Durán-Lara EF, Valderrama A, Marican A (2020) Natural organic compounds for application in organic farming. Agriculture 10:41. https://doi.org/10.3390/agriculture10020041

Edwards DR, Daniel TC (1992) Environmental impacts of on-farm poultry waste disposal—A review. Biores Technol 41:9–33. https://doi.org/10.1016/0960-8524(92)90094-e

Eliot E (2015) Disposal and management of solid waste. In: Disposal and management of solid waste. CRC Press, pp 128–133

Emmanuel-Akerele H, Adamolekun P (2021) Microbiological assessment of poultry droppings water and soil under deep litter (Dl) and battery cage (Bl) systems within Lagos, Nigeria. Malays J Appl Sci 6:80–98. https://doi.org/10.37231/myjas.2021.6.1.279

Epps AV, Blaney L (2016) Antibiotic residues in animal waste: occurrence and degradation in conventional agricultural waste management practices. Curr Pollut Rep 2:135–155. https://doi.org/10.1007/s40726-016-0037-1

Etesami H, Jeong BR, Glick BR (2021) Contribution of arbuscular mycorrhizal fungi phosphate solubilizing bacteria, and silicon to P uptake by plant. Front Plant Sci 12:699618. https://doi.org/10.3389/fpls.2021.699618

Eze JI, Okonkwo TM (2013) Comparative study of composting and anaerobic digestion as a means of animal manure stabilization: a case of cow dung. Int J Sci Eng Res 4

Fernández FGA, Gómez-Serrano C, Fernández-Sevilla JM (2018) Recovery of nutrients from wastewaters using microalgae. Front Sustain Food Syst 2:59. https://doi.org/10.3389/fsufs.2018.00059

Fernandez-Bayo JD, Simmons CW, VanderGheynst JS (2020) Characterization of digestate microbial community structure following thermophilic anaerobic digestion with varying levels of green and food wastes. J Ind Microbiol Biotechnol 47:1031–1044. https://doi.org/10.1007/s10295-020-02326-z

Fredheim L (MGS Johns M) (2017) Design, measurement and verification of abattoir wastewater emissions reduction and biogas capture to offset natural gas/coal consumption. MLA, Sydney, NSW, Australia

Gilchrist MJ, Greko C, Wallinga DB et al (2007) The Potential role of concentrated animal feeding operations in infectious disease epidemics and antibiotic resistance. Environ Health Perspect 115:313–316. https://doi.org/10.1289/ehp.8837

Giola P, Basso B, Pruneddu G et al (2012) Impact of manure and slurry applications on soil nitrate in a maize-triticale rotation: Field study and long term simulation analysis. Eur J Agron 38:43–53. https://doi.org/10.1016/j.eja.2011.12.001

Girija D, Deepa K, Xavier F et al (2013) Analysis of cow dung microbiota—A metagenomic approach. Indian J Biotechnol 12:372–378

Girotto F, Cossu CA (2017) Animal waste and waste animal by-products generated along the livestock breeding and meat food chain. Waste Manage 70:1–2. https://doi.org/10.1016/j.wasman.2017.11.028

Guo R, Li G, Jiang T et al (2012) Effect of aeration rate C/N ratio and moisture content on the stability and maturity of compost. Biores Technol 112:171–178. https://doi.org/10.1016/j.biortech.2012.02.099

Gupta KK, Rana D (2016) Isolation and evaluation of cow dung bacteria for their antimicrobial potential. Biotechnol Int 9:47–54

Hagey JV, Bhatnagar S, Heguy JM et al (2019) Fecal Microbial communities in a large representative cohort of California dairy cows. Front Microbiol 10:6355–6364. https://doi.org/10.3389/fmicb.2019.01093

Haug RT (2018) Composting systems. In: The practical handbook of compost engineering. Routledge, pp 21–93

Hoitink HAJ, Boehm MJ (1999) Biocontrol within the context of soil microbial communities: a substrate-dependent phenomenon. Annu Rev Phytopathol 37:427–446. https://doi.org/10.1146/annurev.phyto.37.1.427

Hong S-W, Park J, Jeong H, Kim M (2021) Evaluation of the microbiome composition in particulate matter inside and outside of pig houses. J Anim Sci Technol 63:640–650. https://doi.org/10.5187/jast.2021.e52

Hutchison ML, Walters LD, Avery SM et al (2005) Analyses of livestock production, waste storage, and pathogen levels and prevalences in farm manures. Appl Environ Microbiol 71:1231–1236

Jacoby R, Peukert M, Succurro A et al (2017) The role of soil microorganisms in plant mineral nutrition current knowledge and future directions. Front Plant Sci 8. https://doi.org/10.3389/fpls.2017.01617

Jambon I, Thijs S, Weyens N, Vangronsveld J (2018) Harnessing plant-bacteria-fungi interactions to improve plant growth and degradation of organic pollutants. J Plant Interact 13:119–130. https://doi.org/10.1080/17429145.2018.1441450

Jang J, Hur H-G, Sadowsky MJ et al (2017) Environmental *Escherichia coli* ecology and public health implications-a review. J Appl Microbiol 123:570–581. https://doi.org/10.1111/jam.13468

Jansa J, Bukovská P, Gryndler M (2013) Mycorrhizal hyphae as ecological niche for highly specialized hypersymbionts or just soil free-riders? Front Plant Sci 4:134. https://doi.org/10.3389/fpls.2013.00134

Jdruchniewicz A (2018) Cyclical fluctuations in the production of polish agriculture. Probl Agric Econ 357:117–140. https://doi.org/10.30858/zer/100714

Jiang X, Deng L, Meng Q et al (2020) Fungal community succession under influence of biochar in cow manure composting. Environ Sci Pollut Res Int 27:9658–9668

Jusoh ML, Manaf LA, Latiff PA (2013) Composting of rice straw with effective microorganisms (EM) and its influence on compost quality. Iranian J Environ Health Sci Eng 10:17

Kim H (2009) Fungal diversity in composting process of pig manure and mushroom cultural waste based on partial sequence of large subunit rRNA. J Microbiol Biotechnol 19(8):743–748. https://doi.org/10.4014/jmb.0807.455

Kim SY, Pramanik P, Bodelier PLE, Kim PJ (2014) Cattle manure enhances methanogens diversity and methane emissions compared to swine manure under rice paddy. PLoS ONE 9:e113593. https://doi.org/10.1371/journal.pone.0113593

Kim E, Lee D-H, Won S, Ahn H (2015) Evaluation of optimum moisture content for composting of beef manure and bedding material mixtures using oxygen uptake measurement. Asian Australas J Anim Sci 29:753–758. https://doi.org/10.5713/ajas.15.0875

Kristiansen A, Saunders AM, Hansen AA et al (2012) Community structure of bacteria and fungi in aerosols of a pig confinement building. FEMS Microbiol Ecol 80:390–401. https://doi.org/10.1111/j.1574-6941.2012.01305.x

Kumar H, Jang Y, Kim K et al (2020) Compositional and functional characteristics of swine slurry microbes through 16S rRNA metagenomic sequencing approach. Animals 10:1372. https://doi.org/10.3390/ani10081372

Laughlin RJ, Stevens RJ, Müller C, Watson CJ (2008) Evidence that fungi can oxidize NH^{4+} to NO^{3-} in a grassland soil. Eur J Soil Sci 59:285–291. https://doi.org/10.1111/j.1365-2389.2007.00995.x
Leffert ML, Clark GA, Hutchinson SL, Barden CJ (2008) Evaluation of poplar trees irrigated with livestock lagoon wastewater. Trans ASABE 51:2051–2060. https://doi.org/10.13031/2013.25408
Leininger S, Urich T, Schloter M et al (2006) Archaea predominate among ammonia-oxidizing prokaryotes in soils. Nature 442:806–809. https://doi.org/10.1038/nature04983
Leung K, Topp E (2001) Bacterial community dynamics in liquid swine manure during storage: molecular analysis using DGGE/PCR of 16S rDNA. FEMS Microbiol Ecol 38:169–177. https://doi.org/10.1111/j.1574-6941.2001.tb00895.x
Leytem AB, Dungan RS, Bjorneberg DL, Koehn AC (2011) Emissions of ammonia methane, carbon dioxide, and nitrous oxide from dairy cattle housing and manure management systems. J Environ Qual 40:1383–1394. https://doi.org/10.2134/jeq2009.0515
Li Y, Zhao J, Krooneman J, Euverink GJW (2021) Strategies to boost anaerobic digestion performance of cow manure: laboratory achievements and their full-scale application potential. Sci Total Environ 755:142940. https://doi.org/10.1016/j.scitotenv.2020.142940
Lim J-S, Yang SH, Kim B-S, Lee EY (2018) Comparison of microbial communities in swine manure at various temperatures and storage times. Asian Australas J Anim Sci 31:1373–1380. https://doi.org/10.5713/ajas.17.0704
Ling ZH, Guo H (2014) Contribution of VOC sources to photochemical ozone formation and its control policy implication in Hong Kong. Environ Sci Policy 38:180–191. https://doi.org/10.1016/j.envsci.2013.12.004
Lovanh N, Cook KL, Rothrock MJ et al (2007) Spatial shifts in microbial population structure within poultry litter associated with physicochemical properties. Poult Sci 86:1840–1849
Lu J, Sanchez S, Hofacre C et al (2003) Evaluation of broiler litter with reference to the microbial composition as assessed by using 16S rRNA and functional gene markers. Appl Environ Microbiol 69:901–908. https://doi.org/10.1128/aem.69.2.901-908.2003
MacSAFLEY LM, Dupoldt C, Geter F et al (1992) Agricultural waste management system component design. Agricultural waste management field handbook. Department of Agriculture, Soil Conservation Service, Washington, DC, pp 1–85
Maeda K, Hanajima D, Toyoda S et al (2011) Microbiology of nitrogen cycle in animal manure compost. Microb Biotechnol 4:700–709. https://doi.org/10.1111/j.1751-7915.2010.00236.x
Maillard É, Angers DA (2013) Animal manure application and soil organic carbon stocks: a meta-analysis. Glob Change Biol 20:666–679. https://doi.org/10.1111/gcb.12438
Malińska K, Zabochnicka-Świtek M (2013) Selection of bulking agents for composting of sewage sludge. Environ Protect Eng 39. https://doi.org/10.37190/epe130209
Malomo GA, Madugu AS, Bolu SA (2018) Sustainable animal manure management strategies and practices. In: Agricultural waste and residues. InTech
Mamun MAA, Sandeman M, Rayment P et al (2019) The composition and stability of the faecal microbiota of Merino sheep. J Appl Microbiol 128:280–291. https://doi.org/10.1111/jam.14468
Mao S, Zhang R, Wang D, Zhu W (2012) The diversity of the fecal bacterial community and its relationship with the concentration of volatile fatty acids in the feces during subacute rumen acidosis in dairy cows. BMC Vet Res 8:237. https://doi.org/10.1186/1746-6148-8-237
Maroušek J, Strunecký O, Stehel V (2019) Biochar farming: defining economically perspective applications. Clean Technol Environ Policy 21:1389–1395. https://doi.org/10.1007/s10098-019-01728-7
Martinez J, Dabert P, Barrington S, Burton C (2009a) Livestock waste treatment systems for environmental quality, food safety, and sustainability. Bioresour Technol 100:5527–5536
Martinez J, Dabert P, Barrington S, Burton C (2009b) Livestock waste treatment systems for environmental quality food safety, and sustainability. Biores Technol 100:5527–5536. https://doi.org/10.1016/j.biortech.2009.02.038
Martín-Marroquín JM, Hidalgo D (2014) Livestock waste: fears and opportunities. In: Environment energy and climate change I. Springer International Publishing, pp 341–373

Massé DI, Croteau F, Masse L (2007) The fate of crop nutrients during digestion of swine manure in psychrophilic anaerobic sequencing batch reactors. Biores Technol 98:2819–2823. https://doi.org/10.1016/j.biortech.2006.07.040

Matassa S, Batstone DJ, Hülsen T et al (2015) Can direct conversion of used nitrogen to new feed and protein help feed the world? Environ Sci Technol 49:5247–5254

Mawdsley JL, Bardgett RD, Merry RJ et al (1995) Pathogens in livestock waste their potential for movement through soil and environmental pollution. Appl Soil Ecol 2:1–15. https://doi.org/10.1016/0929-1393(94)00039-a

May S, Romberger DJ, Poole JA (2012) Respiratory health effects of large animal farming environments. J Toxicol Environ Health Part B 15:524–541. https://doi.org/10.1080/10937404.2012.744288

McMullen J, Fasina OO, Wood CW, Feng Y (2005) Storage and handling characteristics of pellets from poultry litter. Appl Eng Agric 21:645–651. https://doi.org/10.13031/2013.18553

Mehta CM, Sirari K (2018) Comparative study of aerobic and anaerobic composting for better understanding of organic waste management: a mini review. Plant Arch 18:44–48

Millner P, Ingram D, Mulbry W, Arikan OA (2014) Pathogen reduction in minimally managed composting of bovine manure. Waste Manage 34:1992–1999. https://doi.org/10.1016/j.wasman.2014.07.021

Mingyan (2011) Screening of complex thermophilic microbial community and application during municipal solid waste aerobic composting. Afr J Biotechnol 10. https://doi.org/10.5897/ajb10.2559

Möller K (2015) Effects of anaerobic digestion on soil carbon and nitrogen turnover N emissions, and soil biological activity. A Review. Agron Sustain Dev 35:1021–1041. https://doi.org/10.1007/s13593-015-0284-3

Möller K, Müller T (2012) Effects of anaerobic digestion on digestate nutrient availability and crop growth: a review. Eng Life Sci 12:242–257. https://doi.org/10.1002/elsc.201100085

Müller ZO (1980) feed from animal wastes: state of knowledge. Food and Agriculture Organization of the United Nations

Munshi SK, Roy J, Noor R (2019) Microbiological investigation and determination of the antimicrobial potential of cow dung samples. Stamford J Microbiol 8:34–37. https://doi.org/10.3329/sjm.v8i1.42437

Nakai Y (2001) Animal waste management and microorganisms. Nihon Chikusan Gakkaiho 72:1–13

Narula R, Grimberg SJ, Rogers S, Mondal S (2011) Pathogen reduction and factors responsible for pathogen reduction in dairy farm operations treating dairy manure. Biol Eng Trans 4:115–131. https://doi.org/10.13031/2013.39813

Nauanova A, Yerpashev D, Shakhabayeva G et al (2020) Identification and screening of microorganisms common in poultry manure. Syst Rev Pharm 11:1582–1588

Navone L, Speight R (2018) Understanding the dynamics of keratin weakening and hydrolysis by proteases. PLoS ONE 13:e0202608

Neher DA, Weicht TR, Bates ST et al (2013) Changes in bacterial and fungal communities across compost recipes preparation methods, and composting times. PLoS ONE 8:e79512. https://doi.org/10.1371/journal.pone.0079512

Nimmi G, Animal waste: potential, characteristics and disposal. Waste management

Nodar R, Acea MJ, Carballas T (1990) Microbial composition of poultry excreta. Biol Wastes 33:95–105. https://doi.org/10.1016/0269-7483(90)90150-q

Novinscak A, Filion M, Surette C, Allain C (2008) Application of molecular technologies to monitor the microbial content of biosolids and composted biosolids. Water Sci Technol 57:471–477. https://doi.org/10.2166/wst.2008.019

Nwoba EG, Moheimani NR, Ubi BE et al (2017) Macroalgae culture to treat anaerobic digestion piggery effluent (ADPE). Bioresour Technol 227:15–23

Ozi FZ, Boutaleb N, Hadidi M et al (2022) Production of bio-fertilizer by biotransformation of poultry waste enriched with molasses and algae. Environ Qual Manage. https://doi.org/10.1002/tqem.21868
Pan I, Dam B, Sen SK (2011) Composting of common organic wastes using microbial inoculants. 3 Biotech 2:127–134. https://doi.org/10.1007/s13205-011-0033-5
Pandey P, Chiu C, Miao M et al (2018) 16S rRNA analysis of diversity of manure microbial community in dairy farm environment. PLoS ONE 13:e0190126. https://doi.org/10.1371/journal.pone.0190126
Patton J, Turner JC (2008) Effect of animal waste amendments on the chemical composition of surface stable and non-stable aggregates in a cultivated calcareous silt loam. Trans Missouri Acad Sci 42:39–44. https://doi.org/10.30956/0544-540x-42.2008.39
Penakalapati G, Swarthout J, Delahoy MJ et al (2017) Exposure to animal feces and human health: a systematic review and proposed research priorities. Environ Sci Technol 51:11537–11552. https://doi.org/10.1021/acs.est.7b02811
Peng X, Wilken SE, Lankiewicz TS et al (2021) Genomic and functional analyses of fungal and bacterial consortia that enable lignocellulose breakdown in goat gut microbiomes. Nat Microbiol 6:499–511. https://doi.org/10.1038/s41564-020-00861-0
Phillips J, Huhnke R, Atiyeh H (2017) Syngas fermentation: a microbial conversion process of gaseous substrates to various products. Fermentation 3:28. https://doi.org/10.3390/fermentation3020028
Puyol D, Batstone DJ, Hülsen T et al (2017) Resource recovery from wastewater by biological technologies: opportunities challenges, and prospects. Front Microbiol 7:2106. https://doi.org/10.3389/fmicb.2016.02106
Qin H, Lang H, Yang H (2013) Characterization of the methanogen community in a household anaerobic digester fed with swine manure in China. Appl Microbiol Biotechnol 97:8163–8171. https://doi.org/10.1007/s00253-013-4957-z
Ramachandra TV, Bharath S, Gupta N (2018) Modelling landscape dynamics with LST in protected areas of Western Ghats Karnataka. J Environ Manage 206:1253–1262. https://doi.org/10.1016/j.jenvman.2017.08.001
Ramirez J, McCabe B, Jensen PD et al (2021) Wastes to profit: a circular economy approach to value-addition in livestock industries. Anim Prod Sci 61:541. https://doi.org/10.1071/an20400
Randhawa GK, Kullar JS (2011) Bioremediation of pharmaceuticals pesticides, and petrochemicals with Gomeya/cow dung. ISRN Pharmacol 2011:1–7. https://doi.org/10.5402/2011/362459
Rastogi M, Nandal M, Khosla B (2020) Microbes as vital additives for solid waste composting. Heliyon 6:e03343. https://doi.org/10.1016/j.heliyon.2020.e03343
Sawant AA, Hegde NV, Straley BA et al (2007) Antimicrobial-resistant enteric bacteria from dairy cattle. Appl Environ Microbiol 73:156–163. https://doi.org/10.1128/aem.01551-06
Schievano A, D'Imporzano G, Salati S, Adani F (2011) On-field study of anaerobic digestion full-scale plants (Part I): an on-field methodology to determine mass carbon and nutrients balance. Biores Technol 102:7737–7744. https://doi.org/10.1016/j.biortech.2011.06.006
Schiffman SS, Walker JM, Dalton P et al (2000) Potential health effects of odor from animal operations wastewater treatment, and recycling of byproducts. J Agromedicine 7:7–81. https://doi.org/10.1300/j096v07n01_02
Schloss PD, Hay AG, Wilson DB, Walker LP (2003) Tracking temporal changes of bacterial community fingerprints during the initial stages of composting. FEMS Microbiol Ecol 46:1–9
Schneider LK, Einsle O (2016) Shewanella denitrificans nitrous oxide reductase reduced apo form
Shabana II, Albakri NN, Bouqellah NA (2020) Metagenomic investigation of faecal microbiota in sheep and goats of the same ages. J Taibah Univ Sci 15:1–9. https://doi.org/10.1080/16583655.2020.1864930
Shah FA, Mahmood Q, Shah MM et al (2014) Microbial ecology of anaerobic digesters: the key players of anaerobiosis. Sci World J 2014:1–21. https://doi.org/10.1155/2014/183752
Sobsey MD, Khatib LA, Hill VR et al (2006) Pathogens in animal wastes and the impacts of waste management practices on their survival, transport and fate

Sorathiya LM, Fulsoundar AB, Tyagi KK et al (2014) Eco-friendly and modern methods of livestock waste recycling for enhancing farm profitability. Int J Recycl Organic Waste Agric 3:1–12. https://doi.org/10.1007/s40093-014-0050-6

Sun L, Han X, Li J et al (2020) Microbial community and its association with physicochemical factors during compost bedding for dairy cows. Front Microbiol 11:254. https://doi.org/10.3389/fmicb.2020.00254

Taiwo LB, Oso BA (2004) Influence of composting techniques on microbial succession, temperature and pH in a composting municipal solid waste. Afr J Biotechnol 3

Tan H, Cao L (2013) Fungal diversity in sheep (Ovis aries) and cattle (Bos taurus) feces assessed by comparison of 18S 28S and ITS ribosomal regions. Ann Microbiol 64:1423–1427. https://doi.org/10.1007/s13213-013-0787-6

Tanca A, Fraumene C, Manghina V et al (2017) Diversity and functions of the sheep faecal microbiota: a multi-omic characterization. Microb Biotechnol 10:541–554. https://doi.org/10.1111/1751-7915.12462

Tasho RP, Cho JY (2016) Veterinary antibiotics in animal waste its distribution in soil and uptake by plants: a review. Sci Total Environ 563–564:366–376. https://doi.org/10.1016/j.scitotenv.2016.04.140

Teira-Esmatges MR, Flotats X (2003) A method for livestock waste management planning in NE Spain. Waste Manage 23:917–932. https://doi.org/10.1016/s0956-053x(03)00072-2

Thilagam L, Nayak BK, Nanda A (2015) Isolation and enumeration of saprophytic and coprophilous fungi from country cow dung. J Chem Pharm Res 7:74–477

Toljander JF, Santos-GonzÃlez JC, Tehler A, Finlay RD (2008) Community analysis of arbuscular mycorrhizal fungi and bacteria in the maize mycorrhizosphere in a long-term fertilization trial. FEMS Microbiol Ecol 65:323–338. https://doi.org/10.1111/j.1574-6941.2008.00512.x

Tuan NN, Chang Y-C, Yu C-P, Huang S-L (2014) Multiple approaches to characterize the microbial community in a thermophilic anaerobic digester running on swine manure: a case study. Microbiol Res 169:717–724. https://doi.org/10.1016/j.micres.2014.02.003

Vadiveloo A, Nwoba EG, Moheimani NR (2019) Viability of combining microalgae and macroalgae cultures for treating anaerobically digested piggery effluent. J Environ Sci (china) 82:132–144

Vanotti M, Szogi A, Bernal MP, Martinez J (2009) Livestock waste treatment systems of the future: a challenge to environmental quality food safety, and sustainability. OECD Workshop. Bioresour Technol 100:5371–5373. https://doi.org/10.1016/j.biortech.2009.07.038

Waki M, Yasuda T, Fukumoto Y et al (2018) Treatment of swine wastewater in continuous activated sludge systems under different dissolved oxygen conditions: reactor operation and evaluation using modelling. Biores Technol 250:574–582. https://doi.org/10.1016/j.biortech.2017.11.078

Wan J, Wang X, Yang T et al (2021) Livestock manure type affects microbial community composition and assembly during composting. Front Microbiol 12:621126. https://doi.org/10.3389/fmicb.2021.621126

Wiedemann SG, McGahan EJ, Murphy CM (2016) Environmental impacts and resource use from Australian pork production assessed using life-cycle assessment. 1. Greenhouse gas emissions. Anim Prod Sci 56:1418. https://doi.org/10.1071/an15881

Wilkes TI (2021) Arbuscular mycorrhizal fungi in agriculture. Encyclopedia 1:1132–1154. https://doi.org/10.3390/encyclopedia1040085

Zeng G, Yu Z, Chen Y et al (2011) Response of compost maturity and microbial community composition to pentachlorophenol (PCP)-contaminated soil during composting. Biores Technol 102:5905–5911. https://doi.org/10.1016/j.biortech.2011.02.088

Zhang L, Li L, Pan X et al (2018) Enhanced growth and activities of the dominant functional microbiota of chicken manure composts in the presence of maize straw. Front Microbiol 9:1131. https://doi.org/10.3389/fmicb.2018.01131

Zhang X, Ren J, Niu H et al (2010) Composting of sewage sludge using recycled matured compost as a single bulking agent. AIP

Chapter 2
Slaughter Wastes-A Curse or Blessing: An Appraisal

Kashif Nauman, Atif Nauman, and Muhammad Arshad

Abstract Slaughterhouses are designated premises for slaughtering and processing of animals for meat. During processing, many edible and inedible by-products are produced. On further processing, these by-products could produce extra revenue for the facility vice versa and have a harmful effect on the environment. In developing countries, some edible by-products are utilised differently, while inedible by-products go wasted without technological interventions. Under these processing conditions, these wastes can cause adverse effects on the environment in the short and long term. During meat processing, ruminal contents in red meat, while feathers in poultry processing, are produced in high percentages. Different procedures like biogas production, rendering, composting, and biodiesel production could help deal with these wastes, while wastewater could be processed through physicochemical, biological, and advanced oxidation processes. These processes play a critical role in the BOD, COD, and TSS value of the treated water and make it acceptable for the regulatory institutions. In this chapter, a detailed description of these techniques and technologies is discussed to understand the use of different components of slaughterhouse waste for society's benefit, including revenue generation and, subsequently, environmental protection.

K. Nauman (✉)
Department of Meat Science and Technology, University of Veterinary and Animal Sciences, Lahore 54000, Pakistan
e-mail: drkashif@uvas.edu.pk

A. Nauman
Department of Environmental Sciences, Quaid-I-Azam University, Islamabad, Pakistan

M. Arshad
Jhang-Campus, University of Veterinary and Animal Sciences Lahore, Lahore, Pakistan

M. Arshad (ed.), *Climate Changes Mitigation and Sustainable Bioenergy Harvest Through Animal Waste*, https://doi.org/10.1007/978-3-031-26224-1_2

Graphical Abstract

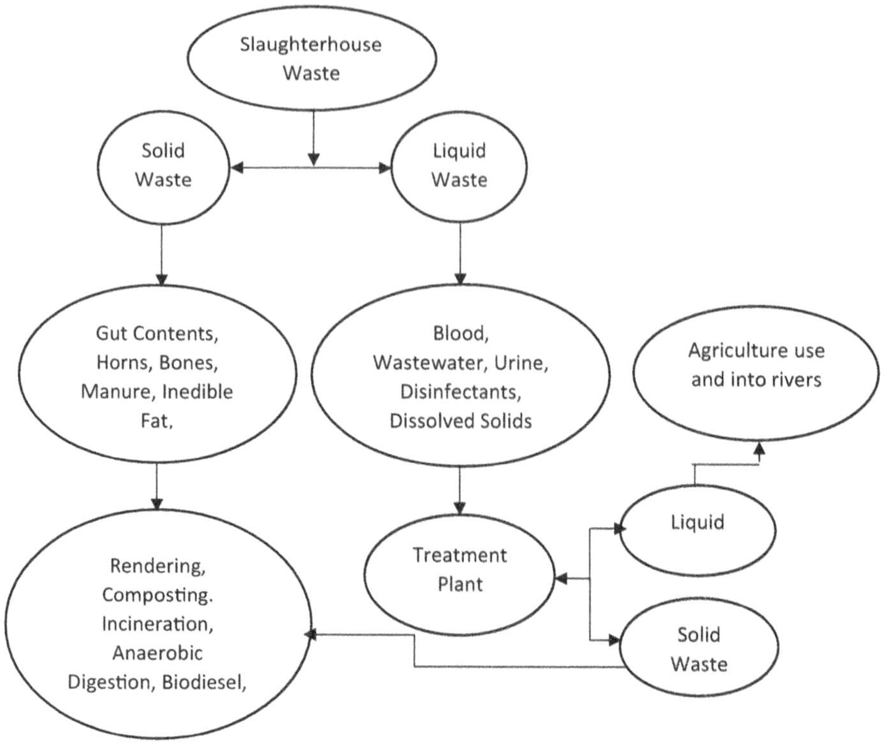

Keywords Slaughterhouse · By products · Economics · Value addition · Waste management · Processing

2.1 Introduction

The waste from the slaughterhouse consists of different parts of the animals generated during meat processing and production; blood and other by-products of animals are among this waste. This inedible waste includes animal tissues like ligaments, blood vessels, organs, integuments, bones, tendons, and feathers, contributing up to 45% to the waste of animals. Pet food manufacturing companies purchase these wastes from slaughterhouses in large amounts, which can be used as a supplement to animal and pet food (Salminen and Rintala 2002). A large amount of organic waste is also generated from the slaughterhouse, including fat, highly suspended solids, and liquids (Adeyemi and Adeyemo 2007). It is assessed that 52% of meat from goats or sheep, 50–54% from cows, 78% from each turkey, and 68–72% of every chicken is consumed, and the remaining animal waste is dumped (Tolera and Alemu 2020).

The solid waste generated from the bovine slaughterhouse is 27.5% of the animal's total live weight (275 kg/ton). During the sheep and goats slaughtering process, 17% of the animal weight and waste is produced, which is about 2.5 kg/head. The poultry shops and slaughtering plants generate about 32.5 to 37% of waste. Moreover, half of the animal by-products are unsuitable for further reuse (Mozhiarasi and Natarajan 2022) (Table 2.1).

A large volume of waste is generated from slaughterhouses globally. Primarily, slaughterhouse waste includes solids like rumen contents, bones, contents of the intestine, dung, feather, ligaments, hooves, and skin. Liquid waste includes urine, blood, internal fluids of the body, and water generated during the washing and cleaning of the animal. If the slaughterhouse waste is not managed correctly, it can contaminate the whole local environment by polluting soil, surface, and undersoil water resources. Poor waste management at landfill sites can raise health issues. Less awareness about slaughterhouse waste management resulted in poor and illegal waste management activities, which can further increase airborne waste, air pollution, highly contaminated wastewater, and infectious stormwater runoff and waste.

Animal by-products are a valuable source of protein at the industrial stage, which can be used in different processed products and applications. Nowadays, such resources are not fully utilised in producing high-value applications, resulting in low-value products. Slaughterhouse waste is highly proteinaceous, a significant source of protein, but currently, there are some hurdles in using feedstock from the valuable protein resource as bio-based product production. The reason is the often mixing of non-homogeneous material with non-proteinaceous material, which results in poor solubility and limited processibility. During the advanced utilisation of the proteinaceous slaughterhouse waste, protein is extracted and combined in the bio-based industrial production of valuable goods.

The waste with its by-products generated from the slaughterhouse can be sold out to increase the revenue of the slaughtering facility and lower operation costs. The practical use of waste trend is expected to increase in coming years because everyday industry searches to find new ways to utilise organic waste like biogas and compost.

Table 2.1 Waste generation (per cent) from animal slaughtering processes (Adhikari et al. 2018a; Mozhiarasi et al. 2020; Jayathilakan et al. 2012; Meeker and Hamilton 2006)

Type of waste	Waste generation (%)
Chicken slaughtering process	
Skin and feathers	57.37
Legs	14.8
Intestines	20.35
Other waste	<1
Lambs/cattle slaughtering process	
Manure	12
Ruminal contents	80
Blood	5
Other waste	3

The waste used for the industrial production of food for pets, animal food, and bone meal by the rendering industry is gaining attractiveness every day. From the rendering processes, poultry meal, meat, blood meal, a hydrolysed meal of feathers, animal fat, and fish meal are among the primary products generated during these processes. This waste can be further used for medical purposes. From the slaughtering facility, the generation of animal by-products and their processing in animal and pet foods production is an effective strategy for adopting the reuse concept. So, in the coming sections, various wastes, their processing, and utilisation are discussed.

2.2 Operations During Slaughterhouse and Waste Production on Each Step

Slaughterhouse operation consists of many steps, from the lairage to initial examination, slaughtering, de-hiding/deskinning, evisceration, post-mortem, deboning, and secondary and tertiary processing steps. In these processing steps, different wastes are produced, which could be divided into two categories as (Table 2.2).

During animal receiving, washing, and staying in lairage, manure, urine, odour, and wastewater are produced. Generally, these wastes are mixed with washing water and go into the sewage system. During slaughtering, animals are usually slaughtered in a killing box, hind limbs used for hanging, and then moved down through the overhead railing for slaughter procedures. During this operational step, a large volume of wastewater containing blood is produced as waste material.

Before starting the de-hiding, the head, hooves, and horns are removed as waste material. After this process, the skin/hide of the animal is removed, which is one of the most valuable wastes produced in a slaughterhouse. In another crucial processing step, evisceration, blood, the visceral organs, liver, intestines, pancreas, heart, lung,

Table 2.2 Edible and inedible parts of animals generated during the slaughtering process (Awan et al. 2015)

Edible parts	Inedible parts
Liver, heart, kidneys, and tails	Horns, hides and hair
Lungs, tongue, and tripe (stomach)	Blood vessels, fats, and bones
Chitterlings, fries (testicles), melt (spleen), natural casing (intestine), sweetbreads, and thymus or pancreatic gland, which depends on animal age	Teeth, blood integuments, and tendons
Lips, rinds, and fats	Feet, trimmings, ligaments, manure, and rumen contents
Trimmings and certain bones	Cartilage, feathers of glands, and glands
Edible by-product yield is around 12% of the live weight	Inedible parts can cover up to 45% or more of the slaughtered animal

stomach contents, paunch manure, digestive system, excretory systems, and wastewater are produced. This process removes white, green, and red offal for further processing or waste.

Deboning and cutting operations are performed in the secondary processing section. In this operation, bones are removed from the carcass and meat is cut into different cuts. During the deboning process, bones, tendons, ligaments, and fat are removed as waste material. Post-mortem inspection examines dressed carcasses and their organs, including head, pluck, and intestines, immediately after slaughter to produce hygienically wholesome meat under adequate light by a qualified meat inspector. During post-mortem, abnormalities and disease conditions noted during the ante-mortem are correlated in deciding to accept and reject the carcass. Only safe and wholesome meat will be provided to the market. In this operation, trimmed carcasses, rejected organs, carcass parts, or carcasses could be presented as waste (Mozhiarasi and Natarajan 2022).

2.3 Types of Waste and Their Handling

2.3.1 Solid Waste Process

In the meat industry, by-products that are in solid form and not further processed can happen are considered solid waste. It includes the remaining material that is in non-edible form and can be sold out, including bones, hoofs, horns, integuments, skin, ligaments, and cartilage tendons, the contents of the gastrointestinal tract and internal body organs. The most significant by-products obtained from the animals are intestinal products used as sausage casing, sports gut, musical string, and surgical sutures. Bone products are used in the gelatine formation, row bone, crushed bone, bone meal, bone ash, handicrafts, livestock feed fertiliser, and biodiesel.

2.3.1.1 Solid Waste Management

Solid waste management consists of several processes such as at the point of waste generation, handling of on-site waste, collection, processing and storage of waste, their transformation, transport to other locations, resource recovery, recycling, reuse, and final disposal (Mozhiarasi and Natarajan 2022). Different processes can manage solids produced in the slaughterhouse, including rendering, composting, incineration, anaerobic digestion, pyrolysis, landfilling, and biodiesel production. Among these processing processes, rendering, composting, and anaerobic digestion are well studied and recommended for application in the field as it destroys many pathogenic and public health significance bacteria, viruses, and parasites.

2.3.1.2 Liquid Process Waste

Liquid waste includes urine, oils, wastewater, fats, sludge, used oil, grease, unsafe household liquids, and gases. Unsafe disposal of this liquid waste is dangerous to human health and the environment. The water requirement on average to produce one-ton cattle meat is 15,500 m^3, for sheep 4000 m^3, for pigs 4800 m^3, and poultry meat production 4000 m^3 (Hoekstra and Chapagain 2006).

2.3.1.3 Liquid Waste Management

A large volume of wastewater is generated while processing meat in the slaughtering facility. Slaughterhouse wastewaters have high organic matter content like fat, protein, and several microorganisms. Mostly, the wastewater from the slaughtering facility is directly discharged to the municipal sewage system without any treatment.

Different processes and technologies are used during the treatment of slaughterhouse wastewater. The first process in the handling of wastewater is preliminary treatment. This step is inexpensive and effectively removes the large particles using specific size screens. These screens stop the large fat particles, feathers, and meat pieces. This step protects the downfall treatment machinery. These steps also help reduce the wastewater biological oxygen demand (BOD), which removes large particles of organic content during this treatment step. The primary treatment method removes the small particles of organic content. During this step, heavy particles are sedimented, and light particles are floated using inexpensive equipment.

In the next step, different techniques like dissolved air flotation, coagulation-flocculation and sedimentation, advanced oxidation process, etc.

2.4 Usage of Slaughterhouse Waste

2.4.1 Blood as a Resource of Bioactive Compounds

In the meat industry, during the slaughtering of animals, the generation of blood is an inevitable by-product and primary waste (effluent). The percentage of blood in the wastewater is 4.5% of live weight, which equals the 10% of the protein available in a slaughtering animal. It is estimated that 75% of blood is allowed to drain during the slaughtering of an animal from the body, which is 4% of the animal's live weight and 6–7% of the carcass meat content. Dried blood contents are rich in lysine, high in protein (80–90%) and have a reasonable reuse price (Wismer-Pedersen 1988). So, it appears as a significant waste that can cause contamination if not removed properly. As per the description of the resultant product, collection methods are adopted. Blood is collected directly from the animal's slaughtering site for medicinal products so contaminants can be minimised.

In contrast, for a blood meal and relevant products, blood is collected through underground pipelines from where blood moves out of the slaughterhouse/industry to blood collection tanks for rendering plants situated nearby or transported to the blood processing units present in other areas through trucks. Blood is a readily available source of protein from the slaughtering animal. From the slaughterhouses, blood supply offers economic, nutritional, and environmental benefits.

There are some religious restrictions on the use of blood; for example, in Jewish shechita and Islamic sharia, there is a prohibition on the use of blood. In some cultures, there are hurdles in using whole blood in food. The hydrolysis process for transforming blood into useable products may present a way for the blood products' acceptability in these cultures. The opinion about using blood products from religious scholars is yet to be presented.

Bioactive food molecules are naturally occurring non-essential elements from plants, animals, and marine that influence biochemical, physiological, and metabolic processes and other valuable effects beyond fundamental nutritious activities (Kris-Etherton et al. 2004). Bioactive compounds include oligosaccharides, peptides, enzymes, biopolymers, fatty acids, and water-soluble minerals. These substances can be naturally present in food sources (gastrointestinal digestion) by microbial fermentation or industrial enzymatic digestion (Hernández-Ledesma et al. 2011). These substances are absorbed primarily through the small intestine or into circulation. These chemicals may function as antioxidants or antimicrobials, cholesterol and blood pressure reducers, and antithrombotic agents. Some bioactive may have many functions (Korhonen and Pihlanto 2003).

The bioactive chemical compounds have the therapeutic and preventative capability for diseases of humans, work as modulation of bio-system function, play the role as substrates for biomolecule and bio-structure creation, antimicrobial properties, the capacity to transport medications, enzymes, and nutrients (Bah et al. 2013).

Due to their various biological effects, such as lowering cardiovascular disease risk factors and antioxidant, antimutagenic, antithrombotic, anticarcinogenic, antiallergenic, anti-inflammatory, antibacterial activity, blood pressure and cholesterol-lowering, and immunoregulatory effect, thus bioactive substances are crucial for the health of humans (Parvathy et al. 2009). In preventing chronic diseases, their role as bioactive substances is significant, including carotenoids, organic acids, vitamins, polyphenols, phytosterols, omega-3 fatty acids, nucleotides, and nucleosides have gained considerable attention.

2.4.1.1 Application of Animal Blood in Food

Animal blood and its components are used as a food or additive in different parts of the world. Some examples of the use of blood as a whole are as the source of protein in blood sausage (Spanish) (Santos et al. 2003), fortification of iron by using bovine haemoglobin in cookies (Walter et al. 1993), use of fat replacer in ham pate by bovine plasma and globin (Viana et al. 2005), fat replacer in Bologna sausage by using animal plasma (Cofrades et al. 2000), production of binder in restructured meat products from thrombin, fibrinogen and porcine transglutaminase (Tseng et al. 2006), protease inhibitor from plasma porcine (Visessanguan et al. 2000), generation of protease inhibitor from porcine plasma and its use in the Surimi (Rawdkuen et al. 2004), plasma of bovine as a white replacer of eggs in cakes (Myhara and Kruger 1998), and plasma of bovine in the cakes as egg white replacer and the use of a stabiliser in the minced meats produced from processed bovine plasma (Furlán et al. 2010).

2.4.1.2 Medicinal Applications of Blood

Several medically significant proteins, such as albumin, plasmin, thrombin, prothrombin, fibrinogen, immunoglobin M, immunoglobin G, haemoglobin, prothrombin, and transferrin, are present in blood as components (Bah et al. 2013). Animals who lose blood or fluids might be given purified bovine albumin to compensate for it. It is a vaccine stabiliser used to screen for the Rh factor in human subjects. Additionally, testing for antibiotic sensitivity employs it (Tanaka et al. 2001). The porcine plasmin enzyme dissolves blood clots in heart attack victims. Bovine thrombin aids in blood coagulation, the healing of wounds, and the retention of skin grafts (Jayathilakan et al. 2012). In serology laboratories, bovine fibrinogen is used as a standard reagent for blood coagulation. Bovine prothrombin is active in topical surgical haemostatic treatments and is a precursor to thrombin generation and purification.

Bovine plasma is utilised as a growing medium in laboratories for porphyrin and probiotic lactobacilli, which are employed in human medicine. Blood products nourish the tissue culture medium (Hyun and Shin 1998). Heme iron polypeptide tonics derived from animal haemoglobin are widely offered by businesses to cure an iron shortage. Bio-peptide tonics assert to lower blood pressure, cholesterol, and blood sugar, improve mineral absorption, boost immunity, and act as opioids (Opioid peptides that bind to opioid receptors as short sequences of amino acids in the brain). The best antigenotoxic impact was shown by bovine blood albumin peptic hydrolysate. Commercial food and nutraceuticals contain peptides from animal blood (non-medicinal nutrients used as supplements). Some businesses openly state that their dietary supplement contains bovine serum, yet many conceal it.

2.4.1.3 Animal Blood in Other Applications

The use of animal blood other than in food can be used in industrial applications. Examples of blood used as a protein source are generated from the slaughtering of sheep, cows, and chickens, which use in the production of pet and animal food. Animal blood has medical field-related applications like plasmin and thrombin, which are recovered from the blood of a slaughtered animal (Tanaka et al. 2001; Pierce et al. 2005; Jayathilakan et al. 2012).

2.4.2 Composting

Organic materials (such as waste from slaughterhouses) are destroyed through the actions of a subsequent set of bacteria during the controlled biological process known as composting (Dees and Ghiorse 2001). The composting slaughterhouse waste may consist of inorganic and organic wastes, paunch, municipal solid waste, pig or poultry manure, sewage sludge, and garden waste (Asses et al. 2019). Sanabria-León et al. (2007) describes that compost made from slaughterhouse waste is often a stable product that may improve soil quality and provide plants with nutrients. Understanding compost's chemical and microbiological characteristics can help to comprehend the composting process that organic leftovers undergo. This method has several advantages, including decreasing environmental pollution, valuable by-products, and eradicating most infections (NABC 2004). The survival and spread of pathogens in animals, people, and plants during the composting process is a common worry.

Composting should not harm people's and animals' health when done under strict care (Franke-Whittle and Insam 2013). Prions and bacteria that generate spores are pathogens that cannot be eliminated. Composting is used to serve as an acceptable, inexpensive, potentially eco-friendly approach to the disposal of was generated from the slaughterhouse. This process significantly lowers the number of pathogens during decomposition due to the temperature rise. The microbial community which takes part in the organic content breakdown is usually fungi and actinomycetes. During breakdown, the microorganisms convert the organic matter into water, carbon dioxide, humus, and heat (Epstein 1997). In this process, anaerobic digestion of waste material takes place, reducing the chance of disease separation by microbes and parasites. Composting technique has certain disadvantages, including space requirements for the remote site, which increases transportation costs. This is challenging to maintain the high-rate conditions of the compost turning ratio; when the required conditions are not met, the compost will be coarser and have a significant oversized fraction of compost. Mainly composting produces leachate, dust water, odour, litter, contamination, vectors, pests, fire, and noise under less control. It is more difficult to achieve consistent results.

There are two types of composting exits; one is windrow composting, while the other is in-vessel composting. Windrow composting is also called aerated (turned) composting (Cekmecelioglu et al. 2005). This composting involves converting organic content into long piles called windrows. Windrows are aerated occasionally by mechanical or manual turning. Windrow composting facility requires equipment, large tracts of land, and labour. In warm, arid climates, windrows are covered to prevent water evaporation. Leachate is a liquid produced during composting that can contaminate groundwater, while odours must also be controlled. There is a modified type of windrow composting called aerated static windrow composting. In this composting type, layers of organic matter are loosely piled so air can pass from top to bottom. Air blowers might be activated. Compost is relatively quickly produced in 3–6 months during the static pile process (da Silva Vilela et al. 2022).

While in other composting types, in-vessel composting involves adding organic content into a silo, drum, or concrete line reactor. It processes a large amount of waste without taking up much space and offers better control and efficiency than windrow composting. This method produces compost within a few weeks. While on the other side, this method is expensive, requires technical expertise to operate correctly, and requires insulation in extremely cold weather (Hoitink and Boehm 1999).

2.4.3 Anaerobic Digestion

Anaerobic digestion is the process in which controlled biological degradation occurs in the absence of oxygen but in the presence of different types of microorganisms inoculated for their specific functions and degradation of organic waste. The products of the procedure are a generation of added value products such as volatile fatty acids, bio manure, biohydrogen, biogas, and alcohols (Moukazis et al. 2018), used as an energy source and fertiliser, respectively. Activities from slaughterhouses generate a large volume of organic wastes in a biodegradable form, such as blood, animal faeces, fat, urine, paunch contents, and animal trimmings. In developing countries, non-availability of technologies and infrastructure, these wastes become part of landfills responsible for water, soil, and air pollution through heavy loads of microorganisms, if untreated or poorly managed. The quantity of these wastes is produced according to the slaughterhouse size and species of the animals slaughtered (Urlings et al. 1992).

Mainly slaughterhouse waste consists of the rumen, stomach, intestine contents, lipids, protein, and blood contents which are ideal for anaerobic digestion, thus elevating the quantity of biogas that could be produced (Chen et al. 2008). Feedstock should be used from different sources to avoid fat flotation, ammonia production, and biomass washout. Certain waste components consist of long-chain fatty acids, primarily responsible for upsetting the digestor working and resulting in desired products.

The technology of anaerobic digestion has been for many years. Renewable energy production and material recovery are aided by anaerobic digestion. Anaerobic digestion occurs throughout four sequential stages: methanogenesis, acidogenesis, hydrolysis, and acetogenesis. Different types of microorganisms' interactions drive and complete these four steps. All collected waste is added at once in the single-stage batch reactors, and these processes are sequentially permitted to take place in the reactor. The final produced compost is then evacuated after a predetermined retention period and when biogas production stops.

Different components of the slaughterhouse waste have different carbon-to-nitrogen ratios (*C*/*N* ratio), which show distinct characteristics like blood has and low carbon-to-nitrogen ratio and high protein contents; due to this, it needs to be processed with high *C*/*N* ratio contents. In contrast, ruminal contents have high *C*/*N* rations along these factors, temperature, pH, volatile fatty acids, alkalinity, carbon-to-nitrogen ratios, solid and volatile solids, solid retention time, and hydraulic retention time that are essential to be considered (Selormey et al. 2021).

Organic material found in anaerobic digesters generally contains polymers in a complex form that are difficult for microbes to degrade without further pre-treatments and hydrolysis. Therefore, hydrolysis breaks down organic macromolecules into smaller parts (Castellucci et al. 2013). Much focus has been placed on ways to speed up hydrolysis in anaerobic digesters because of how crucial it is to the kinetics of anaerobic digestion.

By absorbing hydrolysis by the cell membranes during acidogenic and acidogenesis, bacteria can create fatty acids in volatile form (VFAs) and other compounds. Organic acids like acetates and more powerful organic acids like propionate and butyrate belong to the class of VFAs. Even then, trace levels of lactate and ethanol may exist. Methane is produced during the digestive process of methanogenesis by methanogenic microbes by consuming available intermediates. It was discovered that 99% of Methanococcus vannelli and Methanococcus voltae cells had been killed within ten hours when exposed to oxygen takes place while showing methanogenic microorganisms' rapid sensitivity to oxygen.

Methanosarcina spp. are thought to withstand pH shocks and are relatively resilient and withstand higher levels of sodium, acetate, and ammonia concentrations that can harm methanogenic microorganisms, although methanogenic species in the anaerobic digestion are among the sensitive microbial groups. When in the reactor, biogas production stops, and methanogenesis ends; this process takes up to 40 days. The volatile solids content in the sludge and dewatering capacity can be used to assess how much digestion has taken place.

2.4.4 Rendering

Animal waste that is inappropriate for human consumption is transformed into stable, value-added materials like bone meal and refined fats like lard and tallow through the rendering process. Several drying and separation procedures are primarily used for this. Additionally, the procedure produces bone and poultry meals used to make pet food. The significant steps in the rendering process include removing water from the waste, isolating fats, and applying heat. In the first step, the waste is reduced, put into a cooking vessel, and cooked. The melted fats are subsequently separated from the protein by pressing this waste. Bone fragments and moisture are also taken out. After then, fat can be kept for later use. While the raw material is dried in dry processing, the addition of boiling water or steam causes the fat to surface in wet processing. It is significant to note that the length of the heating process and temperature is essential to the outcome. The rendering benefits include lowering the amount of generated waste dumped directly in landfills, killing protozoa, bacteria, parasites, and viruses, and producing a high-quality product that may be sold to create more cash. Disadvantages of rendering include high energy costs, the need to treat effluent water, and odour annoyance. Removing water from condemned trash also results in high energy expenditures.

Integrated rendering plants work alongside slaughterhouses or poultry processing facilities, whereas independent rendering plants are off-site facilities that gather raw materials from multiple sources. Independent plants can get animal by-product materials from the following places: slaughterhouses, supermarkets, butcher shops, animal shelter restaurants, poultry processors, farms, ranches, fast-food chains, and feedlots. These sources supply complete animal blood, carcasses, offal, grease, feathers, and other by-products.

Food-grade and non-food-grade animal rendering are the two categories of procedures as these facilities are typically run in combination with meat processing facilities, fatty animal tissue converted by edible rendering plants into palatable proteins and fats. Independent renderers or integrated processes that include rendering run inedible rendering plants. Typically, these facilities gather trash from slaughterhouses from facilities nearby. These factories create non-edible grease and tallow that is further used in poultry and animal feed, bone and meat meal, fatty acid production, chicken meat, fish meal, blood meal, hydrolysed feather meal, soap, and biodiesel manufacturing (Meeker and Hamilton 2006). In the nineteenth century, when animal by-products were predominantly used in the large-scale manufacturing of fertiliser, the rendering business underwent tremendous development. They were not necessary economically before it. Nowadays, the rendering sector creates hundreds of valued goods (MLA 2009).

Most solid slaughterhouse wastes were once rendered, giving the facilities an additional source of revenue. However, the commercial worth of such goods has drastically decreased due to the possibility of TSEs and treated as garbage in many situations (Palatsi et al. 2011). As a result, properly disposing of slaughterhouse waste prices has significantly increased lately. This is partly because the bacteria in such wastes pose a health danger.

It is possible to render by using a dry batch system and other technologies, in which the system has a cooker with steam-jacketed walls that keep the steam from coming into direct contact with the material inside. Water and fat are expelled as a result of the procedure. Due to no direct live steam contact, the fat is not substantially damaged. The cooling substance is eliminated, and the freely flowing fat is removed. The humid form material is subsequently squeezed using a continuous screw press, hydraulic press, and decanter centrifuge.

The autoclave system contains a cooker loaded with already-ground raw materials that are sealed prior to a steam injection (about 140 °C). Typically, this procedure lasts three to four hours and begins with high pressures (such as 360 kPa) that eventually drop to roughly 100 kPa of ambient pressure. A continuous dry system operates continuously, much like a dry batch system. Under specific atmospheric pressures, it operates. Heat is generated inside the cooker using steam jackets. To drain the water and fat, the automated process used by a continuous low-temperature system is frequently referred to as a dewatering system (mechanical). In general, the by-products in the uncooked form are squeezed and then placed in a wet or dry cooker at low temperature, where they are held at 60–90 °C for 10–30 min.

2.4.5 Incineration

It is a waste treatment process by destroying something like waste and useless things, burning to convert waste into ash, flue gas, and heat. In this process, waste volume is reduced to less than 5%, which is easy to maintain and reduces the need for landfill space. Heat can be covered during this process; it is the only solution for specific waste types. Most gases are burnt-well-designed systems produced. In the incineration process, relatively simple devices can achieve high removal efficiency. At sufficiently high temperatures and residence time, any hydrocarbon vapour can be oxidised to carbon dioxide and water.

After animal slaughtering to lower the danger of disease, animal by-product incineration is a quick, affordable, and secure disposal option. Abattoir and slaughterhouse waste is a possible breeding ground for pathogens that can infect humans and animals, including bacteria, viruses, prion diseases, and parasites. Heavy metals, dioxins and furans, particulate matter, and acid gases are the pollutants produced by this process. The waste is first received, sorted using an overhead crane, and placed into the incinerator's combustion chamber to begin the burning process. During the burning fuel and waste materials, the heat from the waste produced is collected and can be used to turn water into steam to produce electricity. A high-efficiency filtering system captures emissions 34 from the combustion process, and leftover ash from the system is collected and processed before transportation to a landfill using covered and waste leak-proof vehicles.

Typical waste produced by slaughterhouses and abattoirs is a biosecurity risk and must be disposed of immediately and thoroughly. By-products, including offal, hair, bone, fat, blood, and corpses, can all be efficiently burned in incinerators. It is okay to incinerate any of these waste types. Pollution from this procedure is lower than that from burning. Although it is believed that all viruses and bacteria will be destroyed during incineration, bovine spongiform encephalopathy (BSE) can withstand a high temperature in this incineration procedure not achieved (Franke-Whittle and Insam, 2013).

The incineration process has various advantages, including generating heat to create steam, which generates power and lowers the demand for fossil fuels. Pathogens and waste are destroyed during high-temperature treatment, which results in rising prices for collecting animal by-products but annual cost savings because there are no longer rendering or fallen stock expenses. There is no chance of contamination because trucks carrying potentially hazardous materials are not allowed to enter the site, boosting overall biosecurity. Storage is unnecessary because garbage is burned every day. There are certain obvious advantages that an incinerator provides, such as biosecurity, a decrease in garbage volume, lower expenses, and in some circumstances, a way to produce electricity. Due to the massive amounts of air pollution produced by the smoke released, other drawbacks include the need for high start-up costs, regular maintenance, and proper operation.

2.4.6 Biodiesel Production

The production of biodiesel from vegetable oils, cooking oils, yellow grease, or animal fats is sustainable, biodegradable fuel. Biodiesel is described by the American Society for Testing and Materials (ASTM) as a blend of long-chain monoalkylic esters from fatty acids from renewable resources used in diesel engines. Due to their high lipid content and availability, vegetable fat, animal, or slaughterhouse waste are a possible substitute source for biodiesel generation. One of the most acceptable biofuel options is biodiesel, which is produced using biological materials rather than fossil fuels. Using biodiesel could lessen emissions of hydrocarbons, suspended particulate matter, oxides of sulphur, and carbon monoxide, in addition to potentially extracting valuable products (lipids) from the wastes during this process (Mahyari et al. 2021).

Animal fats used to produce biodiesel include lamb/goat tallow, fish, and chicken fat, lard. Animal fat is the by-product of the rendering process of animal slaughter waste for meal production. Its drawbacks in biodiesel production could be removed, and the transesterification process could be used for production. Animal fat consists of saturated and unsaturated fatty acids, whereas saturated and monosaturated fatty acids are ideal for biodiesel production and polyunsaturated fatty acids make diesel heavy.

Animal waste is heated at 60 °C, and it breaks down the fat elements. Then, the extracted fat will be washed with water. The resultant oil will be filtered/centrifuged and decanted to remove the suspended particles and contamination. The processed fat is separated and stored. The content of free fatty acids, water, iodine number, peroxide value, saponification value, and acid number is vital to achieve high conversion efficiency. The efficiency of the reaction diminishes with the increase of the oil's acidity.

The composition knowledge of the chemical mixture is helpful while performing a reaction. Chemical process reactions take place in the first few minutes. The three stages required for forming esters from triglycerides are supported by the lack of mono- and diglycerides at the start of the chemical reaction and the rise and fall in their concentrations throughout the reaction. Fatty acid methyl esters (FAME) and glycerine are separated by decantation in two phases due to their different densities. In the separator chamber, this occurs in contrast to the majority of fatty acid methyl esters, glycerine, and excess alcohol concentrate in the upper phase.

There are two distinct phases. Decantation is the only method of separation, and because it relies on gravity, it could take several hours to complete. A quicker but more costly alternative is centrifugation. Fatty acid methyl esters have contaminants, including catalyst and methanol residue left over after the separation of glycerine. These raise cloud points; hence, a purification procedure is required. The combination of fatty acids and methyl esters must be purified to meet the appropriate quality standards for biodiesel. Because of this, it is cleaned, neutralised, and dried. The catalyst, methanol, and glycerine remnants are eliminated through repeated washings with water. The production of emulsions during the washing steps must be avoided because these contaminants are water-soluble, decreasing effectiveness. A series of washing procedures are used, with the first washing step using acidic water to neutralise the esters combination. After that, just water is used for two more washing steps. Finally, a drying procedure is required to remove any remaining water residues. The refined product is ready to be used as biodiesel once it has dried.

Glycerine (By-product): Glycerine is also known as glycerol, glycerine, or glycol alcohol. Chemically, it is alcohol with high viscosity at room temperature, odourless, transparent, colourless, low toxicity, and sweet taste. The boiling point of glycerine is high, 290 °C, and its viscosity increases noticeably at low temperatures. Glycerine is hygroscopic and has humectant properties. Glycerine is obtained as a sub-product of soap and biodiesel production. It is purified further to eliminate contaminants.

One of the primary benefits of using biodiesel is energy efficiency. Biodiesel is environment friendly as it emits a low number of sulphur oxide (So_x) and other harmful gases. Biodiesel is highly renewable. Biodiesel is sourced from natural organic matter like plants and animal oils.

Biodiesel is a non-toxic fuel producing lower emissions than fossil fuels when burnt. This lessens the risk of respiratory illnesses due to reduced air pollution. It helps ease the movement of engines as it has a more significant lubricating effect, extending engine lifespan.

On the contrary side, slightly higher fuel consumption than diesel fuel while emitting slightly higher nitrous oxide. Higher freezing point than diesel, while biodiesel is less stable than diesel fuel and therefore cannot be stored for the long term. These disadvantages are significantly reduced when biodiesel is used in blends with diesel fuel. The use of biodiesel as an alternative to diesel fuel has many benefits. After the raw ingredients have been processed, a reaction known as transesterification creates a mixture of fatty acids methyl esters (FAME) and glycerine as a by-product. The glycerine must be removed, and the mixture must be purified for the mixture of methyl esters to meet the specifications defined by global biodiesel standards. Since glycerine is a valuable material with numerous uses in the chemical, cosmetics, and pharmaceutical sectors, it is typically recovered and refined in large-scale production facilities.

2.4.7 Wood Adhesives

Slaughterhouse produces a large quantity of water containing a wide variety of components. Generally, these wastes are utilised for manufacturing meals, pet food other products. However, according to different legislative requirements in developed countries, these materials cannot go into animals' feed as a disease hazard caused by prion protein named bovine spongiform encephalopathy (BSE) is humungous. So, a significant amount of proteinaceous material goes into landfills or is incinerated, causing a financial burden and wastage of limited resources in this circular economy. Different methods are implemented to recover usable and value-added protein-based waste; one is thermal or alkaline-based hydrolysis. This method helps recover hydrolysed-based protein fragments from hydrolysate, which could further be utilised for value-added products, including biocomposites and bioplastics, which could replace formaldehyde-based resins identified as carcinogens (Adhikari et al. 2018a, b).

According to some research publications, the covalent bonds that develop due to the functional group's interaction in the adhesive formulation with those in the wood provide protein-based adhesive systems with their high sticky strength. These adhesives are thought to operate better under low moisture conditions. Peptides combined with synthetic resins or denatured/hydrolysed protein cross-linking are two potential solutions to this problem.

A large portion of the phenol is replaced by the protein component from traditional phenol–formaldehyde resins in the adhesive systems of protein-phenol–formaldehyde. One such formulation, which was created using hydrolysed protein recovered from poultry industry waste, was performed as well as wood glue based on phenol–formaldehyde resin. These applications show a considerable possibility for wood adhesives using a hydrolysed protein recovered from waste streams.

2.4.7.1 Major Types of Adhesive Systems

Adhesive system based on protein-phenol–formaldehyde

In the adhesive systems of protein-phenol–formaldehyde, the components of formaldehyde-based resins are co-reacted with protein through co-polymerisation with the pre-polymers of phenol and formaldehyde or by irreversible incorporation into the phenol–formaldehyde network. This method substitutes phenol from phenol–formaldehyde resins with protein, a sustainable feedstock.

Development of formaldehyde-free adhesive systems by hydrolysed/denatured protein cross-linking chemical modification

The denatured protein is chemically altered or crosslinked in the formaldehyde-free proteinaceous adhesive systems by utilising peptide/protein cross-linking reagents. For these applications, the cross-linking reagents are multifunctional compounds with highly reactive functional groups with peptides and proteins.

In a recent US patent (Patent No. US9522515 B2), Wu and Wang (2016) went into more detail on the embodiments of wood adhesive preparation from spent hen protein and its use for the manufacture of wood specimens from birch veneer. The patent further states that wood glue made from denatured chicken protein had water resistance and greater strength when compared to denatured canola and soy proteins. Under dry conditions, the hydrolysed protein recovered from poultry industry waste used to make wood adhesives was shown to have excellent adhesive strength. However, the formulations created from hydrolysed protein did not achieve water resistance.

By tertiary protein structures partially destroyed with denaturing agents like sodium dodecyl sulphate and urea, water resistance and adhesive strength of protein-based adhesives derived from chicken by-products can be improved. A significant drawback is the water resistance ability of adhesive compositions created only from denatured protein as competitive with commercial wood adhesive resins. In order to identify commercial uses, enhancing the water resistance of such formulations is essential for peptides recovered from the poultry industry in wood adhesives.

2.4.7.2 Bone and Meat Meal

From the meat and bone meal, the disease spread potential is extensive, and relevant institutions strictly monitor this slaughterhouse waste segment and its processing. Although, until now, no such case has been reported concerning paratuberculosis, BSE, and salmonella from slaughterhouse waste, zero tolerance of these disease-causative agents are forcing to explore new value-adding ways (Park et al. 2000). The bone and meat meal consists of 56% crude protein generally. As a constituent of this crud protein, 0.32% tryptophan, 2.87% lysine, and 0.58% cysteine are among the polar amino acids which can further interact with wood (Goedeken et al. 1990).

A patent is the only scientific document developing adhesive formulations for wood bonding from bone and meat meal protein applications (Yang and yang 2010). This patent also describes the potential use of bone and meat meal protein in preparing flocculating agents and bioplastics.

While (Adhikari et al. 2018a, b) worked on the development of plywood adhesives which are hydrolysed protein-based derived from slaughterhouse waste, and their effect on moisture resistance of formulated adhesives by chemical modification of hydrolysed protein (Park et al. 2000) heating suspensions of bone and meat meal protein developed the adhesive formulations at pH values from 5.0 to 9.0, in water for 30 min and at temperature setting ranging from 60 to 90 °C.

Different testing criteria were used using pH values between 6.0 and 7.0 and formulations, in which adhesive performed better than those with lower or higher pH. The best adhesive efficacy was found at pH 6.0 and 7.0, probably due to increased secondary contact between the wood surface and protein in bone and meat meal. Additionally, the effects of partially hydrolysed bone and meat meal protein being chemically crosslinked with glutaraldehyde and glyoxal were investigated. Compared to the control adhesive made from protein concentrate but without cross-linking, the adhesive strength by about 8% is boosted by glutaraldehyde-crosslinked bone and meat meal protein concentrate. This process helps to increase water resistance by up to three times the formulated adhesive. Protein molecules form stiff three-dimensional structures with better water resistance and binding strength due to the addition of cross-linking agents to the formulation.

2.4.7.3 Blood and Blood Meal

Blood-based adhesives are historically highly significant for the adhesive industry since blood and soluble blood meal have been used for ages to manufacture adhesives. Blood albumin glues were the most significant water-resistant glues for the plywood industry until the development of synthetic resins. In the past, blood-based glues were offered as dry powders that could be dissolved in water and other chemicals to create an alkaline, homogenous substance that could be easily dispersed. Sodium hydroxide, lime, sodium silicate, or a mixture of these substances were the chemicals utilised for this.

Typically used as dry extenders, sawdust, wood flour, or other lignocellulosic ingredients were included in conventional blood-based adhesives preparations. Extenders were used to minimise the adhesive price by reducing the quantity of primary binder required per unit area and enhancing the adhesive system's void-filling capabilities. In order to assure uniform loading of the adhesive on the adherend, defoaming agents like terpineol were typically used in wood adhesives. This is because protein solutions generate stable foams that might affect volumetric measurements.

The bond strength and water resistance of the final adhesive system were studied by Yan et al. (2016) after combining acrylic latex- and cow blood-based adhesives. Although they lost market dominance after introducing synthetic adhesives, blood protein-based adhesives from water-soluble blood meal have long been recognised for manufacturing waterproof composite wood panels. However, the emergence of various studies and patents in the recent years suggests increased interest in using blood and blood meal to create proteinaceous adhesives. Some adhesive formulas made from scratch have demonstrated adhesive strength and water resistance on par with industrial resins used to make composite wood goods. Blood protein-based formulations containing up to 70% (w/w) blood protein have been produced by blending and co-reacting the protein with acrylic latex-based glues and partially condensed phenol–formaldehyde resin. These formulations have shown adhesive performance comparable to phenol–formaldehyde resin-based wood adhesives (Adhikari et al. 2018a).

2.4.8 Agriculture Water

After treatment, wastewater from the slaughterhouse could irrigate cultivated land. In many developed and developing countries, effluent water from the slaughterhouse is regulated through different standards to protect the environment, land, and sewerage systems. Slaughterhouse wastewater is heavily loaded with fat, ruminal contents, meat trimmings, disinfectants, cleaners, blood, and disposal from washrooms. For the utilisation of slaughterhouse wastewater as agricultural water, separation of these contaminants through physical and chemical separations needs to be adopted, otherwise irrigated land upper layer respiration will be blocked, and chances are there that it gets barren along with this spoilage of organic matter will cause a nuisance smell for the nearby population. Generally, agriculture after treatment could be a good source of extra income as sweat water availability is getting scared due to a drop in underground water and increasing cost of fetching it, so the mutual benefit for slaughterhouse, environment, and irrigation for cropped land.

2.5 Environmental Challenges Related to Slaughterhouse Waste

The location of the slaughtering facility and its planning is critical issues in an urban setting. The leftovers from animal slaughter end up in our lakes and canals, contaminating the water and becoming a part of our bodies. Untreated slaughterhouse wastewater contains paunch, faeces, urine, blood, lint, fat, carcasses, undigested food, pharmaceuticals, oil and grease, suspended particles, loose meat, disinfectants, and facility cleaners. This results in a high organic matter content and contaminates rivers and drainage systems. Municipal wastewater is not as strong as effluent from slaughterhouses. Such waste products raise phosphorus, nitrogen, sediments, and BOD levels in the receiving water body when dumped directly into it, which may cause eutrophication. All slaughterhouses regularly release enormous amounts of wastewater, an environmental problem.

Land contamination, which occurs due to slaughterhouse waste, is an aesthetic issue rather than one relating to pollution. The slaughterhouse wastewater mainly contains ruminal contents that are a rich source of fibre and could have an effect as a fertiliser and aeration of the internal layers of the land for better root growth and having a positive impact on the nodules of the nitrogenous crops. However, generally, on the negative side, rotten contents could irritate nearby passers. Other contents of the wastewater consist of blood, which will also harm the environment, but the richness of its nutrients may positively impact the irrigated land.

A minute quantity of wasted fat and protein-based trimmings could attract pests and birds in the facility, which could spread disease in the outskirts of the facility. Usually, it is observed that by-products are processed inside the vicinity of the slaughterhouse, in a separate location; this facility is managed with fewer safety measures than the meat processing facility, so resultantly, hazards, particularly biological ones, could slip into the meat processing facility. Slaughterhouses not following local and internationally implemented good hygiene practices because severe environmental and health damage due to discrete waste disposal, highly polluted effluent discharge, burning of bones and hooves, etc. Illegal slaughtering and practices pose a significant environmental risk in developing countries, where the implementation of laws and standards is weak compared to developed countries.

2.5.1 Slaughterhouse Wastewater

Slaughterhouse wastewater contains blood, animal body parts, fat, and animal dug. In a slaughterhouse, no toxic chemicals are used during the operations, but natural material can cause bacterial contamination. Slaughterhouse wastewater contains different compounds such as sulphates, nitrates, and phosphates. These compounds are in high concentrations of wastewater, contaminating the receiving water bodies and leading to environmental contamination (Mees et al. 2009). In developed countries, where water in and out quality is strictly monitored, water treatment plants are installed to minimise the BOD and COD of the disposed water. This water then becomes part of the nearby running canal or rivers, while generally, it is observed in developing countries that slaughterhouses are established near the municipal sewerage system to get rid of environmental obligations.

Such materials are present in slaughterhouse wastewater, which have oxygen demand, i.e. BOD and COD. When discharged to the receiving water bodies, such materials cause severe water quality damage. When discharged, these materials consume dissolved oxygen and cause oxygen deficiency in the streams, killing living organisms (Kundu et al. 2013). Nitrogen and phosphorus also go to the receiving bodies as nutrients and cause eutrophication, resulting in vegetation growth in excessive amounts. In 2019, slaughterhouses in the USA released 28 million pounds of nitrogen and phosphorus into nearby rivers and streams. Excessive vegetation blocks the path of slaughterhouse wastewater and causes overflooding in the path. If from the slaughterhouse, several pathogens are released into the streams, which can be transmitted to humans via water-ground leaching.

Pollutants present in the slaughterhouse can be leached into the underground water resources and cause alteration to the water quality. Humans consume this groundwater, causing several types of diseases like viral diseases in the community. Raw slaughterhouse wastewater is irrigated with polluted groundwater, causing soil pollution again. The presence of pollutants in organic nature can be known by several indications like odour, taste, and foaming. In a study, a soil profile of 4 m was analysed for nitrate concentration irrigated by the slaughterhouse wastewater, and the concentration was 3783 kg in high amounts (López and Borzacconi 2010).

Another vital aspect of slaughterhouses themselves is that they rely heavily on underground water for in-facility utilisation. It is observed that slaughterhouse waste also contaminates the underground water, and resultantly, obtained water is heavily contaminated with pathogenic microorganisms like E-coli. When this water is used for cleaning the facility, equipment, utensils and particularly meat, it will cause contamination and control measures like reverse osmosis plants to be installed. In meat processing facilities, where meat is mainly exported, this condition causes significant losses and sometimes bans the facility from exports.

2.5.1.1 Biodegradable Organic Compound

Activities at the slaughterhouse generate biodegradable material in large volumes, including blood, paunch contents, animal faeces, blood, and urine. Suppose the biodegradable material is not managed correctly and treated and becomes the primary source of water, soil, and air pollution. This organic material is characterised by high concentrations of carbohydrates, mainly in the form of lipids, lactose, and protein. Biodegradable organic material is considered energy rich. Slaughterhouse organic waste comprises long fatty acid chains, ammonia, and hydrogen sulphide (Limeneh et al. 2022).

2.5.1.2 Eutrophication

When released, the direct and indirect discharge of partially and untreated effluent causes eutrophication, a global problem. When the nutrients like phosphorus and nitrogen are in high concentrations and released into the waste bodies, plants and algae growth block the path of sunlight to the water body (Yaakob et al. 2018). The amount of dissolved oxygen in the water bodies consumed by the plants for their excessive growth and, finally, water bodies become water deficient. This process decreases the water quality, and living organisms die, like fishes. Eutrophication causes severe damage to environmental health because water uses become limited and causes outbreaks of diseases when used by living organisms (Sieng 2019).

2.5.1.3 Toxic Compounds

Slaughterhouse wastewater may contain toxic compounds like unionised ammonia, decontaminants, cleaners, disinfectants, surfactants, and steriliser agents, which are highly toxic to aquatic life. Due to improper slaughterhouse wastewater treatment, these surfactants, a component of detergents, can cause short- and long-term effects on the environment, resulting in increased environmental challenges (Yarandi et al. 2021).

2.5.2 Solid Waste

Annually, slaughterhouses produce thousands of tons of solid waste. Slaughterhouse waste generally contains unutilised animal by-products, which need to be accounted. With the growing population in urban areas, the demand for meat is also increasing and putting pressure on slaughterhouses. More solid waste will be generated when more animals are slaughtered, which will be one of the significant problems of handling and dumping. It is estimated that one-third or one-half of the total weight of the slaughtered animal is considered unusable and dumped as solid waste. When the solid waste does not dump with management, this will cause challenges to environmental components, such as degrade the soil, air, and water (Loganath and Senophiyah-Mary 2020).

2.5.2.1 Waste Open Disposal and Landfilling

Solid waste disposal in the open environment is defined as open dumping. When the waste is dumped openly on the empty without following proper waste disposal guidelines, it causes harmful effects on the environment and its components. On the other side, landfilling is the disposal of waste by following the disposal rules, but if proper landfill design is not followed may have a profound effect on the environment

(Omole and Ogbiye 2013). In developing countries, slaughterhouse solid waste is dumped openly in empty spaces. On such dumping sites, organic material degradation contaminates the space, further leaching down and affecting the environment. In the landfills, slaughterhouse waste is dumped properly to generate energy from the organic waste. If dumping guidelines are not followed, flies and insects become a pathway of disease transmission to nearby areas, and viral diseases become common. From the open dumping, nutrients and contaminants with rainfall runoff to the nearby water bodies cause severe environmental conditions such as eutrophication, killing fish, and degrading water quality. Improper slaughterhouse waste disposal when takes place causes the release of greenhouse gases which are responsible for global warming (Selormey et al. 2021).

2.5.3 Air Pollution

Besides, the soil and water pollutions from slaughterhouses also cause air pollution. Greenhouse gases like carbon dioxide and methane are released in high concentrations. These gases are a contributor to climate change. These gases are produced during the degradation of slaughterhouse wastewater and in the slaughtering process. Improper disposal and burning of slaughterhouse waste generate contaminants, affecting the ambient air quality (Mozhiarasi and Natarajan 2022). Slaughterhouse waste contains high concentrations of nitrate, phosphorus, and sulphides, which are released into the air when the waste is burnt out, causing air pollution. When the slaughterhouse works in residential areas, these greenhouse gas emissions create health issues for nearby residents and damage the local air quality. From the disposal of slaughterhouse waste, foul odour is also released, affecting the air quality.

2.5.3.1 Pollution of the Slaughterhouse Environment

The slaughterhouse generates such waste, which generates such odour, which can be a source of local air pollution and disturb the daily tasks in life. Several odorous compounds are stubborn, like mercaptans, organic acids, sulphates, amines, etc. These compounds can attach to clothes, persist for longer, and cause issues (Sweeten 1980).

2.5.3.2 Impact of Solid Slaughterhouse Waste Exposure on the Air Quality

When the slaughterhouse wastewater is disposed of improperly and burnt at the disposed site causes severe damage to the local air quality. Waste burning releases several noxious air pollutants such as carbon dioxide, sulphur dioxide, particulate matter, and nitrogen oxides. When released into the atmosphere, these pollutants affect human health and cause several diseases like cardiovascular diseases, colds, cancer, respiratory diseases, and allergies (Kundu et al. 2013). From the dumpsite, pollutants and pathogens can leach and contaminate the nearby surface and groundwater resources from which water is supplied, which further results in the risk to living beings. When the nutrients-enriched animal faeces and blood is released illegally into the environment cause accumulation of toxic compounds in the environmental components and becomes part of the life cycle of living beings.

2.6 Measures Proposed to Improve the Slaughterhouse Wastewater Management

Proper management of slaughterhouse solid waste and wastewater is vital because it contains a high amount of organic waste that needs to be managed as produced. Wastewater from the slaughterhouse is highly loaded with degradable materials that must be treated before being released to the receiving water bodies. Different wastewater treatment techniques must be applied to meet the National wastewater discharge limits and make wastewater reusable (Bustillo-Lecompte et al. 2016). Different treatment technologies like the coagulation/flocculation process are highly used for the abattoir's wastewater treatment. Abattoir wastewater mainly contains a substantial quantity of solids that can settle down. Settling wastewater has an extended effect on COD and suspended solids reduction, which is helpful for the coagulation and flocculation process. When the wastewater is settled down for 30 min, the 75–79% suspended solids are reduced. The suspended solids settling is further reduced when the settling time exceeds 30 min. On the other side, COD removal in the first 30 min was thirty-two per cent, and when the settling time increased to one hour, COD removal per cent further increased to thirty-eight per cent. In the following treatment stage, biological settling is essential. Chemical treatment of abattoir wastewater is also reported in the literature. Aluminium and ferric salts are also used in the abattoir's wastewater chemical treatment. Different coagulants are used in the treatment (Baker et al. 2021).

The use of common coagulants does not completely flocculate the abattoirs wastewater. The development of alternative treatment methods like anaerobic digestion, media filtration, aerobic and anaerobic sequencing batch reactors, enhanced media, and biofilter systems could be beneficial. On the other side, the coagulation and flocculation methods are energy-saving, easy to operate, and cost-effective compared to other methods for abattoir wastewater. Due to the insufficient treatment facilities, wastewater from the abattoir is deposited on land and finally goes into the water channels, which further causes pollution. In many countries, pollution caused by meat production activities results from the failure of good hygiene practices and good manufacturing practices (Hilares et al. 2021).

To prevent this pollution, it is suggested to seal the gut of slaughtering an animal to avoid the leakage of organic contents. Abattoir wastewater effluent has complex nature, due to which this type of wastewater is harmful to the environment. For example, when the wastewater contains slaughtered animals, blood released into the water channel causes dissolved oxygen (DO) depletion. Due to the paunch manure's improper disposal in the receiving environment, this can exert oxygen, and due to this, large population of decomposers can breed, which can cause a pathogenic effect. Animal waste improper disposal can depletion of oxygen in receiving environment. This situation further causes the enrichment of nutrients in the receiving environment, and toxins in the biological system can accumulate (Musa and Idrus 2021).

The concentration of organic matter in the meat processing plant's effluent is high, and the remaining residues are solubilised, leading to pollution due to the pathogens and organic content in the abattoir's wastewater. Abattoir wastewater is also considered a bulk parameter because various pollutants are derived from the facility and the slaughtering of animals, which fluctuate in the meat industry. In treating the abattoir's wastewater, anaerobic treatment is the preferred biological treatment. On the other side, in the anaerobic treatment, post-treatment requires the discharge to comply with the discharge limits because organic matter stabilisation alone is not possible with the anaerobic treatment. The effluent of anaerobic treatment contains organic matter in solubilised form, which can be done using aerobic processes. So, in contrast with anaerobic processes, aerobic treatment is frequently used because aerobic processes operate at higher rates than conventional processes (Meiramkulova et al. 2020). In the aerobic treatment processes, the treatment time and oxygen demand are directly proportional to the wastewater quantity and pollutant load. Aerobic treatment is used for post-treatment of the effluent of anaerobic treatment but also the removal of nutrients.

The biological processes not only help to produce effluents from the highly organic contain wastewater that complies with local discharge limits, but on the other side, the biological processes like aerobic and anaerobic treatment processes have the potential for resource recovery from the abattoir wastewater with high-level treatment (Filali-Meknassi et al. 2004). Abattoir wastewater also contains bio-resistant, non-biodegradable, recalcitrant, etc. These substances in the abattoir can be removed or transformed using advanced oxidation processes (AOPs) and improve the biodegradability of wastewater. AOPs could be an alternative treatment for the abattoir's effluent and can be attached to biological processes to improve treatment. By combining the AOPs and biological processes, we can achieve an economical and easy operation method with several advantages. For resource recovery, we can use these processes in the abattoir's wastewater (Ng et al. 2022).

2.6.1 Preliminary Treatment

Preliminary treatment is the primary and first step in every wastewater treatment process. This treatment's main objective is to remove large particles and solids from the slaughterhouse wastewater. Effluent quality is improved by primary treatments, including traditional screens, sieves, fat separators, and settlers. In the slaughterhouse wastewater, solids of 10–30 mm diameter are retained on the mesh or sieve screen. In order to avoid clogging and fouling in the other treatment processes, rotary screeners are used to separate solids of a diameter of more than 0.5 mm. To compact and transport the separated solids from the screens, screw screen compactors are used, which further minimise the moisture content and volume of solids treated as solid waste. This process removes about a BOD of 30%. In preliminary treatment, several operations are included, like screening and sieves. Large solids of diameters 10–30 mm are separated while wastewater passes (Bustillo-Lecompte and Mehrvar 2015).

2.6.2 *Physicochemical Treatment*

After the preliminary treatment, it is recommended that the wastewater should be treated with primary and secondary treatment depending on the characteristics of the raw wastewater. To reduce the BOD, fat and total suspended solids in slaughterhouse wastewater (SWW) dissolved air flotation (DAF) are considered a typical method for primary treatment. The solids are separated from the liquid in the physiochemical treatment methods. Physicochemical methods which are mainly used are given below.

Membrane processes: For treating the abattoir industry wastewater, using membranes are an alternative method. To remove the organic matter, pathogens and macromolecules, several membranes can be used, which include reverse osmosis (RO), ultrafiltration (UF), nanofiltration (MF), and microfiltration (MF) (Almandoz et al. 2015). The removal efficiency of membrane processes is up to 90%. To achieve nutrient removal from wastewater, membrane technology must operate or combine with conventional processes. There are several disadvantages of using membranes in abattoir wastewater treatment: the blockage of membranes and the fouling of membranes. These disadvantages restrict the treatment efficiency (Fatima et al. 2021). Yordanov (2010) used ultrafiltration technology for the treatment of slaughterhouse wastewater. Results showed that ultrafiltration is an effective tool to remove fats and total suspended solids with higher removal efficiencies of 99 and 98%. COD and BOD removal efficiency by UF was 94 and 97%. Bohdziewicz and Sroka (2005) evaluated the performance of RO technology in the treatment of slaughterhouse wastewater. Before using RO, raw wastewater was pre-treated by using activated sludge. TN, TP, BOD, and COD removal efficiency was 90, 97.5, 50, and 85.8% using RO treatment.

2.6.2.1 Dissolved Air Flotation (DAF)

Dissolved air flotation (DAF) is considered a primary treatment in wastewater treatment processes. This DAF process includes introducing air at the pressure of 4–6 bar, which aids the liquid and solid separation. The supplied dissolved air escapes from the liquid in the form of bubbles. Bubbles attach to the targeted solids and make a sludge blanket on the surface containing grease and fat with some light solids. This surface blanket can be removed by scraping. Flocculants can be added to the DAF treatment method to denature the protein present in the slaughterhouse wastewater. Before using biological treatment processes, the DAF system was ideal for removing fatty objects and suspended solids. This process addition increases the BOD and COD removal efficiency to 75%. DAF system at a large scale can remove total phosphorus (TP), total nitrogen (TN), grease, and oil by 70%, 55%, 85%, and 70% (Dlangamandla et al. 2018). Combining dissolved air flotation with a membrane reactor (MBR) in slaughterhouse wastewater treatment seems promising to meet discharge limits. If reverse osmosis technology is used as a final step after the previous process, wastewater can be reused in the facility.

2.6.2.2 Coagulation-Flocculation and Sedimentation

In the coagulation process, particles in the form of the colloidal present in slaughterhouse wastewater are grouped to form large particles called flocs. Negatively charged particles that are resistant and stable are present in the abattoir wastewater. For this reason, coagulants are added, and positively charged ions destabilise these colloidal particles and flocs. Aluminium potassium sulphate is mainly used as a coagulant. The flocculation process makes the suspended colloidal particles into flocs after settling down. To increase the efficiency of treatment, coagulation and flocculation are both processes used (Mahtab et al. 2009). Ferric chloride, aluminium chlorohydrate, aluminium sulphate, and ferric sulphate are the coagulants used to treat wastewater. Using poly aluminium chloride as a coagulant, the removal efficiency of COD, TN, and TP was 75, 88.8, and 99%. When the inorganic coagulant is used, the volume of the sludge produced can be reduced by 41.6%. Amuda and Alade (2006) used the coagulation-flocculation technology to treat the slaughterhouse wastewater at the lab scale to remove COD, TSS, and TP. Different coagulants, like alum, ferric chloride, and ferric sulphate, were used. Results showed that the alum coagulant effectively removed the TSS and TP by 34 and 98%. Tariq et al. (2012) used coagulants to treat the slaughterhouse wastewater, i.e. lime and alum. When each coagulant was used in combination, the removal efficiency of COD was 85%, but sludge generation was low in these conditions.

2.6.2.3 Electrocoagulation

In the slaughterhouse wastewater treatment processes, electrocoagulation (EC) is considered cost-effective and advanced treatment. Electrocoagulation is considered effective technology for removing nutrients, organics, heavy metals, and even the removal of pathogens. An electric current is induced without adding any chemicals in the electrocoagulation process. Using different materials electrodes, the EC process generates M^{3+} ions, mainly Al^{3+} and Fe^{3+}. Different types of electrodes are used, like TiO_2, SnO_2, and Pt, which can be used in alkaline and acidic conditions with a high removal efficiency of TSS, BOD, COD, TN, and colour (Emerick et al. 2020). For removing COD from wastewater using electrocoagulation, Bayramoglu et al. (2006) used sacrificial electrodes of Fe and Al with a focus on the operating cost. Results showed that Fe sacrificial electrodes perform better and cost-effectively than the Al electrode. The operating cost of using the Fe electrode is also 50% less than the Al electrodes. Operating costs include sludge handling, electricity, maintenance, etc. Asselin et al. (2008) used the EC process to evaluate its economic cost for removing organic compounds from the slaughterhouse wastewater. In the EC process, Al sacrificial and mild steel electrodes were used for lab scale study. The study result showed that by using steel electrodes, the removal efficiency of BOD, COD, turbidity, TSS, and oil grease was 87, 84, 94, 93, and 99%.

2.6.3 Biological Treatment

To meet the local discharge limits, abattoir wastewater cannot be treated entirely with the primary and physicochemical processes. Biological/secondary treatment methods remove soluble organic compounds to eliminate this limitation. In the biological treatment methods, aerobic, and anaerobic methods include anaerobic digestors, anaerobic lagoons, anaerobic filters, baffled reactors, biological contactors, and sequencing batch reactors. The secondary treatment's primary focus is reducing BOD concentration in slaughterhouse wastewater by removing the remaining organic compounds that are not removed by primary treatment. Biological treatment is considered a secondary treatment, whereas aerobic and anaerobic digestion in combination or individually depends on the characteristics of slaughterhouse wastewater (Mittal 2006a, b). In biological treatment, using microorganisms, organics are removed with pathogens. By using the anaerobic and aerobic processes in the biological treatment, BOD removal efficiency is up to 90%. Biological treatment may include other processes in the combination, like trickling filters and aerobic, anaerobic, and facultative lagoons. Anaerobic treatment is considered ideal in all biological processes when the target is to treat highly contaminated wastewater. In the anaerobic treatment, organic compounds are degraded without oxygen with the help of different bacteria into CH_4 and CO_2. In the aerobic treatment, the organic material is degraded in the presence of oxygen. Aerobic treatment is mainly used after physiochemical treatment to decompose and remove nutrients (Musa and Idrus 2021).

2.6.4 Advanced Oxidation Processes

Advance oxidation processes are becoming alternatives to the complimentary and conventional treatment processes, either pre-treatment or post-treatment to the current biological processes. Compared to using chlorine as a disinfection chemical which can cause the formation of other by-products, AOPs are a cleaner option to inactivate microorganisms. Due to these benefits, AOPs are considered handy for pollution control, water reuse, and advanced degradation processes with better results than complementary processes. Ozonation, gamma radiation, and UV/H_2O_2 are among the AOPs widely used in slaughterhouse wastewater treatment. Wu and Doan (2005) used ozonation in the treatment of slaughterhouse wastewater, and the results showed that the disinfection of wastewater was achieved to 99% by using the ozonation process for 8 min, and the ozone dosage was 23 mg/min per litre. The removal of BOD and COD was low, with the removal efficiency of only 23.06 and 10.7%. The use of UV/H_2O_2 in slaughterhouse wastewater is considered effective in this process; the degradation and oxidation of the pollutants mainly depend on the hydroxyl radicals (*OH), which is a highly reactive species that are produced from the reaction between H_2O_2 and UV (Hamad et al. 2014). Compared with other treatment methods, UV/H_2O_2 is considered effective, with an optimal removal efficiency of up to 50%, but the operating cost is high. To lower the operating cost, it is recommended to use this method with biological processes (Besharati et al. 2020).

2.6.5 Treatment System Maximum Removal Efficiency

Every slaughterhouse wastewater treatment system has separate removal efficiency. The removal efficiency depends on decreased biological content, BOD, COD, and nutrients on the input and output concentration difference. Each treatment system has a minimum and maximum removal efficiency for every pollutant. Figure 2.1 shows the maximum removal efficiency of every treatment system.

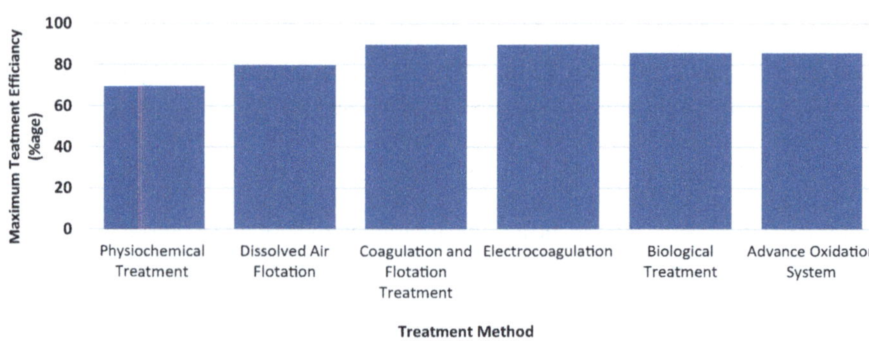

Fig. 2.1 Maximum treatment system efficiency (Mittal 2006a, b; Yordanov 2010)

2.7 Conclusion

Different types of waste are generated in solid and liquid form during the slaughtering process, which needs to be handled timely because the organic material starts degrading quickly and cause environmental pollution, such as the processing of fat which constitutes most of the slaughtering waste. Slaughtering facilities release a high volume of wastewater containing blood and organic content that also needs to be treated; otherwise causes severe environmental challenges. Different products can be produced by processing organic material, blood, and wastewater to increase revenue and solve environmental concerns. In the treatment of slaughterhouse wastewater, different techniques are used with a separate removal efficiency of pollutants like primary and secondary treatment methods, DAF, coagulation/flocculation, advanced oxidation processes, and electrocoagulation. These methods can effectively treat the slaughterhouse wastewater and release the treated water, which can be used as agricultural water or goes into canals/rivers. Solid waste from slaughtering facilities is processed by incineration, composting, rendering, and anaerobic digestion steps and valuable products like wood adhesive and biodiesel are produced, which also increase the return revenue. If we process the slaughtering waste in time, this can be a blessing for us as the source of revenue, but on the other side, if not processed and dumped openly without any treatment can be a curse to the environment and cause serious environmental and health issues and challenges. In developing countries, mostly slaughtering solid and liquid waste is dumped/released without any treatment and causes environmental degradation and health issues in the nearby community.

References

Adeyemi IG, Adeyemo OK (2007) Waste management practices at the Bodija abattoir, Nigeria. Int J Environ Sci 64(1):71–82

Adhikari BB, Chae M, Bressler DC (2018a) Utilisation of slaughterhouse waste in value-added applications: recent advances in the development of wood adhesives. Polymers 10(2):176

Adhikari BB, Kislitsin V, Appadu P, Chae M, Choi P, Bressler DC (2018b) Development of hydrolysed protein-based plywood adhesive from slaughterhouse waste: effect of chemical modification of hydrolysed protein on moisture resistance of formulated adhesives. RSC Adv 8(6):2996–3008

Almandoz M, Pagliero CL, Ochoa NA, Marchese J (2015) Composite ceramic membranes from natural aluminosilicates for microfiltration applications. Ceram Int 41(4):5621–5633

Amuda O, Alade A (2006) Coagulation/flocculation process in the treatment of abattoir wastewater. Desalination 196(1–3):22–31

Asselin M, Drogui P, Benmoussa H, Blais J-F (2008) Effectiveness of electrocoagulation process in removing organic compounds from slaughterhouse wastewater using monopolar and bipolar electrolytic cells. Chemosphere 72(11):1727–1733

Asses N, Farhat W, Hamdi M, Bouallagui H (2019) Large scale composting of poultry slaughterhouse processing waste: microbial removal and agricultural biofertiliser application. Process Saf Environ Prot 124:128–136

Awan ZA, Tariq M, Awan MM, Satti NW, Mukhtar T, Akram W, Yasin MF (2015) Edible by-products of meat. Veterinaria 3(2): 33–36

Bah CS, Bekhit AE-DA, Carne A, McConnell MA (2013) Slaughterhouse blood: an emerging source of bioactive compounds. Compr Rev Food Sci Food Saf 12:314–331. https://doi.org/10.1111/1541-4337.12013

Baker BR, Mohamed R, Al-Gheethi A, Aziz HA (2021) Advanced technologies for poultry slaughterhouse wastewater treatment: a systematic review. J Dispers Sci Technol 42(6):880–899

Bayramoglu M, Kobya M, Eyvaz M, Senturk E (2006) Technical and economic analysis of electrocoagulation for the treatment of poultry slaughterhouse wastewater. Sep Purif Technol 51(3):404–408

Besharati Fard M, Mirbagheri SA, Pendashteh A (2020) Removal of TCOD and phosphate from slaughterhouse wastewater using Fenton as a post-treatment of an UASB reactor. J Environ Health Sci Eng 18(2):413–422

Bohdziewicz J, Sroka E (2005) Integrated system of activated sludge–reverse osmosis in the treatment of the wastewater from the meat industry. Process Biochem 40(5):1517–1523

Bustillo-Lecompte CF, Mehrvar M (2015) Slaughterhouse wastewater characteristics, treatment, and management in the meat processing industry: a review on trends and advances. J Environ Manage 161:287–302

Bustillo-Lecompte C, Mehrvar M, Quiñones-Bolaños E (2016) Slaughterhouse wastewater characterisation and treatment: an economic and public health necessity of the meat processing industry in Ontario, Canada. J Geosci Environ Prot 4(4):175–186

Castellucci S, Cocchi S, Allegrini E, Vecchione L (2013) Anaerobic digestion and co-digestion of slaughterhouse wastes. J Agric Eng 44(s2):e104

Cekmecelioglu D, Demirci A, Graves RE, Davitt NH (2005) Applicability of optimised in-vessel food waste composting for windrow system. Biosyst Eng 91:479–486

Chen Y, Cheng JJ, Creamer KS (2008) Inhibition of anaerobic digestion process: a review. Bioresource Technol 99:4044–4064

Cofrades S, Guerra M, Carballo J, Fernández-Martín F, Colmenero FJ (2000) Plasma protein and soy fiber content effect on Bologna sausage properties as influenced by fat level. J Food Sci 65:281–287

da Silva Vilela RN, Orrico ACA, Junior MAPO, Borquis RRA, Tomazi M, de Oliveira JD, de Ávila MR, dos Santos FT, Leite BKV (2022) Effects of aeration and season on the composting of slaughterhouse waste. Environ Technol Innov 27:102505

Dees PM, Ghiorse WC (2001) Microbial diversity in hot synthetic compost as revealed by PCR-amplified rRNA sequences from cultivated isolates and extracted DNA. FEMS Microbiol Ecol 35(2):207–216

Dlangamandla C, Ntwampe SKO, Basitere M (2018) A bioflocculant-supported dissolved air flotation system for the removal of suspended solids, lipids and protein matter from poultry slaughterhouse wastewater. Water Sci Technol 78(2):452–458

Emerick T, Vieira JL, Silveira MHL, João JJ (2020) Ultrasound-assisted electrocoagulation process applied to the treatment and reuse of swine slaughterhouse wastewater. J Environ Chem Eng 8(6):104308

Epstein E (1997) The science of composting. CRC Press LLC, Boca Ratón, FL, USA

Fatima F, Du H, Kommalapati RR (2021) Treatment of poultry slaughterhouse wastewater with membrane technologies: a review. Water 13(14):1905

Filali-Meknassi Y, Auriol M, Tyagi R, Surampalli R (2004) Treatment of slaughterhouse wastewater in a sequencing batch reactor: simulation vs experimental studies. Environ Technol 25(1):23–38

Franke-Whittle IH, Insam H (2013) Treatment alternatives of slaughterhouse wastes, and their effect on the inactivation of different pathogens: a review. Crit Rev Microbiol 39(2):139–151

Furlán LTR, Padilla AP, Campderrós ME (2010) Functional and physical properties of bovine plasma proteins as a function of processing and pH, application in a food formulation. Adv J Food Sci Tech 2:256–267

Goedeken FK, Klopfenstein TJ, Stock RA, Britton RA, Sindt MH (1990) Protein value of feather meal for ruminants as affected by blood additions. J Anim Sci 68:2936–2944

Hamad D, Mehrvar M, Dhib R (2014) Experimental study of polyvinyl alcohol degradation in aqueous solution by UV/H_2O_2 process. Polym Degrad Stab 103:75–82

Hernández-Ledesma B, del Mar Contreras M, Recio I (2011) Antihypertensive peptides: Production, bioavailability and incorporation into foods. Adv Colloid Interface Sci 165(1):23–35

Hilares RT, Atoche-Garay DF, Pagaza DAP, Ahmed MA, Andrade GJC, Santos JC (2021) Promising physicochemical technologies for poultry slaughterhouse wastewater treatment: a critical review. J Environ Chem Eng 9(2):105174

Hoekstra AY, Chapagain AK (2006) Water footprints of nations: water use by people as a function of their consumption pattern. Water Resour Manag 21:35–48

Hoitink HAJ, Boehm MJ (1999) Biocontrol within the context of soil microbial communities: A substrate-dependent phenomenon. Annu Rev Phytopathol 37(1):427–446

Hyun CK, Shin HK (1998) Utilisation of bovine blood plasma obtained from a slaughterhouse for economic production of probiotics. J Ferment Bioengr 86:34–37

Jayathilakan K, Sultana K, Radhakrishna K, Bawa AS (2012) Utilisation of by-products and waste materials from meat, poultry and fish processing industries: a review. J Food Sci Technol 49:278–293

Korhonen H, Pihlanto A (2003) Food-derived bioactive peptides-opportunities for designing future foods. Curr Pharm Des 9:1297–1308

Kris-Etherton PM, Lefevre M, Beecher GR, Gross MD, Keen CL, Etherton TD (2004) Bioactive compounds in nutrition and health research methodologies for establishing biological function: the antioxidant and anti-inflammatory effects of flavonoids on atherosclerosis. Annu Rev Nutri 24:511–538

Kundu P, Debsarkar A, Mukherjee S (2013) Treatment of slaughter house wastewater in a sequencing batch reactor: performance evaluation and biodegradation kinetics. Biomed Res Int

Limeneh DY, Tesfaye T, Ayele M, Husien NM, Ferede E, Haile A, Gibril M (2022) A Comprehensive review on utilisation of slaughterhouse by-product: current status and prospect. Sustainability 14(11):6469

Loganath R, Senophiyah-Mary J (2020) Critical review on the necessity of bioelectricity generation from slaughterhouse industry waste and wastewater using different anaerobic digestion reactors. Renew Sust Energ Rev 134:110360

López I, Borzacconi L (2010) Modelling of slaughterhouse solid waste anaerobic digestion: determination of parameters and continuous reactor simulation. Waste Manage 30(10):1813–1821

Mahtab A, Tariq M, Shafiq T, Nasir A (2009) Coagulation/adsorption combined treatment of slaughterhouse wastewater. Desalin Water 12(1–3):270–275

Mahyari FZ, Khorasanizadeh Z, Khanali M, Mahyari KF (2021) Biodiesel production from slaughter wastes of broiler chicken: a potential survey in Iran. SN Appl Sci 3:57

Meat and Livestock Australia (MLA) (2009) Co-products compendium. https://www.mla.com.au/contentassets/79c16798add246bfa3162b9411022e93/a.cop.0061_mla_coproducts_compendium.pdf. Assessed on 06 Sep 2022

Meeker DL, Hamilton CR (2006) An overview of the rendering industry In: Meeker DL (ed) Essential rendering. National Renderers Association, Alexandria (VA), pp 1–16

Mees JB, Gomes SD, Boas MAV, Fazolo A, Sampaio SC (2009) Removal of organic matter and nutrients from slaughterhouse wastewater by using Eichhornia crassipes and evaluation of the generated biomass composting. Engenharia Agrícola 29:466–473

Meiramkulova K, Zorpas AA, Orynbekov D, Zhumagulov M, Saspugayeva G, Kydyrbekova A, Inglezakis VJ (2020) The effect of scale on the performance of an integrated poultry slaughterhouse wastewater treatment process. Sustainability 12(11):4679

Mittal GS (2006a) Treatment of wastewater from abattoirs before land application—a review. Biores Technol 97(9):1119–1135

Mittal GS (2006b) Treatment of wastewater from abattoirs before land application—a review. Bioresour Technol 97(9):1119–1135

Moukazis I, Pellera F-M, Gidarakos E (2018) Slaughterhouse by-products treatment using anaerobic digestion. J Waste Manag 71:652–662

Mozhiarasi V, Weichgrebe D, Srinivasan SV (2020) Enhancement of methane production from vegetable, fruit and flower market wastes using extrusion as pre-treatment and kinetic modeling. Water Air Soil Pollut 231:126

Mozhiarasi V, Natarajan TS (2022) Slaughterhouse and poultry wastes: management practices, feedstocks for renewable energy production, and recovery of value-added products. Biomass Convers Biorefin, pp 1–24

Musa MA, Idrus S (2021) Physical and biological treatment technologies of slaughterhouse wastewater: a review. Sustainability 13(9):4656

Myhara RM, Kruger G (1998) The performance of decolorised bovine plasma protein as a replacement for egg white in high ratio white cakes. Food Qual Pref 9:135–138

NABC (2004) Carcass disposal: a comprehensive review. Report written for the USDA Animal and Plant Health Inspection Service. National Agricultural Biosecurity Centre, Kansas State University, USA

Ng M, Dalhatou S, Wilson J, Kamdem BP, Temitope MB, Paumo HK, Kane A (2022) Characterisation of slaughterhouse wastewater and development of treatment techniques: a review. Processes 10(7):1300

Omole D, Ogbiye A (2013) An evaluation of slaughterhouse wastes in southwest Nigeria. Am J Environ Prot 2(3):85–89

Palatsi J, Viñas M, Guivernau M, Fernandez B, Flotats X (2011) Anaerobic digestion of slaughterhouse waste: main process limitations and microbial community interactions. Bioresour Technol 102(3):2219–2227

Park SK, Bae DH, Hettiarachchy NS (2000) Protein concentrates and adhesives from meat and bone meal. J Am Oil Chem Soc 77(11):1223–1227

Parvathy KS, Negi PS, Srinivas P (2009) Antioxidant, antimutagenic and antibacterial activities of curcumin-β-diglucoside. Food Chem 115(1):265–271

Pierce JL, Cromwell GL, Lindemann MD, Russell LE, Weaver EM (2005) Effects of spray-dried animal plasma and immunoglobulins on the performance of early weaned pigs. J Anim Sci 83:2876–2885

Rawdkuen S, Benjakul S, Visessanguan W, Lanier TC (2004) Chicken plasma protein: proteinase inhibitory activity and its effect on surimi gel properties. Food Res Intl 37:156–165

Salminen E, Rintala J (2002) Anaerobic digestion of organic solid poultry slaughterhouse waste–a review. Bioresour Technol 83(1):13–26

Sanabria-León R, Cruz-Arroyo LA, Rodríguez AA (2007) Chemical and biological characterisation of slaughterhouse wastes compost. J Waste Manag 27:1800–1807

Santos EM, González-Fernández C, Jaime I, Rovira J (2003) Physicochemical and sensory characterisation of Morcilla de Burgos, a traditional Spanish blood sausage. Meat Sci 65:893–898

Selormey GK, Barnes B, Kemausuor F, Darkwah L (2021) A review of anaerobic digestion of slaughterhouse waste: effect of selected operational and environmental parameters on anaerobic biodegradability. Rev Environ Sci Biotechnol 20(4):1073–1086

Sieng S (2019) Optimisation of struvite precipitation in poultry slaughterhouse effluent. Universitas Gadjah Mada

Sweeten JM (1980) Water pollution control in slaughterhouses and meat processing plants. Bulletin/Texas Agricultural Extension Service, no 1291

Tanaka K, Sawatani E, Shigueoka EM, Dias GA, Nakao HC, Arashiro F (2001) Isolation of bovine plasma albumin by liquid chromatography and its polymerisation for use in immunohematology. Brazilian J Medical Bio Res 34:977–983

Tariq M, Ahmad M, Siddique S, Waheed A, Shafiq T, Khan MH (2012) Optimisation of coagulation process for the treatment of the characterised slaughterhouse wastewater. Pakistan J Sci Ind Res Ser A Phys Sci 55(1):43–48

Tolera ST, Alemu FK (2020) Potential of abattoir waste for bioenergy as sustainable management, Eastern Ethiopia, 2019. J Energy 6761328

Tseng T, Tsai C, Yang J, Chen M (2006) Porcine blood plasma transgluataminase combined with thrombin and fibrinogen as a binder in restructured meat. Asian Austra J Anim Sci 19:1054–1064

Urlings HAPV, Logtestijn JG, Bijker PGH (1992) Slaughter by-products: problems, preliminary research and possible solutions. Veterinary Quart 14(1):34–38

Viana FR, Silva VDM, Delvivo FM, Bizzotto CS, Silvestre MPC (2005) Quality of ham pâté containing bovine globin and plasma as fat replacers. Meat Sci 70:153–160

Visessanguan W, Benjakul S, An H (2000) Porcine plasma proteins as a surimi protease inhibitor: effects on actomyosin gelation. J Food Sci 65:607–611

Walter T, Hertrampf E, Pizarro F, Olivares M, Llaguno S, Letelier A, Vega V, Stekel A (1993) Effect of bovine-hemoglobin-fortified cookies on iron status of schoolchildren: a nationwide program in Chile. Am J Clin Nutr 57:190–194

Wismer-Pedersen J (1988) Use of haemoglobin in foods—a review. Meat Sci 24(1):31–45

Wu J, Doan H (2005) Disinfection of recycled red-meat-processing wastewater by ozone. J Chem Technol Biotechnol Clean Technol Environ Policy 80(7):828–833

Wu J, Wang C (2016) Adhesives derived from agricultural proteins. 9522515 B2. U.S. Patent. 2016 Dec 20

Yaakob MA, Mohamed RMSR, Al-Gheethi AAS, Kassim AHM (2018) Characteristics of chicken slaughterhouse wastewater. Chem Eng Trans 63:637–642

Yan J, Lin HL, Feng GZ, Gunasekaran S (2016) The effect of acrylic latex-based polymer on cow blood adhesive resins for wood composites. In: Proceedings of the 2016 global conference on polymer and composite materials (PCM 2016), Hangzhou, China, 20–23 May 2016, pp 305–312

Yang G, Yang B (2010) Wood adhesive and method of preparing Thereof. US Patent 0258033 A1, 14 Oct 2010

Yarandi MS, Mahdinia M, Barazandeh J, Soltanzadeh A (2021) Evaluation of the toxic effects of ammonia dispersion: consequence analysis of ammonia leakage in an industrial slaughterhouse. Med Gas Res 11(1):24

Yordanov D (2010) Preliminary study of the efficiency of ultrafiltration treatment of poultry slaughterhouse wastewater. Bulg J Agric Sci 16(6):700–704

Chapter 3
Sustainable Recycling of Manure and Reuse to Mitigate Climate Change

Sadia Javed, Amreen Aftab, Sher Zaman Safi, Ameer Fawad Zahoor, Nazima Anwar, Saboor Gul, and Muhammad Arshad

Abstract This chapter's primary goal is to discuss the current problem of manure wastes, which are dangerous to the environment and agricultural systems due to their rapid growth. Using scientific and eco-friendly methods can assist to reduce waste by reusing manure. However, one of the big challenges to reuse is pathogenic risk. Additionally, vermicomposting and composting's usefulness in recycling crop nutrients, notably nitrogen, is assessed (N). It covers the guidelines and practices of advances for the control of tainting, starting from normal wastes like human waste, serious areas of strength for wastewater, animal fecal matter, and agro-current waste, and the reusing of these regular wastes into critical things like fertilizer, biofuels, algal and fish protein, and immersed crops. It can likewise address various issues connected with the decrease of natural contamination, squander the executives, and human well-being gambles.

S. Javed (✉) · A. Aftab · N. Anwar · S. Gul
Department of Biochemistry, Government College University, Faisalabad, Pakistan
e-mail: sadiajaved@gcuf.edu.pk

S. Z. Safi
Faculty of Medicine. Bioscience and Nursing, MAHSA University, Jenjarom, Selangor, Malaysia

A. F. Zahoor
Department of Chemistry, Government College University, Faisalabad, Pakistan

M. Arshad
Department of Basic Sciences, Jhang-Campus, University of Veterinary and Animal Sciences Lahore, Lahore, Pakistan

M. Arshad (ed.), *Climate Changes Mitigation and Sustainable Bioenergy Harvest Through Animal Waste*, https://doi.org/10.1007/978-3-031-26224-1_3

Graphical Abstract

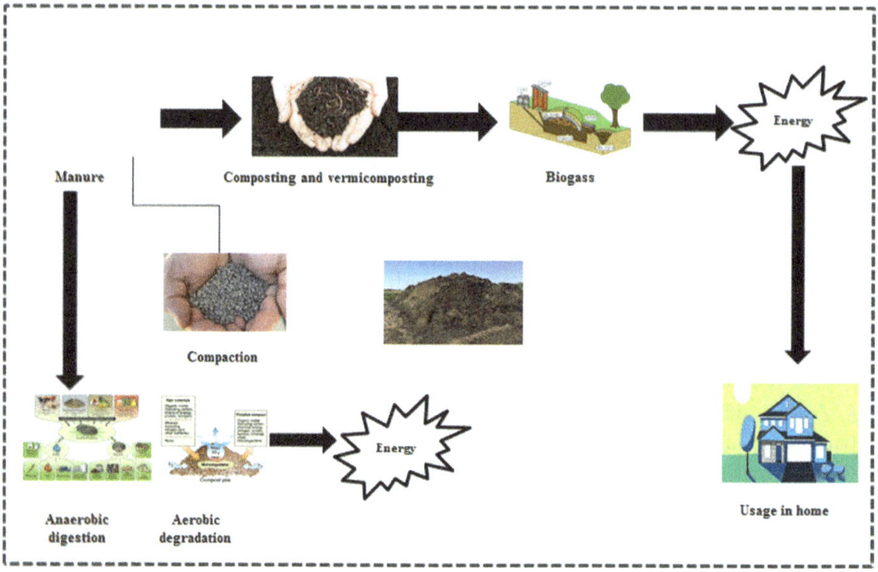

Keywords Animal manure · Biofuels · Organic waste · Water · Waste materials · Composting · Technology · Wastewater treatment

3.1 Introduction

Animal farming is essential for food supply, nourishment, and economic stability. Ruminants, non-ruminants, and aquatic species are the most commonly bred domestic animals in many countries. Cattle, pigs, poultry, and pets are examples. Animal agriculture contributes to the supply of nutritious food, job creation, income generation and family income, the economy of funds, economic output and taxes, divestment agricultural production, wildlife traction, reproduction, and soil transfer, as well as the economic system of resources, economic production and taxes (Swanepoel et al. 2010). One of the significant issues confronting domesticated animals cultivating all over the country is meeting the food needs of a developing total populace, as most would consider to be normal to reach north of 9 billion by 2050. It isn't quite as basic as growing assembling ability to increment food yield. Terrains and water utilization, the ecological outcomes of creature creation, and limitations are factors that can restrict makers' ability to just add more creatures to fulfill future need for animals. Creature horticulture ought to be completed in a manner that doesn't endanger future use of regular assets while endeavoring to satisfy the dietary needs of people and creatures. In some creation frameworks, creatures are raised basically for food, as opposed to for friendship, stows away, or even compost (Romney et al. 1994). Side-effects are made during the creation, handling, transportation and selling of creatures, and while possibly not accurately taken care of, they can become squander. Feed waste or, sewage, incubator squander, slaughterhouse squander, are a portion of the potential squanders created during creature creation exercises. Bedding, pee, wash water, precipitation, spilled feed and spilled water are normal outside inputs in creature excrement. Before the presentation of natural composts, creature excrement assumed a focal part in further developing soil fruitfulness. Job of natural composts in horticultural creation, excrement stays a significant manure asset, particularly in regions where natural manures are not promptly accessible or open to ranchers.

Current ecological issues are because of high creation and neighborhood gatherings of natural squanders for the essential corruption processes in nature. With satisfactory application rates, creature compost is an important asset as a dirt manure, as it gives high full scale and micronutrient content for crop development and is a cheap and harmless to the ecosystem elective. with mineral manures (Swanepoel et al. 2010). Every year more than 1500 million tons compost has been genearted in EU 27 due to escalation of creature cultivation (Romney et al. 1994), which should be reused productively because of the natural issues related with their aimless and untimely application to farming fields. Exorbitant utilization of hurtful follow metals, inorganic salts, and microorganisms; expanded loss of supplements, principally nitrogen and phosphorus, from soils through filtering, disintegration, and overflow brought about by inability to represent the wholesome requirements of harvests; and vaporous emanations of scents, hydrogen sulfide, alkali, and other harmful gases are possible adverse consequences of such unpredictable applications (Malomo et al. 2018). To be sure, agribusiness represents around 10% of generally speaking ozone-depleting substance discharges, with creatures contributing altogether through methane outflows from intestinal maturation and compost age. The overall creature creation chain represents around 65% of anthropogenic N_2O and 64% of anthropogenic NH_3 (Zhang et al. 2017). The presentation of proper administration advancements could in this way alleviate the well-being and natural dangers related with the overproduction of natural waste from domesticated animals by balancing out it before its utilization or removal. Adjustment includes the deterioration of natural material with the end result of disposing of risks and typically brings about diminishes in microbial biomass and action and in groupings of labile mixtures (Malomo et al. 2018). Animals are raised on pastureland, field, yearly grass, and bought grain, with non-domesticated animals cultivating activities representing under 10% of absolute creation costs. Non-animals exercises incorporate families with a portion of those creatures took care of natural waste, clippings from road borderlines, and other non-domesticated animals organizations. Touching, blended, and modern structures are separated into three classes by the Food and Agribusiness Association (FAO).

- **Grazing systems**: The animals are fed in accordance with the animal manufacturing techniques described above. The graze machine is defined by animals that go to get food (mobile), rely on regional common pasture (nutrient deficient), or get enough food inside the farm's boundaries (ranching and grassland). Every hectare of agriculture productivity, stock quotes are fewer than 10 cattle gadgets per year. Grazing manufacturing techniques are used in dry, semi-arid, sub-humid, humid areas, as well as tropical and temperate highlands. In terms of total production rate, grazing systems only provide 9% of world wide meat consumption.

- **Half and half frameworks:** There non-domesticated animals rural creation contributes for over 10% of the absolute expense of creation, or where farming items (for example dinner from modern food fabricating) and additionally stubble represent definitely over 15% of the dry matter took care of to creatures. Eating publicly, rural squanders and harvests, cut-and-convey exercises, on-ranch result, and outside feed are completely used to take care of dairy cattle in these arrangements. In Europe, ranch proprietors should possess more than 1.5 ha of land per animal unit to guarantee that enough developed land is accessible to utilize the micronutrients created in excrement. The creation comes into the blended class of animals fabricating strategies in light of the fact that these fields give most of the feed for the cows.
- **Industrial systems**: The average stocking charge per hectare of agricultural land in industrial plants is larger than 10 animal units, and so less than 11% of the drying count given to farm animals is produced on the farm. The manufacturing techniques discussed here are the poultry industry (meat and eggs), pig productivity, animal feeding meat manufacturing, and dairy product manufacturing. Because commercial cattle production practices rely heavily on outside nutrition, power, and other inputs, the use of those inputs may have an impact on the environment in areas other than where production takes place. More than half of the world's hog and chicken meat, as well as 10% of the world's red meat and mutton, is produced by agricultural infrastructure. Those manufacturing structures are characterized by landless pig and chicken farms. In addition to bright orange meat and dairy farm that have almost a hundred thousand head of animals, pigs farm with slurry sprays on landscapes of several hectares that acquire waste from animal dwellings supporting manufacture of thousands of pigs come into this magnificence in America. Feces is sprinkled on plains to fertilize crops, used in fish ponds to feed algal and fresh water (which plant-eating fishes devour), or used in other ways. Climate, energy generation traditions, and farming methods all influence how manure is collected, stored, and used.

3.2 Historical Background

Occasionally American agriculture commonly contains a concept of a farms some of animals and of chickens, in adding to unique labour wishes of the crop productivity sector. The farm partner was expected to hold the fowl flock and sell eggs and fryers on a normal basis as a deliver of revenue (Miner et al. 2000). The average dairy farm in USA is a small scale family farm with 20 cows. On the farm's grounds, hen flocks graze. Manure was used to offer high nutrition values in a slow-release way. The farm's feed production reassets are placed to the take a look at manner of draught animals. Agricultural waste was no longer an essential issue due to the highest price of the operations. Manure was regarded as an agricultural useful resource at a few level withinside the early year, from the 18th to the nineteenth centuries. Manure treatment aids to generate the soil herbal matter (SOM), which brought about better water invasion and bounty considerably less soil disintegration. This made it less challenging for the rancher to raise feed and one-of-a-kind subjects for off-ranch send out. As a surrender final product of The Second Great War, ways of life and horticulture withinside America changed immensely. Ranch work was in a nutshell supply at a couple of level withinside the conflict.

Additional fuel and nitrogen-based weapons creation capacity stayed after the conflict, which may be changed over into inorganic compost creation capacity. In view of the commitment of ordinary hours and better compensation, city populaces developed at amazing expenses withins ide the twentieth 100 years. Urbanization, horticultural work deficiencies, American ranchers' resourcefulness helped through way of Land Award Rural Trial Stations, copious energy and supplements, great soils, and a reasonable environment prepared for excellent generally speaking execution withinside the creation of feed, food, and fiber. American ranchers and clients lately have a normal and tremendous stockpile of meat, dairy, and hen items in light of the truth to unpracticed creation techniques.

- The excrement interest that results has the ability to hurt the biological system. These worries were issue withinside the establishment of the Perfect Water Act in 1972. In light of the enormous nitrogen unevenness in fertilizer comparative with adjoining crops, it's miles currently broadly seen as a removal danger in a lot of farming locales. Conflicts over water and air high-pleasant, similarly to odors, often arise because of metropolis boom in historically animal-producing areas. Some have confused whether or not or now no longer land software program is a remarkable manure manipulate method because of the ones concerns. By some distance, the most principal byproduct of animal production is animal feces. The production of manure within America is expected to be over 60 million, roughly 1.12 million Mg of nitrogen (N) and 0.60 million Mg of phosphorus (P) in each ton. The whole amount of amassed manure generated withinside America is expected to be over sixty a million tons steady with year, with each tonne containing form of 1.12 million Mg of nitrogen (N) and 0.60 million Mg of phosphorus (P) (Gollehon et al., 2001). According to the Department of Agriculture and the Environmental Protection Agency, land software program of manure at maximum extraordinary agronomic costs is the most favored method for utilizing manure reassets withinside America. While there are benefits to using manure, along with better soil fertility and high-pleasant, reduced runoff and soil erosion, and the possibility of C sequestration, flawed software program may grow to be worse water high-pleasant and convey odour and air high-pleasant problems. According to a research of manure's capacity fertiliser price, recoverable nitrogen in manure money owed for about 15% of the N and 42% of the P consumed withinside the US. These pollutants contaminate water resources and reduce recreational potential of lakes and rivers, destroy wildlife habitat, and eliminate drinking water supplies for people and livestock. Manure, on the other hand, can impair water high-pleasant. This chapter summarizes the benefits of recycling of manure and also discusses the problems related to manure application to mitigate the climate change.

3.3 Accommodations and Activity Area

In different climates like cold and damp, livestock homes or sheds provide a hot indoor living, but in the equator, a cool and dry atmosphere. Housing design is influenced by a variety of factors, including the environment, the animal category being housed, and the production goal.

3.3.1 Cattle

Calves, heifers, bulls, and cows are the four types of cattle. These classifications are based on the animal's age, gender, and role in farm productivity. Air flows via gaps in the walls or through doors opening in most livestock houses, providing natural ventilation. In hot regions, ventilation can be pushed through the use of fans to provide an open air flow, or through the use of tunnel ventilation systems with fans of vantilation located at the end of the house. These produce a continuous air flow throughout the cattle home, with air entering through slots in outer edge (Pain and Menzi 2003). The cattle classification is given as

- Cow: A female cattle delivered; a calf becomes a cow after birth to her 1st animal.
- Calf is newly born cow child
- Feminine calf is considered to as a heifer calf.
- An in-calf heifer is a heifer which is pregnant.
- Dairy cow: A dairy cow is an animal that is raised to produce milk and to rear calf for a dairy animals. It's important to remember what cow needs to calves in order to generate milk.
- A bull: one male cattle is called as bull.
- A male newborn is referred to as a bull kid.
- Beef cattle: Livestock retained for slaughtered at a body mass of 450–550 kg, which can be as young as 14–15 months for intensive grassland or 18–29 months for grazed animals, and which can be as early as 13–16 months for intensive and 17–30 months for grazing living creatures.
- A bull that has been rendered sterile, often known as a steer or bullock.

A loose housing, during which animals are absolve to wander around within the same house, could be a way more frequent style for cow housing. Manure in these dwellings may be managed with suspension (mixed feces) gathered to a lower place the horizontal plane. A slatted floor is typically created out of one to 1.5 m long concrete slabs with many centimeter-wide gaps. Excrement, piddle samples, and spilt potable run through the perforations and roll up a entice or funnel as liquid beneath the slatted surface. A residence with wood floors throughout is referred to as "completely perforated." In residences wherever the perforated floor is reduced to the walking alleys, a part perforated floor is gift. These barns are separated into rows of individual stalls or cubicles where placental will sleep however not be confined.

The compartments have a high top constructed of stone or other sturdy, waterproof material. Solid soil can be sprinkled with straw, sawdust, wood chips, sand, or peat. Walkways can have solid concrete floors, asphalt concrete floors, or rubber-coated concrete floors. In general, once a day, pedestrian paths with solid floors are scraped (Monteny and Erisman 1998). Dairy cows in tethered systems are tied to a neck feed fence with cables, chains, or plates and have limited mobility in their residential area, which is usually a hard floor lined with litter like grain, dust, or dirt. The back of the animal, there may be an iron grill tube or perforated cement floor. For large cattle, such procedures are becoming increasingly popular in Europe due to animal welfare concerns (heifers, beef cattle, and dairy cows).

In Scandinavian nations, private frameworks that hold both watery and strong excrement are being eliminated because of creature government assistance issues. Since the deck on which the creatures walk is shrouded in a weighty covering of dung, strong floor houses are frequently planned with the residing space subterranean. Steers meander beneath the superficial on a litter layer of 30–35 cm toward the beginning of a creation period, yet when extra trash is stored regularly, the creatures are probably going to wind up going on a thick layer of profound waste compacted by their hooves. The profound litter gives solace to the pets while likewise holding water. Destroyed straw, dust, wood shavings, and dirt are the materials used for the cattle bedding (Fig. 3.1).

When the deep waste layer is thin, the submerged floor with steps allows easy access to the feed area. Cattle farms are closed, external animal production systems with no infrastructure to collect liquids or rainwater. Each farm has a number of fenced pens where livestock or dairy farming is housed at the rate of 2.5 animals in all forms of animal husbandry. The animals, that may variety withinside the thousands, are unfastened to stroll at the ground (McGinn et al. 2010). Straw may be scattered across the feedlot to offer resting regions. Most feedlots are determined in semi-arid regions of North America, Antarctica, and numerous components of Europe (Examples consist of regions outdoor the milking chamber in which dairy farm animals congregate earlier than being pressed, in addition to an exercising backyard for dairy calves raised in tethered stalls, that is required in some nations for animal rights safety reasons.

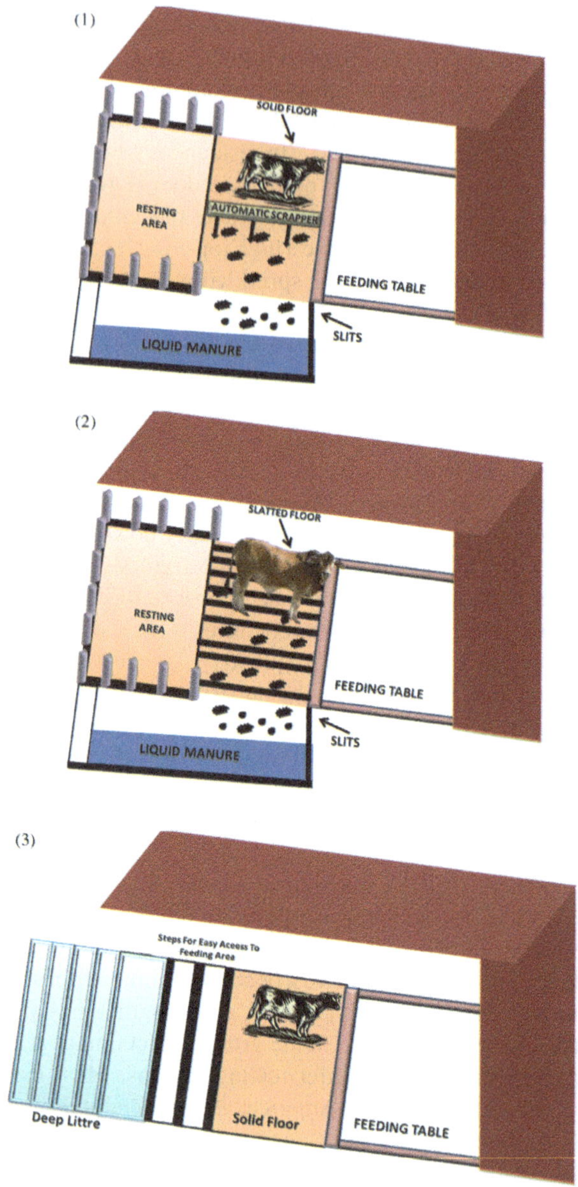

Fig. 3.1 (1) A dairy cows house with consistent deck in sitting regions, pathways, and disposal paths, as well as the taking care of region, where the dung/pee combination is scratched to the channel on the right peak. (2) Yearling home with stable ground in resting region and punctured ground in exercise path and taking care of region. (3) Calf house with strong floor with free moving calves on profound litter

3.3.2 *Poultry*

Sow homes will be worked with semi-escalated courses of action, like how bulls are housed, in which body fecal matter and are kept inside the natural material that covers the floor. Pigs uncover and fabricate homes in various regions of the serious litter while crapping in one part of the pen. Shallow water bowls are implicit the rear of the pig pens in Asian nations where water is copious. They are used by the pig to resit and discharge waste. Pigs enjoy to cool themselves in pools in their natural environment, so the layout is ideal for them. The liquid is pipeline or gravity-fed into lagoon from these watersheds. The resulting liquid excrement is extremely diluted and hard to handle. The remainder of the area should be used as a perforated pit to catch droppings, with at least a third of it coated with feed ingredient. To keep the trash dry, giving birth eggs, feed, and a water supply are placed over the perforated area. The bodily collection of waste in a completely watertight hole beneath the bottom surface. Laying hens are kept in layer cells that are commonly made of steel cable and arranged in rows upon rows in battery cage homes. Excrement is either expelled very cheaply from the cells and captured and processed in an extremely deep hole beneath the cells, or they are retrieved using a transport belt or manual tool technique.

3.4 Integrated Manufacturing Technology

Animal waste material is recycled to fish ponds in Asian farming methods that involve crop cultivation, gardening, fish farming, and animal breeding, whereas gardening, fish farming, and husbandry give the principal outputs for consumption or sale, waste from one system is employed as an input within the others, eliminating the necessity for external chemicals and pollution. Fish are fed directly from the dung, or indirectly through phytoplankton, zooplankton, and zoobenthos, which are after fed to baccivorous fish (Vu et al. 2007). One example is pigs housed in concrete-floored pens, wherever excretion is scraped off the ground and excretion and water are channelled into fishponds. Solid manures are often thrown into the ponds without being processed or composted but it should be processed before putting in the fishponds for good harvesting (Fig. 3.2; Table 3.1).

3.5 Types of Animal Manure

Feces and urine, in addition to bedding, spilt feed, spilt consuming water, and washing water from animal housing, make up animal manure. As an end result, the composition is quite variable, and manure elimination generation has an impact. Manure classifications are generally associated with the shape of the animal domestic or the

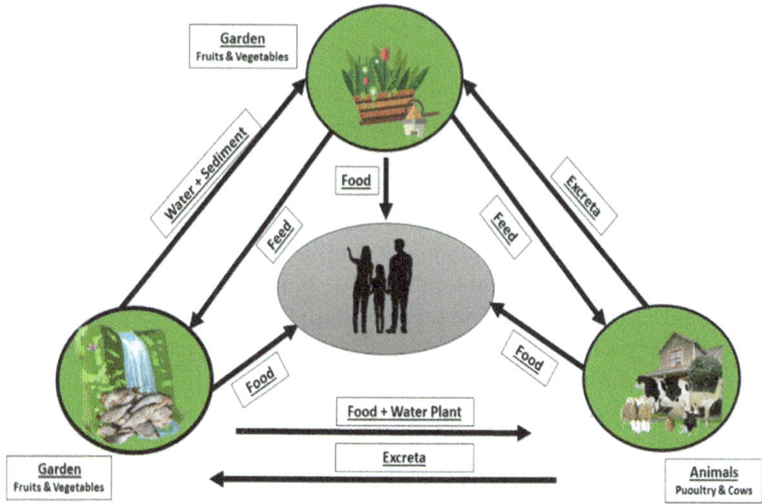

Fig. 3.2 VAC system (garden–pond–animal) is depicted

Table 3.1 Total amount of nutrient elements, carbon, and liquid in excretion and crop residues in manure production within the animals were housed and in manure transmitted to the sphere from Bos taurus deep litter management and animal suspended management: one farm animal produces fifteen tons of deep waste per year1 and one pig place in the pig house (producing three finished pigs per year1) produces 1.5 tons slurry per year1

		Dry matter (kg)	Phosphorous (kg)	Nitrogen (kg)	TAN (kg)	Potassium (kg)	Carbon (kg)	Water (kg)
Dairy cow deep	In house	4600	23.1	128	29	35	175	1765
Litter	To field	2500	22.2	69	15	14	175	739
Pig slurry	In home	100	3.8	11.2	9	8	6.5	38
	To field	73	3.8	7.9	4.8	5.7	5.8	33

out of doors manufacturing device in which the manure is accumulated, so definitions of farm animals manure may also fluctuate through country. Manure may be controlled in ways: liquid or stable. Manure with a dry be counted percent of much less than 12% is stated as "liquid manure." This may be moved the use of drainage or pumps. Solid manure is described as manure that can be stacked and has a dry be counted content material of more than 12%. Due to deposition of washing water, slurries recovered from under slatted floors have a low dry be counted concentration. Cattle slurry may also have a bigger dry be counted content material than pig slurry in general, but there is lots of range among farms inside every animal category. As an end result of the excessive water intake in Asia, the dry be counted of pig slurry is best 0.21%.

Urine and washwater drain from the ground and collect as slurry, an aggregate of washwater, urine, and dissolved fecal components. These facilities produce farmyard manure (FYM) with excessive dry matter content and slurry with low dry matter content material.

In Asia, liquids and part of the excretions are transported to a pond where the fattening pigs are raised in solid-bottomed berries and the pee is deposited in a deep, high-dry-matter well-organized bed. Build nests and dig, which makes it easier to start composting in the barn. Composting is the aerobic microbial breakdown of organic material that creates heat. The garbage in the upper layers (1015 cm) of the barns can start to compost, and eventually the garbage will cover and bury the new litter, which over time and regularly will be compacted. Adding straw and excrement. A conveyor belt transports the manure to an external dry manure deposit (Koerkamp 1994). The dry matter composition of manure and dry chicken manure is significantly different (i.e., 5–27% in manure and 31–67% in dry solid manure). Cause of change in dry matter content of solid manure (Kroodsma and Scholtens 1988). After removal from the barn, the manure content changes depending on storage and processing. Most systems transport manure and animal manure to a concrete or steel pond or tank.

Anaerobic storage conditions result from a lack of air in manure, which means that the composition of the population changes relatively slowly over time. Chicken droppings have a tall structure and begin to compost during storage. This allows solid manure to be naturally composted during storage or by artificial ventilation. The nutritional value of cow manure is higher than that of manure. Cow manure has a higher concentration of dry matter, which explains the difference in manure. Liquids pigs are monogastric animals that do not absorb nutrients from plants as efficiently or use energy from feed as efficiently as ruminants.

Liquid pig manure is diluted by washing or watering pigs with water in warmer climates (which reduces the nutrient content).In addition, the liquid in Asia is poor in vitamin A and plant vitamins, because the sediment is removed before miles from the ground. Cleaning the house is justified by the fact that it reduces the risk of disease and odor emissions. Broiler chicken manure is often extremely dry due to the low moisture content of chicken droppings, the large amount of sludge eaten up, and internal drying processes. Therefore, of all animal fertilizers, chicken dung has the best attention of vegetable vitamins. In some countries, the combination of manure classes has various preferred values. Some versions are due to variations in animal production procedures, as well as the composition of the feeding plan, water use, and waste management.

3.6 Livestock Manure Attributes

Numerous accommodating and reusable parts can be tracked down in fertilizer. Creature compost' physical and synthetic characteristics can impact its expected use, essentially as a manure, additionally that can be dealt with. Creature excrement is separated into compost (five% solids), suspended and semi-strong fertilizer (somewhere in the range of 5 and 25% solids), and strong fertilizer (over 25% solids) with a helped consistency or dampness content. A definitive properties of excrement created by normal creature creation exercises are examined. Given the extraordinary changeability in consistency, body, and substance synthesis of creature excrement, starting with one area then onto the next, inclination ought to be given to the properties of compost of homegrown beginning (Table 3.2).

3.6.1 Particle Size

Since no rules have been laid out for this request, investigations of molecule size in creature slurry have utilized an extensive variety of channel sizes. Accordingly, the biohazardous material molecule orders are erratic. A progression of strainers with diminishing cross-section sizes is many times layered on top of each other, with the best molecule sizes on top. The granules are riveted down the rack by a progressive splash of water of reused fluid until they're gathered on a sifter with a lattice size more modest than the molecule size. Fertilizer is poured to the highest sieves, and as a result, Because particles with a diameter less than 1 m one m are considered soluble dispersed solids, a very affordable sieve with a size of one m is appropriate. Brownian (diffusive) motion is a problem for colloids in a very liquid environment, because they can either not descend or resolve very slowly by gravity. Because colloidal particles are difficult to remove during solid–liquid extraction, the number of colloids is crucial. Despite the fact that there are a few strategies for police work molecule size (for instance, different light-weight dispersing methods), movement is trying because of the expansive pore size dissemination and sporadic construction of the particles. Pig suspension has a more prominent level of dry matter in the division underneath 0.17 mm than cow suspensions, with 66–70% and 50–55% dry matter, correspondingly. Microorganism change of the natural pool, like the transformation of natural parts to ozone-harming substance carbon dioxide (CO_2), methane (CH_4), and ammonium [NH_4 + (aq)], alters the particle size distribution. the particle size distribution. As a result, fermentation process in a biogas decreases particle concentration in animal slurry. Particles smaller than 10 m can account for 64% than ten m can account for sixty-four percent of dry matter in raw slurry, but in anaerobically digested suspension, this proportion jumps to eighty-four percent. Information on the plant dietary organization of molecule size divisions is basic once action no homogeneity of hang on creature slurry happens (for instance, almost 70% of unmelted N and P occurs inside the 0.45-to 250-m molecule size part of bulls

Table 3.2 Farm animal manure characteristics (per 1000 lb animal unit per day)

Manure	Uses	Advantages
Nutrients	Nutrients include compost, fertilizer, and bioenergy conversion	Fertilizer savings and revenue generated from sales of fertilizer
Organic matter	Soil reorganization	Improves the soil's structure and water-holding capacity. Effects on agricultural yield
Solids	Thick bedding	Savings of up to $50 per cow per year on bedding supplies, for example
Energy	Energy Syngas and bio-oil are all examples of renewable energy sources	Farm-specific energy; less dependency on fossil fuels sales of energy; fossil fuels; revenue generation

Animals category	Weight (lb)	Moisture (percent)	Solids (lb)	Volatile solid (lb)	O (lb)	N (lb)	P (lb)	K (lb)
Lactating cow	100–130	87	15–17	8.2–15	3.1	1.0	0.22–0.25	0.41–0.45
Calf	90	85	8.2	6.6	1.2	0.5	0.04	0.22
Heifer	60	85	8.6	6.2	1.3	0.3	0.04	0.13
Dry cow	55	90	5.5	6.5	1.0	0.4	0.053	0.20
Beef cow	108	70	12	12	3.5	0.4	0.09	0.32
Growing cow	88	77	8.8	8.3	2.7	0.5	0.09	0.35
Finishing cattle	68	95	5.4	5.0	3.2	1.2	0.4–0.5	0.3
Gestating sow	28	95	2.8	3.0	0.9	0.2	0.04	0.3
Layers	58	75	18	3.2	3.2	1.2	0.4	0.4
Broiler	89	76	23	5.4	5.0	0.8	0.3	0.6
Turkey toms	35	76	8.0	2.0	2.3	0.4	0.2	0.50
Turkey hen	45	76	15	3.5	3.0	0.8	0.4	0.31
Duck	100	76	26	4.0	5.5	0.4	0.45	0.45

slurry). More than 80% of the absolute N and P in dairy cattle slurry is found in the division under 0.125 mm (Meyer et al. 2007).

3.7 Technologies for Animal Waste Treatment

Animal manures are commonly a combination of excrement, urine, wasted bedding, and waste feed, with various quantities of water. As a result, relying on whether the manure is solid, semi-solid, slurry, or liquid, a few manure remedy structures can be appropriate for dealing with manure than others. The technology mentioned on this segment are opportunity withinside the feel that they minimize pollutants greater correctly than conventional untreated manure land application. These changes in waste control technology can also additionally necessitate extra remedy strategies to growth the system's overall performance and acquire on-farm nutrient discount goals. These extra nutrient discount and recuperation procedures are mentioned in addition in this text below the segment on nutrient discount and recuperation strategies.

3.7.1 Compaction

Actual procedures, for example, granulation and bundling can assist with working on the capacity and treatment of mass compost solids. These methodologies plan to convey supplements to fertilizer in a more practical and without dust way, as well as condition the manure before change to bioenergy. The granulation of a mass material, for example, poultry litter, extraordinarily builds its thickness. Granulation of poultry litter blended in with 3% vegetable oil, for instance, quadrupled the mass thickness of oven litter, from 200 to 790 kg m^3 (McMullen et al., 2005).

The water content material and strength expected to conservative chicken litter the utilization of pelletizing gear choose the most proficient compaction of chicken litter (Bernhart and Fasina 2009). Notwithstanding, the high energy utilization and gear costs make the buy and activity of a pellet plant on a homestead unreasonable for grill makers. Chicken litter squander bundling was created as an energy-effective option in contrast to granulation as a compaction technique. Pressure and wrapping are consolidated in this strategy (Bernal et al. 2009).

3.7.2 Aerobic Degradation

Vigorous biodegradation is the breakdown of regular contamination via microorganisms when oxygen is available. All the more explicitly, it alludes to being or dwelling exclusively within the sight of oxygen. Mechanisms of treating the soil and vermicomposting under cardio conditions, regular depend debasement is an exothermic strategy wherein oxygen highlights as a terminal electron acceptor, revising normal particles into more prominent secure mixtures, delivering carbon dioxide and water, and delivering heat. In the field, cardio corruption happens gradually on the dirt surface without accomplishing inordinate temperatures; in any case, this home grown breakdown technique might be hurried through heaping the texture into windrows to confine warmth misfortunes and as needs be license for temperature increments. Albeit each cardio strategies, fertilizing the soil and vermicomposting, had been broadly utilized for handling various sorts of creature fertilizer, both consistently or joined, practically all the exploration aren't far and wide thought about, because of varieties in exploratory plans, regular count content, worm species, test length, and assessment boundaries, among various things. In spite of those restricts, those explorations have considerably supported a higher data of the changes that emerge withinside the texture at some stage in those natural adjustment strategies, that is significant for their improvement and, eventually, the assembling of an extraordinary stop item. In like manner, a couple of synthetic capacities of creature fertilizers, comprising of an extra of dampness, low porosity, an unreasonable N consideration rather than the regular C substance, or unnecessary pH values, can limit the viability of those strategies (Bernal et al. 2009).

3.7.3 Composting

It is a bio-oxidative method that involves the mineralization and fractional humification of normal matter; following in a balanced out absolute last item this is liberated from phytotoxicity and microbes and has positive humic homes and might be utilized to improve and keep soil brilliant and richness (Zucconi 1987). Ranchers have generally treated the soil animal composts after series with a reason to improve taking care of, transportation, and the executives (Bernal et al. 2009). The squanders have been much of the time heaped high, with little respect for the strategy conditions (air circulation, temperature, alkali misfortune, and so on) and the use of crude procedure. Consistent fertilizing the soil tasks might be thought of as a succession of constant societies, each with its own special arrangement of physical (temperature), synthetic (to be had substrate), and biological (composition of the microbial population) capabilities and comments effects. These fluctuations make it hard to observe the procedure, which is almost not possible to duplicate withinside the lab due to the fact temperature, moisture, aeration, and different elements are all depending on the surface/quantity ratio. Composting, in general, may be defined as a four-segment procedure wherein fungus and microorganism (known as number one decomposers) ruin strength-rich, plentiful, and without difficulty degradable materials consisting of sugars and proteins in the course of an preliminary segment referred to as the mesophilic segment (25–40 °C).

Despite the truth that each microbial agencies combat for with ease to be had substrates, fungus are speedy outcompeted due to the fact microorganism's most precise boom costs are one order of importance more than fungi's (Griffin 1985). Because mycelial organisms are extra visible, the significance of microorganism (except for Actinobacteria) at some stage in the composting procedure has lengthy been overlooked offers a precis of the microbial agencies engaged withinside the preliminary mesophilic segment (Ryckeboer et al. 2003). Manure fauna which integrates night crawlers, vermin, and millipedes can likewise in addition do as impetuses, adding to mechanical breakdown and providing a gastrointestinal home for particular microorganisms, provided that mechanical powers (which consolidates turning) are unobtrusive. The commitment of these creatures might be immaterial or critical, as withinside the instance of vermicomposting The colossal style of mesophilic life forms withinside the exact substrate is three sets of importance higher than the immense style of thermophilic organic entities; be that as it may, essential decomposers' advantage thought processes a temperature climb, and resulting thermophiles debase the mesophilic microbiota, further to the last without inconvenience degradable builds. During this 2nd segment of composting, referred to as the thermophilic segment, the temperature rises swiftly and speedy hurries up as much as a temperature of approximately 62 °C. When a compost pile reaches a temperature of over 55 °C. Fungal improvement is typically hindered, and thermophilic microorganism and Actinobacteria are the number one degraders in the course of this peak-heating period. Furthermore, fungi are extra laid low with oxygen transport than microorganism, or even in force-aerated systems, short anoxic conditions can arise. As a result, besides for the composting of lignocellulosic wastes, fungi play a bit function at some stage in this segment. When the temperature is among 50 and sixty-five tiers Celsius, Bacillus microorganism are commonly dominating. Furthermore, contributors of the Thermus/Deinococcus institution had been found in biowaste composts (Beffa et al. 1996), with most excellent improvement temperatures of sixty-five to seventy-five tiers celsius. Composts have yielded a whole lot of autotrophic microorganism that get their strength from sulfur or hydrogen oxidation. Their most excellent temperature is seventy-five tiers Celsius, and that they extensively resemble Hydrogenobacter species formerly found in geothermal settings.

What's more, the fundamental anaerobic microorganisms are notable withinside the fertilizer, but there might be regardless a comprehension opening roughly this microbial species. It is accepted that the more extended increment occurrences of archaea, when contrasted with microorganism, made them defective for the always changing over environmental elements of the treating the soil strategy. Be that as it may, a huge assortment of cultivable (Methanosarcina termophila, Methanothermobacter sp., Methanobacterium formicicum, among others) and regardless crude archaea had been found in treating the soil strategies in most recent exploration the utilization of the legitimate methods (Thummes et al. 2007a, b). The very last temperature upward push can be extra than eighty levels celsius, because of the impact of abiotic exothermic techniques related to thermostable actinobacteria enzymes. Such excessive temperatures are required for sanitizing compost to kill weed seeds and bug larvae, in addition to human and plant pathogens (Vinneras et al. 2010). Temperatures above 70 levels Celsius have the drawback of killing maximum mesophiles, delaying the restoration of the decaying populace after the temperature peak. Inoculation with mesophilic first-degree materials, on the alternative hand, can be capable of clear up this problem. When thermophilic organisms stop their pastime because of the shortage of substrates, the temperature starts off evolved to drop. This marks the begin of the cooling segment, additionally referred to as the second one mesophilic segment, of composting. The recolonization of the substrate via way of means of mesophilic organisms, each from surviving spores, via way of means of diffusion of blanketed microchials, or via way of means of outside inoculation characterizes it. During this segment, the variety of organisms that could degrade cellulose or starch will increase, specifically microorganism inclusive of Cellulomonas, Clostridium, and Nocardia, in addition to fungi inclusive of Aspergillus, Fusarium, and Paecilomyces

(Ryckeboer et al. 2003). Finally, the fungus/microorganism ratio will increase for the duration of the maturation segment, way to the aggressive benefit of fungi in situations of decreased water capacity and much less availability of the substrate. Some researchers have proposed a 5th segment of composting, referred to as the hardening segment (or garage segment), wherein the physicochemical parameters stay regular, however, the microbial groups change (Danon et al. 2008). Therefore, the chemical and microbiological modifications that arise withinside the substrate at some stage in the one-of-a-kind levels of the composting method will mostly outline the stableness and adulthood of the very last product, in addition to its suitability to be used as a natural soil amendment. Due to the aggressive benefit of fungi in conditions of decreased water capacity and decrease substrate availability, the connection among fungi and microorganism emerges.

Compounds which could not be damaged down, inclusive of lignin humus complexes, increase and end up dominant. Some researchers have proposed a 5th segment of composting, referred to as the seasoning segment (or garage segment), wherein the physicochemical parameters stay regular, however, the microbial groups change (Danon et al. 2008). Therefore, the chemical and microbiological modifications that arise withinside the substrate at some stage in the one-of-a-kind levels of the composting method will mostly outline the stableness and adulthood of the very last product, in addition to its suitability to be used as an natural soil amendment (Dominguez et al. 2010).

3.7.4 Mechanism of Vermicomposting

Detritivore earthworms react directly with bacteria and other members of the anaerobic microbial community during the organic process of vermicomposting, which significantly modifies the biochemical and biological characteristics of organic matter and boosts its stability. Their home-grown ability to colonize normal waste, exorbitant levels of regular material utilization, processing and osmosis, resistance to various natural variables, fast presence cycles, unnecessary conceptive rates, and strength and flexibility to dealing with (Dominguez et al. 2010). The homes are shared through the night crawler species *Eisenia andrei, Eisenia fetida, Perionyx excagus, and Eudrilus eugeniae*, that have been used appreciably in trojan horse composting plants. Reciprocity, predation, and alleviation are examples of biotic interactions. The dating among decomposers (microorganism and fungi) and earthworms in addition to the speedy adjustments in each the useful variety and the substrate first-rate are the primary functions of those structures (Sampedro and Domínguez 2008). Microbes are particularly chargeable for the biochemical decomposition of natural be counted, Since they can influence microbial decomposer movement by brushing on microorganisms straightforwardly and extending the surface region accessible for microbial assault after the comminution of natural matter, night crawlers are critical to the cycle (Monroy et al. 2009) and developing the area of the microbial area attack. Certain microbial organizations respond another way to the gastrointestinal climate

(Schönholzer et al. 1999). For instance, a couple of microorganisms develop to be energetic at some stage in the stomach. They visit by means of the digestion tracts even as others are unaffected and others are processed withinside the gastrointestinal system, primary to a general population decline (Monroy et al. 2009). These outcomes substantiate a new report (Gómez-Brandón et al. 2011a, b) that found solid proof for a bottleneck impact brought about by worm processing (E. andrei) on the microbial populaces of the underlying eaten material. This demonstrates that the worm stomach is a vital participant in molding microbial networks. It fills in as a specific channel for microorganisms in the substrate and supports the improvement of a microbial local area that is capable at processing synthetics created or delivered by the night crawlers in the materials they ingest.

Since digestive tract organisms send off normal be counted into the dung, wherein the discharged regular be included is moreover harmed down, such specific outcomes on microbial gatherings as a result of gastrointestinal entry can change the breakdown pathways at some stage in deception fertilizing the soil, greatest likely through changing over the structure of the microbial gatherings concerned withinside the arrangement of rot. For sure, as prior expressed, purple diversion projects consolidate totally exceptional microorganism populaces than the ones found withinside the decide fabric, and it is guessed that security the ones bunches into present day regular be counted can reason changes like the ones noticeable when worms are available, cleansing microbial organization levels of leisure activity and upgrading the practical assortment of microbial populaces in vermicomposting structures (Aira and Domínguez 2011). Past assessment has incontestable that vermicompost capabilities a greater microbial assortment than the fundamental substrate (Sen and Chandra 2009). Following the delegated magnificence of Holes, the accompanying night crawler projects get through projected related strategies (Covers), which may be various eagerly concerning aging, the existence of unworked fabric, and area change of the egested material (weeks to months; Fig. 3.1). At some stage in those methodology, the impacts of purple worms are for the most part sideways and gotten from the Holes (Aira et al. 2007). In general, there are two segments to worm side interest in vermicomposting: I a lively present that night crawlers framework the normal substrate, dynamic its substantial nation and microorganism organization and (ii) a development area wherein worms are uprooted to first-year recruit layers of undigested substrate, permitting microorganisms to expect over the decay of the worm-handled substrate (Gómez-Brandón et al. 2011a, b).

The time of the development area is variable and is overwhelmed through the productivity with which the enthusiastic period of the procedure is completed, not entirely set in stone through the purple diversion species Partner in Nursing, additionally because of the reality the rhythm at which the buildup can be applied (Domínguez et al. 2010). Vermicompost should accomplish a most satisfying in expressions of nutritionary substance material and unhealthful burden at some stage in this aging period, helping plant improvement and predominant plant sicknesses (Domínguez et al. 2010). Vermicomposting, then again, is a mesophilic system (Monroy et al. 2008). This changed into in sync with the genuine truth that totally remarkable microorganism bunches respond to the viscus environmental elements in some other case contingent upon the night crawler species. Another key feces regulation the unhealthful burden markdown at some stage in the procedure is that the irresistible specialist being thought of. Worms neglected to diminish the recognition of enterobacteria pneumoniae Partner in Nursingd Morganella morganii; however, exceptional microbes like enterobacter aerogenes and Enterobacter cloacae have been totally killed, in sync with (Parthasarathi et al. 2007). At some stage in an ongoing report (Aira et al. 2011) the wealth of waste enterococci, waste coliforms, and E. coli diminished all through the layers of a modern scale vermireactor took care of with cow fertilizer, while general coliforms, Enterobacteria, and clostridia stayed equivalent. In spite of the spearheading examinations of (Riggle 1996) and industrialist (Eastman et al. 2001), practically zero is idea with respect to the vermicomposting framework in modern scale structures, this is, vermicomposting structures intended to take care of significant amounts of junk. This microbespecific impact shows that night crawlers alter now presently not clearly how much unhealthful microorganism, yet in addition their one of a kind organization. in sync with, unaffected microbes ought to very much like the markdown in microorganism and plant biomass all through the layers of the reactor, ability to diminish helpful asset rivalry (Aira et al. 2011).

3.7.5 Anaerobic Digestion

In the absence of oxygen, microorganisms utilize anaerobic assimilation to separate natural squanders including creature fertilizer, wastewater natural squanders, and food scraps is called anaerobic processing. The technique for anaerobic processing (Promotion) has been concentrated on eminently in plant-essentially based absolutely and designed environments for more than 100 years. In home grown propensities, anaerobic corruption of home grown addictions happens in residue, waterlogged soils, and the digestive system of creatures, wherein get right of passage to oxygen is restricted; simultaneously as in designing conditions (Teglia et al. 2011). Biotechnology strategy the utilization of counting techniques wherein normal medication (for example spice waste, sewage, and/or a sustainable helpful asset) is corrupted into the shortfall of oxygen for the minimal expense assembling of biogas which might be utilized as an inventory of unpracticed energy (Insam et al. 2010). According to Massé et al. (2011), biogas changed into created at typical spot charges of 0.30, 0.25, and 0.48 L/g dangerous solids from the anaerobic processing of pig, livestock, and chicken excrement, separately. since, comparably to creating biogas and bringing down nursery gas outflows (Insam and Wett 2008), anaerobic processing is transforming into uncommonly well known as a systemic opportunities for reusing excrement, the sort of anaerobic bioreactors at scale rural assembling achieving 4200 in important and northerly Europe. Microbes make up around 80% of the general assortment of anaerobic digesters (Krause et al. 2008), anaerobic eukaryotes, principally parasites and protozoa, have obtained a huge amount significantly less leisure activity than microorganisms, due to their more slow blast expense and diminishing pervasiveness in anaerobic reactors.

The recurrence and by and large execution of cellulose hydrolysis are very relying upon the microbial biodiversity included, which happens gradually beneath anaerobic circumstances on account of non-public adaptations in cellulose administrative work saw in nature, notwithstanding the assortment of lignin and hemicellulose grids wherein it's miles coordinated for a significant distance (Lynd et al. 2002). Protein hydrolysis gradually creates peptides and amino acids, but debasement of lipids speedy produces glycerol and extensive chain unsaturated fats, as contrasted and next aging or oxidation. Clostridium, Acetivibrio, Bacteroides, Selenomonas or Ruminococcus are magnificent examples of not entirely settled in anaerobic reactors (Ueno et al. 2001). Microorganisms are subsequently responsible for most extreme hydrolytic procedures. The phones of the microorganism will take up the monomeric synthetic compounds produced because of the substance response of the biopolymer and age or oxidize them anaerobically to alcohols, short chain unsaturated fats, CO_2, and atomic H (H_2). Maturation (acidogenesis) is the most common way of delivering an energy-rich moderate which is then used to consolidate ATP, bringing about an aging item which is ousted from the phone. Acidogenesis, the method involved with assembling a maturation item that is released from the cell, is the most common way of making an energy-rich middle that is then used to orchestrate ATP. Since these side-effects are much of the time acidic, the extracellular pH diminishes related to the fermentative responses. This reality, along with an ascent in short-chain unsaturated fats, are the most well-known explanations behind reactor disappointment. Hence, it is fundamental to reestablish the harmony between acidogenic maturation microorganisms and corrosive disposing of organisms to keep the pH of the framework adjusted.

3.8 Animal Manure Management Systems

The expression "creature squander the executives framework" alludes to an arranged framework with fitting parts created and kept up with to direct and involve side-effects of creature creation in a manner that keeps up with and improves the nature of the air, water, soil, plant, and creature assets (adjusted from). The rural waste control machine comprises of a creature excrement control machine. Creatures are brought up in a lot of ways, which has an impact at the compost control systems and strategies utilized. Fertilizer delivered through creatures raised on assortment and field lands is routinely taken care of another way from compost delivered through creatures brought up in constrainment. Excrement control is fundamental as it diminishes the dangers related with compost overseeing and use. Compost or its parts may be limited or deflected from having unfortunate get right of section to the more extensive environmental elements in the event that a wonderful excrement control machine is set up. Sound compost control works on human wellness notwithstanding the environmental elements, monetary framework, and society. By the utilization of garbage as an asset, drawing out the ways of life pattern of valuable materials, and expanding utilizing optional materials, an asset green, socially comprehensive, and low-carbon financial framework might be achieved. The quests for creature compost control structures must be connected sooner than they might be purposeful and completed effectively. The fantasies of a fertilizer control machine should assortment from bringing down the ecological impact of compost making due, confining excrement supplement misfortunes, and offering its green use to administrative consistence, managing the planning of purpose close by various utilizes compost resources, and delivering income (Moreki and Chiripasi 2011) (Table 3.3).

Each manure control gadget has its specific set of challenges, extensively in phrases of nitrogen control. The major ingredient to be concerned about when it comes to animal feces are nitrogen, phosphorous, and potassium because of their importance in soil application. Potential nutrient losses in garage and for the duration of handling, in addition to ability nutrient overload for the duration of land application, are the reassets of situation. Because of the shortage of land and the shortage of nutrient trying out to evaluate necessities previous to application, manure-carried out soils have extra nitrogen and phosphorus (Moreki and Chiripasi 2011). There is lots of proof that manure vitamins are misplaced plenty in garage (Rotz 2004). The primary manure management system operations of production, collection, storage, treatment, transport, and utilization must be managed holistically in order to prevent pollution, reduce nutrient losses, and avoid other potential hazards.

Considering the model withinside the circumstances wherein the waste control instrument is consolidated, as an aide, the leaders' interests, wants and targets should be respected in making arrangements the creature squander control contraption; the turns of events and yearly assembling of the waste that could require control notwithstanding conceivable predetermination changes withinside the size of activity not entirely settled; the decisions the chief is leaned to remember for utilization need still up in the air; the landowner's decision for hardware and district of the capacity not set in stone; and the chart of the device need to cowl from the assembling to the use highlight degree and must be introduced area. These issues are fitting to making arrangements and planning the waste control structures for dairy, hamburger, pig, fowl and various creatures (Table 3.4).

3.9 Benefits of Manure Application

- Manure software at the land has many blessings for each farmers and society.
- Manure is a splendid supply of each essential and secondary vitamins that plants want to grow. Furthermore, land software complements universal soil first-class, which gives oblique blessings to farmers withinside the shape of multiplied crop response, decrease inorganic fertilizer liming, and pesticide inputs, and decreased soil and water losses.
- Green agronomic use of manure, society can advantage different blessings together with greater water first-class and carbon sequestration. This phase carries

Table 3.3 Effective manure handling has climatic, financial, and welfare benefits

Pilla of long production	Effective manure managing significant advantages
Environment	• Stops adverse effects on the environment, including air, water, soil, animals, and the marine environment • Ensures the safety of people in communities and at waste management facilities • Reduces the possibility of waste contamination • Enhances workplace well-being • Minimize waste-related greenhouse gas emissions • Waste and stink are reduced • Defends against flood dangers
Economy	• Increases the amount of business opportunities • Makes a significant contribution to the Gross Domestic Product • Pollution avoidance, recovering, and/or recycling initiatives save money for companies, especially in terms of resource extraction and use • Improves human health and the environment to save money, resulting in improved productivity, fewer medical expenses, better environmental quality, and the preservation of ecological systems
Social	• Provides jobs for low, medium, and high-skilled workers • Integrates and professionalizes informal sector jobs (the route to addressing equity and poverty issues) • Provides more appealing and comfortable urban areas, as well as improved social comfort • Promotes community members to adopt new habits

Table 3.4 Organic matter process management examples

Type of system	Description	Associated nutrient loss challenges
Feeding	During grazing, animals deposit feces freely on the specific field	Draining and evaporation cause significant loss of nutrients, particularly nitrogen
Kraals	Livestock are housed in a fenced landmass that will be used for different crops in the long term	Losses of nutrients by the process of leaching
Dry lot storage	Bedding substances are used to capture excrement and pee	Significant nutritional depletion could result, notably by urine. Draining and water run are further considerations
Preservation of slime	Human wastes are mixed with each other in the storage bin, and the excrement is frequently semi-liquid	Evaporation loss are affected by airflow, holding tank depth, and duration of storage
Lagoon	Solids are removed from animal manure and processed in anaerobically lagoons	Draining from the lagoon's bottom, emission into the water, and stink. There could be a lot of nitrogen, as well as some methane (CH_4) oxide emissions
Energy or fuel	Manure is either burned as fuel or processed anaerobically to produce biogas	Burning results in ammonia, carbon, and sulphur losses. Slurry's presence of water makes it tough to control
Others	Covering for residential construction and cattle feed are two examples. These possibilities are limiting, and using dung as livestock is discouraged	Livestock use for the process of agricultural construction is a total loss

the medical proof for those blessings. Manure serves as a fertilizer manure is a high-quality supply of fundamental plant vitamins, together with N, P, and potassium (K), and additionally gives most of the secondary vitamins that vegetation require (Table 3.5).

Table 3.5 Analysis of nutrient for exclusive structures first three non-liquid and subsequent three liquid gadget kg consistent with Mg and Kg consistent with 1000 L, respectively, Bates and Gagon

Animal species	N	P_2O_5	K_2O	N	P_2O_5	K_2O
Dairy	4–7	1–9	1–18	0.4–5.9	0.3–2.4	0.2–5.8
Beef	3–11	2–8	2–18	0.6–4.5	0.2–3.4	0.4–3.4
Swine	2–10	2–32	2–18	0.2–6.9	0.1–6.9	0.2–4.9
Poultry	3–68	2–47	2–28	4.3–8.9	1.5–10.4	1.7–2.5

From a specific operation will vary extensively because of the kind of animal, its meals ration, approach of series and storage, approach of software and climate

- During treatment, storage, and handling, vitamins in manure can be misplaced or altered, decreasing their availability for utilization through developing plant life.
- Animal waste are key reassets of nutrients for plant life, but software costs are regularly primarily based totally on the quantity of macronutrients like nitrogen, potassium, and phosphorous. Inorganic vitamins are the most effective ones that plant life can utilize.
- Manure N and P may be located in each natural and inorganic paperwork, and that they are not constantly one hundred% to be had to plant life. Before plant life might also additionally appoint natural paperwork, they ought to be mineralized or converted into inorganic paperwork over time.
- Because of potassium in waste is withinside the inorganic, its availability is akin to that of industrial fertilizer (Motavalli et al. 1989).

3.9.1 Available Phosphorus

Inorganic and mineralizable P make up the accessible P in fertilizer. A few examinations have concluded that the convey accessibility of P in excrement is comparing to or better than that of inorganic composts (May and Martin 1966); others have concluded that excrement has lower reactions than manure P. Notwithstanding the absence of proof on excrement P accessibility, it is as of now imagined that 80 to 90% of fertilizer P is plant available on the grounds that inorganic P represents 60 to 90% of all out P (Sharpley and Moyer 2000). Since fertilizer is many times done put together absolutely generally with respect to establish N prerequisites, which prompts over-programming system of P and diminishes the need to exactly gauge plant to be had P, how much examine on compost P accessibility has been restricted. As current issues about P defilement of ground streams achieve compost programs put together absolutely generally with respect to P necessities, more exact assessments of plant to be had P in creature fertilizer can be required.

3.9.2 Crops Response to Manure

Creature fertilizer is a top-notch supplement supply as it integrates limit of the plant essential components. The capacity expense of compost as an inventory of plant nutrients for crop fabricating is astounding in spite of the way that the centralizations of nutrients withinside the fertilizer tends to be low. Over the excess 10 years, expanded quantities of obliged creature taking care of tasks have finished in a blast in the amount of excrement accessible. Many explores have demonstrated that land utility of compost will deliver crop yields equivalent or progressed to the ones got with synthetic manures (Motavalli et al. 1989). Crop quality has furthermore been progressed through compost application (Pimpini et al. 1992). At the point when crop improvements with compost had been extra than the ones accomplished with business compost, response changed into typically ascribed to excrement gave nutrients or to cutting edge soil circumstances currently as of now not provided through business manure (Zhang et al. 1998) found that 2 kg of fertilizer N had been equivalent to no less than one kg of urea-N in expressions of plant take-up and yield response at some stage in the livestock feedlot fertilizer application. Fertilizer advances the real situation of the dirt and will expand P and natural movement (Chang et. al. 1990). The regular matter, general N and micronutrient content material of the floor soil are improved due to fertilizer application. More exploration are difficult to measure the benefits of fertilizer nutrients beside N, and the financial increase connected with those improvements.

Since it comprises of the main part of the plant's significant nutrients, creature squander is an extraordinary compost supply. Although convergences of nutrients in fertilizer have a low propensity, it has various ability to supply nutrients to the plants for crop creation. The wide assortment of controlled creature taking care of tasks has expanded emphatically withinside the previous ten years, following in a vertical push in the amount of fertilizer open. Many examinations have demonstrated that excrement carried out to the floor produces crop yields which may be much the same as or higher than the ones delivered with engineered manures (Motavalli et al. 1989). Fertilizer cure has moreover expanded crop quality (Eck et al. 1990).

3.9.3 Manure Maintains Soil pH

The meaning of compost utilization of plant nutrients supply or soil supplement is broadly perceived, but the limit of excrement, exceptionally chicken tangle, to kill soil sharpness and improve soil pH is considerably less well perceived. Creature compost's liming influence on corrosive and unbiased soils has been introduced in long-lasting period subject and nursery studies. The Magruder Plots are experimental winter wheat field plots at Oklahoma State University established in 1892. Researchers have conducted hundreds of experiments comparing various fertilizer treatments with manure applications and without such treatments. The plots became the center for wheat soil research in the region and provided constant data which helped farmers get maximum yield from their arid climates and naturally dry soils (Sharpley et al. 1993). As indicated by (Eghball 1999), red meat domesticated animals feedlot waste and manure improve soil pH, but inorganic N compost programming brings down soil pH. The impact of long-lasting period (18 years) land programming oven obfuscate on natural associated soil boundaries became explored through Alabama specialists, who verified that dirt pH became 0.5 unit better to a power of 0.6 m under littered soils than unschooled companions (Kingery et al. 1994). In an unpracticed home test, specialists in Hawaii tried the increment response of a tropical rummage vegetable to lime and regular fertilizer as corrosive soil added substances. Chicken excrement became demonstrated to be essentially basically as strong as lime in improving soil pH and diminishing aluminum (Al) harmfulness.

This reviews found that tropical field vegetables can likewise also absorb additional calcium from compost than from lime. The fundamental reason excrement increments soil pH is that it incorporates minerals like calcium (Ca) and magnesium. On a dry weight premise, fowl tangle, for instance, incorporates cycle 50 kg of calcium as per Mg.

3.9.4 Manure Enhance Soil Organic Matter

According to research, the remedy of manure has a full-size effect at the chemical, bodily and organic features of soil (Haynes and Naidu 1998). After 3 years of including hen bedding, observed will increase in soil organic matter (SOM) SOM of fifty-five to eighty percentage in loamy soils of northern Alabama. Most research display that manure additions ought to be made for at the least years to peer will increase in SOM. Manure remedy also can assist lessen losses of SOM in tillage systems (Kapkiyai et al. 1999). In small macroaggregates and microaggregates, the software of manure will increase the covered swimming pools of carbon. Fertilizer is additionally more prominent strong than plant deposits in recharging particulate SOM (Kapkiyai et al. 1999) this is connected with settled regular include in bunches of farming frameworks. Natural count is remembered to convince some of compound homes of the dirt. One renowned effect is a pH laid out extrade in cation substitute capacity (CEC) through the separation of carboxyl, phenolic, and hydroxyl organizations at the regular particles that form SOM (Tisdale et al. 1993). The shape and size-part of SOM influences its commitment to CEC. Research from sandy sub-Saharan soils shows a development in CEC on account of fertilizer utility handiest while dirt estimated particulate regular count is available (Guibert et al. 1999). Because of the ability to buffer of normal count, in acidic soils, excrement tends to development soil and diminish pH in soluble soils (Wahid et al. 1998). In regions wherein P is lacking, excrement assets P and moreover makes P more noteworthy to be had with the guide of utilizing complexing Al. The normal complexing of Al furthermore lessens Al poisonousness. Excrement additionally can be utilized to improve crop fabricating in saline and sodic soils underneath sure circumstances (Haynes and Mokolobate 2001).

3.9.5 Manure Improves Physical Soil Properties

The inclination of composts to invigorate the assembling of water-evidence totals (WSA) has a sizeable effect at the type of the dirt and thus at the real places of the dirt (Haynes and Naidu 1998). Various explorations have demonstrated that the helpfulness of fertilizer advances WSA.

Roberts and Clanton (2000) determined that too excessive a share of waste stockpile area (WSA) will increase infiltration, porosity, and water retention Kirchmann and Gerzabek (1999) determined that too excessive a share of WSA will increase infiltration, porosity, and water retention capacity. WSAs also are associated with reduced compaction and erosion. Even in silty clay soils with an immoderate content material of herbal bacterial counts, the addition of fertilizer will increase macro aggregation and consequently prevents structural degradation, in step with Angers (silty clay marl) has been examined in numerous research to limit compaction and enhance accessibility. In a 90–12 months fertilization strive in Denmark, Schjonning et al. determined that fertilized soils below excessive hundreds display a far decrease compaction than fertilized or unfertilized inorganic soils with comparable water contents and bulk densities. According to this study, the quantity of SOM withinside the fertilized soil accelerated the fragility of the soil. Because of those modifications withinside the bodily houses of the soil, it's far viable to color the ground in wetter conditions. The slurry minimizes the attempt required for soil cultivation and the resistance to seedling emergence and rooting with the assist of the development of the soil frame houses.

3.9.6 Manure Pesticide Vulnerability

Form and intricacy of soil food net are inspired through manner of way of the presence and forms of herbal C withinside the soil, which influences nutrient cycling similarly to plant ailments and parasites. The majority of research suggests that addition of manure may enhance microbial biomass (Estevez et al. 1996). Shifts in nutrient cycling rise up as microbial populations grow. Manure additives were demonstrated in many studies to enhance bacteria participating withinside the nitrogen cycle. After utilizing liquid swine manure, Lalande et al. (2000) observed a boom withinside the N mineralizer population. On a fallow field, Kubat et al. decided that manure addition prolonged nitrification greater than mineral fertilizers. These will growth can be high-quality in terms of giving crop nutrients, but if manure is performed at immoderate charges, they may be capable of make a contribution to nitrate leaching. Plant ailments can also additionally lower as microbial biomass and species range rise, due to the fact pathogen boom might be restricted with the aid of using opposition among microbial purchasers and a growth in predatory species.

Modern investigations have decided that manure suppresses contamination through this route.

Composted cow, sheep, or horse manure did not continuously reduce contamination in this investigation. Extraordinary research shows that excess nitrogenous manure, including that of poultry and pigs, can reduce disease in soil by producing inordinate concentrations of NH_3 and HNO_2.ammonia and/or nitrous acid. Anhydrous ammonia and synthetic nitrite additions no longer offered the same treatment for the disease, according to preliminary research. Fertilizer unpredictable unsaturated fats and acidic corrosive have likewise been displayed to lessen verticillium shrink and potato scab (Lazarovits 2001). Of these mixtures to restrain the illness. The consequences of fertilizer on plant parasitic nematode populaces have been referred to in more than one way. Both as vermin of plants and as people fundamental for supplement remobilization, nematodes assume a significant part in agrarian frameworks (Coleman et al. 1984) concluded that chicken compost diminished plant parasitic nematodes, while dairy cattle excrement had no effect. In the most effective treatment frameworks, Neher and Olsen (1999) settled on a blast in plant parasitic nematodes. They said a selection of elements added to the overflow of plant parasitic nematodes, for example, the enormous measure of natural corrections, phenomenal soil ripeness factors, the utilization of pesticides, crop revolution and remarkable dealing with rehearses. A few examinations have related compost inputs with a blast in bacterivorous nematodes; this ought to expand how much supplements accessible for plant development. Side interest, debasement, and protection from pesticides are completely answered to be impacted by dependence on and positive expenditure of the product fundamentally founded on information on SOM. Treated the soil excrement decreased the volatilization of methyl bromide and methyl isocyanate fungicides. Disarray in poultry has been displayed to cost two times the expense of atrazine debasement (Gupta and Baummer 1996). A microbial side interest, it tends to be of top notch as far as guarding against draining into groundwater. Excrement decreased draining of atrazine from coarse-finished soils, however is presently not generally so viable as actuated C from squander or processed metropolitan sewage slop, as per (Guo et al. 1991). According to a few investigations, treated the soil fertilizer enjoys preferable agronomic benefits over crude fertilizer. Enormous mixtures give humus to the dirt because of their steady spice content. Treated the soil compost, rather than smooth fertilizer, can possibly decrease the practicality of grass seeds and consequently the requirement for pesticides (Edwards et al. 2019). When contrasted with control plots took care of with same charges of N, P, and K found that consistently bundles of spent mushroom fertilizer and hen excrement manure supported the yields of eight types of vegetables.

The style of microbial and regular energizers in the manure likewise truly vaccinate soils, which can be one of the advantages it gives.

3.10 Principles Associated with Manure Management

Waste control and, with the aid of using extension, manure control are ruled with the aid of using some of principles. When growing manure control strategies and interventions, it is crucial to preserve those thoughts in mind. The following are a number of the principles:

- Proximity precept: The proximity precept states that wastes must be dealt with as near wherein they may be produced as possible.
- Sufficiency precept: The precept of sufficiency states that, if practicable, every country, and doubtlessly every condition, area must manipulate its personal garbage. If idea is carried out to animal manufacturing facilities, it method that farms should manipulate the waste they produce.

- The polluter-will pay precept: states that folks that motive or make contributions to pollutants must go through the monetary burden. In this perspective, folks that produce manure must incur the fee of managing it as a way to keep away from capacity fitness and environmental problems.
- Precautionary precept: This method is carried out primarily based totally at the impacted states' capacities. According to the precautionary precept, the shortage of medical self-belief must now no longer be used as a justification for delaying fee-powerful efforts to preclude environmental deterioration, particularly whilst important or irreversible risks exist.
- Sustainable improvement: This concept states that improvement projects aimed toward addressing modern-day wishes should now no longer jeopardize destiny generations' cap potential to fulfill their personal wishes. As a result, manure must be dealt with and controlled in a manner that doesn't damage the environment.
- Intergenerational fairness precept: The precept of intergenerational equality states that waste must now no longer be controlled in this type of manner that destiny generations are accountable for the problems.

3.11 Conclusion

The significance of long-term animal manure management cannot be overstated. Consequently, the influence of manure created on the farm outweighs its basis of production. Organic matter pollution has been linked to a number of major global health outbreaks. A multi-pronged strategy to waste management is required for long-term success. These methods are being used. Dietary techniques, legislative and legal framework, as well as physical, biological, and environmental factors organic matter treatment using chemicals manure policy, laws, and regulations that are effective will encourage. Especially with adequate enforcement and compliance, efficient and sustainable manure management procedures are possible. Manure management solutions should effectively reduce the amount of manure produced. Organic matter has a harmful impact on the environment and the general populace. There are several advantages resulting from long-term fertilizers and manures (Fig. 3.3).

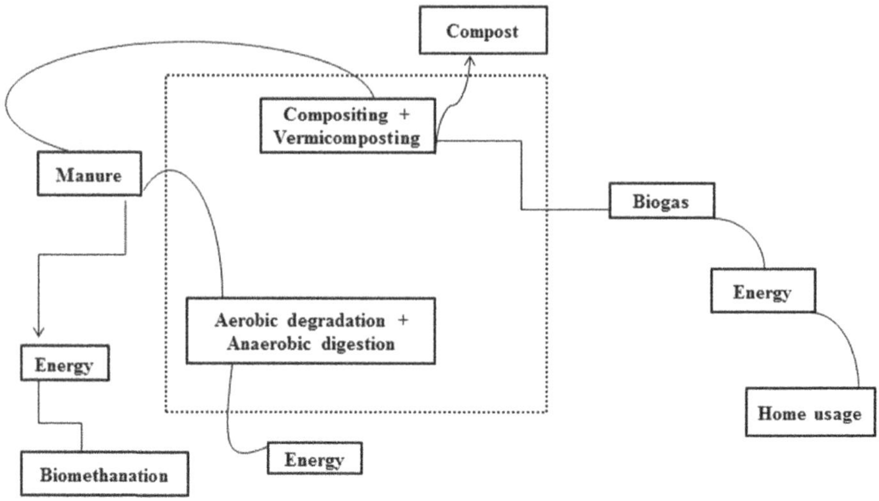

Fig. 3.3 Flow sheet diagram of manure recycling and reuse

References

Aira M, Domínguez J (2011) Earthworm effects without earthworms: inoculation of raw organic matter with worm-worked substrates alters microbial community functioning. PLoS ONE 6(1):e16354

Aira M, Monroy F, Domínguez J (2007) Eisenia fetida (Oligochaeta: Lumbricidae) modifies the structure and physiological capabilities of microbial communities improving carbon mineralization during vermicomposting of pig manure. Microb Ecol 54(4):662–671

Aira M, Sampedro L, Monroy F, Domínguez J (2008) Detritivorous earthworms directly modify the structure, thus altering the functioning of a microdecomposer food web. Soil Biol Biochem 40(10):2511–2516

Aira M, Gómez-Brandón M, González-Porto P, Domínguez J (2011) Selective reduction of the pathogenic load of cow manure in an industrial-scale continuous-feeding vermireactor. Biores Technol 102(20):9633–9637

Beffa T, Blanc M, Lyon PF, Vogt G, Marchiani M, Fischer JL, Aragno M (1996) Isolation of Thermus strains from hot composts (60 to 80 degrees C). Appl Environ Microbiol 62(5):1723–1727

Bernal MP, Alburquerque JA, Moral R (2009) Composting of animal manures and chemical criteria for compost maturity assessment. A Review. Bioresour Technol 100(22):5444–5453

Bernhart M, Fasina OO (2009) Moisture effect on the storage, handling and flow properties of poultry litter. Waste Manage 29(4):1392–1398

Bernhart M, Fasina OO, Fulton J, Wood CW (2010) Compaction of poultry litter. Biores Technol 101(1):234–238

Brady NC, Weil RR, Weil RR (2008) The nature and properties of soils, vol 13. Prentice Hall, Upper Saddle River, NJ, pp 662–710

Briones A, Raskin L (2003) Diversity and dynamics of microbial communities in engineered environments and their implications for process stability. Curr Opin Biotechnol 14(3):270–276

Burn RS (1889) Farming and farming economy: notes, historical and practical, on farming and farming economy. Crosby Lockwood and Son, London, UK

Chang C, Sommerfeldt TG, Entz T (1990) Rates of soil chemical changes with eleven annual applications of cattle feedlot manure. Can J Soil Sci 70(4):673–681

Coleman DC, Anderson RV, Cole CV, Mc Clellan JF, Woods LE, Trofymow JA, Elliott ET (1984) Roles of protozoa and nematodes in nutrient cycling. Microbial-Plant Interact 47:17–28
Danon M, Franke-Whittle IH, Insam H, Chen Y, Hadar Y (2008) Molecular analysis of bacterial community succession during prolonged compost curing. FEMS Microbiol Ecol 65(1):133–144
Darwin C (1892) The formation of vegetable mould, through the action of worms: with observations on their habits. J Murray
Demirel B, Scherer P (2008) The roles of acetotrophic and hydrogenotrophic methanogens during anaerobic conversion of biomass to methane: a review. Rev Environ Sci Bio/technol 7(2):173–190
Dohrmann AB, Baumert S, Klingebiel L, Weiland P, Tebbe CC (2011) Bacterial community structure in experimental methanogenic bioreactors and search for pathogenic clostridia as community members. Appl Microbiol Biotechnol 89(6):1991–2004
Domfnguez J (2004) 20 State-of-the-art and new perspectives on vermicomposting research. In: Earthworm ecology. CRC press, pp 401–424
Dominguez J, Edwards CA (2010) Vermiculture technology-earthworms, organic wastes, and environmental management
Domínguez J, Aira M, Gómez-Brandón M (2010) Vermicomposting: earthworms enhance the work of microbes. In: Microbes at work. Springer, Berlin, Heidelberg, pp 93–114
Drake HL, Küsel K, Matthies C (2006) Acetogenic prokaryotes in the prokaryotes, vol 2. In: Dworkin M et al (eds), pp 354–420
Dubey SK, Mondal RC (1994) Effect of amendments and saline irrigation water on soil properties and yields of rice and wheat in a highly sodic soil. J Agric Sci 122(3):351–357
Eastman BR, Kane PN, Edwards CA, Trytek L, Gunadi B, Stermer AL, Mobley JR (2001) The effectiveness of vermiculture in human pathogen reduction for USEPA biosolids stabilization. Compost Science & Utilization 9(1):38–49
Eck HV, Winter SR, Smith SJ (1990) Sugarbeet yield and quality in relation to residual beef feedlot waste
Edwards DR, Daniel TC (1992) Environmental impacts of on-farm poultry waste disposal—A review. Biores Technol 41(1):9–33
Edwards CA, Arancon NQ, Sherman RL (eds) (2019) Vermiculture technology: earthworms, organic wastes, and environmental management. CRC press
Eghball B (1999) Liming effects of beef cattle feedlot manure or compost. Commun Soil Sci Plant Anal 30(19–20):2563–2570
Estevez B, N'Dayegamiye A, Coderre D (1996) The effect on earthworm abundance and selected soil properties after 14 years of solid cattle manure and NPKMg fertilizer application. Can J Soil Sci 76(3):351–355
Forshell LP (1993) Composting of cattle and pig manure. J Vet Med Ser B 40(1–10):634–640
Goberna M, Gadermaier M, García C, Wett B, Insam H (2010a) Adaptation of methanogenic communities to the cofermentation of cattle excreta and olive mill wastes at 37 C and 55 C. Appl Environ Microbiol 76(19):6564–6571
Goberna M, Schoen MA, Sperl D, Wett B, Insam H (2010b) Mesophilic and thermophilic co-fermentation of cattle excreta and olive mill wastes in pilot anaerobic digesters. Biomass Bioenerg 34(3):340–346
Gollehon NR, Caswell M, Ribaudo M, Kellogg RL, Lander C, Letson D (2001) Confined animal production and manure nutrients, No 1474-2016-120868
Gómez-Brandón M, Aira M, Lores M, Domínguez J (2011a) Epigeic earthworms exert a bottleneck effect on microbial communities through gut associated processes. PLoS ONE 6(9):e24786
Gómez-Brandón M, Aira M, Lores M, Domínguez J (2011b) Changes in microbial community structure and function during vermicomposting of pig slurry. Biores Technol 102(5):4171–4178
Gottschalk G, Peinemann S (1992) The anaerobic way of life. In: The prokaryotes: a handbook on the biology of bacteria: ecophysiology, isolation, identification, applications, vol I, 2 edn, pp 300–311
Griffin DM (1985) A comparison of the roles of bacteria and fungi. In: Bacteria in nature. Springer, Boston, MA, pp 221–255

Guibert H, Fallavier P, Roméro JJ (1999) Carbon content in soil particle size and consequence on cation exchange capacity of alfisols. Commun Soil Sci Plant Anal 30(17–18):2521–2537
Guo L, Bicki TJ, Hinesly TD, Felsot AS (1991) Effect of carbon-rich waste materials on movement and sorption of atrazine in a sandy, coarse-textured soil. Environ Toxicol Chem Int J 10(10):1273–1282
Gupta G, Baummer J III (1996) Biodegradation of atrazine in soil using poultry litter. J Hazard Mater 45(2–3):185–192
Haynes RJ, Mokolobate MS (2001) Amelioration of Al toxicity and P deficiency in acid soils by additions of organic residues: a critical review of the phenomenon and the mechanisms involved. Nutr Cycl Agroecosyst 59(1):47–63
Haynes RJ, Naidu R (1998) Influence of lime, fertilizer and manure applications on soil organic matter content and soil physical conditions: a review. Nutr Cycl Agroecosyst 51(2):123–137
Holm-Nielsen JB, Al Seadi T, Oleskowicz-Popiel P (2009) The future of anaerobic digestion and biogas utilization. Biores Technol 100(22):5478–5484
Hue NV (1992) Correcting soil acidity of a highly weathered Ultisol with chicken manure and sewage sludge. Commun Soil Sci Plant Anal 23(3–4):241–264
Insam H, Franke-Whittle I, Goberna M (2010) Microbes in aerobic and anaerobic waste treatment. In: Microbes at work. Springer, Berlin, Heidelberg, pp 1–34
Insam H, Wett B (2008) Control of GHG emission at the microbial community level. Waste Manage 28(4):699–706
Kapkiyai JJ, Karanja NK, Qureshi JN, Smithson PC, Woomer PL (1999) Soil organic matter and nutrient dynamics in a Kenyan nitisol under long-term fertilizer and organic input management. Soil Biol Biochem 31(13):1773–1782
Kingery WL, Wood CW, Delaney DP, Williams JC, Mullins GL (1994) Impact of long-term land application of broiler litter on environmentally related soil properties, vol 23, no 1. American Society of Agronomy, Crop Science Society of America, and Soil Science Society of America, pp. 139–147
Kirchmann H, Gerzabek MH (1999) Relationship between soil organic matter and micropores in a long-term experiment at Ultuna, Sweden. J Plant Nutr Soil Sci 162(5):493–498
Knapp BA, Podmirseg SM, Seeber J, Meyer E, Insam H (2009) Diet-related composition of the gut microbiota of Lumbricus rubellus as revealed by a molecular fingerprinting technique and cloning. Soil Biol Biochem 41(11):2299–2307
Koerkamp PG (1994) Review on emissions of ammonia from housing systems for laying hens in relation to sources, processes, building design and manure handling. J Agric Eng Res 59(2):73–87
Krause L, Diaz NN, Edwards RA, Gartemann KH, Krömeke H, Neuweger H, Pühler A, Runte KJ, Schlüter A, Stoye J, Szczepanowski R (2008) Taxonomic composition and gene content of a methane-producing microbial community isolated from a biogas reactor. J Biotechnol 136(1–2):91–101
Kroodsma W, Scholtens R (1988) Ammonia emission from poultry housing systems. Jordbrukstekniska Inst
Kubát J, Nováková J, Mikanová O, Apfelthaler R (1999) Organic carbon cycle, incidence of microorganisms and respiration activity in long-term field experiment. Rostlinna Vyroba-UZPI (Czech Republic), 3–269
Lalande R, Gagnon B, Simard RR, Cote D (2000) Soil microbial biomass and enzyme activity following liquid hog manure application in a long-term field trial. Can J Soil Sci 80(2):26
Lazarovits G (2001) Management of soil-borne plant pathogens with organic soil amendments: a disease control strategy salvaged from the past. Can J Plant Path 23(1):1–7
Liedl BE, Bombardiere J, Chatfield JM (2006) Fertilizer potential of liquid and solid effluent from thermophilic anaerobic digestion of poultry waste. Water Sci Technol 53(8):69–79
Lores M, Gómez-Brandón M, Pérez-Díaz D, Domínguez J (2006) Using FAME profiles for the characterization of animal wastes and vermicomposts. Soil Biol Biochem 38(9):2993–2996
Lynd LR, Weimer PJ, Van Zyl WH, Pretorius IS (2002) Microbial cellulose utilization: fundamentals and biotechnology. Microbiol Mol Biol Rev 66(3):506–577

Magnusson U (2016) Sustainable global livestock development for food security and nutrition including roles for Sweden. Ministry of Enterprise and Innovation, Swedish FAO Committee, Stockholm

Malomo GA, Madugu AS, Bolu SA (2018) Sustainable animal manure management strategies and practices. In: Agricultural waste and residues, p 119

Massé DI, Talbot G, Gilbert Y (2011) On farm biogas production: a method to reduce GHG emissions and develop more sustainable livestock operations. Anim Feed Sci Technol 166:436–445

May D, Martin W (1966) Manures are good sources of phosphorus. Calif Agric 20(7):11–12

McGinn SM, Flesch TK, Chen D, Crenna B, Denmead OT, Naylor T, Rowell D (2010) Coarse particulate matter emissions from cattle feedlots in Australia. J Environ Qual 39(3):791–798

McMullen J, Fasina OO, Wood CW, Feng Y (2005) Storage and handling characteristics of pellets from poultry litter. Appl Eng Agric 21(4):645–651

Meyer D, Ristow PL, Lie M (2007) Particle size and nutrient distribution in fresh dairy manure. Appl Eng Agric 23(1):113–118

Miner JR, Humenik FJ, Overcash MR (2000) Managing livestock wastes to preserve environmental quality. Iowa State University Press

Monroy F, Aira M, Domínguez J (2008) Changes in density of nematodes, protozoa and total coliforms after transit through the gut of four epigeic earthworms (Oligochaeta). Appl Soil Ecol 39(2):127–132

Monroy F, Aira M, Domínguez J (2009) Reduction of total coliform numbers during vermicomposting is caused by short-term direct effects of earthworms on microorganisms and depends on the dose of application of pig slurry. Sci Total Environ 407(20):5411–5416

Monteny GJ, Erisman JW (1998) Ammonia emission from dairy cow buildings: a review of measurement techniques, influencing factors and possibilities for reduction. Netherlands J Agric Sci, pp 225–247

Moore JC, Berlow EL, Coleman DC, de Ruiter PC, Dong Q, Hastings A, Johnson NC, McCann KS, Melville K, Morin PJ, Nadelhoffer K, Rosemond AD, Post DM, Sabo JL, Scow KM, Vanni MJ, Wall DH (2004) Detritus, trophic dynamics and biodiversity. Ecol Lett 7(7):584–600

Moral R, Paredes C, Bustamante MA, Marhuenda-Egea F, Bernal MP (2009) Utilisation of manure composts by high-value crops: safety and environmental challenges. Biores Technol 100(22):5454–5460

Moreki JC, Chiripasi SC (2011) Poultry waste management in Botswana: a review. J homepage: http://www.ojafr.ir 285:292

Motavalli PP, Kelling KA, Converse JC (1989) First-year nutrient availability from injected dairy manure, vol 18, no 2. American Society of Agronomy, Crop Science Society of America, and Soil Science Society of America, pp 180–185

Neher DA, Olson RK (1999) Nematode communities in soils of four farm cropping management systems. Pedobiologia 43(5):430–438

NRCS U (1999) Agricultural waste management field handbook. US Department of Agriculture, Natural Resources Conservation Service, Washington, DC. Available at: http://www.nrcs.usda.gov/wps/portal/nrcs/detailfull/national/technical/alphabetical/mnm

Pain B, Menzi H (2003) Glossary of terms on livestock manure management. Pain B, MenziH

Parthasarathi K, Ranganathan LS, Anandi V, Zeyer J (2007) Diversity of microflora in the gut and casts of tropical composting earthworms reared on different substrates. J Environ Biol 28(1):87–97

Pimpini F, Giardini L, Borin M, Gianquinto G (1992) Effects of poultry manure and mineral fertilizers on the quality of crops. J Agric Sci 118(2):215–221

Porter GA, Bradbury WB, Sisson JA, Opena GB, McBurnie JC (1999) Soil management and supplemental irrigation effects on potato: I. Soil properties, tuber yield, and quality. Agron J 91(3):416–425

Riggle D (1996) Worm treatment produces Class A biosolids. Biocycle 37(10):67–68

Roberts RJ, Clanton CJ (2000) Surface seal hydraulic conductivity as affected by livestock manure application. Trans ASAE 43(3):603

Romney DL, Thorne PJ, Thomas D (1994) Some animal-related factors influencing the cycling of nitrogen in mixed farming systems in sub-Saharan Africa. Agr Ecosyst Environ 49(2):163–172
Rotz CA (2004) Management to reduce nitrogen losses in animal production. J Anim Sci 82(suppl_13):E119–E137
Ryckeboer J, Mergaert J, Vaes K, Klammer S, De Clercq D, Coosemans J, Insam H, Swings J (2003) A survey of bacteria and fungi occurring during composting and self-heating processes. Ann Microbiol 53(4):349–410
Sampedro L, Domínguez J (2008) Stable isotope natural abundances (δ13C and δ15N) of the earthworm Eisenia fetida and other soil fauna living in two different vermicomposting environments. Appl Soil Ecol 38(2):91–99
Sawyer JE, Schmitt MA, Hoeft RG, Siemens JC, Vanderholm DH (1991) Corn production associated with liquid beef manure application methods. J Prod Agric 4(3):335–344
Schink B (1997) Energetics of syntrophic cooperation in methanogenic degradation. Microbiol Mol Biol Rev 61(2):262–280
Schönholzer F, Hahn D, Zeyer J (1999) Origins and fate of fungi and bacteria in the gut of Lumbricus terrestris L. studied by image analysis. FEMS Microbiol Ecol 28(3):235–248
Sen B, Chandra TS (2009) Do earthworms affect dynamics of functional response and genetic structure of microbial community in a lab-scale composting system? Biores Technol 100(2):804–811
Sharpley A, Moyer B (2000) Phosphorus forms in manure and compost and their release during simulated rainfall, vol 29, no 5. American Society of Agronomy, Crop Science Society of America, and Soil Science Society of America, pp 1462–1469
Sharpley AN, Smith SJ, Bain WR (1993) Nitrogen and phosphorus fate from long-term poultry litter applications to Oklahoma soils
Stat FAO (2016) Food and agriculture organization of the United Nations: statistics division. Crop. Available online: http://www.fao.org/faostat/en/#data/QC. Accessed on 31 Jan 2020
Steinfeld, H., Gerber, P., Wassenaar, T. D., Castel, V., Rosales, M., Rosales, M., & de Haan, C. (2006). *Livestock's long shadow: environmental issues and options*. Food & Agriculture Org..
Swanepoel FJC, Stroebel A, Moyo S (2010) The role of livestock in developing communities: enhancing multifunctionality. University of the Free State/CTA
Tambone F, Scaglia B, D'Imporzano G, Schievano A, Orzi V, Salati S, Adani F (2010) Assessing amendment and fertilizing properties of digestates from anaerobic digestion through a comparative study with digested sludge and compost. Chemosphere 81(5):577–583
Teglia C, Tremier A, Martel JL (2011) Characterization of solid digestates: part 1, review of existing indicators to assess solid digestates agricultural use. Waste Biomass Valorization 2(1):43–58
Thummes K, Kämpfer P, Jäckel U (2007a) Temporal change of composition and potential activity of the thermophilic archaeal community during the composting of organic material. Syst Appl Microbiol 30(5):418–429
Thummes K, Schäfer J, Kämpfer P, Jäckel U (2007b) Thermophilic methanogenic Archaea in compost material: occurrence, persistence and possible mechanisms for their distribution to other environments. Syst Appl Microbiol 30(8):634–643
Tisdale SL, Nelson WL, Beaton JD, Havlin JL (1993) Soil fertility and fertilizers, 5th edn. Prentice Hall, NJ
Ueno Y, Haruta S, Ishii M, Igarashi Y (2001) Changes in product formation and bacterial community by dilution rate on carbohydrate fermentation by methanogenic microflora in continuous flow stirred tank reactor. Appl Microbiol Biotechnol 57(1):65–73
Vinnerås B, Agostini F, Jönsson H (2010) Sanitation by composting. In: Insam H, Franke-Whittle I, Goberna M (eds) Microbes at work. From wastes to resources
Vivas A, Moreno B, García-Rodríguez S, Benítez E (2009) Assessing the impact of composting and vermicomposting on bacterial community size and structure, and microbial functional diversity of an olive-mill waste. Bioresour Technol 100:1319–1326
Vu TKV, Tran MT, Dang TTS (2007) A survey of manure management on pig farms in Northern Vietnam. Livest Sci 112(3):288–297

Wahid A, Akhtar S, Ali I, Rasul E (1998) Amelioration of saline-sodic soils with organic matter and their use for wheat growth. Commun Soil Sci Plant Anal 29(15–16):2307–2318

Wilkinson KG (2011) A comparison of the drivers influencing adoption of on-farm anaerobic digestion in Germany and Australia. Biomass Bioenerg 35(5):1613–1622

Xie R, MacKenzie AF (1986) Urea and manure effects on soil nitrogen and corn dry matter yields. Soil Sci Soc Am J 50(6):1504–1509

Zhang B, Tian H, Lu C, Dangal SR, Yang J, Pan S (2017) Global manure nitrogen production and application in cropland during 1860–2014: a 5 arcmin gridded global dataset for Earth system modeling. Earth Syst Sci Data 9(2):667–678

Zhang H, Smeal D, Tomko J (1998) Nitrogen fertilization value of feedlot manure for irrigated corn production. J Plant Nutr 21:287–296

Zucconi FD (1987) Compost specifications for the production and characterization of compost from municipal solid waste. In: Compost: production, quality and use, pp 30–50

Chapter 4
Anaerobic Digestion for Bioenergy Production Using Solid Animal Waste: New Avenues

Iram Liaqat, Nazish Mazhar Ali, Muhammad Nauman Aftab, Sikander Ali, and Muhammad Arshad

Abstract Waste production around the globe has become an essential topic of concern since this accumulated waste has resulted in environmental hazards. Anaerobic digestion produces biogas and increases methane production using an optimum substrate. The biogas production from anaerobic digestion of different wastes is highly dependent on process of biodegradation and operating at optimum conditions increases the process efficiency. In biogas production, feedstock composition is the key factor—more methane yields depending on the feedstock type. The digestion rate of organic wastes depends on the relative number of key components. Also, the quantity of the mixture includes physical factors like temperature and pressure. Little information is available for optimum conditions of anaerobic digestion. Therefore, it is suggested that optimum conditions for anaerobic digestion and co-digestion should be explored. Anaerobic digestion provides an alternative green and efficient solution for toxic waste management and energy production. So, this review emphasizes the anaerobic process, enhancement of biogas production, fermentation efficiency, and economic and environmental advantages. Further, the factors influencing anaerobic digestion and the effect of key and trace elements will be discussed.

I. Liaqat (✉) · N. M. Ali
Microbiology Lab, Department of Zoology, Government College University, Lahore, Pakistan
e-mail: dr.iramliaqat@gcu.du.pk

M. N. Aftab · S. Ali
Institute of Industrial Biotechnology, Government College University, Lahore, Pakistan

M. Arshad
Jhang-Campus, University of Veterinary and Animal Sciences Lahore, Lahore, Pakistan

M. Arshad (ed.), *Climate Changes Mitigation and Sustainable Bioenergy Harvest Through Animal Waste*, https://doi.org/10.1007/978-3-031-26224-1_4

Graphical Abstract

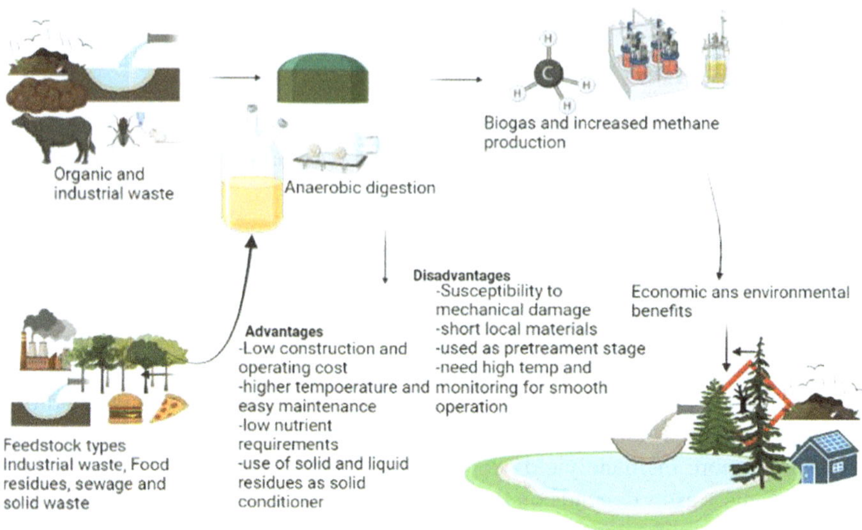

Keywords Biogas · Anaerobic digesters · Biofuel · Free energy · Feedstock

4.1 Introduction

As a feedstock for biogas production, anaerobic digesters utilize various organic resources. However, there are scientific, technical, and legal restrictions adhere to some restrictions whether feedstock is animal manures, food waste, or wastewater effluents (Algapani et al. 2018; Tabatabaei et al. 2010). Furthermore, the feedstock should be a liquid mixture with high moisture content. Common digesters, such as mesophilic complete mix tank digesters, work best with a solids-in-water ratio of 4–8%. Depending on the system's functional architecture, different moisture contents are necessary. Anaerobic digestion (AD) has grown in popularity as more renewable energy sources have been available worldwide. AD results in the generation of biogas in two phases. Hydrolysis phase is first phase in which organic matter is transformed of into CO_2, fatty acids, and hydrogen. Second phase is methanogenic phase which involves decomposition of fatty acids into methane. Biogas is made up of methane, CO_2, and other trace components. The essential technique is the same for both large plants and tiny reactors. Pre-treatments and substrate co-digestion are becoming more common to improve biogas output. The installation location also influences reactor design and substrate choices. Biogas upgrading aids in boosting the gas's utility for various applications. The economic advantage is determined by multiple parameters, including location worldwide and the quality and amount of accessible substrate. To remain lucrative, AD processes rely substantially on government subsidies. AD profitability is especially important in developing countries because this technology improves human life in these locations. The chapter takes a detailed look at AD technology, discusses AD economics, and suggests future studies to improve the technology (Arshad et al. 2018).

4.1.1 Anaerobic Digestion

Anaerobic digestion is a sequential process in which microorganisms degrade biodegradable material in an anaerobic environment. It is used in both industries and domestic purposes for the production of fuels or to manage waste materials. Primarily fermentation is used in industries to produce food products. For yoghurt production in home fermentation, anaerobic digestion is used. Anaerobic digestion naturally occurring in soil lakes and ocean basins is usually termed anaerobic activity. In 1776, Alessandro Volta discovered that anaerobic digestion is the primary source of methane that begins with the hydrolysis of the input material by bacteria. During this process, carbohydrates and other insoluble polymers are degraded to their soluble derivatives and made assessable to other bacteria. Acidogenic bacteria convert sugars an amino acids into hydrogen, organic acids, carbon dioxide, and ammonia. The bacteria also convert the organic acids into acetic acid with hydrogen, carbon dioxide and hydrogen as additional products in a process called acetogenesis. At last, these products are converted into methane and carbon dioxide by methanogenic bacteria. Methanogenic archaea play an essential part in the anaerobic treatments of wastewater. Biodegradable sewage and waste sludge are treated by anaerobic digestion. The integrated system of waste management reduces the atmospheric emanation of methane gas. The biogas produced comprises carbon dioxide, methane, and trace amount of other contaminant gases. It is directly used as a fuel in power and heat gas engines. Additionally, the digestate is enriched with nutrients and can be used as a fertilizer (Aslam et al. 2018; Azzahrani et al. 2018).

4.1.2 Utilization of Animal Wastes for Bioenergy Production

If effectively recycled, many beneficial elements are found in animal waste and could be used as a good fertilizer for crop and energy production. Animal dung is a significant source of potassium, phosphorous, and nitrogen. Biogas is a renewable energy source produced by the anaerobic degradation of organic waste. Modern anaerobic digesters use air-tight chamber, where bacteria convert the solid manure into biogas that can be used for electricity production. Recently, the most frequently used biofuels are biogas and bioethanol. Biogas is generally produced by degrading

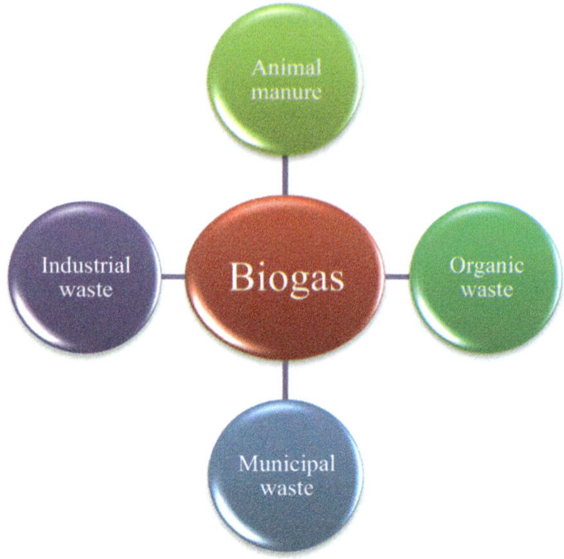

Fig. 4.1 Production of biogas from different types of wastes

organic waste such as manure and industrial and municipal waste. Corn and sugarcane are primarily used for the production of bioethanol (Berghuis et al. 2019).

4.1.3 Biogas

Biogas mainly contains methane and carbon dioxide. Methane forms 40–60% of biogas, and the remaining component is carbon dioxide and traces of water vapors. Biogas can be used as fuel for vehicles if compressed. Biogas could be a suitable replacement for natural gas if it is cleaned and standardized up to natural gas. It is then called biomethane, which could be used as a replacement for methane for cooking and heating purposes. Biogas can be produced by anaerobic digestion from food scraps, animal manure, sewage, and wastewater. Typically, biogas contains 50–75% methane, has a deep blue flame, and could be a good energy source (Fig. 4.1, Table 4.1).

4.1.4 Uses of Biogas

Properly cleaned and upgraded to the standard of natural gas, biogas can be used as a replacement of methane gas for cooking and heating purposes.

Table 4.1 Biogas composition

Components	Percentage (%)
Methane (CH_4)	50–80
Carbon dioxide (CO_2)	20–50
Nitrogen (N_2)	< 1
Hydrogen (H_2)	< 1
Ammonia (NH_4)	< 1
Hydrogen sulfide (H_2S)	< 1

- Compressed biogas can be used as a replacement for compressed natural gas for vehicles.
- Biogas is used for electricity production and water heating, etc.
- Biogas is used to displace CO_2 in combined heat and power (CHP) plants (Boll et al. 2020; Liu et al. 2018a, b).

4.2 Feedstocks for Anaerobic Digestion

The most readily biodegradable organic materials are accepted as feedstocks for anaerobic digestion. Commonly used feedstocks include waste from food processing, sewage sludge, and livestock manure. Feedstocks possess a high potential for energy production, which depends on the level, type of processing, and concentration of the biodegradable material. Feedstock includes any kind of "bio" option, including crop residues, energy crops, plant oils, and waste streams like a municipal waste comprise of production, harvesting, storage, and transportation costs (Tilley et al. 2014).

4.2.1 Types of Feedstocks

- Agricultural residues include all kinds of agricultural waste, such as bagasse, straw, stems, stalks, leaves, husks, shells, pulp, peels, etc., (Fig. 4.2)
- Animal waste, such as manure, is a suitable source for producing energy
- Industrial wastes

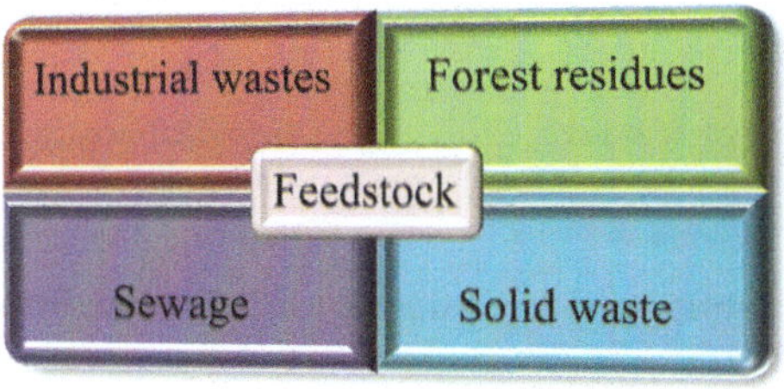

Fig. 4.2 Types of feedstocks

- Forest residues
- Solid waste
- Sewage.

4.2.2 *Feedstock for Biomass*

Renewable resources of biomass can be used directly as a fuel, or it can be converted to any other form of energy products that are commonly termed feedstocks (Table 4.2).

4.2.2.1 Biomass Feedstocks

Biomass feedstocks mainly include residues of crops, dedicated energy crops, algae, forestry residues, wood processing, municipal solid waste, and waste of urban wood (Fig. 4.3).

4.2.2.2 Dedicated Energy Crops

Non-food crops grown on land unsuitable for the commonly cultivated crops like rice, wheat, and corn to produce biomass are called dedicated energy crops. These crops are generally divided into two categories herbs and woody plants. Herbs are perennial and harvested annually; it takes 2–3 years to give maximum productivity. These crops include miscanthus, bamboo, tall fescue, sweet sorghum, Kochia, wheatgrass, and switchgrass. Short rotation woody crops are harvested after 5–8 years of their plantation. These woody trees include hybrid willow, poplar, eastern cottonwood, silver maple, green ash, sycamore, sweetgum, and black walnut. These species aid in improving soil and water quality, habitat for wildlife-related agricultural crops, diversifying the income source, and enhancing farm productivity (Benner 1989).

Table 4.2 Types of biomass feedstock

Biomass type	Examples
Forests stuffs	Sawdust, bark, wood, shrubs residues
Bio renewable energy wastes	Crop residues, agricultural wastes, urban wood wastes, mill wood wastes
Organic wastes	Industrial wastes, municipal wastes, municipal sewage, and sludge
Lichens	Crustose, foliose and fruticose lichens
Mosses	Polytrichales, Bryophyta
Algae	Prokaryotic, eukaryotic algae, and kelps

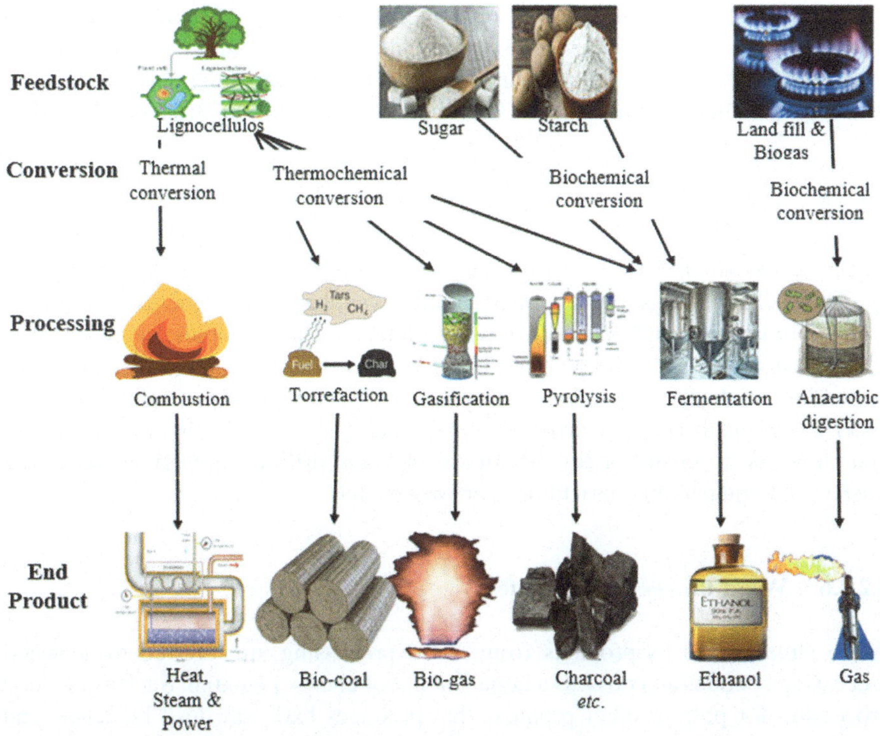

Fig. 4.3 Applications of biomass

4.2.2.3 Agricultural Crop Residue

On existing lands, there are numerous opportunities to maximize agricultural resources without compromising the production of feed, food, and fiber. It includes leaves and stalks that are abundant and widely distributed worldwide. Major residue crops include wheat straw, corn stover, oat straw, rice straw, barley straw, and sorghum stubble. These residues can be sold to the local refineries, which create an additional income source for the farmers.

4.2.2.4 Forestry Residues

Forest residues left over from the logging of timber or whole-tree biomass specifically harvested for biomass make up the two main types of forest biomass feedstocks. After logging, dead, diseased and other unsaleable trees are frequently left in the forest.

The woody waste is collected for bioenergy production with left over waste material to nourish habitat and support nutritional and hydrologic aspects. Additionally, surplus biomass can be used on millions of vast acres of forestland. In addition, forest vitality, restoration, resilience, productivity, and overharvesting of woody biomass will lessen pest infestation. Without harming the health and stability of the forest's biological structure and function, the biomass might be collected for bioenergy production (Niu et al. 2013; Svoboda and Carcluie 2003; Zhao et al. 2018).

4.2.2.5 Algae

Algae as feedstock for bioenergy generally refers to various highly productive species, such as macroalgae, microalgae, and cyanobacteria. Algae use sunlight to produce biomass, which has all necessary elements such as lipids, carbohydrates, and proteins, which can be converted into biofuels and other bioenergy products. Depending on the type of strain, algae can possibly grow in salty, fresh, or brackish water. They can also flourish in water from second-use sources, including the production of water from the operations of oil and gas drilling, municipal, industrial wastewater, aquaculture, agricultural, or wastewater.

4.2.2.6 Wood Processing Residues

Waste streams and by-products from wood processing are referred to as wood processing residues and possess a large amount of energy potential. Such as in wood processing for pulp or other products that produces bark, sawdust, branches, and needles/leaves. Bioproducts or biofuels can be formed by converting these residues as these residues are the waste of wood processing and prove to be an inexpensive and convenient source for biomass production.

4.2.2.7 Sorted Municipal Waste

MSW includes yard trimmings, paperboard, plastics, paper, rubber, food, textiles, and leather wastes are examples of mixed residential and commercial waste. MSW is used for bioenergy production by redirecting substantial amounts of MSW from landfills to the refinery. It also provides an opportunity to lower household and commercial waste.

4.2.2.8 Wet Waste

Institutional, Residential, and Commercial food wastes, organically rich biosolids, manure slurries, organic wastes of industries, and biogas produced from any of the aforementioned feedstock streams are examples of wet waste feedstocks. Rural economies can generate more income, and trash-disposal issues can be resolved by turning these "waste streams" into biofuels (Bastian et al. 2009; Beale et al. 2016; Stowhas et al. 2018).

4.2.3 Feedstock for Biofuel

A sustainable fuel known as biodiesel is created from various feedstocks, such as animal fats, cooking oils, and vegetable oils. Biodiesel produced by different feedstocks possesses different qualities, which should be considered before mixing biodiesel with petroleum diesel (Bertucci et al. 2019).

4.2.4 Chemical Feedstock

A feedstock is sometimes referred to as a raw material or an unprocessed substance. It is an alternative term for biomass which is used to produce or process other products. In carbon-based chemical industries, natural gas, and crude oil accounted for 87% of feedstocks in 2016. During utilization and manufacturing, carbon dioxide is emitted.

Examples of feedstock include crude oil, which is utilized in the production of gasoline, corn, which is utilized in the production of soybean oil and ethanol, which is used in the production of biodiesel (Bharagava et al. 2019; Bingol et al. 2015).

4.3 Best Feedstock for Anaerobic Digestion

- Crops, sewage, slurries, and plant waste are all examples of biodegradable material that can be utilized as fuel for anaerobic digestion. Animal manures
- Spent feed
- Waste products from the food industry
- Waste from slaughterhouses
- Farm fatalities
- Corn silage
- Glycerin; a byproduct for the manufacturing of biodiesel.

4.3.1 Biodegradable Biomass Materials

Biodegradable waste is a waste that can be broken down by other living organisms and often comes from botanical and animal sources. Wastes are considered non-biodegradable if other living things cannot break them down. Biodegradable plastic, green garbage, food, and paper waste are all examples of biodegradable waste frequently found in municipal solid waste also known as BMW (biodegradable municipal waste) (Campanaro et al. 2016, 2019; Cardinali-Rezende et al. 2016).

Following are some biodegradable wastes

- Food waste
- Human feces
- Paper scraps
- Manure
- Sewage
- Medical waste
- Sludge from sewage
- Waste from slaughterhouses.

4.4 Benefits of Anaerobic Digestion

Anaerobic digestion is exciting because it is a streamlined and natural approach to convert a wide range of complex waste into nature friendly fuel gas. There are certain advantages and disadvantages which are summarized in Table 4.3.

There are two main benefits of anaerobic digestion (Fig. 4.3).

- Environmental Benefits
- Economic Benefits.

4.4.1 Environmental Benefits

The natural environment provides many benefits that are difficult to quantify in monetary terms. Natural process assist in reducing greenhouse gas emission, cleaning the air we breathe, clean the water we drink, create food and medicines, decrease noise and chemical pollution, slow floods, and calm streets. This is referred as "ecosystem services" (Date et al. 2012; De Vrieze et al. 2017, 2016).

Table 4.3 Advantages and disadvantages of anaerobic digestion in developing countries

Advantages	Disadvantages
Low construction cost	Susceptibility to mechanical damage
Low sludge production	Lack of locally available materials
Ease of transportation	Low gas pressure requires extra weight
Higher digester temperatures in warm climates	Scum cannot be removed from digester
Easy emptying and maintenance	Used as a pre-treatment stage
Low nutrient (phosphorus and nitrogen) requirements	Need high temperature for effective operation
Use of solid and liquid residues as solid conditioner	Requires monitoring for smooth operation

Anaerobic digestion of organic matter will reduces the organic matter load and associated oxygen demand on manure handling process. This will results in processed components to be more ecofriendly and smaller with less harmful environmental effects. Compared to mechanical aeration, anaerobic pre-treatment being an economical method by converting an anaerobic lagoon to an aerobic lagoon. Digested elute is more operational than raw manure due to the presence of more stable organic load with less volatile odorants.

4.4.2 Economic Benefits

Economic advantages of anaerobic digestion are measured in terms of increased productivity and yield. Recycled nutrients will produce an ecofriendly and sustainable food products resulting in increased net income and revenues, etc. Additionally, produced heat, fuel, or electricity from biogas will be used on-farm, reducing the dependence of agriculture sector's on fossil fuel energy. It will save money while debating a proposal to cut expenditures. Net income and revenues are two examples of economic benefits, which will be managed wisely by effective operation of anaerobic digestion. Profit and cash flow are economic gains as well. Reducing something, e.g., cost, can be considered an economic benefit. Lowering labor and raw material costs are the economic benefits (Campanaro et al. 2016, 2019; Cardinali-Rezende et al. 2016; Castellano-Hinojosa et al. 2018; Chaleckis et al. 2019; De Vrieze, Pinto, et al. 2018a, b).

4.5 Recent Trends in Anaerobic Digestion Technology

Environmental studies for producing biogas to reduce greenhouse gas emissions, focusing on the engineering and microbiological factors involved, have been the most popular topics in the field throughout the past five years of publications. Integrating feedstock pre-treatment with the core processing of the substrates by AD to improve biogas quality and production is the key trend and opportunity. One alternative to achieve the appropriate Sustainable Development Goals is to produce biofuels and bioenergy using AD methods. Finally, understanding feedstock pre-treatment concerning process modeling, optimization, and operation is essential to establishing AD as a successful management system that reaps benefits for both the environment and the economy (De Vrieze, Ijaz, et al. 2018a, b; DeLong et al. 1989; Dione et al. 2016).

4.5.1 Development of Anaerobic Digestion Units

Anaerobic digestion is broken down into four main phases: acidogenesis, hydrolysis, methanogenesis, and acetogenesis. The whole process can be represented by the chemical reaction, in which anaerobic microbes biochemically eat organic material like glucose to produce methane and carbon dioxide.

Most anaerobic digesters use heat exchangers (HXs), which don't mix the liquids as they transport heat from hot water to sludge. A boiler or combined power and heat engine is used to heat the water. The latter process, also referred to as cogeneration, converts biogas into renewable electrical energy (Fang 2010; Ferguson et al. 2016).

4.6 Anaerobic Digestion System and Its Economic Analysis

Even though, the anaerobic digestion of modest amounts of food and organic waste was once thought unprofitable, it is now expanding because of a novel method of producing biogas: small-scale digestion facilities. One hundred thirty micro-scale digestion units were running throughout Europe as of 2016. These manufacturing units, which are smaller, less expensive, and more easily self-sufficient, draw farmers and investors in eco-neighborhoods who want to create new clean energy sources (Martin Alexander Fischer et al. 2019a, b; Martin A Fischer et al. 2019a, b; Franke-Whittle, Goberna, and Insam 2009a, b; Franke-Whittle, Goberna, Pfister, et al. 2009a, b).

4.6.1 Small to Large-Scale Digestion System

Biogas is produced on a small scale in a farm or small community by micro-scale digestion. The production units for small-scale digestion are under 80 kW. While some industrial units have a power capacity of over 1000 kW, most agricultural units are between 100 and 300 kW. Starting with 100 dairy cows, 200 cows, or between 200 and 5000 tons of organic waste per year, a small-scale digestion operation is carried out. Micro-scale digestion enables the system's independence by supplying the digester with agricultural products, which is the attraction of small-scale digestion because it avoids the need to invest in huge facilities (Franke-Whittle et al. 2014; Giacomoni et al. 2015; Grohmann et al. 2018).

4.7 Anaerobic Digestion System and Economic Impact

Small-scale digestion is an excellent, reasonably priced solution for farmers to diversify their businesses. After 2010, on-farm micro-scale digesting plants started to develop. For several reasons, agricultural wastes by farmers are well suited to be used for anaerobic digestion (Fig. 4.4). It will

- Generate their heat and electricity, saving money.
- Reduce greenhouse gas emissions associated with livestock waste.
- Lessen the powerful smells connected with using untreated manure as fertilizer.
- Reduce the distance that organic inputs must be transported to on-site facilities for treatment.
- Take advantage of the digestate's benefits, such as more liquid material that is simpler to distribute, fewer weeds, mineralized nitrogen, etc.

4.7.1 Small-Scale Anaerobic Digestion System and Its Economic Benefits

Israel-based home biogas has created an anaerobic digester for residential use. It uses food waste to generate biogas and fertilizer. It requires six liters of food waste per day to sustain it. Thus, one can use the generated biogas for cooking, lighting, and heating (Gysi et al. 2018; Hagen et al. 2017).

Since 2015, around 1000 family-sized biogas systems have been introduced to over 90 countries. Home biogas systems have already been implemented in several eco-districts in Great Britain. With its ability to digest up to 12 L of kitchen waste and up to 36 L of animal manure, the new model, home biogas 2.0, provides a bigger micro-scale of biogas production. Another example of a micro-technology

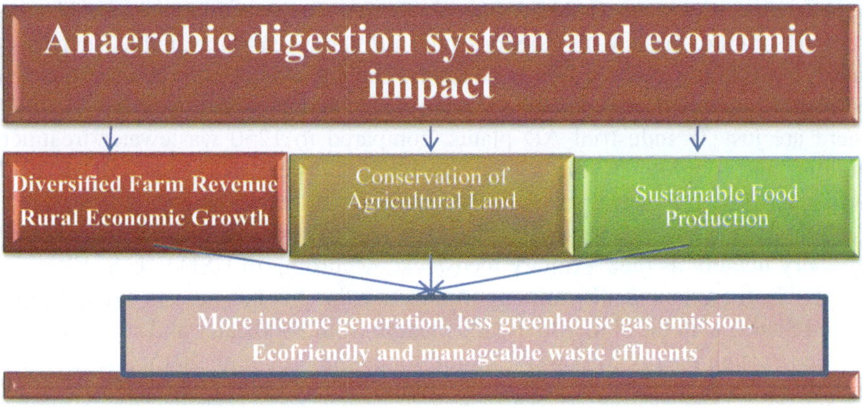

Fig. 4.4 Anaerobic digestion system and economic impact

that aids consumers and businesses in better managing their food waste is the small-scale, fully automated and insulated anaerobic digester offered by MyGug. In home biogas, food waste is converted into biogas and fertilizer. Six liters of food waste per day are needed to feed it. They can hold up to 0.5 tons of food waste annually and come in various sizes suitable for residential or commercial use (Hanreich et al. 2013; Hao et al. 2015; Hassa et al. 2018; Heyer et al. 2015, 2019). The biogas produced can thus be used for cooking, heating, and lighting (Ho et al. 2013; Hori et al. 2014; Iwai et al. 2016; Jia et al. 2019, 2018). Its design explicitly for domestic use,

However, a regulatory framework should be implemented to promote micro-small digestion in cities. Even while the French government continues to support the growth of renewable energies, the legislative framework must permit small-scale digestion facilities in urban areas.

4.7.2 Large-Scale Anaerobic Digestion System and Its Economic Benefits

Large-scale digesters have historically been more common in most developed nations since they require more extensive infrastructure and substantial financial commitment. The biogas that is produced is mostly utilized for heating and power, while it is occasionally upgraded to become a transportation fuel. Europe's two main digester operation modes are "centralized" systems and "farm-scale" digesters. A centralized or combined system codigests agricultural leftovers, food waste, the organic portion of municipal solid waste (MSW), and animal manure from multiple farms. A portion of the digestate is returned to the farms in this model to be utilized as fertilizer, while the remainder is sold to other farms. These centralized facilities have enormous digesters with a capacity of up to 300,000 ft^3 (Jun et al. 2015; Kampmann et al. 2012; Khelaifia et al. 2016; Kim et al. 2017, 2015).

Farm-scale AD facilities are often constructed in big swine or dairy farms, and their digester capacities range from 7000 to 42,000 ft^3. They combine the animal waste from one or more farms with other organic stuff that is readily available, such as the energy crops that were raised on those farms. The AD industry is well-established in the United States and is used to treat sewage sludge in wastewater treatment facilities. There are just 38 industrial AD plants, compared to 1250 wastewater treatment facilities and roughly 250 farm-scale anaerobic digesters (Kirkegaard et al. 2017; Kohrs et al. 2014; Lagier et al. 2012; Langer et al. 2015; Langille et al. 2013).

In the past ten years, around 90% of the AD plants have been built, and 86% use dairy manure as their primary feedstock. According to USDA, U.S. EPA, and U.S. DOE, there is a significant opportunity for the AD business to expand in the United States, with the ability to use the manure from 8000 dairy and swine farms to produce enough energy to power 1.09 million homes (Li et al. 2015, 2017; Limam et al. 2014; Lu et al. 2013). Additionally, almost 2500 wastewater treatment facilities have the capacity to make biogas, many of which are actively producing methane but not using it (Lü et al. 2014; Lv et al. 2010).

Large-scale AD systems use a variety of feedstocks, including waste from agricultural or livestock farms, food waste, and wastewater/sewage sludge, and require considerable capital investment and upkeep. The economics of such plants vary widely because of the nature of these systems. The primary source of production costs for AD systems is capital expenditures. Operating costs range from $18 to $100 per ton of feedstock handled by the facility, depending on the size of the AD plant. According to a survey of 38 AD systems in the United States, the equipment used to generate energy accounts for about 36% of the entire capital expense. The price ($/kWh) for producing energy at AD plants ranges from $0.06 to $0.23. The AD plant and the fuel influence the cost of producing energy. Due to economies of scale, electricity generation costs are often lower for higher plant sizes (Manor and Borenstein 2017; Marchand et al. 2017; Maus et al. 2016).

4.8 Conclusion and Future Prospects

Due to the production of biogas and its various uses, additional products (such as digestate that could be used as biofertilizer), the collection of dumping fees, and government subsidies, AD has several economic advantages. AD has gained popularity worldwide, from modest household digesters in impoverished rural countries to expansive systems in wealthy nations. The primary cost factor for AD plants is capital expenditures. The choice and accessibility of the feedstock are vital factors in the economics of large-scale systems. Although prices have stabilized in Asia, where the technology has been in use for a more extended period of time, small-scale digesters' capital costs vary significantly in countries where their introduction is relatively recent. Future technological developments in digestate management and biogas utilization may improve the economics of AD plants.

References

Algapani DE, Qiao W, Pumpo F et al (2018) Long-term bio-H_2 and bio-CH_4 production from food waste in a continuous two-stage system: Energy efficiency and conversion pathways. Bioresour Technol 248:204–213.

Ammar M, Korai RM, Shahbaz M et al (2019) An Insight into the anaerobic co-digestion of municipal solid waste and food waste: Influence of co-substrate mixture ratio and substrate to inoculum ratio on biogas production. Appl Biochem Biotechnol 187:1356–1370

Angelidaki I, Treu L, Tsapekos P et al (2018) Spatial distribution and diverse metabolic functions of lignocellulose-degrading uncultured bacteria as revealed by genome-centric metagenomics. Appl Environ Microbiol 84(18):1–14

Aquino SF, de Gurgel LVA, Adarme OFH et al (2018) Production of biogas (methane and hydrogen) from anaerobic digestion of hemicellulosic hydrolysate generated in the oxidative pre-treatment of coffee husks. Bioresour Technol 263(3):601–612

Arshad M, Bano I, Khan N, et al (2018) Electricity generation from biogas of poultry waste: an assessment of potential and feasibility in Pakistan. Renew Sust Energ Rev 81:1241–1246

Aslam M, Yang P, Lee PH et al (2018) Novel staged anaerobic fluidized bed ceramic membrane bioreactor: Energy reduction, fouling control and microbial characterization. J Membr Sci 553:200–208

Azzahrani IN, Davanti FA, Millati R et al (2018) Effect of hydraulic retention time (HRT) and organic loading rate (OLR) to the nata de coco anaerobic treatment eficiency and its wastewater characteristics. Agritech. 38(2):160–166

Bastian M, Heymann S, Jacomy M (2009) Gephi: an open source software for exploring and manipulating networks. In: International AAAI conference on weblogs and social media

Batstone DJ, Tait S, Jensen PD et al (2018) Humic acid inhibition of hydrolysis and methanogenesis with different anaerobic inocula. Waste Manag 80:130–136

Beale DJ, Karpe AV, McLeod JD (2016) An 'omics' approach towards the characterization of laboratory scale anaerobic digesters treating municipal sewage sludge. Water Res 88:346–357

Benner R (1989) Book review: biology of anaerobic microorganisms. Limnol Oceanogr 34(3):647

Berghuis BA, Brian YF, Schulz F et al (2019) Hydrogenotrophic methanogenesis in archaeal phylum Verstraetearchaeota reveals the shared ancestry of all methanogens. PNAS 116(11):s5037-5044

Bertucci M, Calusinska M, Goux X (2019) Carbohydrate hydrolytic potential and redundancy of an anaerobic digestion microbiome exposed to acidosis, as uncovered by metagenomics. Appl Environ Microbiol 85

Betenbaugh MJ, Bouwer EJ, Phan D et al (2018) Synergistic co-digestion of wastewater grown algae-bacteria polyculture biomass and cellulose to optimize carbon-to-nitrogen ratio and application of kinetic models to predict anaerobic digestion energy balance. Bioresour Technol 269:210–220

Bharagava RN, Purchase D, Saxena G et al (2019) Applications of metagenomics in microbial bioremediation of pollutants: from genomics to environmental cleanup. In: Das S, Dash HR (eds) Microbial diversity in the genomic era. Elsevier Inc, pp 459–477

Bingol K, Bruschweiler-Li L, Yu C (2015) Metabolomics beyond spectroscopic databases: a combined MS/NMR strategy for the rapid identification of new metabolites in complex mixtures. Anal Chem 87:3864–3870

Boll M, Estelmann S, Heider J (2020) Anaerobic degradation of hydrocarbons: Mechanisms of hydrocarbon activation in the absence of oxygen. In Boll M (ed) Anaerobic utilization of hydrocarbons, oils, and lipids. Handbook of hydrocarbon and lipid microbiology. Springer, Cham, Switzerland

Campanaro S, Treu L, Kougias PG (2016) Metagenomic analysis and functional characterization of the biogas microbiome using high throughput shotgun sequencing and a novel binning strategy. Biotechnol Biofuels 9:26

Campanaro S, Treu L, Rodriguez-R LM (2019) The anaerobic digestion microbiome: a collection of 1600 metagenome-assembled genomes shows high species diversity related to methane production. bioRxiv. 2019:680553

Cardinali-Rezende J, Rojas-Ojeda P, Nascimento AMA et al (2016) Proteolytic bacterial dominance in a full-scale municipal solid waste anaerobic reactor assessed by 454 pyrosequencing technology *Chemosphere* 146:519–525.

Castellano-Hinojosa A, Armato C, Pozo C (2018) New concepts in anaerobic digestion processes: recent advances and biological aspects. Appl Microbiol Biotechnol 102:5065–5076

Chaleckis R, Meister I, Zhang P (2019) Challenges, progress and promises of metabolite annotation for LC-MS-based metabolomics. Curr Opin Biotechnol 55:44–50

Date Y, Iikura T, Yamazawa A et al (2012) Metabolic sequences of anaerobic fermentation on glucose-based feeding substrates based on correlation analyses of microbial and metabolite profiling. J Proteome Res 11:5602–5610

De Vrieze J, Regueiro L, Props R (2016) Presence does not imply activity: DNA and RNA patterns differ in response to salt perturbation in anaerobic digestion. Biotechnol Biofuels 9:244

De Vrieze J, Christiaens MER, Walraedt D (2017) Microbial community redundancy in anaerobic digestion drives process recovery after salinity exposure. Water Res 111:109–117

DeLong EF, Wickham GS, Pace N (1989) Phylogenetic stains: ribosomal RNA-based probes for the identification of single cells. Science 243:1360–1363

De Vrieze J, Pinto AJ, Sloan WT et al (2018a) The active microbial community more accurately reflects the anaerobic digestion process: 16S rRNA (gene) sequencing as a predictive tool. Microbiome 6:63

De Vrieze J, Ijaz UZ, Saunders AM (2018b) Terminal restriction fragment length polymorphism is an "old school" reliable technique for swift microbial community screening in anaerobic digestion. Sci Rep 8:16818

Dione N, Khelaifia S, La Scola B (2016) A quasi-universal medium to break the aerobic/anaerobic bacterial culture dichotomy in clinical microbiology. Clin Microbiol Infect 22:53–58

Douglas GM, Maffei VJ, Zaneveld J et al (2019) PICRUSt2: an improved and extensible approach for metagenome inference. bioRxiv. 672295

Fang HH (2010) Imperial College Press; Covent Garden, London, UK: Environmental Anaerobic Technology

Ferguson RMW, Coulon F, Villa R (2016) Organic loading rate: a promising microbial management tool in anaerobic digestion. Water Res 100:348–356

Fischer MA, Ulbricht A, Neulinger SC (2019a) Immediate effects of ammonia shock on transcription and composition of a biogas reactor microbiome. Front Microbiol 10:2064

Fischer MA, Gullert S, Refai S (2019b) Long-term investigation of microbial community composition and transcription patterns in a biogas plant undergoing ammonia crisis. J Microbial Biotechnol 12:305–323

Franke-Whittle IH, Goberna M, Pfister V et al (2009a) Design and development of the ANAEROCHIP microarray for investigation of methanogenic communities. J Microbiol Methods 79:279–288

Franke-Whittle IH, Goberna M, Insam H (2009b) Design and testing of real-time PCR primers for the quantification of *Methanoculleus*, *Methanosarcina*, *Methanothermobacter*, and a group of uncultured methanogens. Can J Microbiol 55:611–616

Franke-Whittle IH, Walter A, Ebner C (2014) Investigation into the effect of high concentrations of volatile fatty acids in anaerobic digestion on methanogenic communities. Waste Manag 34:2080–2089

Giacomoni F, Le Corguille G, Monsoor M et al (2015) Workflow4Metabolomics: a collaborative research infrastructure for computational metabolomics. Bioinformatics 31:1493–1495

Grohmann A, Fehrmann S, Vainshtein Y et al (2018) Microbiome dynamics and adaptation of expression signatures during methane production failure and process recovery. Bioresour Technol 247:347–356

Gupta S (2010) Biogas comes in from the cold. New Scientist. Sunita Harrington, London, p 14. Retrieved 4 Feb 2011

Gysi DM, Fragoso TM, Buskamp V et al (2018) Comparing multiple networks using the co-expression differential network analysis (CoDiNA) arXiv. Preprint arXiv. 1802:00828

Hagen LH, Frank JA, Zamanzadeh M et al (2017) Quantitative metaproteomics highlight the metabolic contributions of uncultured phylotypes in a thermophilic anaerobic digester. Appl Environ Microbiol 83

Hanreich A, Schimpf U, Zakrzewski M et al (2013) Metagenome and metaproteome analyses of microbial communities in mesophilic biogas-producing anaerobic batch fermentations indicate concerted plant carbohydrate degradation. Syst Appl Microbiol 36:330–338

Hao L, Lu F, Mazeas L et al (2015) Stable isotope probing of acetate fed anaerobic batch incubations shows a partial resistance of acetoclastic methanogenesis catalyzed by Methanosarcina to sudden increase of ammonia level. Water Res 69:90–99

Hassa J, Maus I, Off S et al (2018) Metagenome, metatranscriptome, and metaproteome approaches unraveled compositions and functional relationships of microbial communities residing in biogas plants. Appl Microbiol Biotechnol 102:5045–5063

Heyer R, Kohrs F, Reichl U et al (2015) Metaproteomics of complex microbial communities in biogas plants. J Microbial Biotechnol 8:749–763

Heyer R, Schallert K, Siewert C et al (2019) Metaproteome analysis reveals that syntrophy, competition, and phage-host interaction shape microbial communities in biogas plants. Microbiome 7:1–17

Ho DP, Jensen PD, Batstone DJ (2013) Methanosarcinaceae and acetate-oxidizing pathways dominate in high-rate thermophilic anaerobic digestion of waste-activated sludge. Appl Environ Microbiol 79:6491–6500

Hori T, Akuzawa M, Haruta S et al (2014) Involvement of a novel fermentative bacterium in acidification in a thermophilic anaerobic digester. FEMS Microbiol Lett 361:62–67

Iwai S, Weinmaier T, Schmidt BL et al (2016) Piphillin: improved prediction of metagenomic content by direct inference from human microbiomes. PLoS One 11

Jia Y, Ng SK, Lu H et al (2018) Genome-centric metatranscriptomes and ecological roles of the active microbial populations during cellulosic biomass anaerobic digestion. Biotechnol Biofuels 11:117

Jia Y, Leung MHY, Tong X et al (2019) Rare taxa exhibit disproportionate cell-level metabolic activity in enriched anaerobic digestion microbial communities. mSystems 4:e00208–e00218

Jun SR, Robeson MS, Hauser LJ et al (2015) PanFP: pangenome-based functional profiles for microbial communities. BMC Res Notes 8:479

Kampmann K, Ratering S, Kramer I et al (2012) Unexpected stability of bacteroidetes and firmicutes communities in laboratory biogas reactors fed with different defined substrates. Appl Environ Microbiol 78:2106–2119

Khelaifia S, Lagier JC, Nkamga VD et al (2016) Aerobic culture of methanogenic archaea without an external source of hydrogen. Eur J Clin Microbiol Infect Dis 35:985–991

Kim TG, Jeong SY, Cho KS (2015) Development of droplet digital PCR assays for methanogenic taxa and examination of methanogen communities in full-scale anaerobic digesters. Appl Microbiol Biotechnol 99:445–458

Kim MS, Kim DH, Yun YM (2017) Effect of operation temperature on anaerobic digestion of food waste: performance and microbial analysis. Fuel 209:598–605

Kirkegaard RH, McIlroy SJ, Kristensen JM et al (2017) The impact of immigration on microbial community composition in full-scale anaerobic digesters. Sci Rep 7:9343

Kohrs F, Heyer R, Magnussen A et al (2014) Sample prefractionation with liquid isoelectric focusing enables in depth microbial metaproteome analysis of mesophilic and thermophilic biogas plants. Anaerobe 29:59–67

Lagier JC, Armougom F, Million M et al (2012) Microbial culturomics: paradigm shift in the human gut microbiome study. Clin Microbiol Infect 18:1185–1193

Langer SG, Ahmed S, Einfalt D et al (2015) Functionally redundant but dissimilar microbial communities within biogas reactors treating maize silage in co-fermentation with sugar beet silage. J Microbial Biotechnol 8:828–836

Langille MGI, Zaneveld J, Caporaso JG et al (2013) Predictive functional profiling of microbial communities using 16S rRNA marker gene sequences. Nat Biotechnol 31:814–821

Li Y, Zhang Y, Yang Y et al (2017) Potentially direct interspecies electron transfer of methanogenesis for syntrophic metabolism under sulfate-reducing conditions with stainless steel. Bioresour Technol 234:303–309

Li YF, Wei S, Yu Z (2013) Feedstocks affect the diversity and distribution of propionate CoA-transferase genes (pct) in anaerobic digesters. Microb Ecol 66:351–362

Li YF, Nelson MC, Chen PH et al (2015) Comparison of the microbial communities in solid-state anaerobic digestion (SS-AD) reactors operated at mesophilic and thermophilic temperatures. Appl Microbiol Biotechnol 99:969–980

Limam RD, Chouari R, Mazeas L et al (2014) Members of the uncultured bacterial candidate division WWE1 are implicated in anaerobic digestion of cellulose. Microbiology 3:157–167

Liu Z, Si B, Li J et al (2018a) Bioprocess engineering for biohythane production from low-grade waste biomass: technical challenges towards scale up. Curr Opin Biotechnol 50:25–31

Liu H, Singh L, Mishra P et al (2018b) Impacts of nano-metal oxides on hydrogen production in anaerobic digestion of palm oil mill effluent—a novel approach. Int J Hydrog Energy 43(5):2666–2676

Lu X, Rao S, Shen Z et al (2013) Substrate induced emergence of different active bacterial and archaeal assemblages during biomethane production. Bioresour Technol 148:517–524

Lü F, Bize A, Guillot A, Monnet V et al (2014) Metaproteomics of cellulose methanisation under thermophilic conditions reveals a surprisingly high proteolytic activity. ISME J 8:88–102

Lv W, Schanbacher FL, Yu Z (2010) Putting microbes to work in sequence: recent advances in temperature-phased anaerobic digestion processes. Bioresour Technol 101:9409–9414

Manor O, Borenstein E (2017) Systematic characterization and analysis of the taxonomic drivers of functional shifts in the human microbiome. Cell Host Microbe 21:254–267

Marchand J, Martineau E, Guitton Y et al (2017) Multidimensional NMR approaches towards highly resolved, sensitive and high-throughput quantitative metabolomics. Curr Opin Biotechnol 43:49–55

Maus I, Koeck DE, Cibis KG et al (2016) Unraveling the microbiome of a thermophilic biogas plant by metagenome and metatranscriptome analysis complemented by characterization of bacterial and archaeal isolates. Biotechnol Biofuels 9:171

Niu Q, Qiao W, Qiang H et al (2013) Mesophilic methane fermentation of chicken manure at a wide range of ammonia concentration: stability, inhibition and recovery. Bioresour Technol 137:358–367

Stowhas T, Verdejo J, Yáñez C (2018) Zinc alleviates copper toxicity to symbiotic nitrogen fixation in agricultural soil affected by copper mining in central Chile. Chemosphere 209:960–963

Svoboda I (2003) Anaerobic digestion, storage, olygolysis, lime, heat and aerobic treatment of livestock manures, scotland.gov.uk. Retrieved 17 Aug 07

Tabatabaei M (2010) Importance of the methanogenic archaea populations in anaerobic wastewater treatments (PDF). Process Biochem 45(8):1214–1225

Tilley E, Ulrich L, Lüthi C et al (2014) Compendium of sanitation systems and technologies, 2nd edn. Swiss Federal Institute of Aquatic Science and Technology (Eawag), Duebendorf, Switzerland

Zhao C, Mu H, Zhao Y et al (2018) Microbial characteristics analysis and kinetic studies on substrate composition to methane after microbial and nutritional regulation of fruit and vegetable wastes anaerobic digestion. Bioresour Technol 249:315–321

Chapter 5
Techniques and Strategies for Bioenergy Production from Manure

Neelma Munir, Sher Zaman Safi, Zirwa Sarwar, Muhammad Arshad, Maria Hasnain, and Rukhama Haq

Abstract The rapid increase in population and industrialization initiates the heavy dependency on the imported fossil fuel. Pakistan is facing severe energy crisis. The utilization of alternative energy resources reduces this problem. The search for alternate energy resources, including biowastes, is consequently gaining impetus. The consumption of the animal manure is considered as a most abundant source of energy generation. The successful implementation of animal manure through biochemical conversion technologies, as a primary source of energy in the existing ecosystem, is essential for the development of sustainable environment of the country. By treating the biowaste scientifically enables to do the successful conversion of biowaste to bioenergy. Manure is pretreated chemically, physically, mechanically, heat, and biologically to extract the biogas from it. The development of sustainable environment of human society is only fulfilled by using the concept of clean and green energy. Thermochemical processes are valuable in the maximum generation of bioenergy from manure and high activation energy is required to proceed the reaction of energy formation. The designs of bioreactor allow the efficient functioning of microorganism or cells to perform their desired function under optimum conditions. Gasifier reactor design and anaerobic digester design designs of bioreactors are utilized for the biological and electrochemical gas conversion of manure. Volatile and non-volatile components of the biomass are converted to gaseous compounds at an optimized organic rate. The industrialization of these bioprocesses provides effective applications of biogas as an energy source to meet the future energy needs.

N. Munir · Z. Sarwar · M. Hasnain · R. Haq
Department of Biotechnology, Lahore College for Women University, Lahore, Pakistan

S. Z. Safi
Faculty of Medicine, Bioscience and Nursing, MAHSA University, Selangor, Malaysia

M. Arshad (✉)
Jhang-Campus, University of Veterinary and Animal Sciences Lahore, Lahore, Pakistan
e-mail: muhammad.arshad@uvas.edu.pk

M. Arshad (ed.), *Climate Changes Mitigation and Sustainable Bioenergy Harvest Through Animal Waste*, https://doi.org/10.1007/978-3-031-26224-1_5

Graphical Abstract

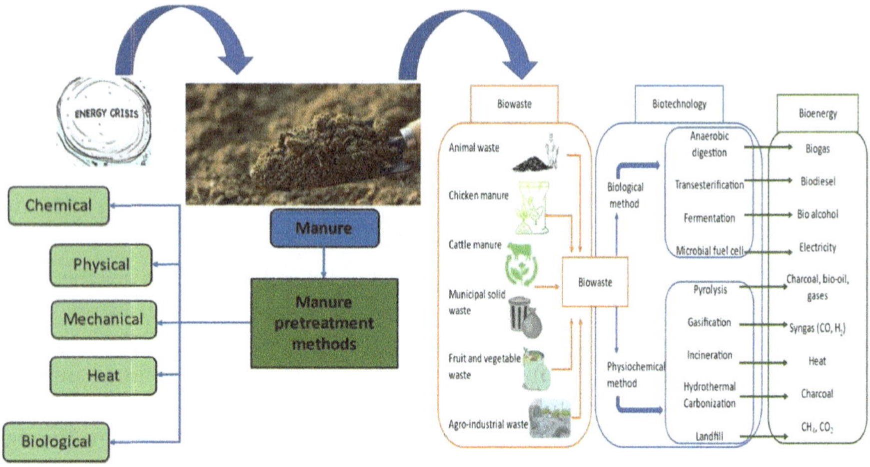

Keywords Anaerobic digestion · Anaerobic digester · Biogas · Biowaste · Manure

5.1 Introduction

5.1.1 Energy Crisis

A huge deficiency of electricity results from the heavy dependence on the imported fuels, which significantly effects the economic development of Pakistan. This situation creates an increase in the prices of local fuel and restricts the establishment of the new industrial sectors. The boost in the greenhouse gases emission threats the global environment. As the population rises, industrial and domestic activities went parallel to fulfill all needs (Li et al. 2021). Hence, the availability of more renewable and sustainable energy sources is required to overcome this electricity problem. The search for alternate energy resources, including biowastes are consequently gaining impetus. Solar, wind, hydro, and biomass are renewable energy sources. These alternative energy sources contribute to the production of sustainable energy development for the future (Raheem et al. 2016).

5.1.2 How Much Energy is Generated from Biowaste

As the need of the energy increases, it has become feasible to convert the municipal solid waste into the electrical energy. The growth of population and industrialization have become a crucial factor of raising the municipal waste. The usage of bio-waste aims to maximize the generation of energy and minimize the emission of greenhouse gases (Bilen et al. 2021). Biological wastes can be transformed to useful energy sources for contributing in three-fourth of global energy emissions. The biowastes were processed through biological and thermo-chemical route for the useful energy resources. The utilization of the biowastes enhances the global economy which in turn yields value added products (Awasthi et al. 2021).

Depleting energy resources, increase in population, and global warming are the major ecological problems for sustainable environmental protection and development. Several tons of organic waste are disposed by the United States. The non-edible sources including the manure, agricultural wastes, and wastewater were managed properly for public health. The conversion of waste to energy optimizes the over increasing amount of waste and depleting energy resources. The energy conversion is eco-friendly and aims to reduce the environmental pollution. The conversion of biowaste to biogas and fertilizer is essential to fulfill the domestic consumption of energy (Iqbal and Kang 2021).

5.1.3 Manure

A huge amount of manure is generated by livestock industry. The organic and animal waste combines to form an organic fertilizer known as manure. Manure is rich in the organic matter and other nutrients like nitrogen, that is consumed by fungi and bacteria present in the soil. The application of manure to the agricultural soils is widely considered as a source of soil organic carbon. The organic carbon concentrations in the soil increases as the quantity of manure increases. The organic matter present in this fertilizer is easily degraded due to the high accessibility of nitrogen content in manure (Gross and Glaser 2021).

5.1.4 Chemical Composition of Manure

Manure consists of heavy metals, organic wastes, nutrients, and microbes that facilitate the process of recycling of nutrients in the soil. The conversion of organic wastes to ecofriendly alternative by composting provides an organic fertilizer. Anaerobic digestion of animals manure show the presence of cellulose and lignin (Abou-Sreea et al. 2021). Changes in the physiochemical characteristics like organic carbon, nitrogen, and pH modifies the structural properties of the compost. There are several spectroscopic techniques such as Fourier transform infrared and nuclear magnetic resonance that were applied to examine the chemical nature of the manure. Low carbon-to-nitrogen ratio decomposes the organic waste and increase the absorbency of the soil (Liu et al. 2021).

Table 5.1 Production of biogas from various biowastes deriving from several sources

Substrate	Biogas l/day	Methane l/kg	References
Animal waste	0.45	–	Karim et al. (2005)
Chicken manure	1.08	–	Belostotskiy et al. (2013)
Cattle manure	12.1	184	Omar et al. (2008)
Municipal solid waste	13.6	350	Martín-González et al. (2010)
Fruit and vegetable waste	2.5	–	Bouallagui et al. (2009)
Agro-industrial waste	16.4	620	Álvarez et al. (2010)

The organic bulking agents like straw, increase the porosity of the topsoil during composting. Different manures constitute different chemical properties and their changes during composting are also distinct. There are several microbial communities that play an important role in decomposing the organic waste. Changes in the chemical properties and the temperature of organic matter governed the process of assembly of bacterial communities. The results of this bacterial assembly predict the better understanding of composting driven by microbes (Bao et al. 2021). Table 5.1 illustrates the production of biogas from different wastes that are originating from various sources.

5.1.5 How Much Manure Produced Annually?

Due to heavy dependency on the imported fossil fuel, Pakistan is facing a severe energy crisis. Partly due to chronic losses and lack of energy from cheap sources, the supply of electricity from utilities is well short of the demand. Pakistan is mainly an agricultural territory with wide variety of production of livestock waste by farming communities. Poultry manure held the largest share at 45.8% livestock population according to the research data of 2016–2018. This livestock population consists of cattle, goats, sheep, mules, horses, and camels. Different amounts of manure are produced by different animals based on their feed and size (Younas et al. 2016).

Large quantities of manure of about 10–20 kg per day are produced by cattle from each province of Pakistan. The annual livestock manure production in Pakistan was 417.3 million tons, recorded in 2018. This annual manure could generate the 26,871.35 million m^3 of biogas. Due to the favorable conditions for the technologies of the biodigester and suitable development of the anaerobic digestion, the current energy crisis can be eliminated in Pakistan by adopting this eco-friendly energy alternative (Khan et al. 2021).

5.2 Manure Pretreatment

The organic waste present in the cattle manure facilitates the alternative biogas production. The biofibers in manure effect the biodegradability of manure. In order to increase the biodegradability of biofibers, manure is pretreated to extract the biogas from it. The application of circular economy in which maximum extraction from the resources is done with the minimal waste disposal to preserve the sources of energy for economic use. A low-cost treatment system to recover resources from waste includes several solid–liquid separation modules. Enough energy can be produced from fertilizers with a positive impact on sustainability of environment (Cândido et al. 2022).

The combination of pretreatments was effective for the anaerobic digestion of the manure fibers. Chemical, physical, mechanical, heat, and biological pretreatments were combined on the anaerobic digestion of manure. All of them showed the positive response in methane yield (Khan et al. 2021).

5.2.1 Chemical

To enhance the anaerobic digestion of lignin and cellulose components of the manure, the combination of mechanical and chemical pretreatments was employed. This anaerobic digestion boosts the lignin removal for the efficient yield of methane and also decomposes the cellulose components of manure. Also, the acidic pretreatment of manure improves the hydrolysis of hemicellulose. This hydrolysis results in the crystallinity of cellulose and removal of lignin. Moreover, the acidic pretreatments generate several types of inhibitors that to promote the digestion process in manure (Khan et al. 2021). The chemical pretreatment showed the improvement of the methane yield when it was fed with 6% NaOH. Highest methane yield was obtained by treating the manure chemically with the addition of calcium oxide and sodium hydroxide to the biofibers manure (Zeng et al. 2021).

5.2.2 Physical

The physical pretreatments demand high energy to break the cells present in the manure by applying the physical force. This force increases the biomass surface area and reduces the particle size to enhance the microbial and enzymatic attack. Microbial attack set off the process of digestion of biomass to improve its accessibility (Orlando and Borja 2020). The production of secondary inhibitory substances was prohibited in physical pretreatments. The combination of physical with mechanical pretreatments reduces the cellulose crystallinity (Victorin et al. 2020).

5.2.3 Mechanical

Several technologies have been developed for treating the agricultural waste. The mechanical pretreatments were present in all industries that consume organic waste to generate biogas. Mechanical mixing homogenizes the manure with the recycled digested matter in order to enhance the solubilization in biodegradable compounds. At laboratory scale, three mechanical pretreatments were applied to recycle waste. First, the organic waste was shredded, then mixed, and finally blended. Shredding was done in rotary at a low rotational speed and reduced the size of the large particles in biowaste (Coarita Fernandez et al. 2020).

Shredding pretreatment was not efficient to obtain a higher methane yield. While blending improved the yield of methane from organic waste. These mechanical pretreatments were efficient in lowering the element size and increasing the biogas production rate. Also, the exposure of organic compounds to enzymatic hydrolysis become more (Zeng et al. 2021). The alternation of cell walls evaluates the enzyme accessibility for degrading the biomass. Several hammer mills were used to grind the manure and to decrease cell size. Combination of aluminum, sandpaper, and stainless steel was applied to achieve the smaller particles during grinding phase (Nabi et al. 2019).

5.2.4 Heat

Heating of biofibers in the organic waste resulted in an increase biomethane production and removal of chemical oxygen demand. In the heat pretreatment, the removal efficiency of chemical oxygen demand is higher. Heat pretreatment involves the influence of hydraulic retention time. The methane production rates can be achieved by pretreating the manure with heat. Due to complete removal of chemical oxygen demand, the digestion was anaerobic. Also, the heat induction stimulates a sterile environment for the system (Orlando and Borja 2020).

Hydraulic retention time was applied to treat the high lignin and organic substrates. In addition, the continuous thermal pretreatment is applicable at industrial scale to achieve maximum enzymatic hydrolysis with the low cost of energy. The production of hydrogen in thermal treatment influences the methane and ammonia production in manure. Low hydraulic retention rate and hydrogen production were unfavorable for methanogenesis (Qian et al. 2019).

5.2.5 Biological

The biological pretreatment of organic waste is efficient on large industrial scale. By consuming various enzymes and microorganisms in biological pretreatment of manure had attracted more attraction due to its eco-friendly nature and lower cost. The abundance of microorganisms was effective in degrading the components in the manure that are unable to break (Shen et al. 2018). Among the biological treatments, the usage of microbial cultures was effective in treating the organic wastes, as a large colony of microbes can be organized with only a single strain. The anaerobic microbial digestion also facilitates the catalytic substrate binding of enzymes. The production rate of methane depends upon the anaerobic efficiency of microbes (Ali et al. 2020a, b).

5.3 Techniques to Convert Manure into Bioenergy

Rapid increases in industrialization and population effects the energy demand. By treatment the biowaste scientifically, enables to do the successful conversion to bioenergy (Bhatia et al. 2018). Different types of manure especially the horse manure has ability to recover for the generation of heat and electricity. Due to the high volatile nature and low ash content in the horse manure, indicated its bioenergy recovery potential. High activation energy was required to proceed the reaction of energy formation. Thermochemical processes were valuable in the maximum generation of bioenergy from manure (Chong et al. 2019). Figure 5.1 gives the complete scheme of production of bioenergy from various biowastes.

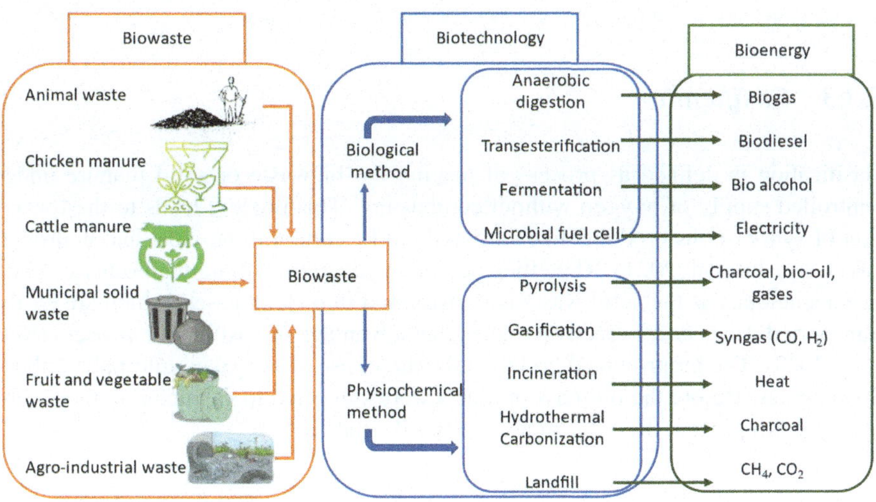

Fig. 5.1 Complete scheme of bioenergy production from various biowastes

5.3.1 Thermochemical Processes

The thermochemical conversion of manure to bioenergy is highly complex and energy sensitive (Perera et al. 2021). The successful commercialization of this reaction depends upon the optimized reactor and process designs. Thermochemical process includes combustion, gasification, pyrolysis, incineration, and carbonization (Guo et al. 2020). The release of gaseous products in the combustion process had a great impact on the ash content (Kirch et al. 2020).

5.3.2 Combustion

The sustainability of energy production was increased by combusting the solid wastes to convert the surplus manure into the renewable fuels (Mboumboue and Njomo 2018). Their combustion behavior was stimulated by blending manure with fossil coal. The hydrocarbons present within this mixture undergoes two stages of combustion: active and fossil fuel combustion. The difference in the combustion techniques depends upon the differences in the characteristics combustion temperatures and kinetic parameters (Kirch et al. 2020). The simulation of blends of animal manure derived char was similar to the coal. Various studies represent the high potential of combusting the small animal manure with coal has high power generation facilities (Sharara and Sadaka 2018).

5.3.3 Gasification

Gasification is defined as process of heating the biowaste or solid manure under controlled supply of oxygen without combusting. This process leads to the formation of syn-gaseous products like carbon dioxide, carbon monoxide, and hydrogen gas (Cerinski et al. 2021). The efficiency of gasification is more than other conversion techniques as the solid waste is transformed into clean gases, which are easily transported. High energy recovery is another advantage of gasification process (Ren et al. 2020). The tar produced prohibits the direct use of syngas in internal combustion engines. Before the biomass products were subjected to gasification, they were pyrolyzed to prevent the tar formation (Gyurik et al. 2019).

5.3.4 Pyrolysis

Pyrolysis is defined as the process of degradation of organic waste in inert environment by applying high heat. The organic waste can be generated from animal manure or agricultural waste can be processed by pyrolysis, and the yield of this process depends upon moisture, temperature, and heating rate. The increase in the moisture content of the reactant resulted in the increase lipid production and decrease in the solid and gaseous products (Czajczyńska et al. 2017). The temperature rates effect the production of lipid fuel. The production of gas and lipid fuel was reduced at high temperature, and the production of charcoal was increased at low heating rate (Pulka et al. 2019).

Based on the operating conditions, the pyrolysis was further classified into the fast and conventional type. The conventional pyrolysis was usually done at low heating temperature in order to maximize the production of charcoal from manure. In contrast, the fast pyrolysis was done at high temperature rates, and it directly depolymerizes the biomass into vapors. After the vapors formation, the biomass was directly converted into the bio-oil (Cai et al. 2021). Very high heating temperature was required to do the fast pyrolysis. The combustion temperature of biowaste and particle size of the animal manure made pyrolysis challenging (Mong et al. 2020).

5.3.5 Incineration

Incineration is the process of production of heat by combusting the material. The heat generated from the process of combustion have ability to run the power turbines for the electricity production. By incinerating 1 ton of waste, nearly 500–600 kWh of electrical energy can be generated (Trindade et al. 2018). The technique of incineration kills pathogens and also helps in the reduction of waste volume. Many developed countries use the process of incineration to generate electricity (Antelava et al. 2021).

Incineration is the complete oxidation of combustible materials with a waste steam. Incineration deals with waste streams on large scale as compared to other processes which deal with waste on rather small scale. Thus, because of this ability of dealing with high degree of waste variety, incineration has advantage over all other processes (Foster et al. 2021). Considering the efficiency, product percentage that we obtain after processing is more in case of dry ash as compared to the fly ash. But fly ash can also be used on roads in the form of asphalt. The bottom ash that we obtain from incineration requires much further purifications in order to obtain useful products (Shanableh et al. 2021).

5.3.6 Carbonization

The thermal conversion of biomass to carbonaceous residues defines the process of carbonization. This process is helpful in decomposing biomass or animal manure to biochar, which can be used as a solid fuel to generate electrical energy (Czerwińska et al. 2022). Cattle manure is more applicable in carbonization energy transformation. Solid fuel is generated from this thermal reaction is clean and contains fewer aromatic compounds. Hydrogen-rich gas and bio-oil are the byproducts of carbonization by consuming cattle manure (Merzari et al. 2020).

In carbonization, the pre-dry feedstock is not necessary for its energy conversion. This technique is efficient in energy as the energy required is much lower than comparison with pyrolysis and incineration (Khoo et al. 2020). In addition, the presence of water is a beneficial factor in improving hydrophobicity of waste and fuel properties. The amount of inorganic substances were also reduced in the biomass and produced high ash yield (Wilk et al. 2019). The conditioning environment of the hydrothermal takes elevated temperature for a specific time with aqueous environment. The conditions degrade the cellulose, lignin, and hemicelluloses. The carbon materials and solid fuels of high energy density can be formulated with carbonization for various applications of energy production (Sharma et al. 2020).

5.4 Biochemical Processes

By interlinking biochemical and thermochemical process for energy production from different waste products, a high product yield is obtained as compared to the past when biochemical and thermochemical processes were used separately (Patel et al. 2021). By the aid of anerobic breakdown of substances, we can obtain a large amount of energy that is previously stored in the form of biowaste. In this regard, methane is equally as useful as other biogases as it can be used as fossil fuel as well. On the other hand, excreta containing nitrogen and phosphorous can be successfully used in the formation of organic fertilizers (Ahorsu et al. 2018).

With the passage of time, the world population has increased in an astonishing rate, which results in increase in waste generation. The dependence of the renewable resources has developed to an extent as an energy source. Bioenergy from the waste of various of substances by providing them preferred anaerobic environment digestion environment over aerobic digestion, which ensures the waste management in a sustainable manner (Kim et al. 2018). By the consumption of biochemical processes, the production of biogas can be done for energy generation in metropolitan and countryside areas (Vyas et al. 2021).

5.4.1 Fermentation

When we come across a list of technologies for organic waste treatment, it would not be false to say that anaerobic fermentation is one of most efficient technology for organic waste treatment and that it can play a fundamental role in meeting the upcoming world energy crises efficiently (Zhu et al. 2020). One of the fermentation processes, the dry organic fermentation, is a quite efficient in obtaining energy as it consumes less water and practically less pollution production is seen, which is one of the many factors for which dry organic fermentation is the preferred one (Wongthanate and Mongkarothai 2018).

On the other hand, it has some shortcomings as well, as we can see that there is less production of energy as compared to the other biowaste management technologies. Thus, in order to overcome this shortcoming, we treat straw in order to enhance its bioavailability and decrease its crystalline nature. But, these pretreatments cost us additional processes which again are not much feasible (Najafi et al. 2021). Dark fermentation technology formulates the generation of biological hydrogen from waste. This process is eco-friendly and has capacity to utilize the organic wastes as renewable energy feedstocks. Pure carbohydrates and carbohydrate rich waste act as feasible substrates for high biohydrogen production. The yield of biohydrogen from the manure waste was elevated after its pretreatment with temperature than that of non-preheated manure (Deepanraj et al. 2021).

5.4.2 Transesterification

The process recovery of energy from organic waste promotes renewable energy production by minimizing the usage of fossil fuels. Transesterification is the chemical conversion of triglycerides into usable biofuel. The biofuel generated by the process of transesterification has lower viscosity (Ali et al. 2020a, b). The thermal degradation of biomass is performed in the presence of the oxygen to maximize the recovery of biodiesel, biochar and syngas. Thermally induced transesterification process transformed the lipid fraction in the cattle manure into biodiesel (Jung et al. 2020).

Biochar for the formation of biofuel acted as a porous medium in transesterification process. Transesterification process can also be enhanced by the usage of basic metal catalysts like potassium and sodium carbonates (Ahmed et al. 2021). The formation of syngas is also helpful in enhancing the properties of metals catalysts. So, the cattle's manure could be considered as a useful resource that can be valorized into value added fuels and chemicals. The transesterification of biomass provides quicker and more stable techniques to improve the production of energy from natural resources (Torres et al. 2021).

Catalyst based and without catalyst are the two types of transesterification process. With the consumption of catalysts, the improve yield of biofuel can be achieved (Chi et al. 2021). Nano-catalysts are widely used in the process of transesterification process, due to better catalytic activity and increase surface area, which results in the enhanced biodiesel yield compared to the solid catalysts (Banerjee et al. 2019).

5.4.3 Anaerobic Digestion

The consumption of biomass generates bioenergy in various forms such as ethanol, methanol, electricity, and biofuels. In anaerobic digestion, many microbes decompose the organic materials in the absence of energy (Bijarchiyan et al. 2020). This process converts the biomass into energy-controlled conditions, and the organic matter becomes agitated and liberates biogas. The biogas produced can be used to generate heat and electricity. Factors effecting the production of biogas include the type, density, and compositions of raw material (Baetge and Kaltschmitt 2018). Following Tables 5.2, 5.3 and 5.4 represents the yield of methane by applying anaerobic digestion to different kinds of manure.

Animal manure contains numerous microorganisms that facilitates the process of anaerobic digestion. Electricity generation through an anaerobic digestion of biomass consumes usable organic waste (Kulkarni and Ghanegaonkar 2019). The transformation of organic wastes starts with hydrolyzing complex natural polymer capability of microorganisms. The unstable and short chained unsaturated compounds like hydrogen and carbon dioxide are released by fermentative microorganisms. In many countries, the plants of biogas run with cow muck (Bharathiraja et al. 2018). As the cow waste comprises of higher concentrations of degradable carbon. The recycling of animal manure for the energy production is also effective in decreasing the ecological contamination. The improved production of methane content is achieved co-digestion of animal manure and microbial biomass (Kulkarni and Ghanegaonar 2020). Figure 5.2 displays the anaerobic digestion process for biogas production from organic waste.

Table 5.2 Methane yield attained with anaerobic digestion of organic dry solids (ODS)

Type of ODS	Methane yield (m^3/kg ODS)	References
Sow	0.28	Møller et al. (2004)
Dairy cattle	0.15	Møller et al. (2004)

Table 5.3 Methane yield attained with anaerobic digestion of municipal solid waste (MSW) (Owens and Chynoweth 1993)

Type of MSW	Methane yield (m^3/kg MSW)
Mechanically sorted (fresh)	0.22
Mechanically sorted (dried)	0.22
Hand sorted	0.21
Grass	0.21
Leaves	0.12
Branches	0.13
Mixed yard waste	0.14
Office paper	0.37
Printed newspaper	0.10

Table 5.4 Methane yield obtained through anaerobic digestion of fruit and vegetable wastes (FVW) (Gunaseelan 2004)

Type of FVW	Methane yield (m^3/kg FVW)
Mango peels	0.37–0.52
Banana peels	0.24–0.32
Orange peels	0.46
Lemon pressings	0.47
Grape pressings	0.28
Pomegranate peels	0.31
Tomatoes	0.21–0.38
Onion exterior peels	0.40
Garden beet leaves	0.23
Carrot leaves	0.24
Cabbage leaves	0.31

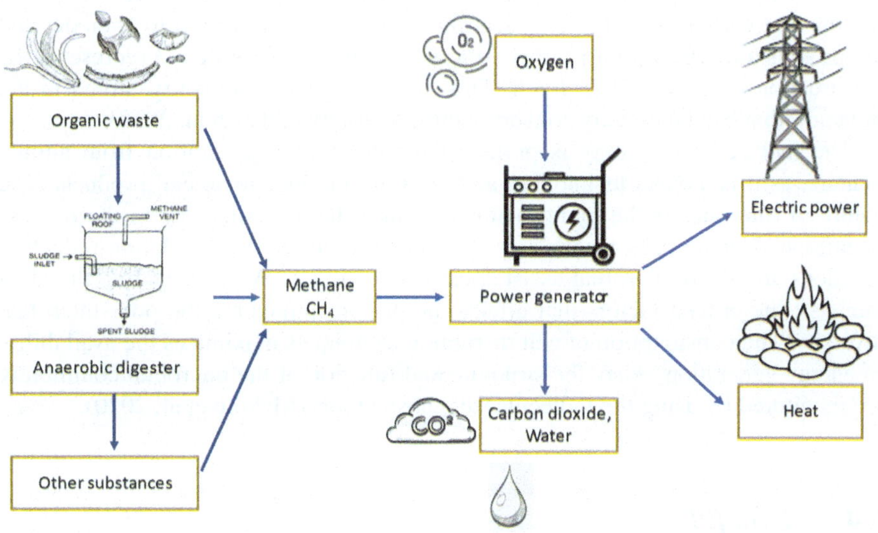

Fig. 5.2 Anaerobic digestion process for biogas production from organic waste

5.4.4 Microbial Fuel Cell

With the aid of microbial catalytic reactions, the conversion of organic matter into electric energy is attained by applying microbial fuel cells technique. These fuel cells generate electricity directly from organic rich wastes and biomasses by treating them (Tao et al. 2020). The apparatus of microbial fuel cells consists of an anode and

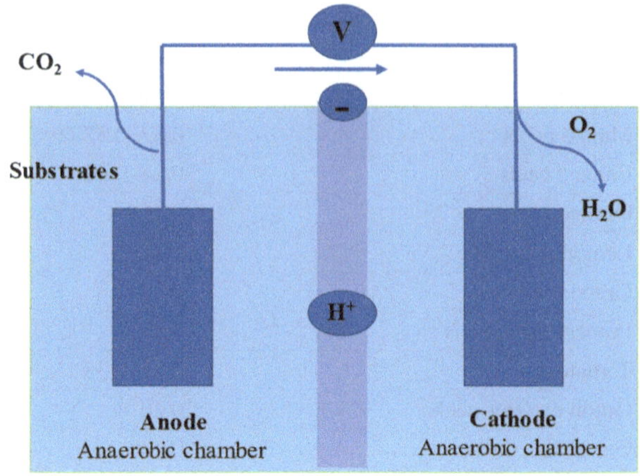

Fig. 5.3 Construction of microbial fuel cell

cathode that are separated by a proton exchange membrane. Isoelectronic microorganisms act as biocatalysts in anode and oxidize the organic matter into electron, proton, and carbon dioxide. The electrons formed are transferred to cathode, and the protons are diffused into cathode through membrane. Figure 5.3 represents the construction of microbial fuel cell. This diffusion generates energy, stored in the organics, that can be directly converted into electricity (Zhao et al. 2018).

The fuel cell technology is preferred for the energy generation from animal manure, as it is a clean technology and does not produce toxic end products. The nature of microbes is the main factor to achieve higher efficiency in the process of microbial flame cells technology (Greenman et al. 2019). Oxidation reduction reactions are aided microbial assisted cathode facilities. While recycling the algal biomass, the crucial factors that effects the power outputs are the poor microbial diversity. The consumption of cell disruption techniques maximizes the availability of energy generation. Also, the structure and function of the microbial community are monitored by using the molecular diagnostic tools (Mekuto et al. 2020).

5.4.5 Landfill

Landfilling is defined as the disposal of waste material by burying it or filling it the excavated pits. Food waste at landfill is either compacted or used to make compost for the reduction of greenhouse gas emission from the coal usage (Pahla et al. 2018). The food waste collected form the landfills is wet and heterogeneous which lovers its heating value and makes it difficult to store and transport. By mixing the coil with this wet waste could lower the combustion efficiencies and increase the emission of toxic gases. Therefore, the utilization of the animal manure is effective as it reduces the emission of toxic gases from combustion (Daiem and Said 2022).

The usage of agriculture residues and microalgae is applicable in reducing the carbon emissions as well as reducing the pressure at landfill sites. The moisture content in the waste is eliminated through evaporation which results in increased temperature of the wet waste (Dogaris et al. 2020). The release volatile compounds are due to the degradation of the hemicellulose in the biomass constituents. By further raising the temperature of biomass releases protein and carbohydrates. The reduction in the ratios of O/C and H/C enhances the production of carbon and diminishes the bulk density of biomass (Xu et al. 2018).

5.5 Bioreactor Design

Bioreactor design allows the efficient availability of microorganism or cells to perform their desired function under optimum conditions. Various designs of bioreactors are utilized for the biological and bio-eletro-chemical gas conversion. The performance of these bioreactors depends upon the gas–liquid mass transfer. Volatile and non-volatile components of the biomass are converted to gaseous compounds at an optimized organic rate. The industrialization of these bioprocesses provides more feasible and economical assessment of gas fermentation (Ayol et al. 2021).

5.5.1 Gasifier Reactor Design

The biomass gasification is the thermal decomposition of the biomass for the production of combustible gases like hydrogen, carbon monoxide, and methane. This mixture of gases is known as pyrolysis gas and is used to run the combustion engines. The production of gases is known as gasification and the reactor used in this process is gasifier (Gomaa et al. 2020). Another type of gasification that converts the wet animal manure to syngas through combined thermal decomposition and hydrolysis is the supercritical water gasification. The wet animal manure has high water affinity as compared to conventional gasification methods (Lee et al. 2021).

The main goal of the water gasification is to obtain syngas rich in hydrogen with the minimal production of char. Factors effecting the yield of hydrogen production from gasification are reaction pressure, temperature, and concentration of biomass. The homogenous and heterogeneous catalysts also enhance the energy production from the organic waste (Ren et al. 2019). The process of gasification prevents the air pollutants and supplies sustainable energy production. The coupling of kinetics and thermochemical equilibrium improves the efficiency of gasification process (Mazaheri et al. 2019).

5.5.2 Anaerobic Digester Design

Bioenergy is normally produced from the biological processes by fertilization of effective microbial microorganisms. Desired production of microorganisms is done by modifying their genetic or metabolic pathways (Srivastava 2019). The main aim of digester is to convert the organic waste into biogas by providing the optimal conditions. Anaerobic digestion is the most suitable way of converting the organic waste into biogas. Energy in the form of methane is recovered in anaerobic reactor. Anaerobic digester suffers from the problem of stability due to accumulation of volatile fatty acids in the manure and also due to drop in pH value (Ekama and Brouckaert 2022).

Many researchers find the inhibitory effect of the unionized volatile fatty acids on the methanogen bacteria. As these fatty acids limits the substrate formation at lower concentrations and substrate inhibition at higher concentrations for microorganisms. The co-digestion of cow and sheep manure under cold climatic conditions helps in improving the biogas generation (Singh et al. 2020). Several types of anaerobic digesters are used alone or in combination to treat and manage the manure. Manure is collected and placed in a centralized location in a way to treat and manage manure with the recovery of the biogas. The captured biogas from the digester directly used to run gas device and the temporary biogas storage may needed to balance the production and utilization system capacity (Sevillano et al. 2021).

5.6 Environmental Impacts

The biogas production from manure is eco-friendly and also inhibits the release of toxic gases. Manure biogases with their combination with the trace elements, avoid the emission of greenhouse gases. The pretreatment of manure with different nanomaterials and applying anaerobic digestion increase the production of biogas and methane (Hijazi et al. 2020). Manure of poultry and cattle is used in various cropping systems. Due to manure bulk density, infiltration, and water holding capacity exhibits beneficial impacts on the various soil properties (Khoshnevisan et al. 2021; Rayne and Aula 2020). Substituting the mineral fertilizers with manure maintains the filed predictability in the production of agricultural systems. Many treatment technologies of manure have potential to manage the solid fraction of it as compared to other waste-to-energy solutions. (Li et al. 2020).

Bio-oil production through the hydrothermal liquefaction is also an environmental promising solution to convert the wet manure to biofuel. The management of livestock manure is inevitable result for the sustainable future development, under the approach of biorefinery, to meet the economical requirements of the population (Montemayor et al. 2019). Anaerobic digestion of animal manure forms compressed biogas, which has been proven to reduce greenhouse gas (van den Oever et al. 2021). The management of the animal manure is challenging to minimize its environmental effects and also ensuring the product viability (Fangueiro et al. 2021).

5.7 Status, Challenges, and Perspective of Biowaste to Bioenergy Technology

The usage of anaerobic technology converts the manure or organic waste into energy to stabilize the economic status of the country (Glivin et al. 2021). The economic feasibility of anaerobic digestion technology is that it uses the pre-treatment techniques for the production of biogas and ethanol. Other parameters like the operating and maintaining costs of the pretreatment of manure are found to creating hurdles for the commercial sustainability (Banu et al. 2022).

The biogas systems and their energy usage and the environmental effects are governed through the lifecycle evaluation. In this evaluation process, the food stuff surplus is used as feedstock for biogas production. Reduction in the energy usage and the environmental effect goes parallel with the increase in the oil content of the food waste and the amount of the bioenergy generated from the approach (Kavitha et al. 2015). By combining the thermo and chemo-sonic disintegration of waste is economical, and the combination of internal biogas with the excess of heat release is beneficial in environmental aspects. The forceful amendment in the bioenergy production technologies and economic aspiration cause vital analysis of these factors for the successful disposal of biowaste to energy origination (Glivin and Sekhar 2020).

5.8 Policy and Government Incentives

The socio-political and regulatory conditions under which the bioenergy produced identifies the governance mechanisms (Ludlow et al. 2021). Despite of being massive ability, the implementation of biogas technology is still not being fully applied in Pakistan due to lack of government initiatives and policies. There is a need of initiating awareness in the government of Pakistan regarding these policies. The farmers should be encouraged to adopt the bioenergy production technologies in the country. The proper handling of organic waste to use a an alternative energy source should be added in these policies (Yaqoob et al. 2021).

The introduction of control mechanisms encourages the waste recycling techniques. Like the open disposal of waste to the landfills promotes these control mechanisms. The financial support programs should be offered by the government in order to attract the attention of farmers to promote biogas technology in Pakistan (Purkus et al. 2018). The involvement of private sectors in the biogas production industry will

Table 5.5 Electricity generation per day from municipal solid waste (Pakistan 2005)

City	Waste generated (t/d)	Generation ratio (kg/capita/d)
Sibi	57	1.30
Bannu	79	0.70
Hyderabad	314	0.32
Bahawalpur	405	0.60
Peshawar	792	0.73
Quetta	1362	0.67
Faisalabad	1860	0.60
Lahore	5120	0.67
Karachi	12,142	0.50

play an important role in the economy of the country. Proper coordination of private and public sectors and also lowering the tax price, reduce the biogas technology cost (Gustafsson and Anderberg 2021). Table 5.5 represents the generation of electricity per day in the major cities of Pakistan from municipal solid waste.

5.9 Conclusion

A huge deficiency of electricity results from the heavy dependence on the imported fuels, which significantly effects the economic development of Pakistan. The successful implementation of animal manure as a primary source of energy in the existing ecosystem through biochemical conversion technologies is essential for the development of sustainable environment of the country.

The heavy dependence on the imported fuels effects the economic development of Pakistan. The application of manure to the agricultural soils is widely considered as a source of soil organic carbon and the organic matter present in this fertilizer is easily degraded due to the high accessibility of nitrogen content in it. The organic and animal waste combine to form an organic fertilizer known as manure. Also, the design of bioreactor modifies the efficiency of biological and bio-electro-chemical gas conversion practices. The biogas production from manure is eco-friendly, and the pretreatment of manure with different nanomaterials increases the production of biogas and methane by applying anaerobic digestion.

References

Abou-Sreea AI, Rady MM, Roby MH, Ahmed SM, Majrashi A, Ali EF (2021) Cattle manure and bio-nourishing royal jelly as alternatives to chemical fertilizers: potential for sustainable production of organic *Hibiscus sabdariffa* L. J Appl Res Med Arom Plants 25:100334

Ahmed M, Abdullah A, Patle DS, Shahadat M, Ahmad Z, Athar M, Vo DVN (2021) Feedstocks, catalysts, process variables and techniques for biodiesel production by one-pot extraction-transesterification: a review. Environ Chem Lett 1–44

Ahorsu R, Medina F, Constantí M (2018) Significance and challenges of biomass as a suitable feedstock for bioenergy and biochemical production. Rev Energ 11(12):3366

Ali S, Shafique O, Mahmood S, Mahmood T, Khan BA, Ahmad I (2020a) Biofuels production from weed biomass using nanocatalyst technology. Biomass Bioenergy 139:105595

Ali SS, Mustafa AM, Kornaros M, Manni A, Sun J, Khalil MA (2020b) Construction of novel microbial consortia CS-5 and BC-4 valued for the degradation of catalpa sawdust and chlorophenols simultaneously with enhancing methane production. Biores Technol 301:122720

Álvarez J, Otero L, Lema J (2010) A methodology for optimising feed composition for anaerobic co-digestion of agro-industrial wastes. Biores Technol 101(4):1153–1158

Antelava A, Jablonska N, Constantinou A, Manos G, Salaudeen SA, Dutta A, Al-Salem SM (2021) Energy potential of plastic waste valorization: a short comparative assessment of pyrolysis versus gasification. Energy Fuels 35(5):3558–3571

Awasthi MK, Sarsaiya S, Wainaina S, Rajendran K, Awasthi SK, Liu T, Binod P (2021) Techno-economics and life-cycle assessment of biological and thermochemical treatment of bio-waste. Renew Sustain Energy Rev 144:110837

Ayol A, Peixoto L, Keskin T, Abubackar, HN (2021) Reactor designs and configurations for biological and bioelectrochemical C1 gas conversion: a review. Int J Environ Res Publ Heal 18(21):11683

Baetge S, Kaltschmitt M (2018) Rice straw and rice husks as energy sources—comparison of direct combustion and biogas production. Biomass Conv Biorefin 8(3):719–737

Banerjee S, Rout S, Banerjee S, Atta A, Das D (2019) Fe_2O_3 nanocatalyst aided transesterification for biodiesel production from lipid-intact wet microalgal biomass. Biorefin Appr Energy Conv Manage 195:844–853

Banu JR, Kavitha S, Kannah RY, Varjani S, Gunasekaran M (2022) Mild hydrogen peroxide interceded bacterial disintegration of waste activated sludge for efficient biomethane production. Sci Total Environ 152873

Bao Y, Feng Y, Qiu C, Zhang J, Wang Y, Lin X (2021) Organic matter-and temperature-driven deterministic assembly processes govern bacterial community composition and functionality during manure composting. Waste Manage 131:31–40

Belostotskiy D, Jacobi H, Strach K, Liebetrau J (2013) Anaerobic digestion of chicken manure as a single substrate by control of ammonia concentration AD13 Recovering (bio) Resources for the World

Bharathiraja B, Sudharsana T, Jayamuthunagai J, Praveenkumar R, Chozhavendhan S, Iyyappan J (2018) Biogas production–a review on composition, fuel properties, feed stock and principles of anaerobic digestion. Renew Sustain Energy Rev 90:570–582

Bhatia SK, Joo HS, Yang YH (2018) Biowaste-to-bioenergy using biological methods–a mini-review. Energy Conv Manage 177:640–660

Bijarchiyan M, Sahebi H, Mirzamohammadi S (2020) A sustainable biomass network design model for bioenergy production by anaerobic digestion technology: using agricultural residues and livestock manure Energy. Sustain Soc 10(1):1–17

Bilen HF, Canakoglu E, Soykan G (2021) The effect of the use of bio-waste on the generation of electrical energy in municipal solid waste incineration. In Paper presented at the 2021 10th international conference on renewable energy research and application (ICRERA)

Bouallagui H, Lahdheb H, Romdan EB, Rachdi B, Hamdi M (2009) Improvement of fruit and vegetable waste anaerobic digestion performance and stability with co-substrates addition. J Environ Manage 90(5):1844–1849

Cai W, Luo Z, Zhou J, Wang Q (2021) A review on the selection of raw materials and reactors for biomass fast pyrolysis in China. Fuel Process Technol 221:106919

Cândido D, Bolsan AC, Hollas CE, Venturin B, Tápparo DC, Bonassa G, Kunz A (2022) Integration of swine manure anaerobic digestion and digestate nutrients removal/recovery under a circular economy concept. J Environ Manage 301:113825

Cerinski D, Ferreiro A I, Baleta J, Costa M, Zimbardi F, Cerone N, Wang J (2021) Modelling the biomass updraft gasification process using the combination of a pyrolysis kinetic model and a thermodynamic equilibrium model. Energy Rep 7:8051–8061

Chi NTL, Anto S, Ahamed TS, Kumar SS, Shanmugam S, Samuel MS, Pugazhendhi A (2021) A review on biochar production techniques and biochar based catalyst for biofuel production from algae. Fuel 287:119411

Chong CT, Mong GR, Ng JH, Chong WWF, Ani FN, Lam SS, Ong HC (2019) Pyrolysis characteristics and kinetic studies of horse manure using thermogravimetric analysis. Energy Conv Manage 180:1260–1267

Coarita Fernandez H, Teixeira Franco R, Bayard R, Buffiere P (2020) Mechanical pre-treatments evaluation of cattle manure before anaerobic digestion. Waste Biomass Valoriz 11(10):5175–5184

Czajczyńska D, Anguilano L, Ghazal H, Krzyżyńska R, Reynolds A, Spencer N, Jouhara H (2017) Potential of pyrolysis processes in the waste management sector. Therm Sci Eng Progr 3:171–197

Czerwińska K, Śliz M, Wilk M (2022) Hydrothermal carbonization process: fundamentals main parameter characteristics and possible applications including an effective method of SARS-CoV-2 mitigation in sewage sludge. Rev Renew Sustain Energy Rev 154:111873

Daiem MMA, Said N (2022) Energetic economic and environmental perspectives of power generation from residual biomass in Saudi Arabia. Alex Eng J 61(5):3351–3364

Deepanraj B, Senthilkumar N, Ranjitha J, Jayaraj S, Ong HC (2021) Biogas from food waste through anaerobic digestion: optimization with response surface methodology. Biomass Conv Biorefin 11(2):227–239

Dogaris I, Ammar E, Philippidis GP (2020) Prospects of integrating algae technologies into landfill leachate treatment. World J Microbiol Biotechnol 36(3):1–25

Ekama G, Brouckaert C (2022) Integration of complete elemental mass-balanced stoichiometry and aqueous-phase chemistry for bioprocess modelling of liquid and solid waste treatment systems—Part 3: measuring the organics composition water SA 48 (1 January)

Fangueiro D, Alvarenga P, Fragoso R (2021) Horticulture and orchards as new markets for manure valorisation with less environmental impacts. Sustainability 13(3):1436

Foster W, Azimov U, Gauthier-Maradei P, Molano L C, Combrinck M, Munoz J, Patino L (2021) Waste-to-energy conversion technologies in the UK: processes and barriers. Rev Renew Sustain Energy Rev 135:110226

Glivin G, Kalaiselvan N, Mariappan V, Premalatha M, Murugan P, Sekhar J (2021) Conversion of biowaste to biogas: a review of current status on techno-economic challenges policies technologies and mitigation to environmental impacts. Fuel 302:121153

Glivin G, Sekhar SJ (2020) Waste potential barriers and economic benefits of implementing different models of biogas plants in a few Indian educational institutions. Bioenergy Res 13(2):668–682

Gomaa MR, Al-Dmour N, AL-Rawashdeh HA, Shalby M (2020) Theoretical model of a fluidized bed solar reactor design with the aid of MCRT method and synthesis gas production. Renew Energy 148:91–102

Greenman J, Gajda I, Ieropoulos I (2019) Microbial fuel cells (MFC) and microalgae; photo microbial fuel cell (PMFC) as complete recycling machines. Sustain Energy Fuels 3(10):2546–2560

Gross A, Glaser B (2021) Meta-analysis on how manure application changes soil organic carbon storage. Sci Rep 11(1):1–13

Gunaseelan VN (2004) Biochemical methane potential of fruits and vegetable solid waste feedstocks. Biomass Bioenergy 26(4):389–399

Guo M, Li H, Baldwin B, Morrison J (2020) Thermochemical processing of animal manure for bioenergy and biochar animal manure. Prod Char Environ Conc Manage 67:255–274

Gustafsson M, Anderberg S (2021) Dimensions and characteristics of biogas policies. Modell Europ Pol Lands Renew Sustain Energy Rev 135:110200

Gyurik L, Egedy A, Zou J, Miskolczi N, Ulbert Z, Yang H (2019) Hydrodynamic modelling of a two-stage biomass gasification reactor. J Energy Inst 92(3):403–412

Hijazi O, Abdelsalam E, Samer M, Amer B, Yacoub I, Moselhy M, Bernhardt H (2020) Environmental impacts concerning the addition of trace metals in the process of biogas production from anaerobic digestion of slurry. J Clean Prod 243:118593

Iqbal MW, Kang Y (2021) Waste-to-energy supply chain management with energy feasibility condition. J Clean Prod 291:125231

Jung S, Kim M, Jung JM, Kwon EE (2020) Valorization of swine manure biochar as a catalyst for transesterifying waste cooking oil into biodiesel. Environ Poll 266:115377

Karim K, Hoffmann R, Klasson T, Al-Dahhan M (2005) Anaerobic digestion of animal waste: Waste strength versus impact of mixing. Biores Technol 96(16):1771–1781

Kavitha S, Kannah RY, Yeom IT, Do KU, Banu JR (2015) Combined thermo-chemo-sonic disintegration of waste activated sludge for biogas production. Bioresour Technol 197:383–392

Khan MU, Ahmad M, Sultan M, Sohoo I, Ghimire PC, Zahid A, Abdeshahian P (2021) Biogas production potential from livestock manure in Pakistan. Sustainability 13(12):6751

Khan MU, Ahring BK (2021a) Anaerobic digestion of digested manure fibers: Influence of thermal and alkaline thermal pretreatment on the biogas yield. Bioenergy Res 14(3):891–900

Khan MU, Ahring BK (2021b) Improving the biogas yield of manure: effect of pretreatment on anaerobic digestion of the recalcitrant fraction of manure. Biores Technol 321:124427

Khoo CG, Lam MK, Mohamed AR, Lee KT (2020) Hydrochar production from high-ash low-lipid microalgal biomass via hydrothermal carbonization: effects of operational parameters and products characterization. Environ Res 188:109828

Khoshnevisan B, Duan N Tsapekos P, Awasthi MK, Liu Z, Mohammadi A, Pan J (2021) A critical review on livestock manure biorefinery technologies. Sustain Chall Fut Perspect Renew Sustain Energy Rev 135:110033

Kim YHB, Ma D, Setyabrata D, Farouk MM, Lonergan SM, Huff-Lonergan E, Hunt MC (2018) Understanding postmortem biochemical processes and post-harvest aging factors to develop novel smart-aging strategies. Meat Sci 144:74–90

Kirch T, Medwell PR, Birzer CH, van Eyk PJ (2020) Small-scale autothermal thermochemical conversion of multiple solid biomass feedstock. Renew Energy 149:1261–1270

Kulkarni M, & Ghanegaonar P (2020) Anaerobic digestion enhancement in biogas production and quality improvement Techno-Societal 2018. Springer, pp 353–361

Kulkarni M, Ghanegaonkar P (2019) Biogas generation from floral waste using different techniques. Glob J Environ Sci Manage 5(1):17–30

Lee CS, Conradie AV, Lester E (2021) Review of supercritical water gasification with lignocellulosic real biomass as the feedstocks: Process parameters biomass composition catalyst development reactor design and its challenges. Chem Eng J 415:128837

Li S Li X, & Ho S -H (2021) Microalgae as a solution of third world energy crisis for biofuels production from wastewater toward carbon neutrality. Updated Rev Chem 132863

Li S, Wu J, Wang X, Ma L (2020) Economic and environmental sustainability of maize-wheat rotation production when substituting mineral fertilizers with manure in the North China Plain. J Clean Prod 271:122683

Liu H, Xu W, Li J, Yu Z, Zeng Q, Tan W, Mi W (2021) Short-term effect of manure and straw application on bacterial and fungal community compositions and abundances in an acidic paddy soil. J Soils Sediments 21(9):3057–3071

Ludlow J, Jalil-Vega F, Rivera XS, Garrido RA, Hawkes A, Staffell I, Balcombe P (2021) Organic waste to energy: resource potential and barriers to uptake in Chile. Sustain Prod Consump 28:1522–1537

Møller HB, Sommer SG, Ahring BK (2004) Biological degradation and greenhouse gas emissions during pre-storage of liquid animal manure. J Environ Qual 33(1):27–36
Martín-González L, Colturato L, Font X, Vicent T (2010) Anaerobic co-digestion of the organic fraction of municipal solid waste with FOG waste from a sewage treatment plant: recovering a wasted methane potential and enhancing the biogas yield. Waste Manage 30(10):1854–1859
Mazaheri N, Akbarzadeh A, Madadian E, Lefsrud M (2019) Systematic review of research guidelines for numerical simulation of biomass gasification for bioenergy production. Energy Convers Manage 183:671–688
Mboumboue E, Njomo D (2018) Biomass resources assessment and bioenergy generation for a clean and sustainable development in Cameroon. Biomass Bioenergy 118:16–23
Mekuto L, Olowolafe AV, Huberts R, Dyantyi N, Pandit S, Nomngongo P (2020) Microalgae as a biocathode and feedstock in anode chamber for a selfsustainable microbial fuel cell technology: a review South African. J Chem Eng 31(1):7–16
Merzari F, Goldfarb J, Andreottola G, Mimmo T, Volpe M, Fiori L (2020) Hydrothermal carbonization as a strategy for sewage sludge management. Influen Process Withdra Point Hydrochar Propert Energ 13(11):2890
Møller H, Sommer S, Ahring BK (2004) Methane productivity of manure straw and solid fractions of manure. Biomass Bioenergy 26(5):485–495
Mong GR, Chong CT, Ng JH, Chong WWF, Lam SS, Ong HC, Ani FN (2020) Microwave pyrolysis for valorisation of horse manure biowaste. Energy Conv Manage 220:113074
Montemayor E, Bonmatí A, Torrellas M, Camps F, Ortiz C, Domingo F, Antón A (2019) Environmental accounting of closed-loop maize production scenarios: Manure as fertilizer and inclusion of catch crops. Resour Conserv Recycl 146:395–404
Nabi M, Zhang G, Zhang P, Tao X, Wang S, Ye J, Wu Y (2019) Contribution of solid and liquid fractions of sewage sludge pretreated by high pressure homogenization to biogas production. Biores Technol 286:121378
Najafi E, Castro E, Karimi K (2021) Biorefining for olive wastes management and efficient bioenergy production. Energy Convers Manage 244:114467
Omar R, Harun RM, Mohd Ghazi T, Wan Azlina W, Idris A, Yunus R (2008) Anaerobic treatment of cattle manure for biogas production. Paper presented at the Proceedings Philadelphia Annual meeting of American Institute of Chemical Engineers
Orlando MQ, Borja VM (2020) Pretreatment of animal manure biomass to improve biogas production. Rev Energ 13(14):3573
Owens J, Chynoweth D (1993) Biochemical methane potential of municipal solid waste (MSW) components. Water Sci Technol 27(2):1–14
Pahla G, Ntuli F, Muzenda E (2018) Torrefaction of landfill food waste for possible application in biomass co-firing. Waste Manage 71:512–520
Pakistan GO (2005) Compendium on environment statistics of Pakistan—2004: Federal Bureau of Statistics Government of Pakistan Statistics Division …
Patel AK, Singhania RR, Chen CW, Tseng YS Kuo, CH, Wu CH, Di Dong C (2021) Advances in micro-and nano bubbles technology for application in biochemical processes. Environ Technol Innov 23:101729
Perera SM, Wickramasinghe C, Samarasiri B, Narayana M (2021) Modeling of thermochemical conversion of waste biomass–a comprehensive review. Biofuel Res J 8(4):1481–1528
Pulka J, Manczarski P, Koziel JA, Białowiec A (2019) Torrefaction of sewage sludge. Kinet Fuel Propert Biochars Energ 12(3):565
Purkus A, Gawel E, Szarka N, Lauer M, Lenz V, Ortwein A, Thrän D (2018) Contributions of flexible power generation from biomass to a secure and cost-effective electricity supply—a review of potentials incentives and obstacles in Germany. Energy Sustain Soc 8(1):1–21
Qian M, Li Y, Zhang Y, Sun Z, Wang Y, Feng J, Zhao L (2019) Efficient acetogenesis of anaerobic co-digestion of food waste and maize straw in a HSAD reactor. Biores Technol 283:221–228
Raheem A, Abbasi SA, Memon A, Samo SR, Taufiq-Yap Y, Danquah MK, Harun R (2016) Renewable energy deployment to combat energy crisis in Pakistan. Energy Sustain Soc 6(1):1–13

Rayne N, Aula L (2020) Livestock manure and the impacts on soil health. Rev Soil Syst 4(4):64
Ren J, Cao JP, Zhao XY, Yang FL, Wei XY (2019) Recent advances in syngas production from biomass catalytic gasification: a critical review on reactors catalysts catalytic mechanisms and mathematical models. Renew Sustain Energy Rev 116:109426
Ren J, Liu YL, Zhao XY, Cao JP (2020) Methanation of syngas from biomass gasification: an overview. Int J Hydrogen Energy 45(7):4223–4243
Sevillano CA, Pesantes AA, Peña Carpio E, Martínez EJ, Gómez X (2021) Anaerobic digestion for producing renewable energy. Evol Technol New Uncert Scen Entropy 23(2):145
Shanableh A, Abdallah M, Tayara A, Ghenai C, Kamil M, Inayat A, Shabib A (2021) Experimental characterization and assessment of bio-and thermo-chemical energy potential of dromedary manure. Biomass Bioenergy 148:106058
Sharara MA, Sadaka SS (2018) Opportunities and barriers to bioenergy conversion techniques and their potential implementation on swine manure. Energies 11(4):957
Sharma HB, Sarmah AK, Dubey B (2020) Hydrothermal carbonization of renewable waste biomass for solid biofuel production: a discussion on process mechanism the influence of process parameters environmental performance and fuel properties of hydrochar. Renew Sustain Energy Rev 123:109761
Shen F, Li H, Wu X, Wang Y, Zhang Q (2018) Effect of organic loading rate on anaerobic co-digestion of rice straw and pig manure with or without biological pretreatment. Biores Technol 250:155–162
Singh B, Szamosi Z, Siménfalvi Z (2020) Impact of mixing intensity and duration on biogas production in an anaerobic digester: a review. Crit Rev Biotechnol 40(4):508–521
Srivastava RK (2019) Bio-energy production by contribution of effective and suitable microbial system. Mater Sci Energy Technol 2(2):308–318
Tao M, Guan L, Jing Z, Tao Z, Wang Y, Luo H, Wang Y (2020) Enhanced denitrification and power generation of municipal wastewater treatment plants (WWTPs) effluents with biomass in microbial fuel cell coupled with constructed wetland. Sci Total Environ 709:136–159
Torres A, Padrino S, Brito A, Díaz L (2021) Biogas production from anaerobic digestion of solid microalgae residues generated on different processes of microalgae-to-biofuel production. Biomass Conver Biorefin 1–14
Trindade AB, Palacio JCE, González AM, Orozco DJR, Lora EES, Renó MLG, del Olmo OA (2018) Advanced exergy analysis and environmental assesment of the steam cycle of an incineration system of municipal solid waste with energy recovery. Energy Convers Manage 157:195–214
van den Oever AE, Cardellini G, Sels BF, Messagie M (2021) Life cycle environmental impacts of compressed biogas production through anaerobic digestion of manure and municipal organic waste. J Clean Prod 306:127–156
Victorin M, Davidsson Å, Wallberg O (2020) Characterization of mechanically pretreated wheat straw for biogas production. BioEnergy Res 13(3):833–844
Vyas S, Prajapati P, Shah AV, Srivastava VK, Varjani S (2021) Opportunities and knowledge gaps in biochemical interventions for mining of resources from solid waste. Spec Focus Anaerobic Digestion Fuel 122–625
Wilk M, Magdziarz A, Jayaraman K, Szymańska-Chargot M, Gökalp I (2019) Hydrothermal carbonization characteristics of sewage sludge and lignocellulosic biomass. Comp Stud Biomass Bioenergy 120:166–175
Wongthanate J, Mongkarothai K (2018) Enhanced thermophilic bioenergy production from food waste by a two-stage fermentation process. Int J Recycl Organ Waste Agric 7(2):109–116
Xu X, Tu R, Sun Y, Li Z, Jiang E (2018) Influence of biomass pretreatment on upgrading of bio-oil: comparison of dry and hydrothermal torrefaction. Biores Technol 262:261–270
Yaqoob H, Teoh YH, Din ZU, Sabah NU, Jamil MA, Mujtaba M, Abid A (2021) The potential of sustainable biogas production from biomass waste for power generation in Pakistan. J Clean Prod 307:127250

Younas U, Khan B, Ali S, Arshad C, Farid U, Zeb K, Vaccaro A (2016) Pakistan geothermal renewable energy potential for electric power generation: a survey. Renew Sustain Energy Rev 63:398–413

Zeng S, Jang HM, Park S, Park S, Kan E (2021) Effects of mechanical refining on anaerobic digestion of dairy manure. ACS Omega 6(26):16934–16942

Zhao N, Jiang Y, Alvarado-Morales M, Treu L, Angelidaki I, Zhang Y (2018) Electricity generation and microbial communities in microbial fuel cell powered by macroalgal biomass. Bioelectrochemistry 123:145–149

Zhu Q, Li X, Li G, Li J, Li C, Che L, Zhang L (2020) Enhanced bioenergy production in rural areas: synthetic urine as a pre-treatment for dry anaerobic fermentation of wheat straw. J Clean Prod 260:121164

Chapter 6
Utilization of Waste Animal Fat for Sustainable Biodiesel Production

Muhammad Arshad, Syed Shatir A. Syed-Hassan, Rehana Masood, Abdur Rahman Ansari, Aqsa Mumtaz, Abdur Rahman, Sadia Batool, Farhan Ahmad Atif, Aun Muhammad, Aziz Ur Rehman, Ammar Tahir, and Iram Saba

Abstract Rising demand of diesel fuel of this high-traffic world compels us to produce more and more fuel which is not eco-friendly practice, so another way to achieve our goal safely is to produce biodiesel which is not only an alternative but also an excellent substitute. In other words, production of conventional diesel severely affects the environment and economy of country. In Pakistan, the major consumer of this imported petroleum fuel is transport sector. In 2010, this sector produced 8.0 $GtCO_2eq$ of direct greenhouse gas emissions which were about 24% of total energy-related CO_2 emissions. This chapter fundamentally reviews the certitudes and anticipation of biodiesel generation along with its effectiveness in reducing the consumption of petroleum-based fuels and resultant contaminated environment in Pakistan. A wide range of feedstock including animal fats and soybean oil can be efficiently used for biodiesel synthesis.

M. Arshad (✉) · A. R. Ansari · A. Mumtaz · S. Batool
Department of Basic Sciences, Jhang-Campus, University of Veterinary and Animal Sciences, Lahore, Pakistan
e-mail: muhammad.arshad@uvas.edu.pk

S. S. A. Syed-Hassan
School of Chemical Engineering, College of Engineering, Universiti Teknologi MARA, 40450 Shah Alam, Selangor, Malaysia

R. Masood
Department of Biochemistry, Shaheed Benazir Bhutto Women University, Peshawar, Pakistan

A. Rahman
Department of Animal Sciences, Jhang-Campus, University of Veterinary and Animal Sciences, Lahore, Pakistan

F. A. Atif · A. Muhammad · A. Tahir
Department of Clinical Sciences, Jhang-Campus, University of Veterinary and Animal Sciences, Lahore, Pakistan

A. U. Rehman
Department of Pathobiology, Jhang-Campus, University of Veterinary and Animal Sciences, Lahore, Pakistan

I. Saba
Department of Chemistry, Government College Woman University, Sialkot, Pakistan

M. Arshad (ed.), *Climate Changes Mitigation and Sustainable Bioenergy Harvest Through Animal Waste*, https://doi.org/10.1007/978-3-031-26224-1_6

Graphical Abstract

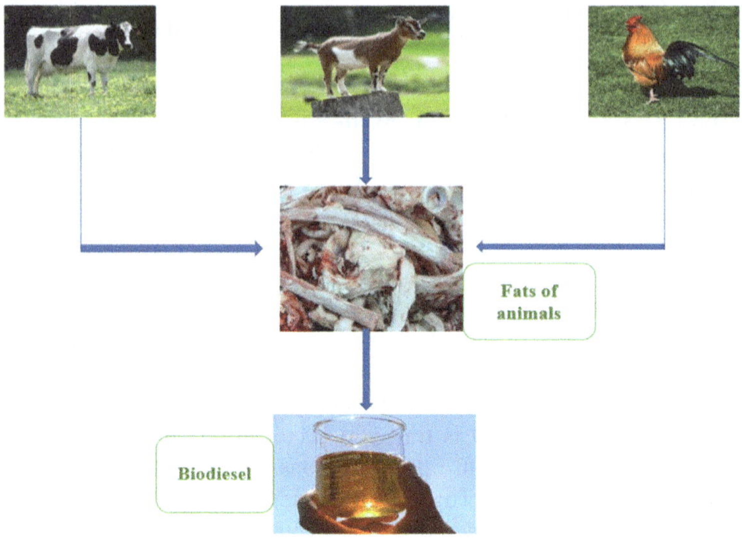

Keywords Biodiesel · Feedstocks · Greenhouse gases · Sustainable energy · Transport sector · Transesterification

6.1 Introduction

Energy constitutes an indispensable entity for the maintenance and uplift of human livelihood standards. The global energy consumption has been doubled over the last three decades, and fossil fuels are the 80% of the total energy (Shirazi et al. 2014). Nevertheless, the energy requirements are gradually increasing owing to rapid economic development and progressively rising human living standards. Consequently, the fossil energy reserves are rapidly running down along with the undesirable emissions of greenhouse gases. Moreover, the abundance of fossil energy resources is confined to only certain regions of the world (Ahmad et al. 2011). Modern transport and communication sector can play a major role in achievement of the economic endeavors and globalization through linked routes. Being essential for the carriage and heavy-duty engines, diesel fuel demonstrated the highest distribution rate among the primary trade products in 2010. Globally, about 30% of the total energy yield is supplied to transportation sector and 80% of it is expended by the road transport. Besides, the road transport receives approximately 60% of global oil supply. In fact, all of the energy used in the transportation sector comes from fossil fuels including gasoline, diesel, LPG, and NG. It mainly derives the fossil fuel-based energy from oil (97.6%), whereas the remaining small amount is contributed by LPG or NG (Ali et al. 2012).

Presently, the world is confronted by a challenging issue of global warming. Therefore, the reduction of progressively increasing CO_2 emissions is recommended for controlling the global warming process and preventing its further exacerbation (Alptekin and Canakci 2010). Historically, fossil fuels have been effectively fulfilling the global energy needs in terms of running vehicles and motor engines and controlling power plants. However, the major emission products of fossil fuel including CO_2, HC, nitrogen oxide (NOx), and volatile organic compound (VOC) have been implicated in causing air pollution, smog, and acid rain (Arbab et al. 2013).

The development of sustainable energy sources offers promising solutions to environmental damage and depletion of decomposing plants and animals. Consequently, there is an awful need to explore authentic, eco-safe, economically feasible, and alternative energy resources. Biodiesel produced from botanical and algal sources is a better substitute to conventional diesel. Chemically, biodiesel is an ester-based oxygenated fuel, comprising of mono-alkyl esters of long-chain fatty acids (Alptekin and Canakci 2010; Atabani et al. 2013a, 2015, 2011).

The unburned total hydrocarbons (HC) and polycyclic aromatic hydrocarbons (PAHs) are withdrawn during biodiesel ignition in the diesel engines (Atabani et al. 2011). Additionally, biodiesel combustion can substantially reduce the emission of carbon monoxide (CO) and particulate matter than fossil-based diesel oil (Atabani and Silva César 2014). In addition being produced using renewable resources, biodiesel saves the environment and people from dangerous air pollution (Atabani et al. 2017). Generally, the efficiency of biodiesel is further improved by mixing 20% v/v or lesser amount of fossil fuel (diesel). Biodiesel has been certified by the Environmental Protection Agency (EPA) based upon the compliance with the qualifications for ASTM D6751 (Atabani et al. 2013b).

Similar to other developing countries, Pakistan also needs vigorous and cost-effective transport and logistic sectors for achieving economic development and enhancing export competitiveness. Therefore, the government is committed to modernize the transport and coordination division through implementing an extensive improvement activity and consistent procedure of change supported by investments in the entirety of its sub-divisions (Atabani et al. 2014a).

Apart from boosting the regional economy, renewable generation of biofuels may also augment the overall energy supply and preclude the environmental deterioration (Arbab et al. 2013; Atabani et al. 2014b). Regardless of a slight diminution in performance, biodiesel can be effectively used in combustion engines and boilers without significant alterations. In addition, the biodiesel nearly release no harmful compounds. Moreover, the physical and synthetic attributes of it are that its utilization either all alone or all blended with oil-based diesel with few specialized changes or no alteration (Atabani et al. 2013c). Several technologically advanced nations used it. For that reason, it is a basic need in the developing countries including Pakistan.

Animal fats are considered as waste and can be obtained from slaughter houses. A minute amount of these fats is used by soap industry, and rest fill up the landfills. It was noticed that percentage of biodiesel and oil produced from mutton fat is higher than those produced from the beef fat. KOH, NaOH, and Na metal were used as a catalyst, and KOH gives higher production of biodiesel as compared to NaOH and Na metal after the transesterification process. Characteristics of biodiesel show that it is an appropriate fuel for vehicles (Balat 2011).

The main sources to obtain animal fats are beef tallow, poultry fat, fish oils, and yellow greases. Beef tallow and poultry fat are the main waste animal fat sources for this purpose. Animal's fat is subtracted from different parts of the animal body, like blood, mesentery, and offal including heart, lungs, and intestines. So the fat can be extracted from these parts of the animal via rendering and then stored in a favorable environment (Balat and Balat 2009). The extracted fats can be converted into the fuel by a reaction called transesterification. The reaction can convert fat into the required product and some other by products on the provision of certain essential conditions like temperature, pressure, and catalyst. The obtained product will be purified from the reaction mixture and stored in a suitable place. The fuel is eco-friendly and sustainable.

6.2 Environment and Climate Change

These days, there is a lot of focus on the environmental issues and to the sustainable exploitation of natural resources. Lack of awareness regarding this matter and proper anticipatory measures may account for high economic and environmental losses in the future. Continuous and widespread contamination is damaging the land, air, and water. Pakistan is predominantly an agricultural country where the availability of natural resources also determines the growth trends of agriculture sector. Therefore, improving the nation's ability is highly essential to accomplish naturally supportable financial advancement and meet the prerequisites of present and future generations.

Pakistan is listed among the highest environmentally vulnerable countries. Urbanization has drastically changed the biological system of cities and provincial zones of the country. The national biodiversity is threatened by prompt depletion of natural resources. However, the government is fully committed to many global conventions and protocols for biodiversity conservation. Moreover, Pakistan has also set the objectives of its economic advancement through Vision 2025.

The most dangerous aspect of nature in the twenty-first century is regarded to be atmospheric changes, which have a variety of effects on the environment and human behavior. Many of these effects, including hurricanes and extremely warm waves, may be life-threatening, whereas spreading of weeds may be less harmful. Regardless of its trivial contribution to the overall emission of GHG, Pakistan may still be influenced by the negative consequences of environmental change.

Excessive ice melting at the glaciers results in dry spells and floods. Extensive rise in the frequency and intensity of unusual climate change coupled with unpredictable storms, recurrent floods, and dry seasons is the major concerns. The projected downturn of Hindu Kush-Karakoram-Himalayan ice sheets on account of perilous atmospheric deviation and carbon residue stores from trans-boundary pollution sources will undermine the natural inflow of water into IRS. Besides, the coastal agriculture, mangroves, and breeding area of aquatic fauna are aggrieved by the entry of aqua saline to Indus delta.

Critical geographical position and socio-economic instability have predisposed Pakistan to the detrimental consequences of climate change (German watch, 2011). Moreover, the deficiency of assets and capabilities for corrective measures will further amplify the circumstances. Atmospheric changes enhance the recurrence and harshness of disastrous incidents, e.g., monsoon rainstorms, swamping, and dry seasons. Terrible dry season of duration of 1999–2003, 2 tornados within thirty days in Karachi/Gwadar coasts in 2008, heavy floods during 2010, 2011, and 2012, land sliding, and GLOFS (chilly lake upheaval floods) in the northern areas represent the obvious manifestations of climate change in Pakistan. Under the 10-year NDMP, the institutional limits of surveillance and forecasting are being improved to combat disasters, by supplanting and introducing the climate observation radars at different regions of the country (Balat and Balat 2010).

6.2.1 *State of Environmental Air*

Rapidly growing human population, increased number of transport vehicles, random infrastructure, and extensive use of low-quality fuels lead to the environmental contamination in Pakistan. The atmospheric data of various cities revealed 2–3.5 times higher concentrations of suspended particulate matter than the threshold limits. The gradually rising expenditure of energy as a result of expanding industrialization, mechanical traffic, and utilization of chemicals tremendously aggravates the urban air quality. Moreover, privatizing the proprietorship of engine vehicles and lack of successfully implementing the vehicle fitness regulations may further augment the air contamination. Besides, the poor fuel consumption efficiency of motorbikes and auto-rickshaws also corresponds to enormous level of noxious gases (Balat and Balat 2010).

6.2.2 *Petro-Fuel Emissions—Health Hazards*

Road transport is promptly developing, particularly in metropolitan areas, and has been recognized as a critical source of air contamination, with consequent ill-effects on human health (Banković-Ilić et al. 2012; Basha et al. 2009; Behçet et al. 2015).

Analyzing the aggregate data of 50 countries and 35 urban areas indicated comparable per capita rise in vehicles and salaries, with faster procurement rate of personal cars than commercial vehicles (Fig. 6.1). More interestingly, the per capita rise in vehicles was twofold higher than salaries in countries like China, Pakistan, and India (Bhale et al. 2009).

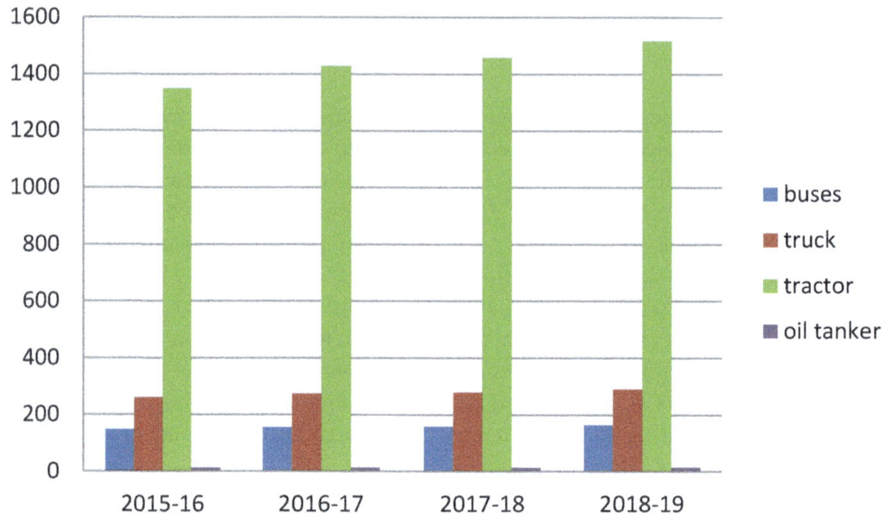

Fig. 6.1 *Source* National Transport Research Centre

Urban transport vehicles are usually fueled by NG, diesel, or gasoline, having different physico-chemical attributes in various regions of the world. Meteorological cosnditions, diverse level of overwhelming polluters (increased number of motorbikes in developing countries), motorways design are regarded as the major confounding factor (Çaynak et al. 2009). Vehicular exhaust emissions are inadequately comprehended and evaluated, particularly in growing nations (Chapagain et al. 2009).

Lead is one of the air contaminants, from human toxicity standpoint. Contrary to industrial regions, the cities of developing countries exhibit relatively higher ambient concentration of fine particulate matter. Diesel-fueled and two-stroke gasoline-powered vehicles are the major sources of fine particulate matter emission. The most noxious air toxins are produced by automobiles. Respiratory problems, pulmonary impairment, pre-term birth, and even mortality are the major consequences of air contamination (Chen et al. 2015). Apart from their diverse toxic effects on human health, these substances are probably carcinogenic (Chhetri et al. 2008) (Table 6.1).

6.3 Potential Feedstock—Biodiesel Achievement in Pakistan

The accessibility and feasibility of feedstock for biodiesel production depend upon the local atmospheric and geological aspects of a country. This wide range of available feedstock can facilitate the advancement of biodiesel industry.

Table 6.1 Health effects of exhaust emissions from automobiles (Cornils et al. 2017; D'Angiola et al. 2010; Demirbas and Biodiesel 2008; Deng et al. 2011)

Exhaust emissions	Detrimental health effects
Carbon monoxide	Besides causing drowsiness, unconsciousness, intra-uterine growth retardation, angina, growth impairment in young children, and death, it also potentiates the toxicity of other pollutants in individuals suffering from respiratory or circulatory disorders
Dioxin	Long-term exposure affects the functions of nervous, endocrine, immune, and reproductive systems
Formaldehyde	Oculo-nasal irritation, coughing, dyspnea, and cancer (as a result of occupational exposure)
Hydrocarbons and other volatile organic compounds	Compounds having low-molecular weight elicit sneezing, coughing, and ocular irritation. High molecular weight compounds may lead to mutagenicity or carcinogenicity
Lead	Inhalation or oral exposure may induce damage to the nervous, circulatory, renal, and reproductive systems; may probably cause hypersensitivity and diminish cognitive ability in children
Nitrogen oxides	Can enhance vulnerability to viral respiratory diseases like influenza and allergic reactions to dust and pollens in asthmatic patients; with typical pathological effects in the form of pneumonia, pulmonary edema, and bronchitis. Most serious health effects are due to their synergistic action with other air pollutants
Ozone	Increased susceptibility to flu and pneumonia; damage to respiratory mucosal membranes associated with coughing and pulmonary dysfunction; exacerbation of asthma, emphysema, bronchitis, and chronic heart disease
PM	Irritation of mucous membranes, pulmonary dysfunction, lung cancer, and death from cardiopulmonary collapse
Polycyclic aromatic hydrocarbons (PAHs)	Lung cancer
Toxic substances	May give rise to birth defects, reproductive problems, and probably cancer. Aldehydes and ketones are ocular irritants, while asbestos and benzene are proven carcinogens

The feedstock for biodiesel generation is selected on the basis of quality, availability, oil contents, physico-chemical properties, and cost. Besides, several biodiesel feedstocks have also been compared in the form of estimated yield and oil contents and have been recorded by several authors (Ali et al. 2012; Atabani et al. 2013c; Falasca et al. 2010; Freedman et al. 1986).

The least expensive one is essential to guarantee low-cost biodiesel production. Generally, the biodiesel feedstocks are categorized into six distinct classes (Gui et al. 2008; Guo et al. 2007; Gurusala et al. 2014).

6.3.1 Edible Vegetable Oils

The utilization of this type of waste results in several issues, like meal versus fuel crisis, significant soil assets destruction, and deforestation. Moreover, the currently raised costs have adversely affected the economic suitability of vegetable oil plants for biodiesel synthesis (Haile 2014; Hajra et al. 2015; Han and Naeher 2006). But their long-term utilization is not feasible in many countries. For example, biodiesel obtained from the entire soybean reserves of USA would hardly meet only 6% of the total diesel requirement (Hoekman and Robbins 2012).

6.3.2 Non-edible Vegetable Oils

Non-edible oils provide potential solutions to diminish the usage of consumable oils for biodiesel generation. Non-edible oil sources are drawing considerable attention owing to their accessibility in numerous areas of the world, particularly the wastelands, lack of competition for food, decreased rate of deforestation, environmental safety, valuable byproducts synthesis, and economic suitability (Hajra et al. 2015; Janaun and Ellis 2010; Kafuku and Mbarawa 2010; Karimi et al. 2015; Karmee and Chadha 2005; Khan et al. 2019; Kondamudi et al. 2009; Kumar and Sharma 2011; Kutzbach 2009; Liaquat et al. 2010; Lim and Teong 2010; Marulanda et al. 2010; Mattingly 2006; Mofijur et al. 2013a, 2013b) (Fig. 6.2).

6.3.3 Waste Cooking Oil

Being an agricultural country, Pakistan depends upon agricultural products for their survival. About 24 kg of edible oil is consumed per capita in Pakistan (Mohammadi et al. 2014). Waste cooking oil is nearly thirty percent effective (Murugesan et al. 2009). Two main suppliers of oil imports to Pakistan are Malaysia and Indonesia (Mohammadi et al. 2014).

Fig. 6.2 Castor oil plant

6.3.4 *Microalgae*

Microalgae are photosynthetic microbes that are more efficient than plants in converting the daylight, water, and CO_2 into algal biomass. In contrast to edible and non-edible sources, microalgae provide relatively greater oil content. When developed in a farm or bioreactor, the oil production of microalgae is nearly 250-folds more as compared to palm oil yields along with that of soybean oil, respectively. Besides, microalgae are also expected to produce sustainable biodiesel. Nevertheless, the development of effective, large-scale bioreactors of microalgae is big constraints in their commercialization. Ongoing studies have shown that algae grown on flue gas can be used for biodiesel generation (Gui et al. 2008; Oanh et al. 2010; Okona-Mensah et al. 2005; Oliveira 2010). Evaluation has been taken for biosafety and subsequently developed as new feedstock for biodiesel production (Omidvarborna et al. 2015).

6.3.5 *Leather Industry Wastes*

Waste products of leather industry are another feedstock for biodiesel generation. A large amount of solid and liquid wastes are not properly utilized (Ong et al. 2011).

6.3.6 Animal Fats

Animal fats of potential significance can be easily collected largely from slaughter houses and meat processing units. Moreover, poultry waste products are preferred than animal fats of other types owing to its low cost, ease of processing, and high availability (Fig. 6.3).

Poultry wastes comprising feathers, blood, offal, and trims can provide a cost-effective feedstock for biodiesel generation. Particularly, chicken slaughterhouses lacking the rendering plants are facing issues of proper waste disposal. Consequently, the waste products can be exploited for the extraction of fats.

Chicken oil obtained from waste skin has been successfully used for biodiesel synthesis in India (Razon 2009). Rendering along with heating yields appreciable amount of oil at comparatively less cost than vegetable feedstocks (Sadhik Basha and Anand 2011). Transesterification of chicken fat at a temperature of 60 °C, 1:30 molar proportion, and 24 h reaction time, resulted in 99.01% yield of biodiesel. Moreover, the physico-chemical properties of chicken fat-derived biodiesel like heating, cetane number, and density were comparable to those of ASTM D 6751 biodiesel standards (Sanjay 2015).

Feather meal also constitutes of 2–12% of chicken fats (Sarma et al. 2005). Chicken fats with 2.3% FFA content are recommended for biodiesel synthesis

Fig. 6.3 Animal fat

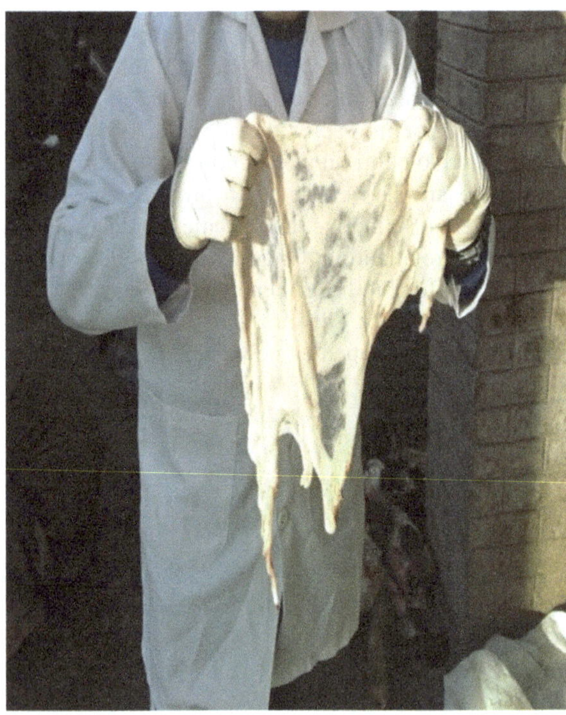

(Schulte 2007). Besides, pre-treatment significantly enhances the yield of biodiesel from chicken fats. The yield of biodiesel reached to about 91%, following the consumption of supercritical methanol (Sharma et al. 2013). Moreover, chicken fats with a high FFA content (13.45%) were successfully used for biodiesel production after the reduction of its FFA level to less than 1% by means of various acid catalysts (Sharma and Singh 2009). Chicken fat is considered suitable for transesterification and biodiesel formation, owing to its lower FFA content and unsaturated fatty acid profile (Sieminski 2014). Diesel blends of fatty acid methyl ester (FAME) of lard, beef tallow, and chicken fat showed lower NOx emanation levels (3.2–6.2%) in comparison with mixture of soybean oil methyl ester and diesel fuel (Sieminski 2014).

Currently, 40% of the total meat requirement is fulfilled by the poultry sector of Pakistan. There are over 25,000 poultry farms, which provide about 1220 million kgs of chicken meat and 10,000 million eggs per annum.

6.4 Biodiesel Production Methods Flowchart

See Fig. 6.4.

6.5 Biodiesel Production Process

The reactor for biodiesel production consists of following components (Fig. 6.5):

- 1L jacketed glass batch reactor,
- reflux condenser (to recover methanol),
- sampling device,
- overhead mechanical stirrer,
- refrigerator,
- circulating water bath for controlling the reaction temperature.

6.5.1 Pre-treatment Process

In this procedure, moisture is removed from crude oil by placing it in a rotatory evaporator for 1 h at 95 °C under vacuum condition.

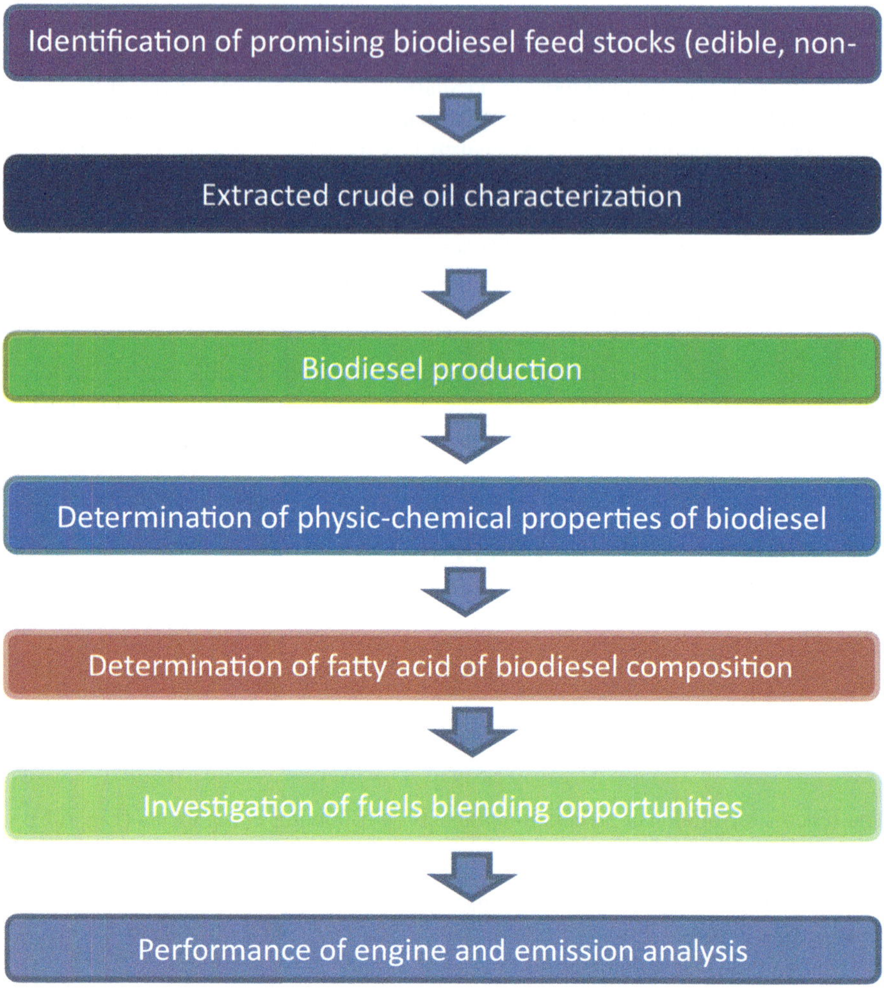

Fig. 6.4 Flowchart of biodiesel production (Arbab et al. 2013; Singh and Singh 2010; Smith et al. 2010; Survey 2014; Usman et al. 2016; Graaf 2012)

6.5.2 *Esterification Process*

The molar ratio 12:1 (50% v/v) of methanol to crude oils with greater values of acid is maintained during this process. Subsequently, 1% (v/v) of sulfuric acid (H_2SO_4) is taken in a glass reactor and added to the pre-heated oils at 60 °C for 3 h along with stirring at 400 rpm. Once the reaction is completed, the excess alcohol is separated from the mixture by separating funnel. The upper layer comprises sulfuric acid and impurities, whereas the lower layer is placed in a rotary evaporator and heated at

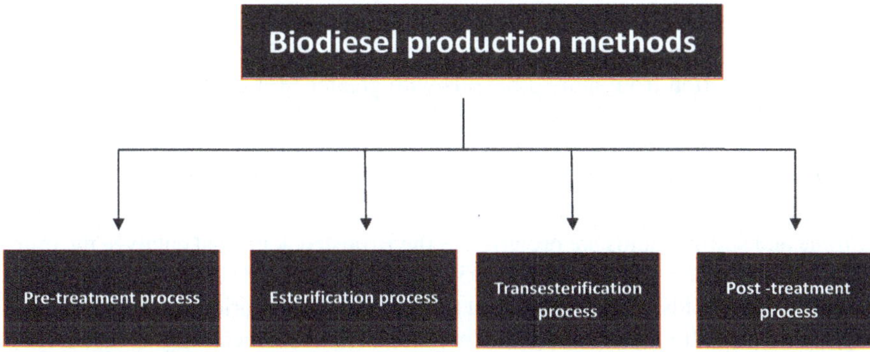

Fig. 6.5 Production of biodiesel can be carried out as follows (Arbab et al. 2013)

95 °C under vacuum conditions for 1 h to remove water and methanol from the esterified oil.

6.5.3 Transesterification Process

Esterified oils and crude oils with low acid values are reacted with 25% (v/v) of methanol and 1% (m/m) of potassium hydroxide (KOH) and kept at 60 °C for 2 h together with stirring at 400 rpm. Upon the completion of reaction, biodiesel is kept in a separating funnel for 12 h to isolate glycerol from biodiesel. The lower layer contains impurities, and glycerol is drawn off.

6.5.3.1 Alkaline-Catalyzed Transesterification

Alkaline-catalyzed transesterification is the most rapid, highly productive, and frequently used procedure for biodiesel production. The abundant methanol is reused and recovered later on. The process of methanol-catalyzed transesterification leads to synthesis of biodiesel and glycerin from free unsaturated fats. The mixing of NaOH with methanol results in the synthesis of sodium methoxide that is subsequently mixed with vegetable oil. The upper layer of the mixture consisting of biodiesel or methyl esters is filtered and washed, whereas glycerin forming the basal layer can be collected and utilized in soap formation.

Moreover, when alkaline catalyst is used, the level of free fatty acid (FFA) declines from a desire limit (running from less than 0.5% to less than 0.3%). Majority of the non-edible oils have greater FFA contents. Therefore, considerable amount of soap is produced by alkaline transesterification method owing to the difficulty in glycerol and ester separation. The typical slow reaction rate of acid-based esterification may necessitate extensive reaction periods to resolve this problem. Consequently, acid-catalyzed transesterification should be followed by alkaline transesterification for biodiesel production from non-edible oils with greater FFA contents.

6.5.3.2 Acid-Catalyzed Transesterification

Sulfonic and sulfuric acids are preferred as the Bronsted acids that catalyze the transesterification process (Wyatt et al. 2005; Yue et al. 2010). These catalysts produce extremely high yields of alkyl esters, but the reactions are laborious and typically take longer than three hours to complete (Zhang et al. 2015). According to Pryde et al., methanolysis of soybean oil at 65 °C with 1 mol% H2SO4 and a 30:1 alcohol/oil molar ratio requires 50 h to complete the conversion of the vegetable oil (> 99%), whereas butanolysis at 117 °C and ethanolysis at 78 °C require 3 h and 18 h, respectively, using the same amounts of catalyst and alcohol (Wyatt et al. 2005).

One of the key elements affecting transesterification is the molar ratio of the alcohol to the vegetable oil. Alcohol in excess promotes the development of the products. However, too much alcohol makes it impossible to recover the glycerol; therefore, the appropriate alcohol to oil ratio must be determined empirically while taking into account each particular process.

6.5.3.3 Enzymatic Transesterification

Enzyme-catalyzed transesterification takes place in the presence of lipase enzyme. Hydrolytic enzymes have been used extensively in organic synthesis because of their wide availability and ease of handling. They are reasonably stable and do not require any coenzymes (Zhou et al. 2010).

These investigations all have one thing in common: they optimize the reaction parameters (solvent, temperature, pH, type of microbe that produces the enzyme, etc.) to provide properties that are appropriate for an industrial use. However, compared to base-catalyzed reaction systems, both the reaction yields and the reaction durations are still unfavorable.

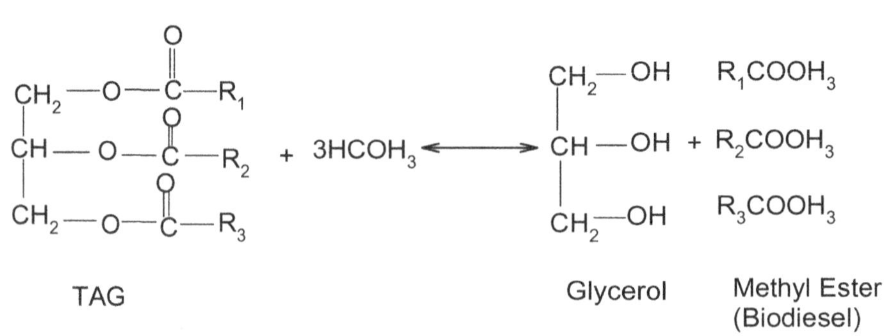

6.5.4 Post-treatment Process

This procedure involves the washing of methyl ester, formed in the upper layer, to eliminate the impurities and glycerol. For this purpose, 50% (v/v) oil of distilled water is showered at 60 °C on the surface of the ester followed by moderate stirring. By repeating the procedure many times, the pH of the distilled water becomes neutral. The lower layer is disposed, and upper layer is dried in a flask containing sodium sulfate Na_2SO_4. Further drying is carried out in a rotatory evaporator for the separation of methanol and water from biodiesel. FTIR spectroscopy has been employed to determine the purity of produced biodiesel and the change of crude oil to methyl ester (Arshad 2017a, b; Arshad et al. 2014, 2019; Bano and Arshad 2018).

6.6 Effects of Additives on the Quality of Biodiesel

Biodiesel emulsion containing 83% jatropha biodiesel, 15% water, and 2% surfactants (Span80 and Tween80) has been manufactured by means of a mechanical agitator (Tahir et al. 2019). The resultant fuel was blended with aluminum nanoparticles in the mass parts of 25, 50, and 100 ppm using an ultrasonicator. Afterward, a steady-speed diesel engine was sequentially fueled with jatropha biodiesel, jatropha biodiesel emulsion, and aluminum nanoparticles-mixed jatropha biodiesel emulsion. The performance of biodiesel emulsion was substantially upgraded, while the harmful emissions were highly reduced than pure biodiesel. In addition, compared to pure biodiesel and biodiesel emulsions, biodiesel emulsions with nanoparticles have high performance and reduced emissions.

Biodiesel acquired from *Madhuca indica* is characterized by low-temperature properties (Sharif et al. 2021). The cold flow characteristics of Madhuca methyl ester fuel are less ideal than those of petroleum-based fuel. The low-temperature properties of biodiesel were determined with and without pour point depressants. Besides, the impact of ethanol, lamp oil, and commercial additives on low-temperature behavior of biodiesel was also evaluated. The pour point of biodiesel was significantly decreased, following the utilization of cold flow improving substances. Moreover, the effect of 2% commercial additive was comparable to blending of 20% ethanol. The cloud point of Madhuca methyl ester was decreased from 291 K (18 °C) to 281 K (8 °C) and 278 K (5 °C) upon mixing with 20% ethanol and 20% lamp fuel, respectively. Likewise, the addition of 20% ethanol and 20% lamp fuel reduced the pour point of Madhuca biodiesel from 280 K (7 °C) to 269 K (−4 °C) and 265 K (−8 °C), respectively. The execution and outflow with ethanol blended Madhuca biodiesel and ethanol–diesel-mixed Madhuca biodiesel have also been examined. The mixing of 20% (by volume) ethanol with Madhuca biodiesel accomplished improved burning with 50% reduction in CO emission without influencing the thermal productivity. Moreover, the emanation of NOx was highly decreased following the addition of 20% (by vol.) ethanol into Madhuca biodiesel. Ethanol-added biodiesel represents a sustainable and feasible alternative fuel, having improved low-temperature behavior and better emanation profile.

Several techniques including the utilization of additives (like traditional oil diesel additives such as wax crystalline modifiers or pour point depressants) and mixing with oil diesel and physico-chemical modification of the oil feedstock or the product of biodiesel have been recommended to enhance the cold flow properties of biodiesel (Arshad et al. 2018). The mixing of oil diesel is effective only at lower concentrations of biodiesel (up to 30% volume). Alternatively, the use of huge moieties for disrupting the organized stacking of ester molecules during nucleation of crystals can also diminish the cloud point of biodiesel. For instance, the cloud point is decreased when large moieties are added to the head-group of alkyl ester or incorporated in the tail-group as a side chain. Fractionation and winterization procedures can be employed for altering the unsaturated fat profile of biodiesel or its feedstock. Particularly, the cold flow properties and oxidation stability of biodiesel can be enhanced through the removal of double bond in the ester group and the addition of side chain. However, this can negatively affect the ignition quality and thickness of biodiesel. Acetone has been successfully used for stabilizing the cold flow properties together with improving the flash point criterion and safety of biodiesel (Arshad and Abbas 2018a).

Organic manganese additives are also known to improve the properties of biodiesel acquired from Pomace oil. It has been documented that doping the fuel at a proportion of 12 mol/l oil methyl ester caused a 20.37% reduction in viscosity, 7 °C (129–122 °C) fall in the flash point, and reduced the pour point from 0 to 15 °C (Arshad and Abbas 2018b).

6.7 Current Challenges of Biodiesel Industry

Concerns regarding the future of biodiesel industry are raised due to the long-term dependence on food-grade vegetable oils (edible oils) as feedstocks for biodiesel synthesis. Besides, it also threatens the supply of edible oils to food industry. Moreover, the gradually increasing prices of feedstocks drastically affect the economic feasibility of biodiesel. Therefore, biodiesel generation from non-edible oils, microalgae, and waste products has become the focus of current research in various parts of the world. Much consideration has been dedicated to biodiesel production through the utilization of easily available, non-customary, and non-palatable feedstock collected from wild plants (Arbab et al. 2013; Atabani et al. 2013c; Janaun and Ellis 2010; Singh and Singh 2010). Effective methodologies targeting the monetary aspects of biodiesel generation are essentially required to fulfill the gradually increasing energy needs (Falasca et al. 2010).

6.8 Conclusion

Reliance upon fossil fuels and global climate changes are the main factors which are responsible for economic issues of Pakistan. In Pakistan, transport sector plays major role in consuming the imported diesel fuel; therefore, largest import bills are due to this sector. To overcome these problems, dependence on sustainable energy resources is an excellent and affordable solution. For the production of biodiesel, animal fat waste is the most promising feedstock because it is sustainable, cost-effective, and easily available. In addition, greenhouse gases emissions are less. It was found that biodiesel produces less emissions in auto vehicles. So, the problems related to air pollution and energy security could be resolved not only by producing biodiesel but also by utilizing it in transportation sector. The government of Pakistan is importing too much diesel and oil in order to fulfill the requirements of people. Therefore, it is the need of the day to produce biodiesel from different renewable sources especially animal fat. Production of biodiesel in Pakistan can be promoted by improving production technology of biodiesel or by making governmental policies to introduce biodiesel in transportation sector as a green fuel.

References

Ahmad A, Yasin NM, Derek C, Lim J (2011) Microalgae as a sustainable energy source for biodiesel production: a review. Renew Sustain Energy Rev 15(1):584–593

Ali A, Ahmad F, Farhan M, Ahmad M (2012) Biodiesel production from residual animal fat using various catalysts. Pak J Sci 64(4):282

Alptekin E, Canakci M (2010) Optimization of pretreatment reaction for methyl ester production from chicken fat. Fuel 89(12):4035–4039

Arbab M, Masjuki H, Varman M, Kalam M, Imtenan S, Sajjad H (2013) Experimental investigation of optimum blend ratio of jatropha, palm and coconut based biodiesel to improve fuel properties, engine performance and emission characteristics (0148–7191)

Arshad M (2017a) Clean and sustainable energy technologies. In: Clean energy for sustainable development. Academic Press, pp 73–89

Arshad M (ed) (2017b) Perspectives on water usage for biofuels production: aquatic contamination and climate change. Springer

Arshad M, Abbas M (2018a) Water sustainability issues in biofuel production. In: Perspectives on water usage for biofuels production: aquatic contamination and climate change, pp 55–76

Arshad M, Abbas M (2018b) Future biofuel production and water usage. In: Perspectives on water usage for biofuels production: aquatic contamination and climate change, pp 107–121

Arshad M, Adil M, Sikandar A, Hussain T (2014) Exploitation of meat industry by-products for biodiesel production: Pakistan's perspective. Pak J Life Soc Sci. 2014a; 12:120–125

Arshad M, Bano I, Younus M, Khan A, Rahman A (2018) Health concerns associated with biofuel production. In: Perspectives on water usage for biofuels production: aquatic contamination and climate change, pp 97–105

Arshad M, Hussain T, Chaudhry N, Sadia H, Aslam B, Tahir U, Abbas M, Qureshi N, Nazir A, Rajoka MI, Iqba M (2019) Enhancing profitability of ethanol fermentation through gamma ray mutagenesis of *Saccharomyces cerevisiae*. Pol J Environ Stud 28(1)

Atabani A, Badruddin IA, Mahlia T, Masjuki H, Mofijur M, Lee KT, Chong W (2013a) Fuel properties of Croton megalocarpus, Calophyllum inophyllum, and Cocos nucifera (coconut) methyl esters and their performance in a multicylinder diesel engine. Energ Technol 1(11):685–694

Atabani A, Badruddin IA, Masjuki HH, Chong W, Lee KT (2015) Pangium edule Reinw: a promising non-edible oil feedstock for biodiesel production. Arab J Sci Eng 40(2):583–594

Atabani A, Badruddin IA, Mekhilef S, Silitonga A (2011) A review on global fuel economy standards, labels and technologies in the transportation sector. Renew Sustain Energy Rev 15(9):4586–4610

Atabani A, da Silva César A (2014) Calophyllum inophyllum L.-A prospective non-edible biodiesel feedstock. Study of biodiesel production, properties, fatty acid composition, blending and engine performance. Renew Sustain Energy Rev 37:644–655

Atabani A, El-Sheekh M, Kumar G, Shobana S (2017) Edible and nonedible biodiesel feedstocks: microalgae and future of biodiesel. In: Clean energy for sustainable development. Elsevier, pp 507–556

Atabani A, Mahlia T, Masjuki H, Badruddin IA, Yussof HW, Chong W, Lee KT (2013b) A comparative evaluation of physical and chemical properties of biodiesel synthesized from edible and non-edible oils and study on the effect of biodiesel blending. Energy 58:296–304

Atabani A, Mofijur M, Masjuki H, Badruddin IA, Chong W, Cheng S, Gouk S (2014a) A study of production and characterization of Manketti (Ricinodendron rautonemii) methyl ester and its blends as a potential biodiesel feedstock. Biofuel Research Journal 1(4):139

Atabani A, Mofijur M, Masjuki H, Badruddin IA, Kalam M, Chong W (2014b) Effect of Croton megalocarpus, Calophyllum inophyllum, Moringa oleifera, palm and coconut biodiesel–diesel blending on their physico-chemical properties. Ind Crops Prod 60:130–137

Atabani A, Silitonga A, Ong H, Mahlia T, Masjuki H, Badruddin IA, Fayaz H (2013c) Non-edible vegetable oils: a critical evaluation of oil extraction, fatty acid compositions, biodiesel production, characteristics, engine performance and emissions production. Renew Sustain Energy Rev 18:211–245

Balat M (2011) Potential alternatives to edible oils for biodiesel production—a review of current work. Energy Convers Manage 52(2):1479–1492

Balat M, Balat H (2009) Recent trends in global production and utilization of bio-ethanol fuel. Appl Energy 86(11):2273–2282

Balat M, Balat H (2010) Progress in biodiesel processing. Appl Energy 87(6):1815–1835

Bano I, Arshad M (2018) Climatic changes impact on water availability. In: Perspectives on water usage for biofuels production. Springer, Cham, pp 39–54

Banković-Ilić IB, Stamenković OS, Veljković VB (2012) Biodiesel production from non-edible plant oils. Renew Sustain Energy Rev 16(6):3621–3647

Basha SA, Gopal KR, Jebaraj S (2009) A review on biodiesel production, combustion, emissions and performance. Renew Sustain Energy Rev 13(6–7):1628–1634

Behçet R, Oktay H, Çakmak A, Aydin H (2015) Comparison of exhaust emissions of biodiesel–diesel fuel blends produced from animal fats. Renew Sustain Energy Rev 46:157–165

Bhale PV, Deshpande NV, Thombre SB (2009) Improving the low temperature properties of biodiesel fuel. Renew Energy 34(3):794–800

Çaynak S, Gürü M, Biçer A, Keskin A, Içingür Y (2009) Biodiesel production from pomace oil and improvement of its properties with synthetic manganese additive. Fuel 88(3):534–538

Chapagain BP, Yehoshua Y, Wiesman Z (2009) Desert date (Balanites aegyptiaca) as an arid lands sustainable bioresource for biodiesel. Biores Technol 100(3):1221–1226

Chen W, Ma L, Zhou PP, Zhu YM, Wang XP, Luo XA, Yu LJ (2015) A novel feedstock for biodiesel production: the application of palmitic acid from Schizochytrium. Energy 86:128–138

Chhetri AB, Tango MS, Budge SM, Watts KC, Islam MR (2008) Non-edible plant oils as new sources for biodiesel production. Int J Mol Sci 9(2):169–180

Cornils B, Herrmann WA, Beller M, Paciello R (2017) Applied homogeneous catalysis with organometallic compounds: a comprehensive handbook in four volumes, vol 4. John Wiley & Sons

D'Angiola A, Dawidowski LE, Gómez DR, Osses M (2010) On-road traffic emissions in a megacity. Atmos Environ 44(4):483–493
Demirbas A, Biodiesel A (2008) A realistic fuel alternative for diesel engines. Springer, London. ISBN: 10, 1846289947
Deng X, Fang Z, Liu Y-H, Yu C-L (2011) Production of biodiesel from Jatropha oil catalyzed by nanosized solid basic catalyst. Energy 36(2):777–784
Doll C, Wietschel M (2008) Externalities of the transport sector and the role of hydrogen in a sustainable transport vision. Energy Policy 36(11):4069–4078
Falasca SL, Flores N, Lamas M, Carballo SM, Anschau A (2010). Crambe abyssinica: an almost unknown crop with a promissory future to produce biodiesel in Argentina. Int J Hydrog Energy 35(11):5808–5812
Freedman B, Butterfield RO, Pryde EH (1986) Transesterification kinetics of soybean oil 1. J Am Oil Chem Soc 63(10):1375–1380
Gui MM, Lee K, Bhatia S (2008) Feasibility of edible oil versus non-edible oil versus waste edible oil as biodiesel feedstock. Energy 33(11):1646–1653
Guo H, Zhang Q, Shi Y, Wang D (2007) On-road remote sensing measurements and fuel-based motor vehicle emission inventory in Hangzhou. China Atmos Environ 41(14):3095–3107
Gurusala NK, Zachariah R, Arul MSV (2014) Effect of EGR on combustion and emissions characteristics of a CI engine fuelled with waste chicken fat biodiesel. Paper presented at the Applied Mechanics and Materials
Haile M (2014) Integrated volarization of spent coffee grounds to biofuels. Biofuel Res J 1(2):65
Hajra B, Sultana N, Pathak AK, Guria C (2015) Response surface method and genetic algorithm assisted optimal synthesis of biodiesel from high free fatty acid sal oil (Shorea robusta) using ion-exchange resin at high temperature. J Environ Chem Eng 3(4):2378–2392
Han X, Naeher LP (2006) A review of traffic-related air pollution exposure assessment studies in the developing world. Environ Int 32(1):106–120
Hoekman SK, Robbins C (2012) Review of the effects of biodiesel on NOx emissions. Fuel Process Technol 96:237–249
Janaun J, Ellis N (2010) Perspectives on biodiesel as a sustainable fuel. Renew Sustain Energy Rev 14(4):1312–1320
Kafuku G, Mbarawa M (2010) Biodiesel production from Croton megalocarpus oil and its process optimization. Fuel 89(9):2556–2560
Karimi K, Tabatabaei M, Sárvári Horváth I, Kumar R (2015) Recent trends in acetone, butanol, and ethanol (ABE) production. Biofuel Res J 2(4):301–308
Karmee SK, Chadha A (2005) Preparation of biodiesel from crude oil of Pongamia pinnata. Biores Technol 96(13):1425–1429
Khan HM, Ali CH, Iqbal T, Yasin S, Sulaiman M, Mahmood H, Mu B (2019) Current scenario and potential of biodiesel production from waste cooking oil in Pakistan: An overview. Chin J Chem Eng 27(10):2238–2250
Kondamudi N, Strull J, Misra M, Mohapatra SK (2009) A green process for producing biodiesel from feather meal. J Agric Food Chem 57(14):6163–6166
Kumar A, Sharma S (2011) Potential non-edible oil resources as biodiesel feedstock: an Indian perspective. Renew Sustain Energy Rev 15(4):1791–1800
Kutzbach MJ (2009) Motorization in developing countries: causes, consequences, and effectiveness of policy options. J Urban Econ 65(2):154–166
Liaquat A, Kalam M, Masjuki H, Jayed M (2010) Potential emissions reduction in road transport sector using biofuel in developing countries. Atmos Environ 44(32):3869–3877
Lim S, Teong LK (2010) Recent trends, opportunities and challenges of biodiesel in Malaysia: an overview. Renew Sustain Energy Rev 14(3):938–954
Marulanda VF, Anitescu G, Tavlarides LL (2010) Investigations on supercritical transesterification of chicken fat for biodiesel production from low-cost lipid feedstocks. J Supercritical Fluids 54(1):53–60

Mattingly BG (2006) Production of biodiesel from chicken fat containing free fatty acids. Paper presented at the Masters Abstracts International
Mofijur M, Atabani A, Masjuki HA, Kalam M, Masum B (2013a) A study on the effects of promising edible and non-edible biodiesel feedstocks on engine performance and emissions production: a comparative evaluation. Renew Sustain Energy Rev 23:391–404
Mofijur M, Masjuki H, Kalam M, Atabani A (2013b) Evaluation of biodiesel blending, engine performance and emissions characteristics of Jatropha curcas methyl ester: Malaysian perspective. Energy 55:879–887
Mohammadi P, Tabatabaei M, Nikbakht AM, Esmaeili Z (2014) Improvement of the cold flow characteristics of biodiesel containing dissolved polymer wastes using acetone. Biofuel Res J 1(1):26–29
Murugesan A, Umarani C, Chinnusamy T, Krishnan M, Subramanian R, Neduzchezhain N (2009) Production and analysis of bio-diesel from non-edible oils—a review. Renew Sustain Energy Rev 13(4):825–834
Oanh NTK, Thiansathit W, Bond TC, Subramanian R, Winijkul E, Paw-armart I (2010) Compositional characterization of PM2. 5 emitted from in-use diesel vehicles. Atmos Environ 44(1):15–22
Okona-Mensah K, Battershill J, Boobis A, Fielder R (2005) An approach to investigating the importance of high potency polycyclic aromatic hydrocarbons (PAHs) in the induction of lung cancer by air pollution. Food Chem Toxicol 43(7):1103–1116
Oliveira EVAD (2010) Síntese de biodiesel a partir da transesterificação do óleo de soja por catálises homogênea e heterogênea
Omidvarborna H, Kumar A, Kim D-S (2015) NOx emissions from low-temperature combustion of biodiesel made of various feedstocks and blends. Fuel Process Technol 140:113–118
Ong H, Mahlia T, Masjuki H, Norhasyima R (2011) Comparison of palm oil, Jatropha curcas and Calophyllum inophyllum for biodiesel: a review. Renew Sustain Energy Rev 15(8):3501–3515
Onursal B, Gautam S (1997) Vehicular air pollution: experiences from seven Latin American urban centers, vol 373. World Bank Publications
Organization WH (2004) Health aspects of air pollution: results from the WHO project. Systematic review of health aspects of air pollution in Europe. Pakistan, e. s. o. (2018–19) (transport and communication)
Ramadhas A, Jayaraj S, Muraleedharan C (2004) Use of vegetable oils as IC engine fuels—a review. Renew Energy 29(5):727–742
Rao MS, Anand R (2015) Production characterization and working characteristics in DICI engine of Pongamia biodiesel. Ecotoxicol Environ Saf 121:16–21
Ravindran B, Sekaran G (2010) Bacterial composting of animal fleshing generated from tannery industries. Waste Manage 30(12):2622–2630
Razon LF (2009) Alternative crops for biodiesel feedstock. CAB Rev Perspect Agric Vet Sci Nutr Natural Resour 4(56):1–15
Sadhik Basha J, Anand R (2011) Role of nanoadditive blended biodiesel emulsion fuel on the working characteristics of a diesel engine. J Renew Sustain Energy 3(2):023106
Salaheldeen M, Aroua MK, Mariod A, Cheng SF, Abdelrahman MA, Atabani A (2015) Physicochemical characterization and thermal behavior of biodiesel and biodiesel–diesel blends derived from crude Moringa peregrina seed oil. Energy Convers Manage 92:535–542
Sanjay B (2015) Yellow oleander (Thevetia peruviana) seed oil biodiesel as an alternative and renewable fuel for diesel engines: a review. Int J ChemTech Res 7(6):2823–2840
Saravanan S, Nagarajan G, Rao GLN (2010) Investigation on nonedible vegetable oilas a compression ignition engine fuel in sustaining the energy and environment. J Renew Sustain Energy 2(1):013108
Sarin R, Sharma M, Khan AA (2009) Studies on Guizotia abyssinica L. oil: biodiesel synthesis and process optimization. Bioresour Technol 100(18):4187–4192
Sarma AK, Konwer D, Bordoloi P (2005) A comprehensive analysis of fuel properties of biodiesel from Koroch seed oil. Energy Fuels 19(2):656–657

Schulte WB (2007) Biodiesel production from tall oil and chicken fat via supercritical methanol treatment. University of Arkansas

Sharma H, Giriprasad R, Goswami M (2013) Animal fat-processing and its quality control. J Food Process Technol 4(8):252

Sharma Y, Singh B (2009) Development of biodiesel: current scenario. Renew Sustain Energy Rev 13(6–7):1646–1651

Shirazi A, Bazgir MJS, Shirazi AMM (2014). Edible oil mill effluent; a low-cost source for economizing biodiesel production: Electrospun nanofibrous coalescing filtration approach. *Biofuel Research Journal, 1*(1), 39-42

Sharif N, Munir N, Hasnain M, Naz S, Arshad M (2021) Environmental impacts of ethanol production system. In: Sustainable ethanol and climate change. Springer, Cham, pp 205–223

Sieminski A (2014) International energy outlook. Energy Inform Adm (EIA), 18:1–24

Singh S, Singh D (2010) Biodiesel production through the use of different sources and characterization of oils and their esters as the substitute of diesel: a review. Renew Sustain Energy Rev 14(1):200–216

Smith PC, Ngothai Y, Nguyen QD, O'Neill BK (2010) Improving the low-temperature properties of biodiesel: methods and consequences. Renew Energy 35(6):1145–1151

Survey PE (2014–15) Environment. Retrieved from http://www.finance.gov.pk/survey/chapters_15/16_Environment.pdf

Tahir A, Arshad M, Anum F, Abbas M, Javed S, Shahzad MI, Ansari AR, Bano I, Shah FA (2019) Ecofuel future prospect and community impact. In: Advances in eco-fuels for a sustainable environment. Woodhead Publishing, pp 459–479

Usman A, Itodo AU, Usman NL, Haruna M (2016) Fatty acid methyl esters composition of Trichilia emetica shell oil. Am Acad Sci Res J Eng Technol Sci 21(1):83–90

Van de Graaf T (2012) Obsolete or resurgent? The international energy agency in a changing global landscape. Energy Policy 48:233–241

Wyatt VT, Hess MA, Dunn RO, Foglia TA, Haas MJ, Marmer WN (2005) Fuel properties and nitrogen oxide emission levels of biodiesel produced from animal fats. J Am Oil Chem Soc 82(8):585–591

Yue J-F, Zuo C-L, Wang Q (2010) Research progress of biodiesel preparation from waste oil. Guangzhou Chem Ind 12(84–85):130

Zhang H, Zhou Q, Chang F, Pan H, Liu X-F, Li H, Yang S (2015) Production and fuel properties of biodiesel from Firmiana platanifolia Lf as a potential non-food oil source. Ind Crops Prod 76:768–771

Zhou Y, Wu Y, Yang L, Fu L, He K, Wang S, Li C (2010) The impact of transportation control measures on emission reductions during the 2008 Olympic Games in Beijing. China Atmos Environ 44(3):285–293

Chapter 7
Biogas from Manure: The Future of Renewable Natural Gas and Its Implications

Charles O. Nwuche, Shruti Gupta, Joseph Akor, Julius Eyiuche Nweze, Justus Amuche Nweze, and Victor U. Unah

Abstract The world's attention to biogas technology has continued to increase as people become more concerned about the environment, especially with the growing problem of waste disposal. Biogas is an ecofriendly energy source and a methane-rich gas produced by anaerobic digestion of organic waste (agricultural, sewage, and landfill) in a bio-digester. Traditionally, anaerobic digestion has been utilized to treat organic waste and digests have also been employed in agriculture as fertilizer. Various studies have enhanced our understanding of the microbial communities' complexity and metabolic pathways involved in the biotechnology of the microbiological process leading to biogas production. Despite its various benefits, biogas technology's potential depends on some criteria, including feedstock type, reactor design, and operation parameters. It has many advantages, including those that translate to improved agricultural profitability and environmental stewardship. The technological and market developments will play a role in determining the future of such renewable natural gas, but its success is also contingent on energy policy fairness in different countries. Biogas production and use can help mitigate many of the impacts of climate change by reducing methane and black carbon emissions, producing renewable energy (cleaner fuels for vehicles, cooking, heating, and electricity), capturing organic wastes, and reintroducing nutrients and organic matter to the soil.

C. O. Nwuche · V. U. Unah
Department of Microbiology, Faculty of Biological Sciences, University of Nigeria, Nsukka, Nigeria

S. Gupta · J. E. Nweze · J. A. Nweze (✉)
Institute of Soil Biology and Biogeochemistry, Biology Centre, Czech Academy of Sciences, Ceske Budejovice, Czech Republic
e-mail: justus.nweze@unn.edu.ng

J. Akor · J. A. Nweze
Department of Science Laboratory Technology, Faculty of Physical Sciences, University of Nigeria, Nsukka, Nigeria

J. E. Nweze · J. A. Nweze
Department of Ecosystem Biology, Faculty of Science, University of South Bohemia in Ceske Budejovice, Ceske Budejovice, Czech Republic

M. Arshad (ed.), *Climate Changes Mitigation and Sustainable Bioenergy Harvest Through Animal Waste*, https://doi.org/10.1007/978-3-031-26224-1_7

Graphical Abstract

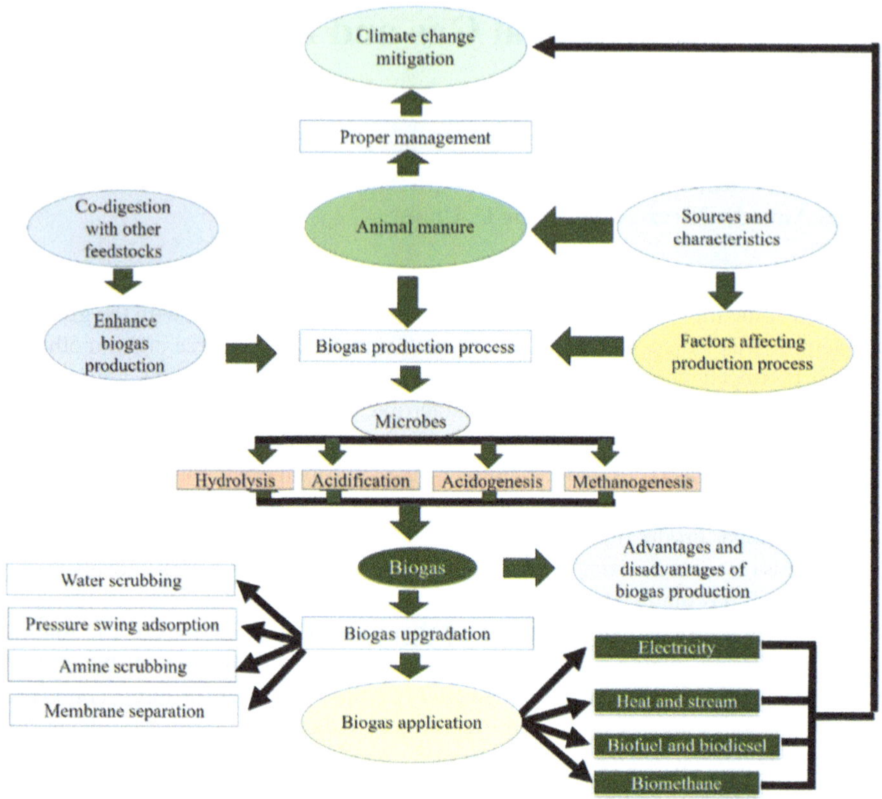

7.1 Introduction

Biogas, unlike fossil fuels, is a renewable energy source because it is produced from biomass and consists entirely of biogenic materials. The main component of this naturally occurring biogas is methane, which when used widely will reduce the burning of fossil fuel and reduce global warming. A large international market for biogas has been promoted in many countries for decades. Biogas is widely used to generate heat, energy, and electricity and has several industrial applications. By the end of 2019, the global capacity of biogas plants is estimated to be 120 GW. A look at international research development over the last decade shows that the biogas industry has grown by 90%, with Europe accounting for over 70% of global biogas production (64 TWh). Organic waste, such as animal manure, is the most popular feedstock for biogas production (Abanades et al. 2021a, b; Damyanova and Beschkov 2020; Arshad et al. 2017).

Animal manures are liquid, semisolid, and solid products from animals raised to produce meat, milk, eggs, and other agricultural products for human consumption. Manure has been used to improve soil since the beginning of civilization, even before chemical fertilizers appeared in the 1940s. Manure is still a valued agricultural resource because it is an essential source of plant nutrients and is known to improve the physical and biological properties of the soil by adding organic matter. However, recent developments in animal production and manure management practices in several countries have raised concerns about the impact on human health and environment (Loyon 2018; Sawyerr et al. 2019).

The production of biogas from manure reduces greenhouse gas emissions (nitrous oxide and methane) while producing a renewable fuel that can replace fossil fuels. For livestock farmers, manure can be a viable alternative energy source. An anaerobic digester converts manure into energy in the form of methane-rich biogas. Biogas is produced when organic materials decompose in a heated, airless environment. This process occurs naturally; mash gas and biogas are nearly identical. Biogas production can be accelerated by enclosing organic waste in a digester, a heated, airtight container without oxygen (Loyon 2018; Paudel et al. 2017).

During anaerobic digestion (AD), a consortium of microbes converts substrates into biogas (40–60% methane) which is used as renewable energy form while the digestate will be applied as a soil conditioner or fertilizer. The anaerobic conversion of organic matter into biogas is a multi-step process in which many different species of bacteria and archaea interact with each other. With the increasing use of AD, engineers and microbiologists are working hard to improve their understanding of the complex microbial interactions that drives the overall process of AD. More technological advances and knowledge is essential for developing better models and designing better AD systems. Despite of apparent benefits of investing in biogas production from animal waste, the technology also has some drawbacks, which is featured in the later part of this article (Abanades et al. 2021a, b; Kadam and Panwar 2017).

The present chapter discusses biogas production from animal manure; the processes involved, microbial activities, factors affecting the production process, the use of biogas, recent developments in production processes, and the advantages and disadvantages of this technology.

7.2 Feedstocks Used for Biogas Production

7.2.1 Sources of Animal Manure

Manure is an unavoidable by-product of poultry and livestock production, consisting primarily of animal faces and urine but also includes bedding, additional water, and unused feed. Cow, pig, chicken, horse, goat, sheep, elephant droppings, and fishery residues are all sources of animal manure. Some animal farm wastes are not suitable substrates with enhanced potential because of their low energy content, so co-digestion with two or more substrates is more beneficial. Also, the availability of some feedstocks might pose a serious challenge. Therefore, it is advocated that organic residues be mixed in order to gain sufficient feedstock loading which could guarantee increase in the yield of biogas during AD (Janni and Cortus 2020; Sommer and Christensen 2013).

7.2.2 Factors Influencing the Amount of Manure Produced

Among several factors, the type of animal (non-ruminant or ruminant), age (which may affect the amount of feed consumed), housing (slatted floor or bedding), feed (grain-based or forage-based), manure management (drying belt, scrapping, flushing, storage pit, etc.), and productivity affect the amount of manure produced. On average, a total of 40 billion tons of manure are produced annually worldwide. Cattle (bulls, cows, steers, heifers, and calves) produce the most manure, followed by pigs (weaners, sows, boars, and market pigs). In contrast, poultry produce the least manure of all livestock species studied, with each animal producing less than 1 kg of manure per day (Otte et al. 2019; Janni and Cortus 2020). Recebli et al. (2015) found that manure from 70 cattle and 1400 chickens produced 25 and 0.036 kg/day, respectively. However, the study eventually found that anaerobic fermentation of 175 kg of cattle manure with 175 kg of water and 50 kg of poultry manure with 325 kg of water produced 6.33 and 0.83 m^3/day of biogas, respectively (Recebli et al. 2015).

7.2.3 Characteristics and Composition of Animal Manure

Animal manure has a variety of characteristics that are mainly determined by solid contents. Depending on the manure's total solids or dry matter content, it is usually classified as liquid, slurry, semisolid, or solid. Feed intake (which affects excreted nutrients), manure storage and management (which affects nutrient yield), and water addition/evaporation (which affects nutrient concentration) all impact the nutrient content of manure (Pagliari et al. 2020; Janni and Cortus 2020).

The chemical composition of the feed supplied to the animal determines the chemical properties of the manure. The feed is metabolized by the animal to provide energy and produce new body tissues and other products. Most of the metabolic waste is collected in the urine and excreted with the faces (which may contain unused feed). The microbial population and activity during the breakdown of manure are determined by the nutrient content of the manure (Huang et al. 2017; Loyon 2018).

7.2.4 Biogas Digester Efficacy and Manure Feedstocks

Manure feedstocks generally contain macro- and micronutrients that promote anaerobic digestion and volatile particles that are broken down to produce biogas. The biogas yields of dairy manure are lower than those of other digestible organic feedstocks. The lower relative potential for biogas production in livestock manure is for some reasons. The most notable reason is the stomach of the animals, which serves as an early digester for the conversion of feed to energy. Although biogas production is lower than other feedstocks, manure has some advantages. It is more suitable for energy conversion on farms, as the disposal of pig and dairy waste is often in liquid form or as slurry, simplifying the necessary transport of slurry. Dairy manure also contains anaerobic bacteria, which increases the number of methane-producing bacteria in the digester and reduces the likelihood of malfunctions (Piekutin et al. 2021; Environmental Protection Agency 2021). Large amounts of litter or other elements such as stones, rocks, and sand should be avoided in the slurry. Small amounts of the fine solids in suspension can affect the reactor capacity and ability to produce biogas and can damage the internal parts of the reactor during agitation. Therefore, the following parameters must be observed for biogas digesters using manure as feedstock to operate effectively. (1) The digester should be constructed so that it can absorb a large portion of the slurry quickly after it is discharged. (2) To avoid losing the capacity of the manure to produce biogas through natural decomposition and to ensure a steady flow of manure/feed into the digester, the manure must be collected regularly. (3) Foreign materials, such as soil and excessive litter, should be kept away from the slurry as they clog the digester and reduce performance. (4) To choose the best digester technology, the total solids content of the manure must be known (Environmental Protection Agency 2021).

7.3 Biogas Production Processes

7.3.1 Anaerobic Fermentation: A Biochemical Process

Anaerobic digestion is a complex biochemical process that has been long exploited in waste management by processing organic wastes into biogas. It provides two essential benefits: waste management and renewable energy (Molino et al. 2018). Anaerobic digestion is an anoxic microbial process in which organic wastes are converted (broken down) to carbon dioxide (CO_2) and methane (CH_4) gases in a sequence of ordered biological and chemical activities. The process is categorized into separate stages involving a diverse community of microorganisms. The most readily degradable organic fractions are first converted to gases, while the rest are spent building biomass through microbial species' growth and development (Kougias and Angelidaki 2018). Anaerobic digestion occurs in four distinct phases, i.e., hydrolysis, acidogenesis, acetogenesis, and methanogenesis (Themelis and Ulloa 2007). Each of the different stages of digestion (or fermentation) is usually undertaken by a group of metabolically dissimilar, physiologically diverse, but syntrophically interrelated, groups of microorganisms often referred to as consortia of microorganisms. The activities of these large communities of microorganisms inevitably generate some amounts of heat. The thermophilic conditions that prevail promote biogas production. The digestate liquor remaining in the digesters after fermentation can be used as fertilizers and soil conditioners due to their richness in organic nutrients. Biogas is composed chiefly of methane and carbon dioxide, but other gases such as ammonia, hydrogen, and hydrogen sulfide are present in minute amounts as contaminants.

Anaerobic digestion technology is simple and finds ready applications in sustainable energy recovery at home and farm settings where wastes generated from municipal and agricultural sources are substantial. Besides the provision of green energy, the technology is useful in pollution control and environmental protection because it creates avenues for the utilization and conversion of abundant biomass produced yearly from forestry, agriculture, animal husbandry, and other agro-industrial activities (Yang and Li 2014; Ighravwe and Babatunde 2018). The appropriation of anaerobic digestion along the lines of clean energy, organic fertilization, and waste management could contribute to a cleaner and healthier environment, advance crop production, and animal husbandry, thereby lifting people's standard of living, particularly in rural communities. The four major stages of anaerobic fermentation are hydrolysis, acidogenesis, acetogenesis, and methanogenesis, which are further discussed below.

a. **Hydrolysis**

Hydrolysis is the foremost step in anaerobic digestion. The process entails the breakdown of complex organic polymers such as carbohydrates, proteins, and lipids into monomers (Kumar and Sharma 2017). During hydrolysis, the resident microflora produces copious quantities of hydrolytic enzymes, which enables them to break down complex organic polymers (present in the original chemistry of the feedstock) into simple, digestible units (or monomers) such as sugars, amino acids, and long-chain fatty acids. Some enzymes released at this phase of digestion include amylase, protease, lipase, cellulase, and xylanase (Raja and Wazir 2017). After hydrolysis, the products in their soluble forms become available for cellular transport and undergo absorption and degradation by both facultative and obligate anaerobes to produce short-chain volatile fatty acids (VFA), which constitute the base materials as the fermentation advances into the secondary (acidogenic) and tertiary (acetogenic) phases. The fatty acids react with alcohol to form acetate, hydrogen, and carbon dioxide (Chandra et al. 2012). The long-chain fatty acids formed are converted into hydrogen, acetate, and other volatile fatty acids (i.e., propionic and butyric acids) by the acetogenic bacteria. The microorganisms involved at this stage of breakdown are composed mainly of strict anaerobes such as species of *Clostridia*, *Bacteroides* and *Bifidobacteria* (Raja and Wazir 2017). They may also include facultative anaerobes, e.g., *Enterobacteriaceae* and *Streptococci*. The kinetics of the hydrolytic process is driven by several parameters such as pH, particle sizes, concentration of enzymes, diffusion gradient, and rate of adsorption of enzymes by the feedstock (Sawyerr et al. 2019).

b. **Acidification (Acidogenesis)**

During acidogenesis, the products of the hydrolytic stage are converted by the fermenting microbes into organic acids such as acetic acid, hydrogen, carbon dioxide, and other secondary products (i.e., propionic, butyric, and lactic acids and alcohols) (Huang et al. 2015). This stage is also characterized by the production of high amount of hydrogen by the acidogenic bacteria as a result of electrons accumulated by the secondary metabolic products and volatile fatty acids. To be able to control this buildup, it has been reported that the levels of hydrogen within the digesters need to be adjusted to low levels to enable the acidogenic reactions to be thermodynamically successful (Sawyerr et al. 2019). Biogas production also depends on the interrelationships between the two dominant microbial groups within the digester—the acetogens and the methanogens. The maintenance of mutual interaction between the acidogenic and methanogenic groups is key to the successful performance of an anaerobic digesters (Wareham et al. 2014) because the more acid is generated during the acidogenic phases, the more the methanogens develop to produce methane gas. Ammonia and hydrogen sulfide are among the products of this phase (Ntaikou et al. 2010).

c. **Symbiosis of Bacteria (Acetogenesis)**

The main highlight of the acetogenic stage is the conversion of alcohols and volatile fatty acids (produced in the acidogenic stage) into acetate by the acetate-forming bacteria and the release of hydrogen and carbon dioxide (Parawira et al. 2008; Kumar and Sharma 2017). Acetogenic bacteria (such as those of the genera *Syntrophobacter* and *Syntrophomonas*) convert products of the acidogenic phase into acetates and hydrogen which may be utilized by the methanogenic bacteria in a later phase. It has been reported also that bacteria such as *Methanobacterium propionicum* break down propionic acid to acetic acid while the activities of *Methanobacterium suboxydans* bring about the formation of propionic acid from pentanoic acid (de Bok et al. 2005). Hydrogen and carbon dioxide produced also undergo chemical reduction to acetate by homo-acetogenic microorganisms, to forestall hydrogen accumulation which could endanger the activities of the acetogenic bacteria (Weiland 2009; Chandra et al. 2012). High hydrogen concentrations have been reported to bring about loss of ability to produce acetate among the acetogens. Therefore, maintaining a low hydrogen partial pressure is important for sustaining the acetogenesis stage because a quiet hydrogen environment drives the acetogenic reaction forward. In order to accomplish this, a symbiotic relationship between acetogens and hydrogenotrophic methanogens must be allowed to flourish. This will allow the methanogens to use the acetate produced by the acetogens as substrates against the next and final stage of anaerobic fermentation, known as methanogenesis (Nges 2012). Acetogenesis determines the success or otherwise of anaerobic fermentation because more than 70% of activities at this stage result in the microbial production of acetates. Acetates are therefore crucial extra intermediates of the anaerobic process.

d. **Methanogenesis**

Methanogenesis is the last phase of anaerobic digestion. It is the phase with the greatest impact on the overall digestion process because it is the major determinant of the state of "health" of the fermentation and the degree of microbial activities occurring within the digester (Vrieze et al. 2012). If the system is stable and running well, it creates the setting for increase in the production of methane. In fact, more than 70% of methane generated during the course of anaerobic fermentation comes from this stage (Raja and Wazir 2017; Sawyerr et al. 2019). Among the signature reactions of this phase is the conversion of hydrogen and carbon dioxide to methane and water by a class of hydrogen oxidizing and carbon dioxide-reducing methanogens, respectively. However, the dominant biological activity is the utilization of acetates by the acetoclastic methanogens to produce methane and carbon dioxide (Parawira et al. 2008). These two pathways for the formation of methane are complementary and can be represented in simple chemical equations hereunder. Other less known breakdown products from the previous digestion phases such as formats, methylamine, methanol, and dimethyl sulfide are also taken up by the methanogens to produce methane and carbon dioxide.

$$CO_2 + 4H_2 \rightarrow CH_4 + 2H_2O$$

$$CH_3COOH \rightarrow CH_4 + CO_2$$

7.4 Co-Digestion and Biogas Potential Enhancement

7.4.1 Co-Digestion of Animal Manure with Other Feedstocks

The simultaneous decomposition of manure and one or more feedstocks from other sources and processed in the anaerobic digester/biogas system is called co-digestion. Feedstocks for co-digestion should be carefully selected to increase rather than decrease methane production. When the feedstocks with better energy potential are combined or co-digested with manure, the amount of biogas produced increases optimally. Many millennia ago, it was believed that anaerobic co-digestion was practiced without knowing the effects of adding high carbon substrates to anaerobic digesters. The nature of the co-substrate used for co-digestion must also be considered, including biogas production per volume of material, potential inhibitors, effects on solid retention time and hydraulic retention time in existing digesters, ease of integration into current anaerobic digester operations, and the fate of the additional liquid volume and nutrients. In addition, manures with high C and low N content are usually preferred. Among the various feedstocks that could be used in anaerobic digestion as co-substrate, lignocellulosic containing wastes seem to be a potential option as they have high carbon content and are easily and cheaply available (Neshat et al. 2017; Lusk 1998). Since animal manure contains high nitrogen but low-carbon, the C/N ratio needs to be optimized before anaerobic digestion can begin. Lignocellulosic materials have a high carbon content that can compensate for the carbon deficit in animal manure, but due to their slow decomposition and low methane production, they are rarely used as the sole substrate in an anaerobic digester. This problem can be overcome by pretreatment of lignocellulosic materials by processes such as enzymatic hydrolysis or steam explosion. However, this approach may not be financially viable (Ebner et al. 2016; Neshat et al. 2017).

Combination of manure from different animals

In the same way, different manures are produced from different animals, so is the diversity and abundance of their rumen microbial communities. Co-digestion of manure from different animals enhances methane production because of the positive synergies created in the medium during digestion, the microbiological diversity of the different animal manures, and the supplementation of nutrients that are lacking (Li et al. 2014). Compared to individual yields, dry anaerobic co-digestions of cow and swine manure in various ratios resulted in higher methane yields (5.10–18.01%), higher volatile solid removals (2.03–12.95%), and 2.98–12.51 percent chemical oxygen demand degradation. Co-digestion improved nutritional balance and diluted high NH3 concentrations in swine manure, resulting in improved digester performance efficiency and increased biogas production (Li et al. 2014). The co-digestion of swine and poultry manure in a ratio of 1:1 (w/w dry basis) was reported to significantly enhance biogas production compared with individual manures (Ogunwande 2013).

Combination of animal manure and agricultural waste

Co-digestion of feedstocks from other sources and animal manure is becoming increasingly popular to increase CH_4 yields and to control organic waste. However, the literature on the advantages of co-digestion compared to mono-digestion varies widely. The co-digestion of other feedstocks and animal manure anaerobically has significantly increased biogas production. The type and amount of substrate, hydraulic retention time, organic loading rate, volatile solids, and carbon/nitrogen ratio were all factors that affect the methane yields (Ma et al. 2020). The co-digestion of pig manure and agricultural wastes in the anaerobic digester (energy crop residues) resulted in a daily rise in biogas production (Cuetos et al. 2011). According to a study that looked at the methane potential production of ternary mixtures of cattle and pig manure and food waste at various mixing ratios, there was no loss of CH_4 potential at any mixing ratio and a potentially antagonistic effect was noticed when the food waste content exceeded 50%. Co-digestion was also found to accelerate the start of anaerobic digestion, implying that the technique could be used to hasten the start of the decomposition of cattle manure in a digester (Baek et al. 2020). The results of co-digestion of cow dung and swine manure with lawn grass suggested that swine manure would be a better co-substrate to combine with lawn grass for a longer period of time if it yielded a higher biogas potential. When compared to a mixing ratio of swine manure: lawn grass (40:20g), digestion of only lawn grass had a higher biogas and biomethane potential buildup over time. Because of the buffering and balanced nutrients in the digester, the best potential biogas and biomethane generation was attained when three substrates were co-digested (Singh et al. 2017). According to a recent meta-analysis of CH_4 yield in co-digestion compared to single-digestion of animal manure with other feedstocks, the former significantly increased CH_4 production. The findings of this research give a solid foundation for realistic economic assessments and control of anaerobic digestion procedures for or with various types of animal manure and digesters (Ma et al. 2020).

7.4.2 Boosting of Biogas Production Potential

The amount of biogas produced from manure is determined by the chemical content of the manure and the efficiency with which volatile solids present in the manure or feedstocks are decomposed. Biogas production has been improved as a result of recent changes to classic digesters. When digesters operate at a medium (mesophilic) or high (thermophilic) temperature, maintaining that temperature consistently increases biogas output compared to digesters with a wider temperature range. The biogas yield per unit mass entering the digester has increased thanks to a two-phase system (acid and methanogenic phase) that provides near ideal conditions for microbial growth and an increase in the number of methane-producing microorganisms. Furthermore, increasing the surface area of feedstocks by reducing particle size during maceration will expose more volatile solids to microorganisms, resulting in higher biogas production. Studies have shown that biological and/or chemical improvement of manure leads to an increase in biogas production. These improvements include bioaugmentation with certainly known microorganisms to accelerate the rate of organic decomposition, the addition of enzymes, the addition of micronutrients such as iron and magnesium to prevent the inhibitory effect of hydrogen sulfide production and ammonia accumulation in the digester, gas stripping of digestate to release dissolved biogas, use of biochar to improve digester performance, and integration of ammonia strips into biogas or anaerobic digester systems to remove the ammonia that can be inhibitory to the microbial community (Environmental Protection Agency 2021; Johari et al. 2020).

7.5 Full-Scale Biogas Digester Microbial Composition

7.5.1 Biologically Based Pretreatment of Manure

The biological pretreatment of animal manure often involves the complex activities of various microbial communities such as bacteria, archaea and fungi. It is much more cost-effective than pretreatment with enzymes because these microorganisms are already present in the feedstock and use their enzymes during the process. Microorganisms are known for converting high molecular weight substances in manure into lower molecular weight products that can be used in fermentation. The production of microbial extracellular enzymes capable of degrading the recalcitrant polymeric structures of the substrate is responsible for this process (Ferdeş et al. 2020; Lim et al. 2018). These various microbial communities use their enzymes such as ligninolytic, cellulolytic, pectinolytic, proteolytic, lipolytic, amylolytic, and other enzymes to meet their nutrient requirements. The degrading microorganisms' proliferation and their enzyme synthesis are involved in this treatment process (Ferdeş et al. 2020). As will be described later, it is critical to consider the optimal conditions for survival and growth of the relevant microorganisms during pretreatment, such as pH, temperature, nutrients, inhibitors, and so forth. The change of each of these parameters impacts the microbial diversity, structure, and composition during the decomposition of the substrate, and these changes can be adjusted according to the desires and requirements of the biogas process. However, unlike the addition of enzymes, which act much faster, the use of microbes requires more time, more stringent operating conditions and the risk of growth of undesirable microorganisms (Huang et al. 2015; Wan et al. 2021).

Metataxonomic profiles of animal manure could thus be used to control and monitor composting processes. As mentioned earlier, animal feeding affects manure's biological and biochemical properties, leading to predictable changes in composition, diversity, and microbial community assembly during decomposition. According to a study examining how the type of livestock manure used by omnivores (chickens and pigs) and herbivores (cattle and sheep) affects compost microbial communities, there were significant changes in all microbial communities during composting, resulting in lower bacterial and fungal diversity and significant shifts in community composition and species dominance. Firmicutes dominated pig and chicken manure composts, while Proteobacteria and Chloroflexes dominated sheep and cattle manure composts (Wan et al. 2021).

7.5.2 Bacteria in Biogas Digester

The bacteria used in the pretreatment may be derived from microbial cultures from previous fermentations, sludge, lignocellulosic substrates, rumen fluid, and various agricultural by-products. The bacteria may be single species, consortia, or combinations with microorganisms. These bacteria can rapidly adapt to new environments and substrates and continue to proliferate after a longer or shorter lag. Several bacterial species, including *Actinomycetes, Nocardia*, *Streptomyces*, and *Eubacteria*, are known for their properties in modifying, solubilizing, and degrading lignin (Ferdeş et al. 2020; Bajpai 2016). As explained earlier in the anaerobic degradation processes, all anaerobic degradation processes, such as hydrolysis, fermentation, acidification, etc., are carried out by a bacterial community. The presence of microorganisms in biogas plants using the hydrolytic anaerobic process varies depending on the type of reactor. The degradability of different polymers is determined by their nature, composition, and complexity. For example, the hydrolysis of carbohydrates takes only a few hours, while the hydrolysis of proteins and lipids can take several days. The degradation of lignocellulose and lignin is also slow and incomplete. According to research, hydrolytic bacteria cannot synthesize enzymes without the presence of a cellulosome (Rehman et al. 2019; Schroyen et al. 2018; Yin et al. 2018).

Bacteria with high hydrolytic activity were used for biological treatments. In one study, cellulose-degrading bacteria from the families *Prevotellaceae* and *Fibrobacteraceae* were found to be abundant in the rumen fluid of cattle but not in the manure, which are predominated by hydrolytic and fermenting bacteria from the family *Ruminococcaceae* and the orders *Clostridiales* and *Bacteroidales*. This explains why anaerobic digesters inoculated with rumen fluid from cattle perform better in biogas production (Ozbayram et al. 2018). Jin et al. (2018) also confirmed the influence of rumen bacteria that produce cellulase and xylanase on anaerobic co-digestion of corn straw with pig manure (Jin et al. 2018). According to Pampillón-González et al., in an anaerobic digestion, reactor loaded with pig manure, Firmicutes dominated, followed by Proteobacteria and Bacteroidetes, and many phylotypes of Firmicutes are known to have enzymes that can ferment plant lignocellulosic materials (Pampillón-González et al. 2017). Firmicutes (50–73%), Bacteroidetes (6–27%), and Proteobacteria (3–8%) were the most abundant phyla in a study examining the microbial communities of six biogas plants connected to respective cattle farms (Fontana et al. 2016). The environmental and operational parameters influence the overall composition of the microbial community in the biogas plant. Analysis of microbial communities of 29 full-scale anaerobic digesters showed a high abundance of phyla Firmicutes (major classes were Bacilli and Clostridia) in all samples, followed by Bacteroidetes (dominance of Bacteroidia class) and Proteobacteria (Vrieze et al. 2015).

7.5.3 Archaea in Biogas Digester

Methanogens are important for biogas production and have been the subject of numerous studies of microbial communities. Using high-throughput sequencing methods, researchers can now better understand the structure and function of the entire microbial community, allowing them to focus on understanding complex relationships to maximize biogas production (Pampillón-González et al. 2017). In contrast to methylotrophs, represented in the rumen sample only by the genus *Methanosphaera*, hydrogenotrophic and acetotrophic methanogens were identified in large numbers in the metagenomics study of the anaerobic digester-fed bovine excreta. The genera *Methanoculleus, Methanosaeta*, and *Methanolinea* were most abundant and changed over time as the digester ran. This appears to change toward the end of the experiment, with hydrogenotrophic *Methanoculleus* (70%) dominating the bio-digester community, while acetotrophic *Methanosaeta* (7%) and hydrogenotrophic *Methanolinea* (5%) are minor components (Senés-Guerrero et al.

2019). Also, according to a study by Pampillón-González et al., methanogens, acetoclastic *Methanosaeta*, hydrogenothrophic *Methanoculleus*, and *Methanospirillum* dominated in the middle of an anaerobic digestion reactor loaded with pig manure (Pampillón-González et al. 2017). Using amplicon sequencing of mcrA genes, Ozbayram et al. studied the methanogenic communities in bovine manure. They found that Methanocorpusculum was the most abundant genus (66%) and *Candidatus methanoplasma* accounted for only 5%, in contrast to what they found in the rumen, where *Methanobrevibacter* and *Candidatus methanoplasma* accounted for 85% of all mcrA genes. Due to the high occurrence of lignocellulose-degrading bacteria such as hydrogenotrophic methanogens and hydrolytic/fermenting bacteria of the orders Bacteroidales and Clostridiales, the rumen microbiome can be used as a seed for anaerobic digesters working with lignocellulosic feedstock or manure (Ozbayram et al. 2018). In a study examining the archaeal communities of six biogas plants connected to cattle farms, it was reported that the predominant archaea are *Methanosarcina*, *Methanobrevibacter* (80%), *Methanobacterium*, and *Methanosaeta*, although these methanogens varied depending on the biogas plant. The genus *Methanosarcina* was associated with ammonium concentration, while the genus *Methanoculleus* was more represented in the thermophilic digesters and the order Thermotogales was associated with hydraulic retention time (Fontana et al. 2016). The existence of the family *Methanosaetaceae* and the order *Methanobacterium* was discovered in the methanogenic communities of 29 full-length Anaerobic Digester Plants analyzed using real-time PCR targeting the 16S rRNA genes. Methanobacteriales were found in all samples and the authors emphasized that the methanogenic community in a given plant remains constant over time, suggesting that they may be the major acetoclastic and highly hydrogenotrophic methanogens in anaerobic digestion, respectively (Vrieze et al. 2015).

7.5.4 Fungi in Biogas Digester

Anaerobic fungi are excellent degraders of lignocellulosic biomass in the digestive tract of their hosts, making them a potential source for bio-fermentation in biogas production. Their unique combination of enzymatic and mechanical attacks on recalcitrant plant structures shows great promise for improving anaerobic digestion (Vinzelj et al. 2020). The addition of anaerobic fungi to plant biomass has resulted in increased biogas production (Procházka et al. 2012), faster initial hydrogen molecule and methane synthesis, and better degradation of volatile fatty acids (Nkemka et al. 2015), suggesting the potential of fungi to promote fiber digestion. The molecular approach used to evaluate the transcriptional cellulolytic activity of fungi from animal and biogas plant-derived materials was able to discriminate between fungi at the species level, as demonstrated for the genera *Neocallimastix* (*N. frontalis* and *N. cameroonii*) and *Orpinomyces* (*O. joyonii* and *O. intercalaris*) (Dollhofer et al. 2016). Dollhofer et al. used qPCR targeting the 18S rRNA gene to screen 10 different biogas plants fed with cattle manure or slurry in Germany for anaerobic fungi and detected the fungi in the majority of the investigated biogas plants. According to the study, several factors influenced the existence and activity of fungi in biogas plants (Dollhofer et al. 2017). In a follow-up study, Young et al. found anaerobic fungi enzyme, endoglucanase that is transcriptional active in the digester tank of a two-stage biogas plant fed with cow manure. This study also examined a variety of biogas plants and found that no such fungi were detected in anaerobic digester reactors fed with a variety of agricultural commodities and wastes, as well as pig and poultry manure. It has shown that fungi such as *Trichoderma capillare, Coprinopsis cinerea*, and Neocallimastigomycetes are potential candidates for improving the use of lignocellulosic raw materials in the practice of biogas production (Young et al. 2018). However, aerobic fungi from the phyla Ascomycota, Basidiomycota, and Zygomycota were found in large numbers in these biogas plants. Previous studies have also detected aerobic fungi in biogas plants (Dollhofer et al. 2017; Bengelsdorf et al. 2012) (Dollhofer et al. 2017; Bengelsdorf et al. 2012).

7.5.5 Microbial Enzyme in Biogas Digester

Even though enzyme treatment requires a short reaction time and is unaffected by some inhibitors (e.g., coumarin-rich plants) or microbial metabolism, the high cost of the enzyme is an economic barrier to the development of biogas production on an industrial scale. Most of the enzymes involved in the fermentation or biodegradation of complex organic polymers, carbohydrates, lipids, and proteins in animal waste in biogas reactors are produced by indigenous microorganisms, as well as exogenous enzymes provided for the degradation of most of these organic materials (Wei 2016). Some of the extracellular enzymes that help in the initial degradation of biomass into an even finer consistency are formate dehydrogenase, cellulase, laccases, lipase, xylanase, amylase, pectinases, proteases, hemicellulases, methyl-CoM reductase, formylmethanofuran transferase, etc. Enzymes involved in organic matter degradation or methanogenesis, the most important step in methane production, include methyl-CoM reductase, formylmethanofuran transferase (FTR), the F420-dependent N5, N10-methylenetetrahydromethanopterin dehydrogenase, the MER, N5, N10-methylenetetrahydromethanopterin reductase, the formylmethanofuran dehydrogenase, and the coenzyme M-HTP heterodisulfide. In addition, pretreatment with enzymes has been shown to be a promising method to increase biomethane production (Hosseini et al. 2019; Ferdeș et al. 2020).

7.6 Chemical and Physical Properties of Biogas

Analysis of the chemical components of the biogas produced from anaerobic digestion of organic matter indicate that methane (50–75%) and carbon dioxide (25–50%) make up the primary constituents of the biogas (Table 7.1) while hydrogen (0–1%), nitrogen (0–2%), ammonia (0–1%), hydrogen sulfide (0–1%), carbon monoxide (0–2%), and water (2–7%) occur in trace amounts as impurities (Raja and Wazir 2017; Sawyerr et al. 2019). A wide range of feedstocks can be coupled with a variety of fermentation processes to produce biogas with varying trace concentrations. It all depends on the feeding rate of the digester, the substrate, and its organic matter load, as well as the internal parameters within the anaerobic digester (pH and temperature) (Herout et al. 2011). For instance, the levels of this trace chemicals in biogas from dairy manure are generally lower, but the combustion products have a significantly higher toxicity than the other feedstocks (Li et al. 2019). A study of five types of cows revealed differences in biochemical properties and, as a result, biogas potential of the produced manure. Higher lipid and protein concentration resulted in reduced biogas potential, but increased carbohydrates content resulted in higher biogas generation (Abdallah et al. 2018).

Methane is a combustible gas, and the amount of methane in biogas determines how much energy it contains. Biogas is a gas that is much lighter than air and provides twice as many calories as natural gas. The characteristics of biogas, like those of any pure gas, are pressure and temperature dependent. Biogas is influenced by changes in calorific value (a function of water vapor content, temperature and pressure), changes in volume (a function of pressure and temperature), and changes in water vapor content (a function of pressure and temperature) (Bharathiraja et al. 2018). Biogas is difficult to compress, liquefy, or store. Methane must be liquefied at a temperature of −83 °C at a pressure of 35 MPa or −162 °C at the standard atmospheric pressure (101,325 Pa). Biogas has a calorific content of roughly 6 kWh/m^3 and other properties (Table 7.1), which is about half a liter of diesel fuel. The efficiency of the burners or appliances determines the net calorific value (5000 kcal per m^3).

Table 7.1 The general composition and properties of biogas

Composition		Properties	
Gas	% age amount in biogas	Chemical properties	Range
Methane (CH_4)	50–70	Density (kg/m^3)	0.94
Carbon dioxide (CO_2)	25–50	Ignition temperature (°C)	700
Hydrogen (H_2)	0–1	Required air for combustion (m^3/ m^3)	5.7
Nitrogen (N_2)	0–2	Net calorific value (KWh/m^3)	6
Ammonia (NH_3)	0–2		
Hydrogen sulfide (H_2S)	0–1		
Carbon monoxide (CO)	0–2		
Water (H_2O)	2–7		

7.7 Factors and Parameters that Can Influence Biogas Production from Manure

A number of factors as well as parameters, such as manure nutrients and content, temperature, pH, nitrogen inhibition and carbon/nitrogen (C/N) ratio, retention time, substrate particle size, and agitation, must be given specific attention in order to achieve optimal biogas generation. These elements have been shown to have a direct impact on the rate of biogas production and the efficiency of the digestive process. In addition, the amount of organic material used as inoculums in the fermentative organic substrate and the size of the inoculums had an impact on the rate of gas production (Nuchdang et al. 2015; Noraini et al. 2017). As a result, we highlighted several key factors as well as parameters that are known to have a direct impact on the efficiency of the manure processing and biogas generation rate, as well as examined the findings of past studies in this sector.

7.7.1 Temperature

One of the most important factors that affect the formation of biogas from manure is temperature. Negligence to correctly manage the reaction temperature can reduce process effectiveness and have an indirect impact on reaction rate, heavy metal as well as carbon dioxide solubility and buffering. The pace of reaction theoretically increase as the ambient temperature rises. As a result, the production of biogas will rise. During anaerobic digestion, there are three temperature ranges: psychrophilic temperatures ranging from 0 to 15 °C, mesophilic temperatures ranging from 15 to 45 °C, and thermophilic temperatures ranging from 45 to 65 °C (Table 7.2) (Kardos et al. 2011). After evaluating the type of inoculums to be employed, temperature ranges are chosen. In most conventional digesters, mesophilic temperatures of around 35 °C are used in the system. Thermophilic temperatures of 55–60 °C, on the other hand, are worth investigating because they produce more biogas in a shorter amount of time. Several studies have shown that thermophilic systems have benefits over mesophilic systems. When it comes to response times, thermophilic temperatures allow for quicker responses in less time, causing a greater gas output (Noraini et al. 2017).

Moreover, because it destroys disease-causing microorganisms as well as weed seeds to a higher degree, the thermophilic system is surely capable of providing improved sanitation for the end-digestate product (Elmashad 2004). Furthermore, thermophilic temperatures can inactivate *Salmonella* as well as *Mycobacterium paratuberculosis* for as little as 24 h, but mesophilic temperatures need at least a couple of weeks or a month to keep the same microbes dormant (Sahlström 2003). On the other hand, other studies have found that the mesophilic temperature range is almost but not quite as effective as the thermophilic temperature range, but only on certain substrates. Arikan and other researchers found that handling dairy dung

Table 7.2 Temperatures at which various methanogenic bacteria thrive best (Noraini et al. 2017)

Range of temperature	Genus of microorganisms	Optimal temperature (°C)
Thermophilic	*Methanohalobium*	50–55
	Methanosarcina	50–55
Mesophilic	*Methanohalophilus*	35–45
	Methanogenium	35–40
	Methanococcoides	30–35
	Methanocorpuseulum	30–40
	Methanobacterium	37–45
	Methanobrevibacter	37–40
	Methanoplanus	20–40

at reduced temperatures, such as 22 as well as 28 °C, produces methane at 70 and 87% of the value reported in a system operating at 35 °C (Dobre et al. 2014). Aside from that, anaerobic digestion of Buffalo dung at thermophilic (55 °C), as well as mesophilic (37 °C) temperatures, revealed that the system with higher temperatures produced more methane. Furthermore, the thermophilic reactor can run smoothly at pH 8.0 (Noraini et al. 2017).

The rate of solubilization was calculated by removing suspended solids, and it was discovered that at temperatures ranging from 25 to 45 °C, the solubilization rate is quite high, ranging between 62.2 and 72.7% This shows that under these temperature ranges, microbial activity was high, which led to a high rate of solubilization. Furthermore, the abrupt transition from mesophilic to thermophilic or vice versa, as well as variation in temperature placed on the system will have a direct impact on the process. Biogas yield will decrease until the appropriate populations have been restored for the process to run smoothly (Ward et al. 2008). In addition, even little variations in temperature in the digestion process, such as from 35 to 30 °C or vice versa, can drastically lower the biogas generation rate (Chae et al. 2008).

The energy requirements for heating are higher in thermophilic systems, which increases the system's operating costs. In addition, there is a need for temperature regulation due to the increased sensitivity of the systems to operational and environmental conditions.

7.7.2 The pH Values

In anaerobic digestion, the pH value is the most important element. The pH influences the growth of bacteria during anaerobic fermentation. Hydrogen, carbon dioxide, and volatile fatty acids influence the pH of digestive contents. The temperature of the reaction medium has an impact on this. A study by Carotenuto et al. reported twice the production of biogas (70%) at a pH > of 8.0 under thermophilic (55 °C) conditions than at a pH of 7.0 under mesophilic (37 °C) conditions during anaerobic digestion of buffalo manure. Their results show that a basic pH at the beginning favors the production of biomethane because it reduces the lag time and thus increases the production rate (Carotenuto et al. 2016). A recent study has shown that pH varies during different stages of anaerobic digestion of livestock manure due to microbial activity. In light of this, the acidic condition could be due to the accumulation of volatile fatty acids caused by the inhibition of ammonia driven by the high nitrogen concentration of chicken manure (Johari et al. 2020). In addition, the results of a study confirmed that alkaline pH, which increased from 8 to 12, was beneficial for pretreatment of manure (alkaline microwave treatment), which could promote the decomposition rate of manure. Although some methanogens are sensitive to pH, few acidophilic methanogens are known to survive in a neutral or weakly alkaline environment. With such manure pretreatment, methanogens will always be the dominant microflora in weakly alkaline environments (Yu et al. 2017). It has been shown that the performance of the bioreactor for the anaerobic digestion of swine manure is highly dependent on pH. This was supported by the fact that during biogas production, methane content was significantly higher (51.81%) at neutral pH (7.0) than at pH 6.0 (42.9%) and 8.0 (35.6%) (Zhou et al. 2016).

7.7.3 The Nutrients and Content of Manure

The nature of the feedstock utilized influences the quality as well as quantity of biogas produced. In addition, feedstock should have essential nutrients as well as carbon sources, which lets microorganisms thrive more sustainably. The kind of feedstock chosen and its nature i.e., solid content, determine the operational considerations of the waste to be processed (Daim et al. 2012) (Daim et al. 2012). "A plug-flow digester is apt for ruminant animal dung with solid concentrations of 11–13%," "a complete-mix digester is apt for manure that is 2–10% solids," and "a covered lagoon digester" is employed for liquid manure with less than 2% solids (Daim et al. 2012). The total quantity, as well as solids' type in the wastes they considered, were such that they could flow on their own or form slurries with water before eventually flowing, allowing them to be employed in a continuous process (Igoni et al. 2008).

7.7.4 Nitrogen Inhibition and Carbon/Nitrogen (C/N) Ratio

Protein, carbohydrate as well as fat make up the bulk of organic solid waste. Nitrogen is required for microbial growth plus replication in anaerobic circumstances. It's critical to keep nitrogen concentrations consistent throughout the procedure to avoid disruption. Microbes involved in first-step biological degradation remove the amino groups from the nitrogenous molecule during hydrolysis, resulting in ammonia as a by-product. Currently, if the rate of hydrolysis is not adequately regulated, NH_3

will build up in the digester, causing NH_3 toxicity or NH_3 inhibition (Yin et al. 2018). According to one investigation, NH_3 inhibition transpires between 1500 and 3000 mg L^{-1} of total ammoniacal nitrogen plus a pH of over 7.4 (Calli et al. 2005). Ammonia concentrations greater than 4 g NH^{4+}–NL^{-1} impede the methanogenesis process. Free NH_3 has also been discovered as one of the primary components impeding methanogenesis, according to the study (Chen et al. 2008). Investigators stated that NH_3 affects methanogenesis by two mechanisms: (1) direct inhibition of CH_4 producing enzyme by NH^{4+} and (2) diffusion of NH_3 through the cell wall of microorganisms, resulting in pH shift. The second mechanism is thought to have a bigger impact on the methanogenesis process. NH_3 is thought to be able to permeate passively into methanogen cell walls, where some of them will transform to NH^{4+}. This transpires when internal pH differences drive ammonia molecules to absorb proton (H^+) as a result of the mechanism (Noraini et al. 2017).

Excess NH_3 could bring about the rise in pH of the digester, triggering digester failure as well as low product yield if not rectified. During the intervening time, differing C/N ratios produce slightly varied results when animal excrement is digested. The biogas production was 0.50 m^3/kg VS for digestion with the lowest C/N ratio of 13. The greatest C/N ratio employed was 24, with a biogas production of 0.70 m^3/kg volatile volids (VS) (Yilmaz et al. 2018). The most ideal C/N ratios in a CH_4 generating process have been reported to be in the range of 20–30 (Wang et al. 2014). It has as well been claimed that microbes deplete carbon 25–30 times faster than nitrogen since carbon is employed as an energy source whereas nitrogen is needed to construct cell structures (Okonkwo et al. 2018). When the C/N ratio is too high, nitrogen is used up first, allowing carbon to accumulate, slowing the process. A low C/N ratio, on the other hand, indicates that nitrogen is abundant in the digester while carbon is scarce. Carbon will become scarce in the near future and fermentation will cease, leaving nitrogen unused. This situation could lead to the production of hazardous NH_3 species (Okonkwo et al. 2018).

7.7.5 Hydrogen Sulfide Concentration

Methane (CH_4) makes up half to 70% of the volume of biogas produced for sustainable energy production. Biogas also contains a lot of CO_2, between 30 and 50% by volume, as well as traces of other gases such as NH_3, H_2, and N_2, as well as other nutrients such as hydrogen sulfide. When stored, combined, or handled in biogas systems, input feedstock alongside output digestate materials (including manure) can produce these gases. These gases are toxic and create a harmful atmosphere if not adequately vented. The most frequent gas is H_2S. It is also the deadliest (Zagorskis et al. 2012).

7.7.6 Substrate Particle Size and Agitation

The size of the feedstock's particles has an impact on how quickly biogas is produced. The rate of hydrolysis, which is limited in the anaerobic digestion process, increases as particle size decreases (Sheridan et al. 2012). Vast particles may cause the digester to clog. Small particles, on the other hand, provide a huge surface area for microbe adsorption, which boosts the activity of microbes and, consequently, biogas output (Wasajja et al. 2021).

7.7.7 Retention Time (Flow-Through Time)

The theoretical period during which a particle or volume of liquid added to a digester remained in it is known as the retention time. The amount of time that volatile solids persist in the reactor is also known as the retention time. The average range of the complex molecule kept in the digesters, in contact with the biomass and decomposes into metabolic products such as monosaccharides, polysaccharides, and amino acids, is referred to as hydraulic retention time (Dobre et al. 2014). Organic matter decomposes slowly in anaerobic circumstances; therefore, digestion will take some time. One of the factors that will affect the retention time is the type of microbe and its temperature range. When compared to a mesophilic temperature system, a thermophilic temperature system in anaerobic digestion will have a shorter retention time. The particle kinetics rate and the reaction rate both increase as the temperature rises. Retention time is one of the most important criteria to consider when developing a cost-effective digester, as well as the other metrics listed above (Alvarez et al. 2006). Ezekoye and other researchers investigated the influence of retention time on biogas production of poultry droppings as well as cassava peels under mesophilic conditions. He concluded that as the retention period is increased, biogas output increases. During the first 5 to 15 days of digestion, cumulative biogas output peaks. The pace remains consistent for the next 20–30 days before gradually decreasing (Ezekoye et al. 2011).

7.8 Biogas Applications

Biogas contains combustible methane, which carries chemical energy. Methane gas is made up of many carbon and hydrogen atoms, and when the gas is burned, the energy stored in these atoms is released. Burning the gas produces hot air that can be used in a number of ways (Fig. 7.1). Biogas can be utilized in all natural gas-based applications such as cooking, drying, cooling, absorption heating, gas turbines, direct combustion, space heating, and water heating. It can also be used to power internal combustion engines and fuel cells for electrical and mechanical operations. When

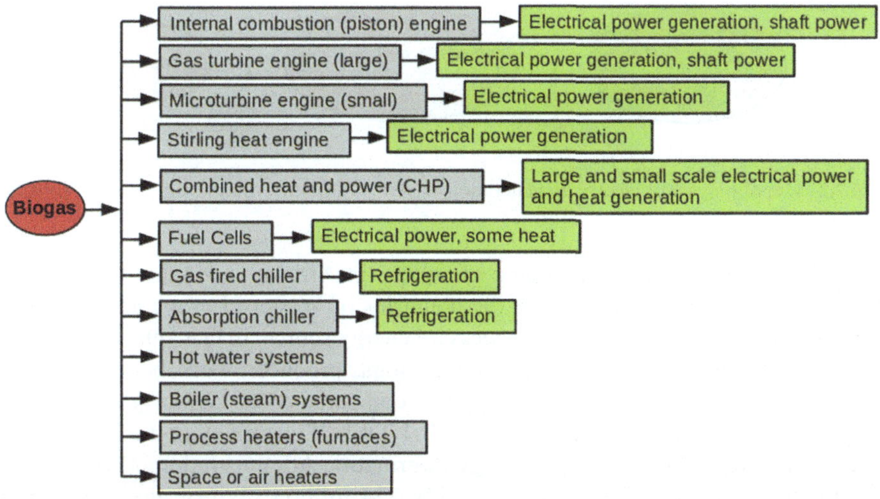

Fig. 7.1 Possible applications of biogas in many types of equipment

properly purified and compressed, it can be used in gas pipelines for steam generation and lighting (Kadam and Panwar 2017).

7.8.1 Electricity Generation

Biogas can be used in a variety of internal combustion engines, including gas turbines, Jenbacher and Caterpillar gas engines, which can convert biogas into both electricity and heat. These combustion engines can convert biogas into mechanical energy, which is used to power an electric generator to produce electricity (Scholwin and Nelles 2013). When the biogas is carefully mixed with the appropriate amount of air, it is drawn into the internal combustion engine by the force of the engine pistons moving downward and creating a vacuum. As the piston moves upwards, the biogas-air mixture is compressed. The compressed biogas-air mixture is ignited by a high-energy spark plug, since biogas is a slow-burning fuel and requires an engine with a higher compression ratio for optimal combustion. This gas mixture heats up quickly, expands, and pushes the piston down, creating torque that turns the engine. The exhaust valve of the biogas engine opens at this point, allowing the spent fuel–air combination to enter a heat exchanger, where any remaining fuel is removed. The generator converts the mechanical energy generated by the internal combustion engine into electricity (Fig. 7.2). To generate an electrical current, this mechanical energy turns an iron core wrapped in copper wire inside a strong magnet (Qian et al. 2017; Mustafi et al. 2006).

Biogas-based electricity generation has seen the fastest increase in the bioenergy sector, with projections of up to 11.2 GW capacity by 2020 (Scarlat et al. 2015).

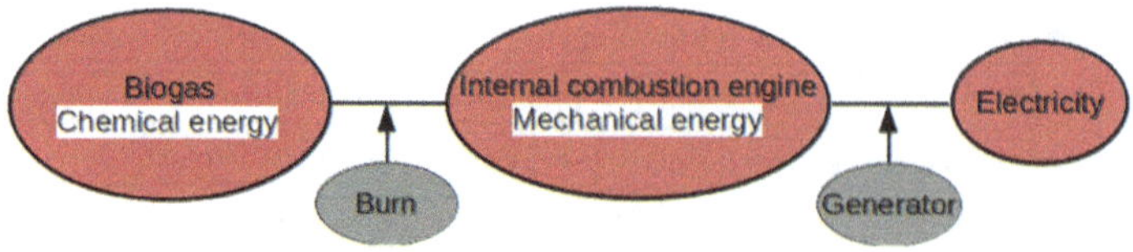

Fig. 7.2 Use of biogas for electricity generation

It has been widely used as a source of energy in African and Asian households, despite its rudimentary nature. For the countries of Western Europe and Northern America, it is a very important and necessary energy source. In fact, due to European Union (EU) countries is the world leading biogas producer, with about 10 GW of installed capacity and 17,400 biogas plants, compared to 15 GW of global biogas capacity. In 2015, biogas generated around 61 TWh of electricity in the EU (Scarlat et al. 2018). According to the US Energy Information Administration, 25 big dairies and livestock enterprises generated roughly 224 million kWh of electricity from biogas in 2019 (EIA 2020). Cattle waste from a dairy barn in Meoqui, Chihuahua, Mexico, was reported to have been used to generate up to 22 kW of power on a small scale (Huerta-Reynoso et al. 2019). In Asia, a few small-scale and low-cost residential biogas systems have been widely used for electricity generation, especially in rural areas (Dimpl 2010). According to a study conducted in Indonesia, biogas from cattle, pig, and poultry waste has an annual electricity potential of 80 TWh, which is more than enough to substitute diesel fuel in the power sector by 2030. In Malaysia, a similar study in 2016 predicted that animal manure would generate 10 TWh by 2020, which is about 15% of the replacement of natural gas in the power sector. The current installed capacity of power injection from biogas in Malaysia is 220.86 MW, generating only 1.7 TWh per year, according to Malaysia's Sustainable Energy Development Authority (Hoo 2019).

7.8.2 Heat or Steam Generation

The most obvious application of biogas is as a source of heat (thermal energy). Biogas simply burns to generate heat through the combustion process. One cubic meter of biogas provides about 2.0/2.5 kWh of heat energy when burned. Part of the heat produced by the plant can be used directly to supply energy to the plant and surrounding buildings, while the remaining heat is not wasted but can be used for domestic heating and transmitting of hot water via a local pipe network. In several Scandinavian countries, the concept of heating water and sending it to households as part of central heating is common. Small biogas systems can offer heat energy for basic cooking (Anderman et al. 2015) and water heating in locations where fuels are scarce. Biogas is used in gas lighting systems and conventional gas burners by simply adjusting the air-to-gas ratio. From the standpoint of energy gain, direct thermal application of biogas is the best option. In 2015, biogas produced 127 trillion joules of heat in Europe, with heat generation accounting for over half of total biogas consumption (Scarlat et al. 2018).

7.8.3 Co-Generation or Combined Heat and Power Generation

Bioelectricity is generated by burning biogas in a combined heat and power plant (CHP), and waste heat is collected from the combustion system's engine exhaust by a thermal storage unit. This heat can be converted into a usable form of thermal energy, such as steam or hot water. Some CHP plants first generate electricity and then use the waste heat to heat domestic hot water (topping cycle). Other CHP systems primarily generate heat, with electricity as a by-product (bottoming cycle). In either instance, the overall (combined) efficiency of the electricity (35%) and heat (50%) produced and utilized is significantly higher than when the fuel (biogas) is used just to generate power or heat. The most commonly used in engine in CHP applications are the internal combustion engines and turbines (micro or large) with comparable efficiencies to spark-ignition engines and low maintenance. In order to employ biogas in any of these systems, hydrogen sulfide and water vapor levels must be significantly reduced (Frazier 2019; Kaparaju and Rintala 2013; Arshad et al. 2022). Co-generation is one of the strategies for increasing the cost-effectiveness of production. When compared to separate power and heat generation, CHP allows for improved utilization of the chemical energy contained in biogas, as well as reduced fuel usage, pollution emissions, and few geographic limitations. CHP is now state of the art in Europe and its share of electricity generation is higher than in other regions, due to the long history of CHP. Despite the obvious environmental benefits, the use of CHP remains some 70–80% below its potential (Kusch, 2017).

7.8.4 Biomethane Production

Biogas purification, also known as biogas upgrading to biomethane (BUB), is a cost-effective alternative to employing a CHP system. In most cases, the biogas produced in a biogas plant is not clean biomethane (it consists mostly of 60% CH_4 and 40% CO_2). Other biogas gasses, except CH_4, are considered as contaminants. The removal of these contaminants, particularly CO_2, will improve the quality of biogas for alternate uses. CO_2 and other unwanted components are filtered out using various biogas upgrading technologies, such as membrane-based separation, pressure-driven membrane absorption, amine purification, organic-physical scrubbing, water purification, or freeze distillation. Because it may be injected into the natural gas system or compressed into CNG (compressed natural gas) for transportation, biomethane is a good form of renewable energy. After being pumped into the natural gas line, it can be used to fuel the central heating system or used in a gas stove for cooking (Adnan et al. 2019; Arshad et al. 2021).

Europe is the world's leading producer of biomethane for injection into the existing natural gas or as a vehicle fuel, with 340 plants putting 1.5 million m^3 into the gas line and 459 plants creating 1.2 billion m^3. Roughly 697 biomethane fuel stations facilitated the use of 160 million m^3 of biomethane as a fuel for transportation in 2015 (Scarlat et al. 2018). The most ambitious activities are now being carried out in Germany. As a result of the German government setting targets for biomethane production and long-term use, 83 biogas upgrading units were operational by the end of 2011 (Beil and Beyrich 2013).

7.8.5 Transportation Fuel

In a number of countries, particularly in Europe, purified biogas, such as compressed (CBG) and liquefied methane (LBG), is used as a transportation fuel. In addition to CBG and LBG, biogas is used to generate syngas, which can then be used to produce sustainable biofuels such as methanol, hydrogen or dimethyl ether. These fuels differ from CBG and LBG in terms of their properties and potential, which may make them ideal for different elements of the renewable energy system that needs to be developed. They, like compressed and liquefied methane, are all currently produced mainly from fossil fuels (Dahlgren 2020). In 2016, Germany, Sweden, Switzerland, the United Kingdom, and the United States were the top producers of biogas as a car fuel. An estimated 500 plants throughout the world produce biogas and upgrade it to biomethane, which is roughly 50 Petajoule (PJ)/a (IRENA 2018). Biogas was expected to account for around 10% of the natural gas fuel pool by 2020, with gasoline gallon equivalent production ranging from 7 million in 2015 to 111 million in 2020. Although biogas accounts for a tiny proportion of total low-carbon fuel volumes generated in the United States, it is an appealing compliance alternative for regulated parties because distinct biogas routes have been rated with a few of the reduced carbon levels. While biogas as a fuel for transportation is still in its infancy, this lower-carbon alternative has a promising future (Rogulska et al. 2018).

7.8.6 Biohydrogen Production

The perfect raw material for a sustainable energy market transition is hydrogen. Biohydrogen has the potential to increase energy efficiency. Biogas is utilized to produce hydrogen as a sustainable resource through a multi-step process that includes biogas reformation, water–gas shift reaction, and hydrogen extraction (Minh et al. 2018). In Germany, biogas facilities have transition to steam reforming instead of generating 31.9 terawatt-hours of electricity per year, producing roughly 58 terawatt-hours, or 1.7 billion kilos of hydrogen per year. However, one of the most significant barriers to using gas, particularly in transportation, is its inefficient conversion efficiency (Schleupen 2020). Currently, there is no substantial facility for providing hydrogen as a transportation fuel and in-vehicle storage capacity is also a challenge. Furthermore, hydrogen fuel cells are expensive to produce and have a limited lifespan. Chemical hydrogen storage is the subject of extensive investigation.

7.8.7 *Fuel Cells*

Fuel cells are devices that directly transform the chemical energy stored in biogas into electrical and thermal energy using essentially an electrolytic substance put between positive and negative electrodes (anode and cathode), with only water and heat as by-products. In contrast to combustion, which drives a mechanical process that powers an electrical generator to produce power, a biogas fuel cell produces direct current electricity via an electrochemical process (Arshad et al. 2018). "Outside" air is pumped through piping to generate power. When oxygen atoms come into touch with a negative electrode (cathode), they gain additional electrons and become ions, which migrate into the electrochemical reaction. The ionized oxygen migrates to the positively charged electrode (anode), where they mix with the hydrogen in the fuel and release more electrons, which are then emitted as a current by the fuel cell. The differential in oxygen partial pressure across the electrolyte drives electrochemical processes (Hanson 2013; Arshad et al. 2018). Fuel cell systems can produce net electric power outputs ranging from 250 kW to 1.4 MW and can run on 100% renewable fuel. The solid oxide fuel cell has been reported to have a 50–60% electrical conversion efficiency with little pollutant emissions and high thermal leftovers. Its high-quality residual heat could be used to boost biogas output by thermally pre-treating anaerobic digestion substrates. The heat could be used for absorption freezing, water heaters, or supplied to local businesses. Fuel cells make it possible for a facility to become energy self-sufficient (Saadabadi et al. 2019). The ability of biogas-fueled fuel cells to directly transform waste streams into power with zero emissions holds a lot of potential (Margalef 2021). Although hydrogen is the ideal fuel for fuel cells, alternative fuels are being investigated due to the associated costs of hydrogen production and storage (Saadabadi et al. 2019).

7.9 Biogas Development Across the Globe

The development of the biogas industry varies across the globe. Due to variances in the environment, energy use, and production methods, some continents have evolved from distinct biogas systems while others have not (Abanades et al. 2021a, b; Arshad et al. 2014). Profiling the biogas growth of particular nations throughout the world's continents with an emphasis on manure as a feedstock could therefore provide some challenges in terms of scale, commercialization, technical perspective, availability of feedstock, and economic policy. This research won't provide any definitive comparisons across the continents because the biogas industries were founded at separate eras and never developed consistently. Rather, it focuses on researching the level of biogas production and advancement that has been made across several continents using dung as a feedstock, as well as potential economic benefits (Gao et al. 2019), potential barrier's identification, digester capacities, standard framework, policies, and operational modalities (Nevzorova and Kutcherov 2019). One noteworthy plan calls for the implementation of zero-grazing systems to increase the supply of manure as a feedstock for biogas production. Since the Domestic Biogas Stove Implementation Program, the first biogas standard, was launched in April 1987, more than 70 biogas standards have been released. These standards have an impact on how biogas facilities are designed, built, operated, and maintained (Giwa et al. 2020).

7.9.1 Africa

Africa has one of the highest potentials for producing biogas, but nothing has been done to advance its development. This shows that despite the continent's abundant natural resources, which serve as a suitable feed stock for biogas production, biogas production in Africa had no economic objectives and did not benefit from advanced industrial technologies (Kemausuor et al. 2018). Despite this, only household biogas systems on the continent have made headway, and a lot has to be done (Clemens et al. 2018). Over the past few decades, small-scale facilities, the most of which are located in remote locations, have produced the majority of the biogas in Africa. But as several nations in sub-Saharan Africa strive to improve their infrastructure for biogas production, a paradigm change is taking place. Within the framework of the African Biogas Partnership Program (BPPA), seventy thousand biogas digesters were constructed in Tanzania, Burkina Faso, Ethiopia, Uganda, Kenya, and Senegal in 2013 (Clemens et al. 2018; Aliyu et al. 2015; Shane and Gheewala 2017). These varied digesters use typical manures as feedstock, including animal excreta and waste from slaughterhouses (Mukumba et al. 2016). The National Domestic Biogas Programs (NDBP) were aided by the African Biogas Partnership Program, which in 2013 ended successfully. This gave rise to increase in biogas installation in Africa (Ghimire 2013). This program's success, along with that of other non-governmental organizations promoting the growth of biogas in Africa, has been made possible by merging farming systems to take advantage of the multiple potential advantages of biogas digesters (Yimen et al. 2018).

7.9.2 Asia

Several Asian nations have launched significant endeavors to spread the use of biogas technology with the aid of their governments and foreign aid (Kemausuor et al. 2018).

The development and future of the biogas business differ significantly between Asian nations. China dominates Asia and the world in terms of residential biogas technology (Chen et al. 2010). It contains a significant amount of agricultural wastes, particularly animal manure utilized as a feedstock. China has devised an internationally acclaimed standard system for its biogas industry. The report states that at the end of 2015, China had 111,000 profit-oriented biogas plants, of which 103,898 were small and medium-sized, 6737 were industrial scale, and 34 were extremely large. The report predicts that the current number will significantly increase over the next six years. Their primary feedstocks are manure and agricultural waste (Chen et al. 2017). With 14.9 gigawatts (GW) of installed capacity and 79.4 terawatts hour (TWh) produced in 2017, China has overtaken all other countries as the world's top producer of bioelectricity (Dwyer and Teske 2018). By the end of 2017, India had established over 300 MW of profit-driven biogas production capacity (Havrysh et al. 2020). In Asia, biogas systems have been promoted in Nepal, Bangladesh, Vietnam, and Thailand. Attempts to expand facilities in Japan using dairy agriculture waste as a feedstock have been reported (Takeuchi et al. 2018).

7.9.3 Europe

The biogas market is most developed in Europe, where laws are much more developed than in other continents (Korbag et al. 2020). The use of biogas in Europe is seen as a possibly exclusive approach for the continent's broad switch to sustainable energy, a course that has gained strong support from the European Union (EU), the continent's top governing body (Achinas and Euverink 2020). In Europe, the United Kingdom produces the most biogas (bioelectric capacity reached 6 GW, generating 31.8 TWh), followed by Italy (Dwyer and Teske 2018). Austria, Bulgaria, the Czech Republic, Denmark, and the Czech Republic are among the European nations having a modest biogas production (Aryal and Kvist 2018).

7.9.4 South and North America

In America, large-scale production has mostly become commercial (Kemausuor et al. 2018). The most well-known biogas producers in South and North America are, respectively, Brazil and the United States. According to the U.S. Biogas Council, the country has about 2100 active biogas systems and more than 11,000 possible locations (Kemausuor et al. 2018). The annual biomethane production from animal waste is 5,128,334.6 million gallons (Pasqual et al. 2018). The strategic incentives provided for commercial biogas facilities in various US states include tax relief, subsidies, and soft loans (Sam et al. 2017). Brazil now feeds its 127 commercial biogas plants in South America mostly with municipal waste and cattle manure (Pasqual et al. 2017; Arshad et al. 2022). Approximately 100 million m^3 of methane might be produced every day from agricultural waste and animal manure, according to calculations (Langeveld and Peterson 2018). The possibility for producing biomethane from animal waste alone in Brazil is estimated to be almost two million gallons yearly (Pasqual et al. 2017). It is now essential to develop inventive and improved biogas production efficiency processes for enhanced quality and productivity because its uses in a variety of industries are quickly developing as a feasible approach to offer renewable energy for industrial and residential consumption. The four main stages of anaerobic digestion are a combined system that consists of a biochemical process involving active microorganisms, metabolism of energy, and the processing of raw materials subject to controlled environments (Luque and Clark 2010). Similarly, microbial communities are sensitive to changes in the working environment. Thus, improper regulation of the anaerobic digestion process will cause it to become unstable and produce less biogas (Achinas et al. 2020). Recent advancements in biology, engineering, and cross-disciplinary collaboration are rekindling optimism that will lead to a greater understanding of AD processes and, consequently, a comeback of these strategies (Prasad et al. 2017).

7.10 Recent Progress in Biogas Production

7.10.1 Choice of Biomass and Use of Additives

One of the main feedstocks in a biogas plant is biomass, which are of varieties. The amount of feedstock used in biogas facilities can be decreased by using highly fermentable waste (Feiz and Ammenberg 2017). How effectively the biogas facility operates and productivity can be improved by enhancing the usage of chemicals additives in the activities of microorganisms under various operating conditions. To provide the ideal nutritional conditions for bacteria, additives are usually used, although the ideal dosage depends on biocenosis (Demirel and Scherer 2011). Calcium and magnesium salts are added, which results in increased methane generation and decreased slurry foaming (TR et al. 2004). Application of chemicals that moderate pH variations lowers the concentration of hydrogen and ammonium sulfides. According to research, adding biological agents with a long retention time of 7 weeks significantly increases the production of biogas and biomethane (Ursa 2017). Gas generation dramatically enhanced when crop leftovers like onion garbage, wheat, and maize straws were treated with partially digested cow dung. Also, iron salts (50 mM-$FeSO_4$ and 70 lM-$FeCl_3$) are claimed to be an inorganic additive for the production of biogas (Dhanya et al. 2013).

7.10.2 Improvement in Digester Designs

The rate of biogas production in current biogas plant designs is being increased. In order to decrease HRT and boost gas production rate, improved biogas plant designs are critically required. An innovative, inexpensive household type biogas plant (Konark Model) with exceptional productivity was recently created by spiraling the biogas plant's design. The storage capacity of the Konark Model, which is often used due to its cost effectiveness, is increased by 50% as opposed to that of the Deenbandhu and Utkal Models, which are increased by just 33%. When using sturdy concrete blocks, construction expenses for the Deenbandhu model were reduced by 15% and by 30–35%, respectively. It had the smallest surface area and was structurally sound due to its spiral form. These variations have more space for storing gas. Another significant obstacle to the production of biogas is ammonia accumulation. Abouelenien et al. claim that using ammonia stripping devices with chicken droppings as a feed stock and digesting processes at an initial pH of 8–9 and temperature of 55 °C can boost biogas output (Abouelenien et al. 2010).

7.10.3 Meta-Omics Tool

The robustness of anaerobic digestion (AD) is influenced by microbial diversity, which is essential. High-throughput sequencing provides improved data on microbiomes and stability in biological systems, but it takes more than just knowing the microbial population to comprehend the dynamics of microbial activities. In order to ascertain the physiological link between organisms in biological processes, sophisticated meta-omic techniques have been created (Hashemi et al. 2021). Meta-genomic approaches offer a variety of data regarding the phylogeny of microbes, despite the fact that they are not ideal for real-time use due to their prolonged duration and cost inefficiency (Cai et al. 2016; Hashemi et al. 2021). To strengthen and advance AD, additional data is necessary about, among other things, the pattern of electron sharing, the rate of microbial metabolic activities, and functionality in recently identified microorganisms (Lamb and Hohmann-Marriott 2017; Gaspari et al. 2021). To close the gaps in our understanding of AD processes, gene amplicon sequencing and meta-omic approaches can be used to correlate the function and activity of the microbial community.

Metagenomics

Metagenomics based on next-generation sequencing is a rapidly growing research area that contributes to our understanding of the functional complexity and biodiversity of biological systems (Jünemann et al. 2017). In AD, a digester's activities can be observed using a metagenomic method. One illustration is the capability to follow the AD cycle from its inception through an acidic condition and back to its regular functioning state (Zhang et al. 2020). Reconstructing significant sections of the genomes of species identified in microbial consortiums is the main objective of metagenomic techniques, especially in less complex environments (Jaenicke et al. 2011). In more complex systems such as AD, gene-centric metagenomes has improved by providing more information on the prevalence of genes (Fontana et al. 2018).

Meta-Transcriptome

The study of how a significant number of transcripts from a culture specimen function is known as meta-transcriptomics. By measuring in situ gene expression, it sheds light on the actions of microorganisms (Mutz et al. 2013). By concentrating on metabolically active species of microorganisms, meta-transcriptomic approaches reduce the challenges of metagenomics (Su et al. 2011). The information obtained from these techniques must be compared to a standard genome data base in order to calculate gene expression. As a result, it is a more effective, affordable, and trustworthy technology that, unlike more traditional techniques like microarrays, could discover new genes. Gene expression profiling has been enhanced by high-throughput transcriptomics, which has also made it possible to find previously unknown sequencing transcripts (Lowe et al. 2017).

Meta-Proteome

Wilmes and Bond define meta-proteomics as the identification of all gene products within an ecosystem at a certain time and in a particular circumstance (Heyer et al. 2016). Heyer et al. discovered that the microbial population is shaped by microbial competition, interaction, and transposon-induced interactions, which cause cell lysis to produce biogas slowly. They discovered enzymes involved in numerous activities, such as pentose phosphate pathways, glycolysis, and CO_2/acetate methanogenesis (Heyer et al. 2019).

Meta-Metabolome

Metabolites are the term for the metabolic process's intermediates and final products. Utilizing meta-metabolome analysis techniques, metabolites from living things, such as enzymes, hormones, and light compounds, are identified and evaluated (Kikuchi et al. 2018). The meta-metabolome can therefore offer a glimpse of recent or ongoing cellular activities (Callaghan 2013). The biggest issue with metabolomics is that identifying metabolites necessitates an understanding of biological processes (Xiao et al. 2012). Gas chromatography-mass spectrometry (GC–MS) has demonstrated encouraging results when combined with chemometrics, despite the fact that numerous techniques for evaluating metabolites have been established (Lu et al. 2019).

7.10.4 Improved Anaerobic Digestion Techniques

Phase Separation Technique

In the most commonly used single-stage stirred tank biogas reactors, all phases of the process are performed simultaneously. Intercellular hydrogen transfer has advantages, but it also has substantial disadvantages. When hydrolytic, acidogenic, or acetogenic phases are more prevalent and proceed more quickly than acid uptake by acetoclastic species, acidification may occur. This could lead to significant fermentation failures and protracted downtime (Moeller and Zehndorf 2016). In a conventional multi AD, hydrolytic and acid-producing microbes predominate in the early stages, whereas acetogenic and methanogenic microbes predominate in the latter stages. By observing and adjusting the pH value, it is simple and requires minimal effort to separate these phases (Chatterjee and Mazumder 2019). The cost-effective production of hydrogen-enriched biogas (10–30% v/v) has been demonstrated to benefit from phase separation (Dahiya et al. 2018).

Cryogenic Separation

Cryogenic Separation is a cutting-edge method for upgrading methane that operates at extremely high pressure and low temperature. It is significantly more effective than traditional options when producing liquefied biomethane (LBM) (Hashem et al. 2019). The biogas mixture is compressed (200 bars), dried, and chilled (160 °C). Although the average pressures are lower under these circumstances, methane becomes a liquid that can be separated from other gases and pollutants.

Advanced Membrane Technology for Biogas Treatment

Biomethane was produced from more than 12% of biogas produced in Europe in 2018. Processes for upgrading biogas to biomethane can be separated using a separation process such as adsorption, absorption, or membrane separation (Khan et al. 2021). The least expensive techniques among the several available are membrane separation and water scrubbing, but chemical scrubbing gives comparatively high biomethane purities with reduced CH_4 losses (Katariya and Patolia 2021). Water scrubbing has typically been used because of its simplicity. On the other hand, membrane separation has stood out in the last 10 years due to its prospective economic advantages (Baena-Moreno et al. 2020). Since 2011, the number of plants has increased in lockstep, and membrane separation in particular has seen a significant increase, from less than 5% of all biomethane plants in 2011 to 34% in 2019. Then, approximately 56% of the biomethane produced in Europe is upgraded using water and chemicals (9EBA 2019).

7.11 Recent Methods Use for Biogas Upgradation

7.11.1 What is Biogas Upgradation?

Biogas upgrading removes impurities from raw biogas and converts biogas into high-quality biomethane that meets the demand of gas specifications of the grid infrastructure operator. The purified gas is then injected into the gas grid (Sahota et al. 2018). As mentioned earlier, biogas obtained directly from anaerobic digestion is not pure. It contains CH_4 (55%), CO_2 (40–45%), air (1%), hydrogen sulfide (H_2S), oxygen, hydrogen, ammonia, volatile organic compounds (VOCs), and siloxanes, depending on factors such as feedstock and the digester technology. The discrepancy between the calorific values of biogas (21.5 MJ/m^3) and natural gas (35.8 MJ/m^3) is primarily due to biogas's non-combustible component which is mainly CO_2. Raw biogas is also freed from unwanted impurities before upgrading to avoid corrosion and mechanical wear on the upgrading equipment. Biogas upgrading is gaining popularity in the United States, Europe, Australia (like Eneraque), China, and India as a sustainable, renewable, and environmentally friendly solution to climate change (Sahota et al. 2018; Chandra et al. 2020).

7.11.2 Biogas Upgradation Technologies

Various biogas upgrading methods or technologies are currently available in the market, all with the same goal: To remove carbon dioxide and other chemical components from methane to meet vehicle fuel standards or to achieve natural gas quality (biomethane) for injection into the natural gas grid, based on the mechanisms of absorption, adsorption, and membrane separation, the commercially available methods for purifying raw biogas are water scrubbing, pressure swing adsorption, chemical or amine scrubbing, membrane separation technology, cryogenic distillation or separation, biological upgrading methods, and in situ upgrading methods (Kapoor et al. 2019; Sahota et al. 2018). Each of these technologies has unique characteristics, efficiencies, and applications that depend on the particular biogas source.

Water scrubbing

Water scrubbing, also known as high-pressure water scrubbing, is the most widely used technology for removing CO_2 and hydrogen sulfide (which are soluble in water) from biogas and landfill gas because it is simple, inexpensive, and low in toxicity. In most cases, this purification process is carried out in a two-column system. In the first column, the raw biogas is passed through a scrubber or absorption column with a pressurized water stream to remove CO_2 and most of the H_2S, resulting in an upgraded gas stream with a high methane concentration. The CO_2 is then removed from the H_2O and returned to the first stage via a second stripper column. Reportedly, this technique can produce up to 97% pure methane with a methane loss of only 5% (Paolini et al. 2021).

Pressure swing adsorption

One of the most common techniques for separating CO_2 from raw biogas is pressure swing adsorption (PSA). In PSA, compressed raw biogas (between 4 and 10 bar) is fed into a vessel (column) where it interacts with adsorbents (zeolites, activated carbon, titanium silicates, etc.) to retain CO2 selectively. At some point, the saturated adsorbent can be regenerated as well as recovering the purified CH_4 at the top of the column by lowering the pressure before reloading. The use of two adsorption tanks allows for near-constant methane production. It also allows pressure equalization, using the gas from the unpressurized tank to partially pressurize the second tank. This is a common industrial practice that results in significant energy savings. Studies using PSA have yielded a final CH_4 product with a purity of 99–88% and a recovery of 91–81% (Augelletti et al. 2017).

Amine scrubbing

This refers to a process for removing H_2S and CO_2 from raw biogas using aqueous solutions of various alkylamines, such as mono-diethanol amine (MDEA). It can also be referred to as chemical scrubbing. The scrubbing chemical does not affect the methane content of the biogas as it passes through the packed bed reactor. As observed in some countries, e.g., Germany, high methane purities (> 99.9%) can be achieved in the recovered biogas with minimal losses. Therefore, it is considered one of the best systems for upgrading biogas (Sahota et al. 2018).

Membrane separation technology

Under high pressure, polymeric membranes are used to separate the CO_2 (highly ionically charged) from the nonpolar methane in biogas. The difference in chemical affinity and particle size of distinct molecules is the basic strategy of the membrane separation process. The gases separated by the membrane separation module are separated using the selective permeation technique (Chen et al. 2015).

Currently, serious progress is being made toward developing new advanced biogas upgrading technologies such as cryogenic separation, in situ methane enrichment, and hybrid membrane cryogenic technologies (Baena-Moreno et al. 2019).

7.12 Biogas Technologies: Their Benefits and Drawbacks

7.12.1 Benefits of Biogas Technologies

Significantly declined GHG production

Significant reductions in GHC emission were recorded after the consumption of biogas over the other household fuel, thus subsequently easing the reduction of biomass fuel and kerosene usage and annual GHC emission (Mengistu et al. 2016).

Reduction in the use of chemical fertilizer

According to the estimation made by Mengistu and co-workers, about 36% from the household biogas users agree that the obtained biogas slurry helped to reduce the usage of chemical fertilizers, consequently resulting in the reduction of GHC as well as declined usage of chemical fertilizer attained from the increased utilization of biogas (Mengistu et al. 2016).

Reduction in the usage of woody biomass

Utilization of biogas helps in minimizing the usage of woody fuels. Thus, it helps in improving the efficiency of energy and reducing the depletion of woody biomass.

Boosting soil fertility

With the usage of biogas, not only it gives energy efficient and pollution free alternative, perhaps, it also contributed in the enhancement of soil fertility. The application of bio-slurry obtained as an end product of biogas production improves the soil fertility and quality that reduces the overall consumption of chemical fertilizer. With regards to manure, bio-slurry provides the readily available nutrients (micro and macro nutrients) to plants (Shireen 2017).

7.12.2 Limitations of Biogas Technology

Production of Methane increases

Biogas production requires organic waste including animal and agriculture wastes. Subsequently, it raises the cattle farming and related human activities. Thus, it leads to the significant methane production. However, this is not having direct impact on human. Nevertheless, methane is a potent greenhouse gas (GHG) and responsible for the several significant damage to the environment.

Less significant technological advancement

Biogas technology is beneficial in several ways yet less technological advancement was reported in this direction. No significant technology is available for the bulk production and low-cost inputs.

Spiked with certain impurities

Biogas itself is not considered as pure combustible fuel, as it contains impurities that need to be clean. However, even after the cleaning and upgradation process, still some impurities are remained in general that may cause the damage to the engine or machine. Further it can be suitable for household usage including kitchen stoves etc.

Not ideal for urban areas

Biogas production demands plenty of agriculture and animal waste to proceed. Thus, this is not considered at larger urban and metropolitan areas (Shireen 2017; Omer 2017).

7.13 Conclusions

Animal manure is a substantial waste product that varies with farm size, nutrient content of the animal feed, and animal species. Because this waste product poses a significant environmental risk if it is not managed correctly, manure cannot be applied to the field immediately after it is produced. It can usually be stored for a while until it is time to apply it to the field. It could be utilized as a feedstock in digesters to produce biogas in the time between production and application to the field. Biogas is produced by anaerobic digestion, in which microorganisms (bacteria, fungi, and enzymes) break down the biodegradable components of manure in a closed chamber. The financial and environmental benefits of dealing with manure produced by livestock farms can be tapped using the biogas produced by this process. All natural gas-based uses, such as cooking, drying, cooling, absorption heating, gas turbines, direct combustion, space heating, and water heating, can be used with biogas. It can also be utilized to power electrical and mechanical processes in turbines, internal combustion engines, and fuel cells. It can be utilized in gas pipelines for steam generation and lighting after being purified and compressed adequately. Furthermore, it also contributes to the reduction of methane-related greenhouse gas emissions. Biogas plants offer a number of advantages and benefits that increase farm profitability while ensuring proper stewardship of resources and the environment.

References

Abanades S, Abbaspour H, Ahmadi A et al (2021a) A critical review of biogas production and usage with legislations framework across the globe. Int J Environ Sci Technol. https://doi.org/10.1007/s13762-021-03301-6

Abanades S, Abbaspour H, Ahmadi A et al (2021b) A critical review of biogas production and usage with legislations framework across the globe. Int J Environ Sci Technol 1–24

Abdallah M, Shanableh A, Adghim M et al (2018) Biogas production from different types of cow manure. In: 2018 Advances in science and engineering technology international conferences (ASET). IEEE

Abouelenien F, Fujiwara W, Namba Y et al (2010) Improved methane fermentation of chicken manure via ammonia removal by biogas recycle. Biores Technol 101:6368–6373

Achinas S, Euverink GJW (2020) Rambling facets of manure-based biogas production in Europe: a briefing. Renew Sustain Energy Rev 119:109566

Achinas S, Achinas V, Euverink GJW (2020) Microbiology and biochemistry of anaerobic digesters: an overview. Bioreactors 17–26

Adnan A, Ong MY, Nomanbhay S et al (2019) Technologies for biogas upgrading to biomethane: a review. Bioengineering 6:92. https://doi.org/10.3390/bioengineering6040092

Aliyu AS, Dada JO, Adam IK (2015) Current status and future prospects of renewable energy in Nigeria. Renew Sustain Energy Rev 48:336–346

Alvarez R, Villca S, Liden G (2006) Biogas production from llama and cow manure at high altitude. Biomass Bioenerg 30:66–75

Anderman TL, DeFries RS, Wood SA et al (2015) Biogas cook stoves for healthy and sustainable diets? A case study in Southern India. Front Nutr 2. https://doi.org/10.3389/fnut.2015.00028

Arshad M (2017) Clean and sustainable energy technologies. In: Clean energy for sustainable development 2017 Jan 1. Academic Press, pp 73–89

Arshad M, Ansari AR, Qadir R, Tahir H, Nadeem A, Mehmood T, Alhumade H, Khan N (2022) Green electricity generation from biogas of cattle manure: an assessment of potential and feasibility in Pakistan. Front Energy Res 2022:1256

Arshad M, Javed S, Ansari AR, Fatima A, Shahzad MI (2021) Biogas: a promising clean energy technology. Bioenergy Resour Technol 91–120

Arshad M, Abbas M (2018) Water sustainability issues in biofuel production. In: Perspectives on water usage for biofuels production. Springer, Cham, pp 55–76

Arshad M, Adil M, Sikandar A, Hussain T (2014) Exploitation of meat industry by-products for biodiesel production: Pakistan's perspective. Pakistan J Life Soc Sci 12:120–125

Aryal N, Kvist T (2018) Alternative of biogas injection into the Danish gas grid system—a study from demand perspective. Chem Eng 2:43

Augelletti R, Conti M, Annesini MC (2017) Pressure swing adsorption for biogas upgrading. A new process configuration for the separation of biomethane and carbon dioxide. J Clean Prod 140:1390–1398. https://doi.org/10.1016/j.jclepro.2016.10.013

Baek G, Kim D, Kim J et al (2020) Treatment of cattle manure by anaerobic co-digestion with food waste and pig manure: methane yield and synergistic effect. Int J Environ Res Public Health 17:4737. https://doi.org/10.3390/ijerph17134737

Baena-Moreno FM, Rodríguez-Galán M, Vega F et al (2019) Biogas upgrading by cryogenic techniques. Environ Chem Lett 17:1251–1261. https://doi.org/10.1007/s10311-019-00872-2

Baena-Moreno FM, le Saché E, Pastor-Pérez L, Reina TR (2020) Membrane-based technologies for biogas upgrading: a review. Environ Chem Lett 18:1649–1658. https://doi.org/10.1007/s10311-020-01036-3

Bajpai P (2016) Pretreatment of lignocellulosic biomass. In: Springer briefs in molecular science. Springer, Singapore, pp 17–70

Beil M, Beyrich W (2013) Biogas upgrading to biomethane. In: The biogas handbook. Elsevier, pp 342–377

Bengelsdorf FR, Gerischer U, Langer S et al (2012) Stability of a biogas-producing bacterial archaeal and fungal community degrading food residues. FEMS Microbiol Ecol 84:201–212. https://doi.org/10.1111/1574-6941.12055

Bhardwaj S, Das P (2017) A review: advantages and disadvantages of biogas. Int Res J Eng Technol 04

Bharathiraja B, Sudharsana T, Jayamuthunagai J, and others (2018) Retraction notice to Biogas production—a review on composition fuel properties, feed stock and principles of anaerobic digestion [Renew Sustain Energy Rev 90(2018) 570–582]. Renew Sustain Energy Rev 94:1229. https://doi.org/10.1016/j.rser.2018.08.010

Cai M, Wilkins D, Chen J et al (2016) Metagenomic reconstruction of key anaerobic digestion pathways in municipal sludge and industrial wastewater biogas-producing systems. Front Microbiol 7:778

Callaghan AV (2013) Metabolomic investigations of anaerobic hydrocarbon-impacted environments. Curr Opin Biotechnol 24:506–515. https://doi.org/10.1016/j.copbio.2012.08.012

Calli B, Mertoglu B, Inanc B, Yenigun O (2005) Effects of high free ammonia concentrations on the performances of anaerobic bioreactors. Process Biochem 40:1285–1292. https://doi.org/10.1016/j.procbio.2004.05.008

Carotenuto C, Guarino G, Morrone B, Minale M (2016) Temperature and pH effect on methane production from buffalo manure anaerobic digestion. Int J Heat Technol 34:S425–S429

Chae KJ, Jang A, Yim SK, Kim IS (2008) The effects of digestion temperature and temperature shock on the biogas yields from the mesophilic anaerobic digestion of swine manure. Biores Technol 99:1–6. https://doi.org/10.1016/j.biortech.2006.11.063

Chandra R, Isha A, Kumar S et al (2020) Potentials and challenges of biogas upgradation as liquid biomethane. In: Biogas production. Springer International Publishing, pp 307–328

Chandra R, Takeuchi H, Hasegawa T (2012) Methane production from lignocellulosic agricultural crop wastes: a review in context to second generation of biofuel production. Renew Sustain Energy Rev 16:1462–1476. https://doi.org/10.1016/j.rser.2011.11.035

Chatterjee B, Mazumder D (2019) Role of stage-separation in the ubiquitous development of anaerobic digestion of organic fraction of municipal solid waste: a critical review. Renew Sustain Energy Rev 104:439–469. https://doi.org/10.1016/j.rser.2019.01.026

Chen B, Hayat T, Alsaedi A (2017) History of biogas production in China. In: Biogas systems in China. Springer, pp 1–15

Chen XY, Vinh-Thang H, Ramirez AA et al (2015) Membrane gas separation technologies for biogas upgrading. RSC Adv 5:24399–24448. https://doi.org/10.1039/c5ra00666j

Chen Y, Cheng JJ, Creamer KS (2008) Inhibition of anaerobic digestion process: a review. Biores Technol 99:4044–4064. https://doi.org/10.1016/j.biortech.2007.01.057

Chen Y, Yang G, Sweeney S, Feng Y (2010) Household biogas use in China: a study of region suitability and sustainability. Renew Sustain Energy Rev 14:545–549

Clemens H, Bailis R, Nyambane A, Ndung'u V (2018) Africa biogas partnership program: a review of clean cooking implementation through market development in East Africa. Energy Sustain Develop 46:23–31

Cuetos MJ, Fernández C, Gómez X, Morán A (2011) Anaerobic co-digestion of swine manure with energy crop residues. Biotechnol Bioprocess Eng 16:1044–1052. https://doi.org/10.1007/s12257-011-0117-4

Dahiya S, Kumar AN, Sravan JS et al (2018) Food waste biorefinery: sustainable strategy for circular bioeconomy. Biores Technol 248:2–12. https://doi.org/10.1016/j.biortech.2017.07.176

Dahlgren S (2020) Biogas-based fuels as renewable energy in the transport sector: an overview of the potential of using CBG LBG and other vehicle fuels produced from biogas. Biofuels 1–13. https://doi.org/10.1080/17597269.2020.1821571

Daim T, Harell G, Hogaboam L (2012) Forecasting renewable energy production in the US. Foresight

Damyanova S, Beschkov V (2020) Biogas as a source of energy and chemicals. In: Biorefinery concepts energy and products. IntechOpen

Demirel B, Scherer P (2011) Trace element requirements of agricultural biogas digesters during biological conversion of renewable biomass to methane. Biomass Bioenerg 35:992–998

de Bok FAM, Harmsen HJM, Plugge CM et al (2005) The first true obligately syntrophic propionate-oxidizing bacterium Pelotomaculum schinkii sp. nov., co-cultured with Methanospirillum hungatei, and emended description of the genus Pelotomaculum. Int J Syst Evol Microbiol 55:1697–1703. https://doi.org/10.1099/ijs.0.02880-0

Dhanya MS, Prasad S, Singh A (2013) Biogas technology for developing countries: an approach to sustainable development. In: Bioenergy production by anaerobic digestion. Routledge, pp 427–451

Dimpl E (2010) Small-Scale Electricity Generation from Biomass. Part II: Biogas. GTZ-HERA – Poverty-oriented Basic Energy Service

Dobre P, Nicolae F, Matei F et al (2014) Main factors affecting biogas production-an overview. Romanian Biotechnol Lett 19:9283–9296

Dollhofer V, Callaghan TM, Dorn-In S et al (2016) Development of three specific PCR-based tools to determine quantity cellulolytic transcriptional activity and phylogeny of anaerobic fungi. J Microbiol Methods 127:28–40. https://doi.org/10.1016/j.mimet.2016.05.017

Dollhofer V, Callaghan TM, Griffith GW et al (2017) Presence and transcriptional activity of anaerobic fungi in agricultural biogas plants. Biores Technol 235:131–139. https://doi.org/10.1016/j.biortech.2017.03.116

Dwyer S, Teske S (2018) Renewables 2018 global status report

EBA (2019) The 'European Biomethane Map 2020' shows a 51% increase of biomethane plants in Europe in two years. https://www.europeanbiogas.eu/the-european-biomethane-map-2020-shows-a-51-increase-of-biomethane-plants-in-europe-in-two-years/

EIA (2020) Biomass explained landfill gas and biogas. US Energy Information Administration

Ebner JH, Labatut RA, Lodge JS et al (2016) Anaerobic co-digestion of commercial food waste and dairy manure: characterizing biochemical parameters and synergistic effects. Waste Manag 52:286–294

Elmashad H (2004) Effect of temperature and temperature fluctuation on thermophilic anaerobic digestion of cattle manure. Biores Technol 95:191–201. https://doi.org/10.1016/j.biortech.2003.07.013

Environmental Protection Agency US (2021) Project development handbook a handbook for developing anaerobic digestion/biogas systems on farms in the United States 3rd Edition AgSTAR Project Development Handbook i. AgSTAR

Ezekoye VA, Ezekoye BA, Offor PO (2011) Effect of retention time on biogas production from poultry droppings and cassava peels. Nigerian J Biotechnol 22:53–59

Feiz R, Ammenberg J (2017) Assessment of feedstocks for biogas production, part I—a multi-criteria approach. Resour Conserv Recycl 122:373–387

Ferdeș M, Dincă MN, Moiceanu G et al (2020) Microorganisms and enzymes used in the biological pretreatment of the substrate to enhance biogas production: a review. Sustainability 12:7205. https://doi.org/10.3390/su12177205

Fontana A, Campanaro S, Treu L et al (2018) Performance and genome-centric metagenomics of thermophilic single and two-stage anaerobic digesters treating cheese wastes. Water Res 134:181–191

Fontana A, Patrone V, Puglisi E et al (2016) Effects of geographic area feedstock, temperature, and operating time on microbial communities of six full-scale biogas plants. Biores Technol 218:980–990. https://doi.org/10.1016/j.biortech.2016.07.058

Frazier RS (2019) Biogas utilization and cleanup. Oklahoma State University, Renewable Energy Extension Engineer

Gao M, Wang D, Wang Y et al (2019) Opportunities and challenges for biogas development: a review in 2013–2018. Curr Poll Rep 5:25–35

Gaspari M, Treu L, Zhu X et al (2021) Microbial dynamics in biogas digesters treating lipid-rich substrates via genome-centric metagenomics. Sci Total Environ 778:146296

Ghimire PC (2013) SNV supported domestic biogas programmes in Asia and Africa. Renew Energy 49:90–94

Giwa AS, Ali N, Ahmad I et al (2020) Prospects of China's biogas: fundamentals, challenges and considerations. Energy Rep 6:2973–2987

Hanson C (2013) Equipment Profile: Biogas-powered Fuel Cells. Biomass Magazine

Hashemi S, Hashemi SE, Lien KM, Lamb JJ (2021) Molecular microbial community analysis as an analysis tool for optimal biogas production. Microorganisms 9:1162

Hashemi SE, Sarker S, Lien KM et al (2019) Cryogenic versus absorption biogas upgrading in liquefied biomethane production an energy efficiency analysis. Fuel 245:294–304. https://doi.org/10.1016/j.fuel.2019.01.172

Havrysh V, Kalinichenko A, Mentel G, Olejarz T (2020) Commercial biogas plants: lessons for Ukraine. Energies 13:2668

Herout M, Malaťák J, Kučera L, Dlabaja T (2011) Biogas composition depending on the type of plant biomass used. Res Agric Eng 57:137–143. https://doi.org/10.17221/41/2010-rae

Heyer R, Benndorf D, Kohrs F et al (2016) Proteotyping of biogas plant microbiomes separates biogas plants according to process temperature and reactor type. Biotechnol Biofuels 9. https://doi.org/10.1186/s13068-016-0572-4

Heyer R, Schallert K, Siewert C, and others (2019) Metaproteome analysis reveals that syntrophy competition, and phage-host interaction shape microbial communities in biogas plants. Microbiome 7. https://doi.org/10.1186/s40168-019-0673-y

Hoo R (2019) Biogas as a sustainable energy solution for Southeast Asia. Energy Studies Institute at the National University of Singapore

Hosseini KE, Dahadha S, Bazyar LAA et al (2019) Enzymatic pretreatment of lignocellulosic biomass for enhanced biomethane production—a review. J Environ Manage 233:774–784

Huang J, Yu Z, Gao H, and others (2017) Chemical structures and characteristics of animal manures and composts during composting and assessment of maturity indices. PLOS ONE 12:e0178110. https://doi.org/10.1371/journal.pone.0178110

Huang W, Wang Z, Zhou Y, Ng WJ (2015) The role of hydrogenotrophic methanogens in an acidogenic reactor. Chemosphere 140:40–46. https://doi.org/10.1016/j.chemosphere.2014.10.047

Huerta-Reynoso EA, López-Aguilar HA, Gómez JA et al (2019) Biogas power energy production from a life cycle thinking. In: New frontiers on life cycle assessment—theory and application. IntechOpen

IRENA (2018) Biogas for road vehicles: technology brief. International Renewable Energy Agency, Abu Dhabi

Ighravwe DE, Babatunde MO (2018) Determination of a suitable renewable energy source for mini-grid business: a risk-based multicriteria approach. J Renew Energy 2018:1–20. https://doi.org/10.1155/2018/2163262

Igoni AH, Ayotamuno MJ, Eze CL et al (2008) Designs of anaerobic digesters for producing biogas from municipal solid-waste. Appl Energy 85:430–438

Jaenicke S, Ander C, Bekel T, Bisdorf R, Droge M, Gartemann KH, Jü nemann S, Kaiser O, Krause L, Tille F et al (2011) Comparative and joint analysis of two metagenomic datasets from a biogas fermenter obtained by 454-pyrosequencing. PLoS ONE 6:e14519

Janni K, Cortus E (2020) Common animal production systems and manure storage methods. In: Animal manure. American Society of Agronomy Crop Science Society of America, and Soil Science Society of America, pp 27–43

Jin W, Xu X, Yang F (2018) Application of rumen microorganisms for enhancing biogas production of corn straw and livestock manure in a pilot-scale anaerobic digestion system: performance and microbial community analysis. Energies 11:920. https://doi.org/10.3390/en11040920

Johari SAM, Aqsha A, Osman NB et al (2020) Enhancing biogas production in anaerobic co-digestion of fresh chicken manure with corn stover at laboratory scale. SN Appl Sci 2. https://doi.org/10.1007/s42452-020-3063-y

Jünemann S, Kleinbölting N, Jaenicke S et al (2017) Bioinformatics for NGS-based metagenomics and the application to biogas research. J Biotechnol 261:10–23

Kadam R, Panwar NL (2017) Recent advancement in biogas enrichment and its applications. Renew Sustain Energy Rev 73:892–903. https://doi.org/10.1016/j.rser.2017.01.167

Kaparaju P, Rintala J (2013) Generation of heat and power from biogas for stationary applications: boilers gas engines and turbines, combined heat and power (CHP) plants and fuel cells. In: The biogas handbook. Elsevier, pp 404–427

Kapoor R, Ghosh P, Kumar M, Vijay VK (2019) Evaluation of biogas upgrading technologies and future perspectives: a review. Environ Sci Pollut Res 26:11631–11661. https://doi.org/10.1007/s11356-019-04767-1
Kardos L, Juhasz A, Palko GYORGY et al (2011) Comparing of mesophilic and thermophilic anaerobic fermented sewage sludge based on chemical and biochemical tests. Appl Ecol Environ Res 9:293–302
Katariya HG, Patolia HP (2021) Advances in biogas cleaning enrichment, and utilization technologies: a way forward. Biomass Conv Biorefin. https://doi.org/10.1007/s13399-021-01750-0
Kemausuor F, Adaramola MS, Morken J (2018) A review of commercial biogas systems and lessons for Africa. Energies 11:2984
Khan MU, Lee JTE, Bashir MA et al (2021) Current status of biogas upgrading for direct biomethane use: a review. Renew Sustain Energy Rev 149:111343. https://doi.org/10.1016/j.rser.2021.111343
Kikuchi J, Ito K, Date Y (2018) Environmental metabolomics with data science for investigating ecosystem homeostasis. Prog Nucl Magn Reson Spectrosc 104:56–88. https://doi.org/10.1016/j.pnmrs.2017.11.003
Korbag I, Omer SMS, Boghazala H, Abusasiyah MAA (2020) Recent advances of biogas production and future perspective. In: Biogas-recent advances and integrated approaches. IntechOpen
Kougias PG, Angelidaki I (2018) Biogas and its opportunitiesA review. Front Environ Sci Eng 12. https://doi.org/10.1007/s11783-018-1037-8
Kumar AK, Sharma S (2017) Recent updates on different methods of pretreatment of lignocellulosic feedstocks: a review. Bioresour Bioprocess 4. https://doi.org/10.1186/s40643-017-0137-9
Kusch S (2017) Cogeneration (combined heat and power production) in Europe. In: Proceedings of The 5th international virtual research conference in technical disciplines. Publishing Society
Lamb JJ, Hohmann-Marriott MF (2017) Manganese acquisition is facilitated by PilA in the cyanobacterium Synechocystis sp. PCC 6803. PLoS One 12:e0184685
Langeveld JWA, Peterson EC (2018) Feedstocks for biogas production: biogas and electricity generation potentials. In: Biogas. Springer, pp 35–49
Li J, Jha AK, Bajracharya TR (2014) Dry anaerobic co-digestion of cow dung with pig manure for methane production. Appl Biochem Biotechnol 173:1537–1552. https://doi.org/10.1007/s12010-014-0941-z
Li Y, Alaimo CP, Kim M et al (2019) Composition and Toxicity of biogas produced from different feedstocks in California. Environ Sci Technol 53:11569–11579. https://doi.org/10.1021/acs.est.9b03003
Lim J-S, Yang SH, Kim B-S, Lee EY (2018) Comparison of microbial communities in swine manure at various temperatures and storage times. Asian Australas J Anim Sci 31:1373–1380. https://doi.org/10.5713/ajas.17.0704
Lowe R, Shirley N, Bleackley M et al (2017) Transcriptomics technologies. PLoS Comput Biol 13:e1005457. https://doi.org/10.1371/journal.pcbi.1005457
Loyon L (2018) Overview of animal manure management for beef pig, and poultry farms in France. Front Sustain Food Syst 2. https://doi.org/10.3389/fsufs.2018.00036
Lu J, Muhmood A, Czekała W et al (2019) Untargeted metabolite profiling for screening bioactive compounds in digestate of manure under anaerobic digestion. Water 11:2420. https://doi.org/10.3390/w11112420
Luque R, Clark J (2010) Handbook of biofuels production: Processes and technologies. Elsevier
Lusk P (1998) Methane recovery from animal manures the current opportunities casebook. Office of Scientific and Technical Information (OSTI)
Ma G, Ndegwa P, Harrison JH, Chen Y (2020) Methane yields during anaerobic co-digestion of animal manure with other feedstocks: a meta-analysis. Sci Total Environ 728:138224. https://doi.org/10.1016/j.scitotenv.2020.138224
Margalef P (2021) Fuel cells powered with biogas

Mengistu MG, Simane B, Eshete G, Workneh TS (2016) The environmental benefits of domestic biogas technology in rural Ethiopia. Biomass Bioenerg 90:131–138. https://doi.org/10.1016/j.biombioe.2016.04.002

Minh DP, Siang TJ, Vo D-VN, and others (2018) Hydrogen production from biogas reforming: an overview of steam reforming dry reforming, dual reforming, and tri-reforming of methane. In: Hydrogen supply chains. Elsevier, pp 111–166

Moeller L, Zehnsdorf A (2016) Process upsets in a full-scale anaerobic digestion bioreactor: over-acidification and foam formation during biogas production. Energy Sustain Soc 6. https://doi.org/10.1186/s13705-016-0095-7

Molino A, Larocca V, Chianese S, Musmarra D (2018) Biofuels production by biomass gasification: a review. Energies 11:811. https://doi.org/10.3390/en11040811

Mugodo K, Magama PP, Dhavu K (2017) Biogas production potential from agricultural and agro-processing waste in South Africa. Waste Biomass Valoriz 8:2383–2392

Mukumba P, Makaka G, Mamphweli S et al (2016) Biogas technology in South Africa, problems, challenges and solutions. Int J Sustain Energy Environ Res 5:58–69

Mustafi NN, Raine RR, Bansal PK (2006) The use of biogas in internal combustion engines: a review. In: ASME 2006 Internal combustion engine division spring technical conference (ICES2006). ASMEDC

Mutz K-O, Heilkenbrinker A, Lönne M et al (2013) Transcriptome analysis using next-generation sequencing. Curr Opin Biotechnol 24:22–30. https://doi.org/10.1016/j.copbio.2012.09.004

Neshat SA, Mohammadi M, Najafpour GD, Lahijani P (2017) Anaerobic co-digestion of animal manures and lignocellulosic residues as a potent approach for sustainable biogas production. Renew Sustain Energy Rev 79:308–322. https://doi.org/10.1016/j.rser.2017.05.137

Nevzorova T, Kutcherov V (2019) Barriers to the wider implementation of biogas as a source of energy: a state-of-the-art review. Energy Strategy Rev 26:100414. https://doi.org/10.1016/j.esr.2019.100414

Nges IA (2012) Anaerobic digestion of crop and waste biomass: impact of feedstock characteristics on process performance. Master's thesis, Department of Biotechnology, Lund University

Nkemka VN, Gilroyed B, Yanke J et al (2015) Bioaugmentation with an anaerobic fungus in a two-stage process for biohydrogen and biogas production using corn silage and cattail. Biores Technol 185:79–88. https://doi.org/10.1016/j.biortech.2015.02.100

Noraini M, Sanusi S, Elham J et al (2017) Factors affecting production of biogas from organic solid waste via anaerobic digestion process: a review. Solid State Sci Technol 25:29–39

Ntaikou I, Antonopoulou G, Lyberatos G (2010) Biohydrogen production from biomass and wastes via dark fermentation: a review. Waste Biomass Valoriz 1:21–39. https://doi.org/10.1007/s12649-009-9001-2

Nuchdang S, Khemkhao M, Techkarnjanaruk S, Phalakornkule C (2015) Comparative biochemical methane potential of paragrass using an unacclimated and an acclimated microbial consortium. Bioresour Technol 183:111–119

Ogunwande GAOJAAAOFOF (2013) Effects of co-digesting swine manure with chicken Manure on biogas production. Ife J Sci 15

Okonkwo UC, Onokpite E, Onokwai AO (2018) Comparative study of the optimal ratio of biogas production from various organic wastes and weeds for digester/restarted digester. J King Saud Univ Eng Sci 30:123–129

Omer A (2017) Biogas technology for sustainable energy generation: development and perspectives. MOJ Appl Bion Biomech 1. https://doi.org/10.15406/mojabb.2017.01.00022

Otte J, Pica-Ciamarra U, Morzaria S (2019) A comparative overview of the livestock-environment interactions in Asia and Sub-saharan Africa. Front Veterinary Sci 6. https://doi.org/10.3389/fvets.2019.00037

Ozbayram E, Ince O, Ince B et al (2018) Comparison of Rumen and Manure microbiomes and implications for the inoculation of anaerobic digesters. Microorganisms 6:15. https://doi.org/10.3390/microorganisms6010015

Pagliari P, Wilson M, He Z (2020) Animal Manure production and utilization: impact of modern concentrated animal feeding operations. In: Animal Manure. American Society of Agronomy Crop Science Society of America, and Soil Science Society of America, pp 1–14
Pampillón-González L, Ortiz-Cornejo NL, Luna-Guido M et al (2017) Archaeal and bacterial community structure in an anaerobic digestion reactor (lagoon type) used for biogas production at a pig farm. J Mol Microbiol Biotechnol 27:306–317. https://doi.org/10.1159/000479108
Paolini V, Tratzi P, Torre M et al (2021) Water scrubbing for biogas upgrading: developments and innovations. In: Emerging technologies and biological systems for biogas upgrading. Elsevier, pp 57–71
Parawira W, Read JS, Mattiasson B, Björnsson L (2008) Energy production from agricultural residues: high methane yields in pilot-scale two-stage anaerobic digestion. Biomass Bioenerg 32:44–50. https://doi.org/10.1016/j.biombioe.2007.06.003
Pasqual JC, Bollmann HA, Scott C (2017) Biogas perspectives in livestock sector in Brazil and the United States: electric, thermal and vehicular energy use. J Agric Sci Technol A 7:258–273
Pasqual JC, Bollmann HA, Scott CA et al (2018) Assessment of collective production of biomethane from livestock waste for urban transportation mobility in Brazil and the United States. Energies 11:997
Paudel SR, Banjara SP, Choi OK et al (2017) Pretreatment of agricultural biomass for anaerobic digestion: current state and challenges. Biores Technol 245:1194–1205. https://doi.org/10.1016/j.biortech.2017.08.182
Piekutin J, Puchlik M, Haczykowski M, Dyczewska K (2021) The efficiency of the biogas plant operation depending on the substrate used. Energies 14:3157. https://doi.org/10.3390/en14113157
Prasad S, Rathore D, Singh A (2017) Recent advances in biogas production. Chem Engin Process Tech 3:1038
Procházka J, Mrázek J, Štrosová L et al (2012) Enhanced biogas yield from energy crops with rumen anaerobic fungi. Eng Life Sci 12:343–351. https://doi.org/10.1002/elsc.201100076
Qian Y, Sun S, Ju D et al (2017) Review of the state-of-the-art of biogas combustion mechanisms and applications in internal combustion engines. Renew Sustain Energy Rev 69:50–58. https://doi.org/10.1016/j.rser.2016.11.059
Raja IA, Wazir S (2017) Biogas production: the fundamental processes. Univ J Eng Sci 5:29–37. https://doi.org/10.13189/ujes.2017.050202
Recebli Z, Selimli S, Ozkaymak M, Gonc O (2015) Biogas production from animal manure. J Eng Sci Technol 10:722–729
Rehman MLU, Iqbal A, Chang C-C et al (2019) Anaerobic digestion. Water Environ Res 91:1253–1271. https://doi.org/10.1002/wer.1219
Rogulska M, Bukrejewski P, Krasuska E (2018) Biomethane as transport fuel. In: Biofuels—state of development. InTech
Rupf GV, Bahri PA, de Boer K, McHenry MP (2016) Broadening the potential of biogas in Sub-Saharan Africa: an assessment of feasible technologies and feedstocks. Renew Sustain Energy Rev 61:556–571
Saadabadi SA, Thattai AT, Fan L et al (2019) Solid oxide fuel cells fuelled with biogas: potential and constraints. Renew Energy 134:194–214. https://doi.org/10.1016/j.renene.2018.11.028
Sahlström L (2003) A review of survival of pathogenic bacteria in organic waste used in biogas plants. Biores Technol 87:161–166. https://doi.org/10.1016/s0960-8524(02)00168-2
Sahota S, Shah G, Ghosh P et al (2018) Review of trends in biogas upgradation technologies and future perspectives. Bioresour Technol Rep 1:79–88. https://doi.org/10.1016/j.biteb.2018.01.002
Sam A, Bi X, Farnsworth D (2017) How incentives affect the adoption of anaerobic digesters in the United States. Sustainability 9:1221
Sawyerr N, Trois C, Workneh T, Okudoh V (2019) An overview of biogas production: fundamentals, applications and future research. Int J Energy Econ Pol 9. https://doi.org/10.32479/ijeep.7375
Scarlat N, Dallemand J-F, Fahl F (2018) Biogas: developments and perspectives in Europe. Renew Energy 129:457–472. https://doi.org/10.1016/j.renene.2018.03.006

Scarlat N, Dallemand J-F, Monforti-Ferrario F et al (2015) Renewable energy policy framework and bioenergy contribution in the European union an overview from national renewable energy action plans and progress reports. Renew Sustain Energy Rev 51:969–985. https://doi.org/10.1016/j.rser.2015.06.062

Schleupen M (2020) Green hydrogen from biogas. Master's thesis, RWTH Aachen, Aachen, Germany

Scholwin F, Nelles M (2013) Biogas biogas for electricity generation biogas for electricity generation hi-tech applications biogas hi-tech applications. Renewable energy systems. Springer, New York, pp 161–169

Schroyen M, Vervaeren H, Raes K, Hulle SWHV (2018) Modelling and simulation of anaerobic digestion of various lignocellulosic substrates in batch reactors: Influence of lignin content and phenolic compounds II. Biochem Eng J 134:80–87. https://doi.org/10.1016/j.bej.2018.03.017

Senés-Guerrero C, Colón-Contreras FA, Reynoso-Lobo JF, and others (2019) Biogas-producing microbial composition of an anaerobic digester and associated bovine residues. MicrobiologyOpen 8. https://doi.org/10.1002/mbo3.854

Shane A, Gheewala SH (2017) Missed environmental benefits of biogas production in Zambia. J Clean Prod 142:1200–1209

Sheridan C, Petersen J, Rohwer J (2012) Technical note on modifying the Arrhenius equation to compensate for temperature changes for reactions within biological systems. Water SA 38. https://doi.org/10.4314/wsa.v38i1.18

Singh S, Matheri AN, Belaid M, Muzenda E (2017) Co-digestion of lawn grass with cow dung and pig manure under anaerobic condition. In: The nexus: energy environment and climate change. Springer International Publishing, pp 221–243

Sommer SG, Christensen ML (2013) Animal production and animal manure management. In: Animal manure recycling. John Wiley & Sons Ltd, pp 5–23

Su C, Lei L, Duan Y et al (2011) Culture-independent methods for studying environmental microorganisms: methods application, and perspective. Appl Microbiol Biotechnol 93:993–1003. https://doi.org/10.1007/s00253-011-3800-7

Takeuchi Y, Andriamanohiarisoamanana FJ, Yasui S et al (2018) Feasibility study of a centralized biogas plant performance in a dairy farming area. J Mater Cycles Waste Manage 20:314–322

Themelis NJ, Ulloa PA (2007) Methane generation in landfills. Renew Energy 32:1243–1257. https://doi.org/10.1016/j.renene.2006.04.020

Ursúa N (2017) Los ocho pecados mortales de la humanidad civilizada. Una relectura de Konrad Lorenz y los problemas de la naturaleza humana. Ludus Vitalis 13:165–180

Valenti F, Zhong Y, Sun M et al (2018) Anaerobic co-digestion of multiple agricultural residues to enhance biogas production in southern Italy. Waste Manage 78:151–157

Vinzelj J, Joshi A, Insam H, Podmirseg SM (2020) Employing anaerobic fungi in biogas production: challenges and opportunities. Biores Technol 300:122687. https://doi.org/10.1016/j.biortech.2019.122687

Vrieze JD, Hennebel T, Boon N, Verstraete W (2012) Methanosarcina: the rediscovered methanogen for heavy duty biomethanation. Biores Technol 112:1–9. https://doi.org/10.1016/j.biortech.2012.02.079

Vrieze JD, Saunders AM, He Y et al (2015) Ammonia and temperature determine potential clustering in the anaerobic digestion microbiome. Water Res 75:312–323. https://doi.org/10.1016/j.watres.2015.02.025

Wan J, Wang X, Yang T et al (2021) Livestock Manure type affects microbial community composition and assembly during composting. Front Microbiol 12. https://doi.org/10.3389/fmicb.2021.621126

Wang X, Lu X, Li F, Yang G (2014) Effects of temperature and carbon-nitrogen (C/N) ratio on the performance of anaerobic co-digestion of dairy manure chicken manure and rice straw: focusing on ammonia inhibition. PLoS ONE 9:e97265. https://doi.org/10.1371/journal.pone.0097265

Ward AJ, Hobbs PJ, Holliman PJ, Jones DL (2008) Optimisation of the anaerobic digestion of agricultural resources. Biores Technol 99:7928–7940. https://doi.org/10.1016/j.biortech.2008.02.044
Wareham DG, Elefsiniotis P, White J (2014) Anaerobic digestion of coconut copra: methane generation potential. J Environ Eng Sci 9:162–170. https://doi.org/10.1680/jees.14.00004
Wasajja H, Al-Muraisy SAA, Piaggio AL et al (2021) Improvement of biogas quality and quantity for small-scale biogas-electricity generation application in off-grid settings: a field-based study. Energies 14:3088
Wei S (2016) The application of biotechnology on the enhancing of biogas production from lignocellulosic waste. Appl Microbiol Biotechnol 100:9821–9836
Weiland P (2009) Biogas production: current state and perspectives. Appl Microbiol Biotechnol 85:849–860. https://doi.org/10.1007/s00253-009-2246-7
Xiao JF, Zhou B, Ressom HW (2012) Metabolite identification and quantitation in LC-MS/MS-based metabolomics. TrAC, Trends Anal Chem 32:1–14. https://doi.org/10.1016/j.trac.2011.08.009
Yilmaz A, Ünvar S, Koçer A, Aygün B (2018) Factors affecting the production of biogas. Int J Sci Eng Res 9:59–62
Yang L, Li Y (2014) Anaerobic digestion of giant reed for methane production. Biores Technol 171:233–239. https://doi.org/10.1016/j.biortech.2014.08.051
Yimen N, Hamandjoda O, Meva'a L et al (2018) Analyzing of a photovoltaic/wind/biogas/pumped-hydro off-grid hybrid system for rural electrification in Sub-Saharan Africa—case study of Djoundé in Northern Cameroon. Energies 11:2644
Yin DM, Westerholm M, Qiao W et al (2018) An explanation of the methanogenic pathway for methane production in anaerobic digestion of nitrogen-rich materials under mesophilic and thermophilic conditions. Bioresour Technol 264:42–50
Young D, Dollhofer V, Callaghan TM et al (2018) Isolation identification and characterization of lignocellulolytic aerobic and anaerobic fungi in one- and two-phase biogas plants. Biores Technol 268:470–479. https://doi.org/10.1016/j.biortech.2018.07.103
Yss TR, Kohli S, Rana V (2004) Enhancement of biogas production from solid substrates using different techniques. J Bioresour Technol 95:1–10
Yu T, Deng Y, Liu H et al (2017) Effect of alkaline microwaving pretreatment on anaerobic digestion and biogas production of swine manure. Sci Rep 7. https://doi.org/10.1038/s41598-017-01706-3
Zagorskis A, Baltrėnas P, Misevičius A, Baltrėnaitė E (2012) Biogas production by anaerobic treatment of waste mixture consisting of cattle manure and vegetable remains. Environ Eng Manage J (EEMJ) 11
Zhang Y, Zhang L, Guo B et al (2020) Granular activated carbon stimulated microbial physiological changes for enhanced anaerobic digestion of municipal sewage. Chem Eng J 400:125838
Zhou J, Zhang R, Liu F et al (2016) Biogas production and microbial community shift through neutral pH control during the anaerobic digestion of pig manure. Biores Technol 217:44–49. https://doi.org/10.1016/j.biortech.2016.02.077

Chapter 8
Greenhouse Gases Emissions Assessments and Mitigation Opportunities from Animal Manure Processing

Muhammad Umar Ijaz, Muhammad Faisal Hayat, Sher Zaman Safi, Ali Hamza, Asma Ashraf, and Muhammad Arshad

Abstract Our atmosphere is continuously depleting due to immense emission of greenhouse gases (GHGs). This chapter will assess the emissions of GHGs and ameliorative opportunities from animal manure processing. Animal manure is a major contributor in the generation of GHGs during different management processes. Animal manure is generated 37% of total GHGs globally. Emission of GHGs occurs during manure collection from animal yards, manure storage and manure spreading in fields for the sake of fertility. Methane (CH_4) and nitrous oxide (N_2O) are two significant GHGs which are generated from animal manure by the process of methanogenesis and denitrification, respectively. Livestock adds 240 metric tons of CO_2 eq. to methane in the atmosphere and it is designated as top-notch contributor of anthropogenic methane emitter. Emissions of GHGs from animal manure are rely on certain environmental conditions viz. temperature and water availability which correlate with microbial processes (Methanogenesis, Nitrification, Dentification, Methane oxidation). The complete assessment of GHGs from livestock may help to understanding the contribution of livestock in climate variations and to develop strategic approaches to ameliorate these emissions. Mitigation of these emissions is strictly based upon the type of manure, management practices and climatic conditions. Phase feeding in livestock yards is tremendous strategy to reduce the emission of GHGs. Limit the losses of ammonia by the process of biofiltration is another approach to diminish the emissions of GHGs. Reduction in the storage time of manure prevents anaerobic decomposition which ultimately reduces the emissions of GHGs. Shuffling of

M. U. Ijaz · M. F. Hayat · A. Hamza
Department of Zoology, Wildlife and Fisheries, University of Agriculture, Faisalabad, Pakistan

S. Z. Safi
Faculty of Medicine, Bioscience and Nursing, MAHSA University, Jenjarom, Selangor, Malaysia

A. Ashraf (✉)
Department of Zoology, Government College University, Faisalabad, Pakistan
e-mail: asmaashraf@gcuf.edu.pk

M. Arshad
Department of Basic Sciences, University of Veterinary and Animal Sciences Lahore, Jhang-Campus, Lahore, Pakistan

M. Arshad (ed.), *Climate Changes Mitigation and Sustainable Bioenergy Harvest Through Animal Waste*, https://doi.org/10.1007/978-3-031-26224-1_8

heaps, impermeable covers and acidification of manure has proved to be efficient in the mitigation of these GHGs except nitrous oxide. Anaerobic digestion (AD) technology uses huge biogas digestor and it is a modern approach to not only limit the emissions of GHGs but also act as alternative source of fossil fuel to overcome the energy crises. As a result of this assessment, it is determined that livestock is the significant contributor in total GHGs emissions and effective management practices can mitigate these GHGs emissions.

Graphical Abstract

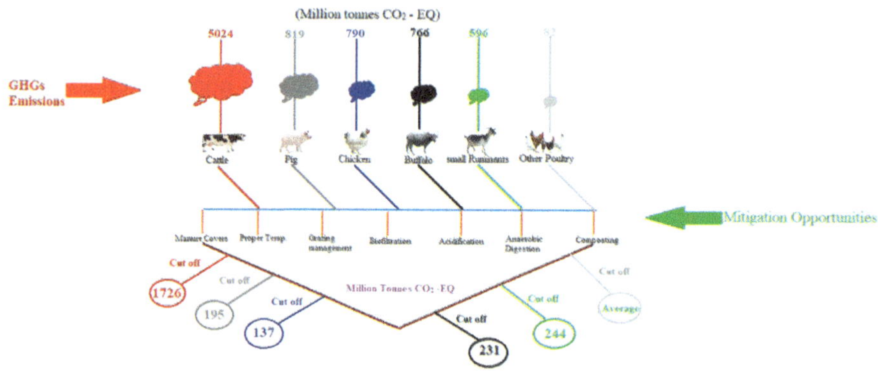

Keywords Greenhouse gases · Methanogenesis · Anaerobic digestion · Biofiltration

8.1 Introduction

The world population is estimated to outreach 9 billion till 2050. Demands for energy, food and lands will increase day by day due to population pressure. Land will be less accessible to cultivate other food or bioenergy if it is employed for animals. The solution will be to intensify production in all regions of the globe by approaching to the usage of fertilizers and other inputs, which will cause a broad range of environmental concerns, such as the contamination of water and soil (IAASTD 2008). When resources become limited, the results will appear in the form of soil deterioration and eventually low productivity everywhere (FAO 2006). The body of evidences on the emission of greenhouse gas (GHG) associated to livestock husbandry is substantial and expanding day by day (Cederberg and Mattson 2000; Cederberg and Stadig 2003; Basset-Mens and van der Werf 2005; Casey and Holden 2005, 2006; FAO 2006; Lovett et al. 2006) (Table 8.1).

Our atmosphere is the core component of our Earth. It directly or indirectly interacts with all the other components of earth such as hydrosphere, lithosphere and biosphere (Jerez et al. 2018). Anthropogenic activities are continuously disrupting

Table 8.1 GHGs emissions on the basis of region (FAO 2015)

Region	GHGs emissions
Eastern Europe	127 million tons CO_2—EQ
North America	604 million tons CO_2—EQ
Caribbean and Latin America	1889 million tons CO_2—EQ
Western Europe	579 million tons CO_2—EQ
Oceania	157 million tons CO_2—EQ
South Asia	1507 million tons CO_2—EQ
Sub-Saharan Africa	416 million tons CO_2—EQ
East and South Asia	1576 million tons CO_2—EQ
Near East and North America	579 million tons CO_2—EQ
Russian Federation	92 million tons CO_2—EQ

this balanced chain by various sorts of actions. The societal actions caused by human beings put devastating impacts on the balance of atmospheric gases which ultimately leads to the disruption of natural equilibrium (Dai 2016). Greenhouse gases are those gases which have potential to absorb and emit infrared radiations. These gases trap the energy (heat) from our atmosphere and warm the planet. We can predict the dynamic state of our atmosphere from the fact that concentration of carbon dioxide in our atmosphere increases about 43% from 1755 to till now (Ciais et al. 2013).

This immense elevation in growth rate was due to anthropogenic actions such as ignition of fossil fuels for energy and cutting of forests for different purposes. This growth rate casted highly devastating impact on global temperature by raising it to 0.85 °C as compared to pre-industrial time (Hansen et al. 2010). Similarly, the concentration of methane, another greenhouse gas, was 722 ppm since pre-industrial time but elevated to 1834 ppm till 2013. This excessive concentration elevated by a factor of 2.5 and highest recorded concentration since 0.8 million years. The most important greenhouse gas nitrous oxide (N_2O) had 300 times more warming abilities than carbon dioxide since last decade (Myhre et al. 2013).

Animal manure is the world second largest root cause of greenhouse gases primarily of methane and nitrous oxide (USEPA 2006a, b). NH_3 in volatilized form by animal manure can constitute up to 70% of total nitrogen loss. It incorporated in the water and other terrestrial bodies to impart devastating effects as well as it becomes the major source of N_2O (Hristov et al. 2002). From the twentieth century, the size of population increased day by day without proper balance which eventually created pressure on world resources. To fulfil the demand of food, milk and other dairy products, the trend for livestock production increased excessively in well-developed nations (Gerber et al. 2005) (Table 8.2).

As a consequence, the continuum of livestock manure produced in immense volume. The direct emission of greenhouse gases results as anaerobic digestion, incomplete metabolism and manure management, while the indirect ammonia release occurs from microbial activities in the animal manure. Animal manure is generally used in agricultural lands to improve fertility, productivity and nutrient balance but

Table 8.2 World population estimation and different sorts of meat consumption from 2010–2050

	2010	2020	2030	2050	References
Population of human (billions)	6.96	7.6	8.6	9.8	(FAO 2009)
Meat demand (millions of tons)					
Poultry	100.2	131.2	143	181	(McLeod A and Food and Organization of the United Nations 2011)
Cows and buffalo	65	70.9	88.9	106	
Pork	108.7	106.3	129.9	143	
Sheep and goat	13.2	15.5	18.5	25	
Total demand	287.1	323.9	380.3	455	

on the other hand its continuous usage set forth severe impacts on the equilibrium of greenhouse gases in the atmosphere (Zhou et al. 2017b). Almost 7 billion tons of animal excreta are used widely in world each year in agriculture land for the sake of high productivity and fertility (Thangarajan et al. 2013).

Because of their impact on the earth's climate, emissions of greenhouse gases have drawn more attention on a worldwide scale. There are various reports of GHG emissions, livestock documented to produce 8.4% total of US GHGs emissions and 11% of global GHGs emissions overall (Smith et al. 2014; EPA 2017). Due to their impacts on the environment and their commitment to GHG emissions, livestock has garnered additional attention. According to recent statistics, livestock produces 5335 Mt of carbon dioxide eq. annually, or 11% of all greenhouse gases caused by humans (Smith et al. 2014).

Animal manure is used to overcome the deficiency of essential nutrients to the soil. It was reported that excreta of animal contributed 22% of total nitrogen and about 38% of total phosphate used globally (United Nations Environment Programme, undated). Animal excreta enhance the soil fertility as well as make it biologically active as compared to fertilizer alone (Fließbach et al. 2007). After application to land, manure emits N_2O and CH_4 as it gets decomposed into soil (FAO 2006). It has been reported that 1/3 of all cultivated cereals crops all over the world are used to feed animals which eventually become the part of their manure. Later on, this manure is used to endure their own existence. Without any doubt, the rearing of livestock for different purposes poses some serious threats in the form of greenhouse gases which overweigh the advantages (Garnett 2009).

Overall, animals manure contributed about 37% of globally produced greenhouse gases emission (Vac et al. 2013). Soil pH, crop type, pore spaces and topography are the key factors for greenhouse gases (GHGs) emission from terrestrial ecosystem (Severin et al. 2015). Almost 5–30% of global methane emission comes from manure of livestock (Svenson et al. 1991; Sommer et al. 2000; Kulling et al. 2002). Emissions of methane from animal manure depend upon (i) type of animal from which manure was being collected such as pig, cattle, (ii) temperature at which manure was stored and (iii) storage condition of the manure, for instance, slurry, paste or pasture (Husted 1994; Amon et al. 2001; Su et al. 2003). CH_4 emission occurs when conditions favour anaerobic decomposition, so decomposition in the presence of aerobic environment can disrupt the peak of this emission. Emission of nitrous oxide (N_2O) accounts for 18% of total GHGs from livestock manure. In this chapter, we will assess the emission of these GHGs from different animal manures and amelioration of these emissions with management potential. The rising concern of greenhouse gases regulated by the Kyoto protocol under the supervision of UNFCCC. The reported outcomes claimed the reduction in GHGs emission by 9% between 2008 and 2012 since 1990.

Ruminants have special digestive track to convert plant material into nutritious food. These animals also synthesize fibres which are further consumed by human. This special type of digestion produces methane as well as other greenhouse gases which directly affect the climate. These gases have different global warming potential (GWP) such as CO_2 has 1 GWP, CH_4 has 25 GWP and N_2O has 310 GWP (Sejian et al. 2011).

8.1.1 Emission of Methane and Nitrous Oxide from Livestock Manure

Methane (CH_4) and Nitrous oxide (N_2O) are considered as major gases which have significant contribution in total GHGs emission. Both of these gases emitted as response of various microbial activity during management of animal manure. Following are the processes which involve in the generation of these gases:

8.1.2 Methanogenesis

Methanogenesis is an anaerobic mechanism to form methane as final product as a result of manure metabolism. Emission of methane occurs from all type of manure but most significantly it is linked with liquid and compact excreta (Osada et al. 2000). Manure contains organic matter, the anaerobic conditions along with other chemical processes break down the content of manure (Valentine 2007). But the process of methanogenesis occurs under strict conditions such as at low temperature and that's why cooling is the ameliorative factor of methane emission from manure. Aerobic bacteria involved in the process of methane oxidation (Hanson and Hanson 1996). Both the processes, methanogenesis and methane oxidation, contain similar bacteria that are present in compact animal manure (Sharma et al. 2011) (Fig. 8.1).

In case of livestock slurry, an organic layer is present on the surface which enhances the oxidation of methane (Ambus and Petersen 2005; Hansen et al. 2009). Another investigation of Nielsen et al. (2013) demonstrated that potency for CH_4 oxidation in two natural pig farms remain stunted prior to late autumn. It is indicated that activity of methanogens and oxidizing bacteria remained low in early autumn and summer when most of the emissions take place (Husted 1994).

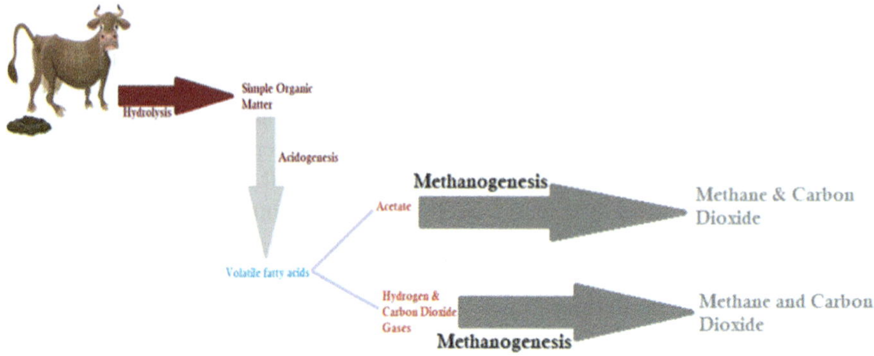

Fig. 8.1 Illustrating the process of methanogenesis

8.1.3 Emissions of Nitrous Oxide (N_2O) by Nitrification and Denitrification

Plant losses nitrogen during conversion of oxidized nitrogen to gaseous form during various sorts of operations viz. manure handling, storing and applying in field as fertilizer (Rotz 2004). Different form of N such as N_2O, NH_3 and NO_3^- contribute to the GHGs during manure management operations. The process of nitrification proceeds in two steps, in the foremost step ammonia is converted into nitrite (NO_2^-) with the help of NH_3 oxidizing bacteria and in later step NO_2^- converts into NO_3^- with the help of nitrite oxidizing bacteria.

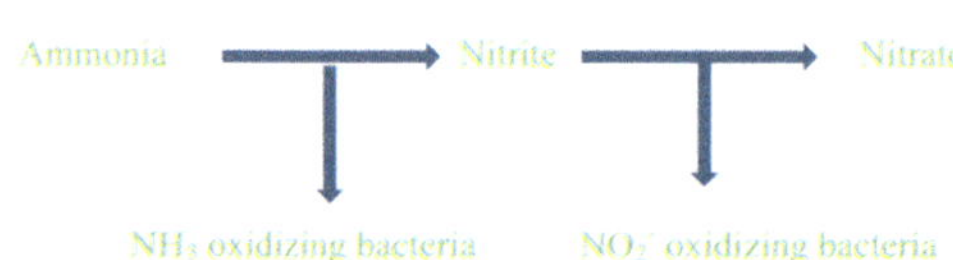

Ammonia oxidizing bacteria most prominently found in compact manure or soil and their contribution is totally uncertain (Jia and Conrad 2009; Yamamoto et al. 2010). Nitrous oxide generated during the process of ammonia formation or by the mechanism of denitrification (Goreau et al. 1980; Kool et al. 2011). The process of denitrification is accomplished by nitrifier bacteria which operate in the presence of oxygen otherwise, they used NO_2^- or NO_3^- as electron receptor (Thomsen et al. 2012). During the storage of manure and application in field, it is enriched in O_2 due to intense decomposing operations which enhance the process of nitrification and denitrification. During these chemical processes, N_2O emissions occur. Fresh manure contains neutral-alkaline pH while contain high quantity of ammonia and ammonium ions. In the atmosphere, CO_2 and NH_3 release in gaseous form. As the ammonia oxidizing is an acidic process, therefore pH of manure decreases. Petersen et al. (1992) documented that application of low pH manure in loam soil is ten times increase the ammonia oxidizing capacity.

8.2 Manure Management and Emission of Greenhouse Gases

Manure management consists of four fundamental steps (i) Generation from livestock, (ii) Manure storage, (iii) Manure treatment and (iv) Application to the fields/spreading. Each step of manure management emits certain amount of GHGs with different potential.

8.2.1 *Emission of N_2O from Animal Yards*

Animal yards without proper bedding material contain slurry, urine and faeces in anaerobic state. In this situation, there is negligible or no opportunity for nitrification of NH_4^+. As a consequence, there is little or almost no nitrous oxide emission from these yards (Zhang et al. 2005). Other investigations showed that animals' yards having straw bedding emit 4–5 mg of nitrous oxide N-m^2/d, while there was little or no emission in those yards which didn't have bedding material/slurry (Thorman et al. 2003). According to sources mentioned by Jungbluth et al. (2001), cow sheds emit 0.14–2.0 g N_2O/livestock unit (LU)/d or 0.05–0.7% of nitrogen excreted. Slurry-based pig homes with fully perforated flooring have been shown to emit between 0.66 and 3.62 g of N_2O per LU per day (Costa and Guarino 2009). Deep litter setups with fattening pigs may emit greenhouse gases that are substantially higher. Groenestein and Van Fassen (1996), for instance, recorded ratios of 4.8 and 7.2 g N_2O/LU/day, which correspond to 50–60% of the total N fluxes from the waste, with ammonia serving as the other prime component of gaseous N loss. Owing to higher fluxes nitrification and subsequent denitrification of NO_3, it has been documented that mechanically stirring of deep litter in the animal building causes N_2O emissions upsurge (Groenestein et al. 1993).

8.2.2 *Emission of Nitrous Oxide During Manure Storage and Treatment*

Animal manure in solid form contains both aerobic and anaerobic conditions and therefore can be a prime stimulator of nitrous oxide emission/generation. Emission of nitrous oxide ranges from < 1 to 4.3% of total stored nitrogen recorded in the heaps of cattle and pig farmyard. However, emission of 9.8% is also recorded in certain farmyard (Webb et al., in press). Other investigations showed the emission of nitrous oxide ranges from 0.2 to 0.8% (Thorman et al. 2006), in those animal heaps which have proper covering. These covering ultimately lessen the emission of NH_3 but did not show pronounced effect on reduction of nitrous oxide emission. Another study conducted by Chadwick (2005) manifested that covering of heaps in farmyard lessen the emission of NH_3 and N_2O. So, aforementioned studies demonstrated that maintaining anaerobic conditions in the farmyard is key factor to limit the emission of N_2O from solid manure heaps. Sommer et al. (2000), Yamulki (2006) demonstrated that addition of straw during solid excreta storage has ability to lessen the emission of GHGs. It has been reported that addition of 50% chopped straw in solid manure lessen the emission of nitrous oxide about 32% (Yamulki 2006).

Storage of slurry remained anaerobic under normal conditions until O_2 is introduced in the process of treatment. Slurry or liquid manures which do not have proper surface covering emit negligible amount of N_2O during storage (Sommer et al. 2000). Additionally, slurry with surface covering provides aerobic condition and thus nitrification takes place along with the emission of nitrous oxide (Sommer et al. 2000). Crust/surface layer in cattle manure is normal process while in the case of pig manure, it only form when content of organic matter will high. Different sorts of materials are used to cover liquid manure for lessen the release of NH_3 but these conditions enhance the process of crust formation which ultimately leads towards the emission of N_2O (Berg et al. 2006) (Fig. 8.2).

Crust formation in slurry has potential to control nitrous oxide emission. Sommer et al. (2000) demonstrated that emission of N_2O became high when water contents were reduced and even at extremely low water content, emissions of nitrous oxide have been reported as high as 25 N_2O-N-m^2/h. A study performed by Sommer et al. (2000) emphasized that N_2O emissions remained negligible during winter when the temperature was low and water contents were high at surface crust. Mechanism of intensive aeration for the sake of nitrogen removal enhances the emission of N_2O from the slurry. Burton et al. (1993) reported that emission of nitrous oxide exceeds to 19% of total nitrogen in the slurry during the process of aeration.

Process of slurry partitioned was also performed to get nutrient enriched solid and liquid fraction as well as to increase the storage capacity (Fangueiro et al. 2008).

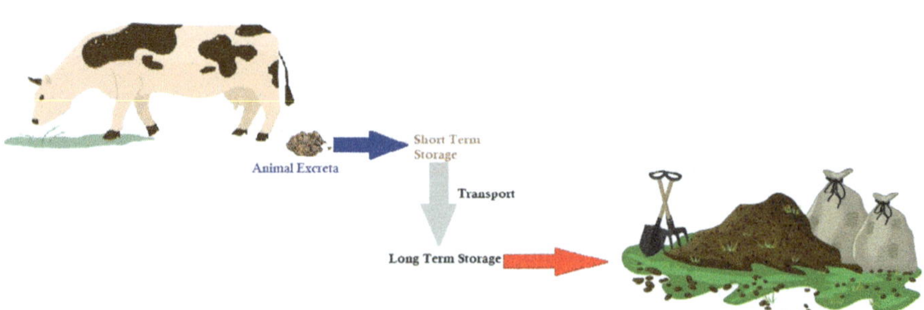

Fig. 8.2 Processing of manure management

Almost 4.8% of total nitrogen content of solid slurry losses in the form of N_2O in first 4 month of storage (Hansen et al. 2006). However, storage of liquid slurry emits negligible amount of N_2O as compared to untreated slurry. But overall, slurry fraction process losses nitrous oxide in the atmosphere during storage phase due to immense emission from solid slurry (Fangueiro et al. 2008; Dinuccio et al. 2008).

8.2.3 *Emission of Nitrous Oxide (N_2O) from Manure Spreading*

The spreading of solid/slurry manure to fields causes the emission of N_2O when ammonium ions (NH_4^+) of manure are subjected to aerobic condition of soil and it is start the process of nitrification which generates flux of nitrate (NO_3^-) (Chadwick et al. 2001). Emission factor ranges from < 0.1 to 3% but in case of pig slurry, it may reach up to 7.3–13.9% (Velthof et al. 2003). Variation in the range of emissions depends upon different factors viz. soil composition, soil conditions such as pH, open pore spaces, temperature and soil type. Manure spreading never causes the emission of nitrous oxide immediately until it already contains the flux of NO_3^-. Emission of nitrous oxide from manure spreading can be outlined by following factors:

8.2.4 *Type of Manure*

Chadwick et al. (2000) showed the emission of N_2O from soil having pig manure and dairy slurry where NH_4-N applied at same rate. The emission of N_2O was noticed substantially higher in dairy slurry as compared to pig. The dairy slurry showed the emission 2.42% of NH_3-N and 0.97% of total applied nitrogen, while the slurry of pig showed the emission of 0.94% of NH_3-N and 0.44% of total applied nitrogen. Difference in the emission pattern was observed due to carbon contents present in the manure as well presence of fine solids in the dairy slurry which block the pores' spaces in the soil and enhance anaerobic conditions. Rochette et al. (2008) reported no difference between the rate of emission by applying solid and liquid manures treatments.

8.2.5 *Type of Soil*

The emission of N_2O was recoded high in clay soil as compared to sandy. In another study conducted in the soils of Netherlands showed that emission rate in clay soil became two times higher as compared to sandy soil (van Groenigen et al. 2004). The main factor behind this difference is that clay soil provides favourable conditions for dentification while sandy soil favours nitrification only; therefore, emission of nitrous oxide was higher in clay soil as compared to sandy soil (Rochette et al. 2008).

8.2.6 *Season of Manure Application*

Agricultural crops demand proper manure application at proper time for better soil fertility and productivity (Anon. 2010). Excessive supply of manure can provide opportunity for NO_3^- by the process of leaching as well as emission of N_2O. The process of N_2O emission takes place by the activity of microbes; however, the rate of emission vary according to the time at which manure was applied and the temperature or water condition of the soil (Dobbie et al. 1999). The leaching of nitrogen as NO_3^- was observed highest during autumn/winter as compared to spring (Chambers et al. 2000).

8.2.7 *Rate of Manure Application in Fields*

Many studies manifested that rate of application of manure was found to be directly proportional to the emission of N_2O. Similar findings showed that high quantity of nitrogen fertilizer result in high emission of nitrous oxide (Cardenas et al. 2010). At high application rate, the oxygen depletion takes place which causes denitrification and eventually became a source of N_2O emission. Velthof et al. 2003 determined that application of manure at optimum rate, temperature and time is effective for proper utilization of nutrients and ceases the direct and indirect emission of N_2O.

8.3 Emission of Methane from Manure Management

Ruminants are generating methane gas as result of normal digestion. According to an estimate, the methane emission from manure contributed about 4% of all anthropogenic methane emission (USEPA 2006a, b). CH_4 emission from livestock estimated to be more than 16% till 2030 as compared to data collected in 2005 (USEPA 2011). This expending trend of methane emission reported in African countries (45%), Asian countries contributed 39% and Middle East account for 10% from 2005 till 2030; however, 3% of decreasing trend was observed between 1995 and 2005 due to reduction in livestock production. Swine contributes 50% methane production because their manure management significantly depends upon liquid-based systems which generate more CH_4 as compared to other systems (USEPA 2011).

8.3.1 Emission of CH_4 from Animal Houses

Manure present in animal yards emit continuously methane but most of the methane emission results as enteric emission. This globe emits more than 7 billion Gt of greenhouse gases every year from livestock. The majority of greenhouse gases emission occurred from cattle (81%), buffaloes (11%), while contribution of small ruminants limit up to 8%. Enteric fermentation from aforementioned ruminants contributes about 90 teragram of methane in the atmosphere while the rest of emission takes place by excrement. It has been reported that enteric emission of methane enhanced when the ruminants were fed with fibrous feed. Different approaches in different places were conducted to determine the enteric emission of methane. One of these approaches was conducted in Indian livestock where enteric emission was observed to be high. But the publish data is not consistence, some showed low emissions while others showed immense enteric emission from livestock yards (Popp et al. 2010). Animal yards are the major source of methane emission when manure was stored in the form of slurry and continuous elimination of slurry can become a potential factor to reduce this emission (Sommer et al. 2009).

8.3.2 Emission of CH_4 from Storage

Animal manure store in the solid form is considered as a major source of methane generation (Sommer and Møller 2000; Amon et al. 2001; Chadwick 2005; Hansen et al. 2006; Yamulki 2006; Szanto et al. 2007). Chadwick (2005) stated that management of heaps directly affects the generation of methane. Yamulki (2006) investigated that incorporation of straw with equal volume of cattle manure reduced the emission of methane. Methane emission can be reduced by promoting or inhibiting the anaerobic conditions. As covering of heap with air tight container inhibits the establishment of aerobic condition but elevation in temperature favours the anaerobic condition which leads to emission of methane. For instance, perfect covering of heaps reduces the emission of methane from 1.6 to 0.2 kg C/t or from 1.3 to 0.17% from initial carbon contents (Hansen et al. 2006). Frequently, turning and shuffling of heaps during management reduce the occurrence of anaerobic circumstances thus reducing the emission of methane (Amon et al. 2001, 2006).

Storage of slurry also caused emission of methane gas. A slight agitation in the stored slurry produces bubbles and dissolved gas which contain methane in the atmosphere (VanderZaag et al. 2010a, b). However, emission through this pathway considered to be very little and non-viable for long term. Covering the slurry with straw having porous surfaces limits the production of methane due to oxidation of carbon dioxide (Sommer et al. 2000; VanderZaag et al. 2009). VanderZaag et al. (2010b) elaborated that covering the slurry with permeable floating synthetic cover could reduce the emission of N_2O, CO_2 but remained unable to limit the emission of methane. Proper clearing of slurry can reduce the emission of methane from an area. For instance, elimination of slurry from animal yard can reduce 40% of methane emission because frequently removal of slurry also removes the bacteria which enhance methanogenesis (Haeussermann et al. 2006).

A pragmatic association was observed between the temperature and emission of methane from manure or slurry (Massé et al. 2003; Møller et al. 2004b; Pattey et al. 2005). Emission of methane was reported to be high when temperature exceeded 15 °C and it became low when temperature reached less than 15 °C (Husted 1994; Khan et al. 1997; Clemens et al. 2006; Sommer et al. 2007). Methane emission was observed high at 20 °C as compared to 10 °C. Vanderzaag et al., indicated that emission of methane has positive correlation with temperature (VanderZaag et al. 2010a). Fangueiro et al. (2008) reported that instead of complete slurry, cattle slurry should be converted into liquid and solid fractions which have potential to reduce the emission of methane more than 35%. On the other hand, Dinuccio et al. (2008) observed minor reduction (3–4%) in methane emission by slurry fraction during storage depending on various temperatures and conditions of slurry. Therefore, it is difficult to say that fraction of slurry can decrease or increase the emission of methane because it depends upon the storage condition as well as the type of slurry being stored.

8.3.3 *Emission of CH_4 by Manure Spreading*

The process of methanogenesis begins as soon as manure is spread on the soil (Chadwick and Pain 1997; Chadwick et al. 2000). These immediate methane emissions are short lived because as the spreading of manure takes place, the O_2 contents start to diffuse and hinder the process of methanogenesis thus ultimately reducing the emission of methane. In case of outdoor storage, the process of fermentation or separation of slurry in biogas digester is tremendous pathway to minimize the emission of methane. Treatment of slurry in the biogas digester reduces the emission during storage, the reported emissions from treated slurry were 30–66% lower in contrast to that slurry which left undisturbed (Clemens et al. 2006; Amon et al. 2006).

Major emissions of GHGs take place by enteric fermentation, management of manure, feed production and energy consumption. Enteric fermentation takes place in the gut of ruminant during the process of digestion and produce methane. Manure management produces both methane and nitrous oxide which depends upon the type of manure management system. Feed processing is another major source of CO_2 which contributed about 41% in GHGs emission as shows in Table 8.3. Energy consumption for fertilizer and other machinery also generates GHGs which accounted for 5% of total emissions (FAO 2015).

Table 8.3 Contribution of different resources in the emission of GHGs (FAO 2015)

Sources	Emissions
A. Enteric fermentation	44.3% CH_4
B. Feed processing	
(i) Deposited and applied manure	13.5% N_2O
(ii) Feed	13% CO_2
(iii) Fertilizer and crop residue	5.9% N_2O
(iv) Pasture expansion	4.8% CO_2
(v) Soy and palm	3.8% CO_2
(vi) Rice	0.5% CH_4
C. Manure management	5.2% CH_4 & 4.3% N_2O
D. Energy consumption	
(i) Post formgate	2.8% CO_2
(ii) Direct entirely use	1.6% CO_2
(iii) Indirectly use	0.3% CO_2

8.4 GHGs Mitigation Opportunities from Animal Manure Processing

The variety of manure conditions and management techniques observed in different regions has been discussed in the earlier sections. The complicated monitoring of microbial processes that cause CH_4 and nitrous oxide emissions makes it evident that GHG mitigation for manure management is a major problem. Regional differences in the socioeconomic and cultural environment of livestock farming can affect the availability of options to enhance manure handling for Mitigation. Number of environmental issues originated due to immense production and accumulation of organic waste in the surrounding and inappropriate way to manage it. Sufficient and balanced supply of animal manure provides essential micro and macro nutrients to enhance soil fertility and productivity (Moral et al. 2009). However, overcrowding of cattle in yards produced huge volume of manure each year. This immense volume needs to be recycled otherwise it cause adverse effects such as leaching of essential nutrients from the soil, leaching of nitrogen and ammonia, emission of greenhouse gases and other harmful gases (Massé et al. 2011). Introduction of appropriate technologies to recycle/process the animal manure are highly and urgently needed to eliminate the risk of environmental/climatic balance.

Although, NH_3 is not actual greenhouse gas but along with its ionized form NH_4^+ these are vital component of manure nitrogen cycle. As the animal urine spread on barn floor and pasture, the process of decomposition initiates immediately in which microbial enzyme urease releases ammonia in the form of NH_4^+ ions. Urease enzyme is abundant in faecal organic matter and therefore converts excreted waste into ammonium ions when existing conditions are favourable. Under aerobic conditions, the ammonium ions are converted into NO_3^-. . Plants have potential to avail

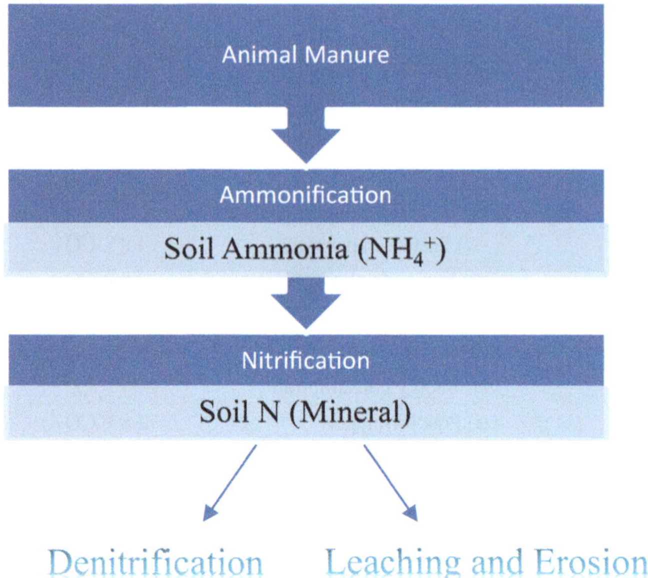

Fig. 8.3 Conversion of animal manure into soil ammonia

both of these forms while cannot intake organic form of manure (Beegle et al. 2008). NH_4^+ ions also act as carrier of soil nitrogen to the atmosphere as well as precursor of nitrous oxide emission from manure applied in the field or stored in pasture. Therefore, ammonia is put under strict consideration while managing the greenhouse gases emissions from animal manure (Fig. 8.3).

8.4.1 Management of Animals and Their Housing

Covering the animal house cannot directly limit the emission of methane or nitrous oxide but the type of material used to cover the house can help to quantify the management, storage and application of manure. Animal houses along with solid floors with straw/hay can accumulate more dry matter manure which enhances the process of nitrification and denitrification thus emission of greenhouse gases takes place. Külling et al. (2001, 2003) argued that slurry can limit the emission of greenhouse gases as compared to solid manure. Amon et al. (2001) reported that slurry-based system produces less greenhouse gases as compared to composted manure where most of the gases emission occurs due to their turning for the sake of aeration.

Hassouna et al. (2010) investigated that emission of gases was high in solid and straw bedding-based system while generation rate was much lower in liquid-based system. Manipulating the route of nitrogen emission is key tool for mitigation. Urea contributed 60–80% of total nitrogen in the urine due to CP. Therefore, reducing the amount of dietary CP is the core strategy to ameliorate the emission of NH_3 in stored manure. Velthof et al. (2005) observed that emission of N_2O and CH_4 decreases as the protein content of swine diet decreases during storage of manure.

Management and feeding are another key factor to mitigate the emission of greenhouses. For instance, phase feeding operations diminish the emission of GHGs. Lessen the amount of dietary protein during production cycle, reduce the emission of N (Cole et al. 2005, 2006; Vasconcelos et al. 2007). Erickson and Klopfenstein (2010) observed 12–21% less emission of N through phase feeding in cattle yards.

8.4.2 Animals Grazing Practices

Improving forage digestibility is wonderful approach to minimize GHGs emissions and volume of manure production. In pasture, forage digestibility improvement means enhancing application rates of nitrogen fertilizer which pose noxious impact on urinary nitrogen excretion thus eventually on emissions of N_2O and NH_3. Emission of N_2O was observed to be high in pasture-based system because of high concentration of nitrogen in the urine due to elevated CP content in diet. De Klein et al. (2001) reported 40–57% reduction in nitrous oxide emission when animals were bound to graze on an area of 3 h/day in late autumn. Reduction in the emissions of GHGs has been reported by adopting following approaches (i) improve the nitrogen use efficiency, (ii) enhance soil management practices, (iii) pasture rehabilitation, (iv) using soil additives to manipulate nitrogen cycling in soil, (v) planation for proper utilization of nitrogen and (vi) modifying grazing and farm management (Luo et al. 2010).

8.4.3 Process of Biofiltration

The process of biofiltration captures the ventilated air of pasture building to control odour, absorb NH_3 and conversion of ammonia to NO_3 by the assistance of biological scrubbers. Limit the losses of ammonia is an indirect approach to diminish the emission of nitrous oxide in the atmosphere by reducing the deposition of ammonium ions in the soil and ultimately cease the process of nitrous oxide generation as discussed above. Different biological filters have distinct potential to eliminate ammonia. For instance, acidic scrubbers have 91% efficiency to remove ammonia while bio trickling filters have 70% (Melse and Ogink 2005). Biofilters along with heat exchangers proved to be efficient in elimination of NH_3. It has been reported that biofilters proved effective in treating high concentration of ammonia with elimination efficiency of 79% (Shah et al. 2011). These biofilters proved to be efficient in trapping the sulphur gases emitted from broiler house. The elimination of ammonia in biofilters is directly linked with the conversion of ammonia to NO_2^- to NO_3^-.. So, in the elimination process through biofilters, the generations of nitrous oxide must be kept under considerations (Maia et al. 2012).

Several studies reported the reduction in CH_4 by passing the swine through biofiltration process. Reduction in CH_4 was observed about 50–60% by Canadian pork council in 2005. Girard et al. (2011) observed the elimination rate up to 40%. Melse and van der Werf (2005) findings revealed that 85% of methane reduction was reported via the biofiltration system which consisted of compost and perlite treated with methane oxidizing bacteria (taken from manure). The elimination potential of methane in biofilters depends upon the concentration of methane present in the biofilters stream. Therefore, to eliminate 50% of methane emission require immense biofilters systems which limit the use of this technology on random basis. Melse and Timmerman (2009) suggested the use of multipollutant scrubbers, biofilters and acid combing filters to not only mitigate the greenhouse gases but also ameliorate the particulate matter from animal yards.

8.4.4 Amelioration of GHGs from Manure Storage and Treatment

Manure storage as aforementioned facilitates anaerobic conditions to favour methane emission, although the emission of nitrous oxide and ammonia also takes place in high quantity. A quick approach to reduce the emission of GHGs particularly methane is reduction in storage time (Philippe et al. 2007; Costa et al. 2012). Long-term storage provides opportunity for emission of methane along with fast rate of emission due to anaerobic condition (Philippe et al. 2007). Regulating the manure temperature is another strategy to reduce the emission of methane during storage condition (Steed and Hashimoto 1994).

Providing the temperature of less than 10 °C to manure by removing it from building and store in cold weather can reduce the emission of methane (Monteny et al. 2006). In poultry manure, the storage treatments along with appropriate moisture and aeration lessen the emission of CH_4 (Li and Xin 2010). Partitioning of swine slurry into liquid and solid and later treated the solid slurry with aerated composting diminish the emission of methane to 99% and nitrous oxide to 75% as compare to undisturbed manure (Vanotti et al. 2008). However, due to opposite relationship of ammonia and nitrous oxide (Petersen and Sommer 2011), this process may surge the emission of nitrous oxide. Amon et al. (2001) observed substantial nitrous oxide emission from turned pile of solid manure as compare to those which left undisturbed.

8.4.5 Different Types of Covers for Manure Storage

Different types of covers mentioned in the literature for manure such as wood cover, oil layer, natural crust, straw cover and many others. How much a cover is affective is correlated with number of factors such as degradability, porosity, permeability of cover and thickness. Semipermeable covers enhance the emission of nitrous oxide because they provide aerobic condition which favour nitrification, while on the downside it creates oxygen deficient environment which further induces the process of denitrification (Hansen et al. 2009; Nielsen et al. 2010). Such sort of covers helps to limit the emission of methane, ammonia and to control odour but elevate the emissions of nitrous oxide (Sommer et al. 2000).

Impermeable covers (oil layers, solid plastic) help to mitigate the emissions of methane, nitrous oxide and ammonia (Sommer et al. 2000; Guarino et al. 2006; VanderZaag et al. 2008). Guarino et al. (2006), VanderZaag et al. (2008) reported that using vegetable oil as cover can limit the emissions of GHGs but in practical not applicable due to its quick degradation. Use of impermeable cover becomes effective to limit the emission of methane, if trapped methane used immediately for other purposes such as running an engine to generate electricity otherwise accumulation of methane under the cover will create pressure which may burst the cover. The success of the impermeable covers depends upon the transfer of trapped gases particularly GHGs for other purposes (Nicolai and Pohl 2004).

8.4.6 Mitigation by the Process of Anaerobic Digestion

Anaerobic digestion is a process that is carried out by archaea in the absence of oxygen to degrade organic material and it cause the emission of CO_2, N_2O, CH_4 and other gases as by-product. A careful practice can assist to ameliorate the emissions of GHGs. When this process is carried out in efficient way, it produces 60–80% methane which is used as a source of energy depending upon the substrate and operation conditions (Roos et al. 2004). Biogas digesters are widely used to capture the emitted methane and direct it for energy uses to prevent its emission into atmosphere. Biogas digester varies in size, functions and other operations. Some digester designed to operate with manure from some animals to many. These digester produce 23–53% less GHGs as compared to house hold biogas digester (Dhingra et al. 2011). Anaerobic digestion (AD) technology adoption can eliminate the emission of greenhouse gases. Results demonstrated in Table 8.3 showed that AD not only offset the emission of greenhouse gases but also act as an alternative of fossil fuels to overcome the downfall of energy. Anaerobic digestion reduced the GHGs annually 177 mg CO_2 eq. emissions from dairy manure, 87.7 mg CO_2 eq. from sow and 125.6 Mg CO_2 eq. from pig farms (Kaparaju and Rintala 2011) (Fig. 8.4).

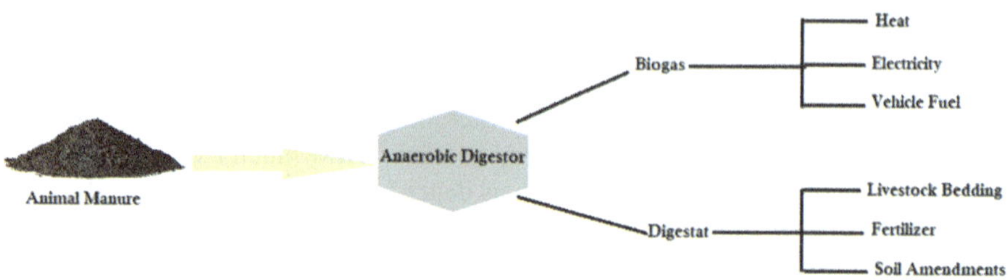

Fig. 8.4 Mechanism of anaerobic digestion

8.4.7 *Acidification of Manure*

Lowering the pH of stored manure is effective way to mitigate the emission of greenhouse gases. Petersen and Sommer (2011) reported that acidification of manure is effective strategy to mitigate the emission of ammonia but its effect on nitrous oxide is not reported so far. Study of 15 different cattle, swine and poultry manure in which emissions of NH_3 were ameliorated successfully (from 14 to 100%) by using hydrochloric acid, sulfuric acid, calcium, phosphoric acid or monocalcium phosphate. These studies concluded that using the strong acid mitigated the emission of ammonia to larger extent rather than weaker one but use of strong acid at farms may be hazardous (Ndegwa et al. 2011).

Petersen et al. (2012) investigated the impact of acidification on emission of methane from stored manure. PH of stored manure was fixed at 5.5 with the help of H_2SO_4 and sample was stored for 95 days. During the storage, emissions of ammonia and methane were monitored carefully. PH of manure gradually increased up to 6.5–7.0. Results showed that the emissions of methane reduce by 67–87% and NH_3 was completely mitigated. Aforementioned study showed that acidification is cost-effective technique to mitigate the emission of greenhouse gases.

8.4.8 *Composting*

Composting is microbial process carried out in the presence of oxygen to decompose the organic matter (OM) which helps to control odour, manure handling and pathogen control. Solid composted manure is used as bedding for the comfort of cows ensuring that their health is not compromised (Husfeldt et al. 2012). The process of composting of manure causes the losses of N which depends upon moisture level, temperature, PH and material stability (Zeman et al. 2002). The authors also claimed that by raising the solids content of the feedstocks and the carbon to Nitrogen ratio, it is conceivable to drastically diminish emissions from compost piles. Consequently, composting can be a useful technique for lowering GHG emissions from a variety of waste products, especially livestock manure (Brown et al. 2008).

8.5 Conclusions

Carbon dioxide (CO_2), nitrous oxide (N_2O) and methane (CH_4) are all key sources of manmade GHG (greenhouse gas) from animal manure (CO_2). About 37% of the world's methane emissions are caused by animals. The accumulation of liquid manure, also known as slurry, and compact solid manure is the prime sources of GHG (Vac et al. 2013). According to IPCC statistics, animal agriculture is accountable for 8–10.8% of the overall of the world's greenhouse gas (GHG) emissions, and according to lifecycle assessment, livestock can account for up-to 18% of these emissions. Cattle emissions account for the majority of these emissions (O'Mara 2011). At every step of the manure management chain, it is evident that manure handling has an influence on the volumes of both direct and indirect nitrous oxide emissions and CH_4 emissions. Anaerobic digestion (AD) processing of liquid manure and other wastes contributes to renewable energy goals by making it a desirable option for greenhouse gas (GHG) mitigation. In order to reuse nutrients for crop production, unprocessed manure is frequently kept in storage for a period. Methane (CH_4), nitrous oxide (N_2O) and ammonia (NH_3) emissions that occur throughout storage influence the balance of greenhouse gas emissions.

References

Amon B, Amon T, Boxberger J, Alt C (2001) Emissions of NH_3, N_2O and CH_4 from dairy cows housed in a farmyard manure tying stall (housing, manure storage, manure spreading). Nutr Cycl Agroecosyst 60:103–113

Amon B, Kryvoruchko V, Amon T, Zechmeister-Boltenstern S (2006) Methane, nitrous oxide and ammonia emissions during storage and after application of dairy cattle slurry and influence of slurry treatment. Agric Ecosyst Environ 112:153–162

Anon. (2010) Fertiliser manual (RB209), 8th edn. The Stationary Office, London, UK, 256 pp. www.defra.gov.uk/foodfarm/landmanage/landsoil/nutrient/documents/rb209-rev-100609.pdf

Ambus P, Petersen SO (2005) Oxidation of 13C-labeled methane in surface crusts of pig- and cattle slurry. Isot Environ Health Stud 41:125–133

Basset-Mens C, van der Werf HMG (2005) Scenario-based environmental assessment of farming systems: the case of pig production in France. Agr Ecosyst Environ 105:127–144

Berg W, Brunsch R, Pazsiczki I (2006) Greenhouse gas emissions from covered slurry compared with uncovered during storage. Agric Ecosyst Environ 112:129–134

Beegle DB, Kelling KA, Schmitt MA (2008) Nitrogen from animal manures. Nitrogen Agricultural Syst 49:823–881

Brown S, Kruger C, Subler S (2008) Greenhouse gas balance for composting operations. J Environ Qual 37:1396–1410

Burton CH, Sneath RW, Farrent JW (1993) Emissions of nitrogen oxide gases during aerobic treatment of animal slurries. Biores Technol 45(3):233–235

Casey JW, Holden NM (2005) The relationship between greenhouse gas emissions and the intensity of milk production in Ireland. J Environ Qual 34:429–436

Casey JW, Holden NM (2006) Quantification of greenhouse gas emissions from suckler-beef production in Ireland. Agric Syst 90:79–98

Cederberg C, Mattson B (2000) Life cycle assessment of milk production—a comparison of conventional and organic farming. J Clean Prod 8:49–60

Cederberg C, Stadig M (2003) System expansion and allocation in life cycle assessment of milk and beef production. Int J Life Cycle Assess 8(6):350–356

Chambers BJ, Smith KS, Pain BF (2000) Strategies to encourage better use of nitrogen in animal manures. Soil Use Manage 16:157–161

Cardenas LMR, Thorman R, Ashlee N, Butler M, Chadwick D, Chambers B, Cuttle S, Donovan N, Kingston H, Lane S, Scholefield D (2010) Emission factors for N_2O fluxes from grazed grassland soils in the UK. Agric Ecosyst Environ 136:218–226

Ciais P et al (2013) Climate change 2013: the physical science basis. Contribution of working group I to the fifth assessment report of the intergovernmental panel on climate change. Cambridge University Press

Chadwick DR, Pain BF (1997) Methane fluxes following slurry applications to grassland soils: laboratory experiments. Agric Ecosyst Environ 63:51–60

Chadwick DR, Pain BF, Brookman SKE (2000) Nitrous oxide and methane emissions following application of animal manures to grassland. J Environ Qual 29:277–287

Chadwick DR, Martinez J, Marol C, Béline F (2001) Nitrogen transformations and ammonia loss following injection and surface application of pig slurry: a laboratory experiment using slurry labelled with 15N-ammonium. J Agric Sci 136:231–240

Chadwick DR (2005) Emissions of ammonia, nitrous oxide and methane from cattle manure heaps: effect of compaction and covering. Atmos Environ 39(4):787–799

Costa A, Guarino M (2009) Definition of yearly emission factor of dust and greenhouse gases through continuous measurements in swine husbandry. Atmos Environ 43:1548–1556

Costa A, Chiarello GL, Selli E, Guarino M (2012) Effects of TiO_2 based photocatalytic paint on concentrations and emissions of pollutants and on animal performance in a swine weaning unit. J Environ Manage 96:86–90

Clemens J, Trimborn M, Weiland P, Amon B (2006) Mitigation of greenhouse gas emissions by anaerobic digestion of cattle slurry. Agric Ecosyst Environ 112:171–177

Cole NA, Clark RN, Todd RW, Richardson CR, Gueye A, Greene LW, McBride K (2005) Influence of dietary crude protein concentration and source on potential ammonia emissions from beef cattle manure. J Anim Sci 83:722–731

Cole NA, Defoor PJ, Galyean ML, Duff GC, Gleghorn JF (2006) Effects of phase-feeding of crude protein on performance, carcass characteristics, serum urea nitrogen concentrations, and manure nitrogen of finishing beef steers. J Anim Sci 84:3421–3432

Dhingra R, Christensen E, Liu Y, Zhong B, Fu C, Yost M, Remains J (2011) Greenhouse gas emission reductions from domestic anaerobic digesters linked with sustainable sanitation in rural China. Environ Sci Technol 45:2345–2352

Dobbie KE, McTaggart IP, Smith KA (1999) Nitrous oxide emissions from intensive agricultural systems: variations between crops and seasons, key driving variables, and mean emission factors. J Geophys Res 104:26891–26899

De Klein CAM (2001) An analysis of environmental and economic implications of nil and restricted grazing systems designed to reduce nitrate leaching from New Zealand dairy farms. II. Pasture production and cost/benefit analysis. N Z J Agric Res 44:217–235

Dai A (2016) Future warming patterns linked to today's climate variability. Sci Rep 6:19110

Dinuccio E, Berg W, Balsaria P (2008) Gaseous emissions from the storage of untreated fractions obtained after mechanical separation. Atmos Environ 42:2448–2459

EPA (2017) Inventory of U.S. Greenhouse gas emissions and sinks: 1990–2015. EPA 430-P-17–001. Environmental Protection Agency (EPA), Washington DC

Fangueiro D, Pereira JL, Chadwick D, Coutinho J, Moreira N, Trindade H (2008) Laboratories estimates of the effect of cattle slurry pre-treatment on organic N degradation after soil application and N_2O and N_2 emissions. Nutr Cycling Agroecosyst 80:107–120

FAO (2006) Livestock report. Food and Agricultre Organization of the United Nations, Rome, p 85

FAO (2009) Global agriculture towards 2050. https://www.fao.org/fileadmin/templates/wsfs/docs/Issues_papers/HLEF2050_Global_Agriculture.pdf

FAO (2015) GLEAM 3.0 Assessment of greenhouse gas emissions and mitigation potential. https://www.fao.org/gleam/dashboard-old/en/

Fließbach A, Oberholzer H-R, Gunst L, Mäder P (2007) Soil organic matter and biological soil quality indicators after 21 years of organic and conventional farming. Agriculture Ecosyst Environ 118(1–4)

Garnett T (2009) Livestock-related greenhouse gas emissions: impacts and options for policy makers 12:491–503

Gerber P, Chilonda P, Franceschini G, Menzi H (2005) Geographical determinants and environmental implications of livestock production intensification in Asia. Biores Technol 96:263–276

Girard M, Ramirez AA, Buelna G, Heitz M (2011) Biofiltration of methane at low concentrations representative of the piggery industry—influence of the methane and nitrogen concentrations. Chem Eng J 168:151–158

Goreau TJ, Kaplan WA, Wofsky SC, McElroy MB, Valois FW, Watson SW (1980) Production of NO_2 and N_2O by nitrifying bacteria at reduced concentrations of oxygen. Appl Environ Microbiol 40:526–532

Groenestein CM, Oosthoek J, Van Faassen HG (1993) Microbial processes in deep-litter systems for fattening pigs and emission of ammonia, nitrous oxide and nitric oxide. EAAP Publ

Groenestein CM, Van Fassen HG (1996) Volatilisation of ammonia, nitrous oxide and nitric oxide in deep-litter systems for fattening pigs. J Agric Eng Res 65:269–274

Husted S (1994) Seasonal variation in methane emission from stored slurry and solid manures. J Environ Qual 23:585–592

Hanson RS, Hanson TE (1996) Methanotrophic bacteria. Microbiol Rev 60:439–471

Hansen MN, Henriksen K, Sommer SG (2006) Observations of production and emission of greenhouse gases and ammonia during storage of solids separated from pig slurry: effects of covering. Atmos Environ 40(22):4172–4181

Hansen RR, Nielsen DA, Schramm A, Nielsen LP, Revsbech NP, Hansen MN (2009) Greenhouse gas microbiology in wet and dry straw crust covering pig slurry. J Environ Qual 38(3):1311–1319

Husfeldt AW, Endres MI, Salfer JA, Janni KA (2012) Management and characteristics of recycled manure solids used for bedding in Midwest freestall dairy herds. J Dairy Sci 95:2195–2203

Hansen J, Ruedy R, Sato M, Lo K (2010) Global surface temperature change. Rev Geophys 48:RG4004

Hassouna M, Robin P, Brachet A, Palliat J, Dollé J, Faverdin P (2010) Development and validation of a simplified method to quantify gaseous emissions from cattle buildings. In: Proceedings of the XVII world congress of the international commission of agricultural engineering (CIGR). Québec City, Canada, 13–17 June 2010

Hristov AN, Zaman S, Vander Pol M, Ndegwa P, Campbell L, Silva S (2002) Nitrogen losses from dairy manure estimated through nitrogen mass balance and chemical markers. J Environ Qual 38:2438–2448. https://doi.org/10.2134/jeq2009.0057

Haeussermann A, Hartung E, Gallmann E, Jungbluth T (2006) Influence of season, ventilation strategy, and slurry removal on methane emissions from pig houses. Agric Ecosyst Environ 112:115–121

IAASTD (2008) Executive summary of the synthesis report. International Assessment of Agricultural Knowledge, Science and Technology for Development. http://www.agassessment.org/

Jerez S, López-Romero JM, Turco M, Jiménez-Guerrero P, Vautard R, Montávez JP (2018) Impact of evolving greenhouse gas forcing on the warming signal in regional climate model experiments. Nat Commun 9:1304

Jia Z, Conrad R (2009) Bacteria rather than archaea dominatemicrobial ammonia oxidation in an agricultural soil. Environ Microbiol 11:1658–1671

Jungbluth T, Hartung E, Brose G (2001) Greenhouse gas emissions from animal houses and manure stores. Nutr Cycling Agroecosyst 60:133–145

Kaparaju P, Rintala J (2011) Mitigation of greenhouse gas emissions by adopting anaerobic digestion technology on dairy, sow and pig farms in Finland. Renew Energy 36:31–41

Khan RZ, Müeller C, Sommer SG (1997) Micrometeorological mass balance technique for measuring CH_4 emission from stored cattle slurry. Biol Fertil Soils 24:442–444

Kool DM, Dolfing J, Wrage N, Van Groenigen JW (2011) Nitrifier denitrification as a distinct and significant source of nitrous oxide from soil. Soil Biol Biochem 43(1):174–178

Külling DR, Menzi H, Krober TF, Neftel A, Sutter F, Lischer P, Kreuzer M (2001) Emissions of ammonia, nitrous oxide and methane from different types of dairy manure during storage as affected by dietary protein content. J Agric Sci 137:235–250

Külling DR, Menzi H, Sutter F, Lischer P, Kreuzer M (2003) Ammonia, nitrous oxide and methane emissions from differently stored dairy manure derived from grass- and hay-based rations. Nutr Cycling Agroecosyst 65:13–22

Kulling DR, Dohme F, Menzi H, Sutter F, Lischer P, Kreuzer M (2002) Methane emissions of differently fed dairy cows and corresponding methane and nitrogen emissions from their manure during storage. Environ Monit Assess 79:129–150

Li H, Xin H (2010) Lab-scale assessment of gaseous emissions from laying-hen manure storage as affected by physical and environmental factors. Trans ASABE 53:593–604

Luo J, de Klein CAM, Ledgard SF, Saggar S (2010) Management options to reduce nitrous oxide emissions from intensively grazed pastures: a review. Agric Ecosyst Environ 136:282–291

Lovett DK, Shalloo L, Dillon P, O'Mara FP (2006) A systems approach to quantify greenhouse gas fluxes from pastoral dairy production as affected by management regime. Agric Syst 88(2–3):156–179

Maia GDN, Day GB, Gates RS, Taraba JL (2012) Ammonia biofiltration and nitrous oxide generation during the start-up of gas-phase compost biofilters. Atmos Environ 46:659–664

Massé DI, Croteau F, Patni NK, Massé L (2003) Methane emissions from dairy cow and swine manure slurries stored at 10 °C and 15 °C. Can Biosyst Eng 45:6.1–6.6

Massé DI, Talbot G, Gilbert Y (2011) On farm biogas production: a method to reduce GHG emissions and develop more sustainable livestock operations. Anim Feed Sci Tech 166–167:436–445

McLeod A, Food and Organization of the United Nations (2011) World livestock 2011 livestock in food security world.

Melse RW, van der Werf AW (2005) Biofiltration for mitigation of methane emission from animal husbandry. Environ Sci Technol 39:5460–5468

Melse RW, Timmerman M (2009) Sustainable intensive livestock production demands manure and exhaust air treatment technologies. Bioresour Technol 100:5506–5511

Møller HB, Sommer SG, Ahring BK (2004) Biological degradation and greenhouse gas emissions during pre-storage of liquid animal manure. J Environ Qual 33:27–36

Moral R, Paredes C, Bustamante MA, Marhuenda-Egea R, Bernal MP (2009) Utilisation of manure composts by high-value crops: safety and environmental challenges. Bioresour Technol 100:5454–5460

Monteny GJ, Bannink A, Chadwick D (2006) Greenhouse gas abatement strategies for animal husbandry. Agric Ecosyst Environ 112:163–170

Myhre G, Shindell D, Bréon FM, Collins W, Fuglestvedt J, Huang J, Koch D, Lamarque JF, Lee D, Mendoza B, Nakajima, Zhang H (2013) Climate change 2013: the physical science basis. In: Contribution of working group I to the fifth assessment report of the intergovernmental panel on climate change, pp 659–740

Nielsen DA, Nielsen LP, Schramm A, Revsbech NP (2010) Oxygen distribution and potential ammonia oxidation in floating, liquid manure crusts. J Environ Qual 39(5):1813–1820

Nielsen DA, Schramm A, Nielsen LP, Revsbech NP (2013) Seasonal methane oxidation potential in manure crusts. Appl Environ Microbiol 79(1):407–410

Nicolai R, Pohl S, Schmidt D (2004) Covers for manure storage units

Ndegwa PM, Hristov AN, Ogejo JA (2011) Ammonia emission from animal manure: mechanisms and mitigation techniques. In: He Z (ed) Environmental chemistry of animal manure. Nova Science Publishers, Hauppauge, NY, pp 107–151

O'Mara FP (2011) The significance of livestock as a contributor to global greenhouse gas emissions today and in the near future. Anim Feed Sci Technol 166–167:7–15

Osada T, Kuroda K, Yonaga M (2000) Determination of nitrous oxide, methane, and ammonia emissions from a swine waste composting process. J Mater Cycles Waste Manage 2:51–56

Pattey E, Trzcinski MK, Desjardins RL (2005) Quantifying the reduction of greenhouse gas emissions as a result of composting dairy and beef cattle manure. Nutr Cycling Agroecosyst 72:173–187

Petersen SO, Nielsen AL, Haarder K, Henriksen K (1992) Factors con-trolling nitrification and denitrification: a laboratory study with gel-stabilized liquid cattle manure. Microb Ecol 23:239–255

Petersen SO, Sommer SG (2011) Ammonia and nitrous oxide interactions: roles of manure organic matter management. Anim Feed Sci Technol 166–167:503–513

Petersen SO, Andersen AJ, Eriksen J (2012) Effects of cattle slurry acidification on ammonia and methane evolution during storage. J Environ Qual 41:88–94

Philippe FX, Laitat M, Canart B, Vandenheede M, Nicks B (2007) Comparison of ammonia and greenhouse gas emissions during the fattening of pigs, kept either on fully slatted floor or deep litter. Livest Sci 111:144–152

Popp A, Lotze-Campen H, Bodirsky B (2010) Food consumption, diet shifts and associated non-CO_2 greenhouse gases from agricultural production. Glob Environ Chang 20(3):451–462

Roos KF, Martin JH, Moser MA (2004) AgSTAR handbook: a manual for developing biogas systems at commercial farms in the United States, 2nd edn. EPA-430-B-97-015. U.S. Environmental Protection Agency, Washington, DC

Rotz CA (2004) Management to reduce nitrogen losses in animal production. J Anim Sci 82:E119–E137

Rochette P, Angers DA, Chantigny MH, Gagnon B, Bertrand N (2008) N_2O fluxes in soils of contrasting textures fertilized with liquid and solid dairy cattle manures. Can J Soil Sci 88:175–187

Sharma R, Ryan K, Hao X, Larney FJ, McAllister TA, Topp E (2011) Real-time quantification of mcrA, pmoA for methanogen, methanotroph estimations during composting. J Environ Qual 40:199–205

Severin M, Fuß R, Well R, Garlipp F, Van den Weghe H (2015) Soil, slurry and application effects on greenhouse gas emissions. Plant Soil Environ 61(8):344–351

Shah SB, Workman DJ, Yates J, Basden TJ, Merriner CT, de Graft-Hanson J (2011) Coupled biofilter-heat exchanger prototype for a broiler house. Appl Eng Agric 27:1039–1048

Smith P, Bustamante M, Ahammad H, Clark H, Dong H, Elsiddig E, Haberl H, Harper R, House J, Jafari M, Masera O, Mbow C, Ravindranath N, Rice C, Robledo Abad C, Romanovskaya A, Sperling F, Tubiello F (2014)

Sommer SG, Møller HB (2000) Emission of greenhouse gases during composting of deep litter from pig production—effect of straw content. J Agric Sci 134:327–335

Sommer SG, Petersen SO, Sørensen P, Poulsen HD, Møller HB (2007) Methane and carbon dioxide emissions and nitrogen turnover during liquid manure storage. Nutr Cycling Agroecosyst 78:27–36

Sommer SG, Olesen JE, Petersen SO, Weisbjerg MR, Valli L, Rohde L, Béline F (2009) Region-specific assessment of greenhouse gas mitigation with different manure management strategies in four agroecological zones. Glob Change Biol 15:2825–2837

Sommer SG, Petersen SO, Sogaard HT (2000) Atmospheric pollutants and trace gases. J Environ Qual 29:744–751

Szanto GL, Hamelers HVM, Rulkens WH, Veeken AHM (2007) NH_3, N_2O and CH_4 emissions during passively aerated composting of straw-rich pig manure. Bioresour Technol 98:2659–2670

Sejian V, Lal R, Lakritz J, Ezeji T (2011) Measurement and prediction of enteric methane emission. Int J Biometeorol 55:1–16

Su JJ, Liu BY, Chang YC (2003) Emission of greenhouse gas from livestock waste and wastewater treatment in Taiwan. Agr Ecosyst Environ 95(1):253–263

Steed J Jr, Hashimoto AG (1994) Methane emissions from typical manure management systems. Bioresour Technol 50:123–130

Thangarajan R, Bolan NS, Tian G, Naidu R, Kunhikrishnan A (2013) Role of organic amendment application on greenhouse gas emission from soil. Sci Total Environ 465:72–96. https://doi.org/10.1016/j.scitotenv.2013.01.031

Thorman RE, Harrison R, Cooke SD, Chadwick DR, Burston M, Balsdon SL (2003) Nitrous oxide emissions from slurry- and straw-based systems for cattle and pigs in relation to emissions of ammonia. In: McTaggart I, Gairns L (eds) Proceedings of SAC/SEPA conference on agriculture, waste and the environment. Edinburgh (UK), 26–28 Mar 2002, pp 26–32

Thorman RE, Chadwick DR, Boyles LO, Matthews R, Sagoo E, Harrison R (2006, July) Nitrous oxide emissions during storage of broiler litter and following application to arable land. In: International congress series, vol 1293. Elsevier, pp 355–358

Thomsen IK, Olesen JE, Møller HB, Sørensen P, Christensen BT (2012) Carbon dynamics and retention in soil after anaerobic digestion of dairy cattle feed and faeces. Soil Biol Biochem 58:82–87

USEPA (2006a) Global anthropogenic non-CO_2 greenhouse gas emissions: 1990–2020. U.S. Environmental Protection Agency, Washington, DC

United States Environmental Protection Agency (USEPA) (2006b) Global anthropogenic non-CO_2 greenhouse gas emissions: 1990–2020, EPA 430-R-06-003. Washington D.C.

USEPA (2011) DRAFT: global anthropogenic non-CO_2 greenhouse gas emissions: 1990–2030. U.S. Environmental Protection Agency, Washington, DC

United Nations Environment Programme, undated. Animal waste management in livestock farms: practical environmental policies and reviews. http://www.agrifoodforum.net/home.asp

VanderZaag AC, Gordon RJ, Jamieson RC, Burton DL, Stratton GW (2009) Gas emissions from straw covered liquid dairy manure during summer storage and autumn agitation. Trans Am Soc Agric Biol Eng 52:599–608

VanderZaag AC, Gordon RJ, Jamieson RC, Burton DL, Stratton GW (2010a) Effects of winter storage conditions and subsequent agitation on gaseous emissions from liquid dairy slurry. Can J Soil Sci 90:229–239

VanderZaag AC, Gordon RJ, Jamieson RC, Burton DL, Stratton GW (2010b) Floating covers to reduce gas emissions from liquid manure storages: a review. Appl Eng Agric 26:287–297

Vac SC, Popiţa GE, Frunzeti N, Popovici A (2013) Evaluation of greenhouse gas emission from animal manure using the closed chamber method for gas fluxes. Not Bot Horti Agrobot Cluj-Napoca 41:576–581

Vasconcelos JT, Greene LW, Cole NA, Brown MS, McCollum FT III, Tedeschi LO (2007) Effects of phase feeding of protein on performance, blood urea nitrogen concentration, manure nitrogen:phosphorus ratio, and carcass characteristics of feedlot cattle. J Anim Sci 84:3032–3038

Velthof GL, Kuikman PJ, Oenema O (2003) Nitrous oxide emission from animal manures applied to soil under controlled conditions. Biol Fertil Soils 37:221–230

Velthof GL, Nelemans JA, Oenema O, Kuikman PJ (2005) Gaseous nitrogen and carbon losses from pig manure derived from different diets. J Environ Qual 34:698–706

Vanotti MB, Szogi AA, Vives CA (2008) Greenhouse gas emission reduction and environmental quality improvement from implementation of aerobic waste treatment systems in swine farms. Waste Manage 28:759–766

van Groenigen JW, Kasper GJ, Velthof GL, van den Pol-van Dasselaar A, Kuikman PJ (2004) Nitrous oxide emissions from silage maize fields under different mineral nitrogen fertilizer and slurry applications. Plant Soil 263:101–111

Valentine DL (2007) Adaptations to energy stress dictate the ecology of the archaea. Nat Revi Microbiol 5:316–323

Yamamoto N, Otawa K, Nakai Y (2010) Diversity and abundance of ammonia-oxidizing bacteria and ammonia-oxidizing archaea during cattle manure composting. Microb Ecol 60:807–815

Yamulki S (2006) Effect of straw addition on nitrous oxide and methane emissions from stored farmyard manures. Agric Ecosyst Environ 112:140–145

Zeman C, Depken D, Rich M (2002) Research on how the composting process impacts greenhouse gas emissions and global warming. Compost Sci Util 10:72–86

Zhou M, Zhu B, Wang S, Zhu X, Vereecken H, Brüggemann N (2017b) Stimulation of N_2O emission by manure application to agricultural soils may largely offset carbon benefits: a global meta-analysis. Glob Change Biol 23:4068–4083. https://doi.org/10.1111/gcb.13648
Zhang G, Strom JS, Li B, Rom HB, Morsing S, Dahl P, Wang C (2005) Emission of ammonia and other contaminant gases from naturally ventilated dairy cattle buildings. Biosyst Eng 92:355–364

Chapter 9
Lifecycle and Risk Assessment of Animal Manure Utilization

Hamid Masood, Sami Ullah Khan, Shujaul Mulk Khan, Aneela Nawaz, Syeda Haseena Wajid, Atiq Ur Rehman, and Abdullah Abdullah

Abstract Animal manure is the main source of biofertilizer, biogas, and bioenergy. It increases soil fertility leading to increased agricultural crop production which ultimately fulfills the food demand. Nowadays, animal manure is used in biogas and bioenergy production throughout the world and by 2050 the world will face a severe energy crisis, so animal manure can be used as a sustainable alternative to non-renewable energy resources. Animal manure is the source of pathogenic microorganisms, which causes severe illness in the human population and studies showed that these pathogens are resistant to most of the currently available antibiotics. Before applying animal manure as fertilizer in the soil, the manure should be processed through various methods including biological, physical, and chemical, to reduce the number of pathogens and their pathogenicity in the manure. Aerobic composting of animal manure reduces the pathogens' burden. Likewise, studies showed that anaerobic digestion of animal manure at various temperatures not only reduces the pathogenic microbes, but it also down-regulates the antibiotic resistance genes (ARGs).

H. Masood · S. U. Khan · A. Nawaz · S. H. Wajid · A. U. Rehman
Department of Microbiology, Quaid-I-Azam University, Islamabad 45320, Pakistan
e-mail: samikhan@qau.edu.pk

S. H. Wajid
e-mail: shahseena@bs.qau.edu.pk

A. U. Rehman
e-mail: atiq_dvm@gu.edu.pk

S. M. Khan (✉) · A. Abdullah
Department of Plant Sciences, Quaid-I-Azam University, Islamabad 45320, Pakistan
e-mail: shuja@qau.edu.pk

A. Abdullah
e-mail: abdullahkhan@bs.qau.edu.pk

M. Arshad (ed.), *Climate Changes Mitigation and Sustainable Bioenergy Harvest Through Animal Waste*, https://doi.org/10.1007/978-3-031-26224-1_9

Graphical Abstract

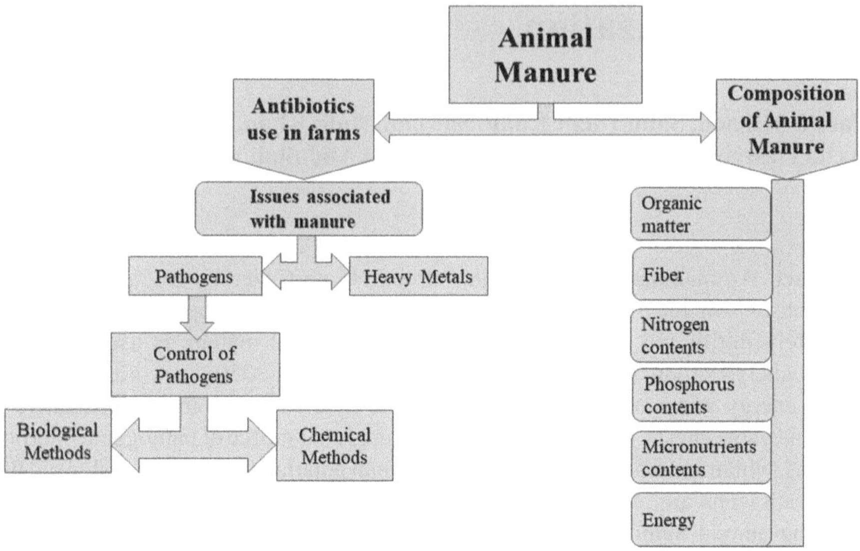

Keywords Biofertilizers · Organic matter · Pathogens · Antibiotics · Anaerobic digestion

9.1 Introduction

Animal manure is the animal excretion that mainly contains feces and urine, but it also contains bedding material, wasted feed, water, scurf, and soil depending upon the animal management system (Zhang et al. 2014). A huge amount of manure is produced in the livestock industry which is the richest source of organic and inorganic nutrients essential for the promotion of plant growth and crop yield (Sharma et al. 2022). The utilization of animal manure as fertilizer recycles nutrients, improves the physiochemical structure of the soil, and increases soil organic content. Animal manure provides essential minerals like potassium (K), phosphorus (P), nitrogen (N), and other minerals (Eghball et al. 2002) to soil that enhances its fertility and ultimately promotes plants growth and productivity. The contents of different nutrients and minerals largely depend upon the feed, animal type, and also on manure collection, how it is stored and the way of application to the soil. Studies showed that the organic content of the soil is reduced after continuous harvesting of crops but the addition of regular manure into the soil will increase its organic contents (Rawal et al. 2022).

On the other hand, animal manure is the major source of heavy metals and pathogens that potentially contaminates the environment (Liu et al. 2021). According

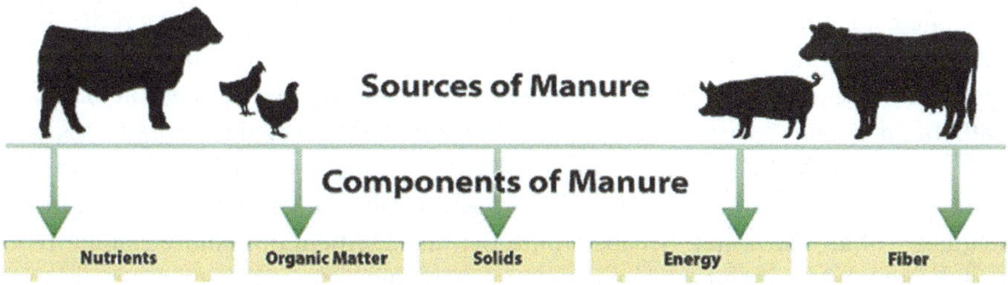

Fig. 9.1 Major sources of manure (Reproduced from Ribaudo et al. 2003)

to World Health Organization (WHO), many pathogens like viruses, bacteria, parasites, and fungi causing many diseases are the second leading cause of deaths worldwide. The animal manure used as a fertilizer enhances soil fertility, but also spreads pathogens in soil and water, which ultimately cause food and water-borne infections (Ayilara et al. 2020). In fields, mostly cow, buffalo, goat, and sheep manure is used to increase fertility (Fig. 9.1). To reduce the spread of these pathogens and to protect the environment, animal manure should be treated before its application to soil. Anaerobic digestion and aerobic composting are the most promising biological methods to reduce or eliminate the pathogens in manure (Agga et al. 2022).

9.2 History

Previously, it was believed that the use of fertilizer had started back 2000–3000 years ago, but now it is thought that it had been more than 8 thousand years ago that the farmers were fertilizing their crops by using manure (Xiong et al. 2010). To find the shreds of evidence about the earlier use of fertilizers, an Archaeo-botanist team at Oxford University conducted a study, and Amy Bogaard led this team. The team observed that in ancient times one of the most logical forms of fertilizer was manure because manure contains a high content of nitrogen (N-15 isotopes) (Szpak et al. 2012). Archeological samples of cereals and pulses like barley, wheat, peas, and lentils were collected by researchers from 13 different ancient field sites in Europe, that were dated back from four thousand and four hundred years to seven thousand and nine hundred years ago. More than 2500 pulse and cereals seeds were analyzed, and it was observed that in the ancient samples that were treated with manure, the concentration of Nitrogen-15 was higher in them than the untreated ones (Reed 2015).

While some studies reported that the use of organic fertilizers (animals and plant residues, human manure, and river humus in the soil) as a rich source of nutrient for crops and agriculture, dated back to the Neolithic period (12,000 years ago) (Goudie 2018).Till the mid of eighteen century, there was a shortage of organic manure especially animal manure due to its high demand in various regions of the world (Huang 2002).The beginning of 1950 was the era of technological advancement in the field of agriculture which resulted in improved capacity, high productivity, and storage of crops. This advancement in technology also resulted in increased animal farming (source of organic manure). But nowadays, the increase in the use of synthetic inorganic fertilizers has reduced the demand for organic fertilizer (Kumar et al. 2019). With the help of modern technology, manure can be used efficiently to protect our environment from the effect of greenhouse gases and also to protect the quality of air and water, because improper management of manure can harm our environment, while the proper management of manure can make it a renewable and valuable resource (Chiumenti et al. 2007).

Animal manure utilization can be preferred over the utilization of inorganic synthetic fertilizers because it not only provides nitrogen, potassium, and phosphorous to the soil to increase its quality but it is also a good source of macro-nutrients, it acts as a complete package of nutrients, can reduce the risk of soil erosion, and help in the improvement of soil quality, and it is cheaper than synthetic fertilizer (Tabitha et al. 2018).

9.3 Composition of Animal Manure

Manure contains phosphorous, nitrogen, and potassium and considered is an efficient source of micronutrients that are essential for plant growth (Table 9.1). Manure can be applied either directly to the soil or after proper processing such as nutrient extraction, pelleting, and composting. Manure can be securely recycled through modern agricultural systems (Logan and Visvanathan 2019).

For making all the nutrients available in the manure to the plants, there should be proper management plans defining the required concentration mandatory for the proper growth of plants. To prevent over-application, avoid runoff and protect water quality all the specifications of manure should be mentioned in the manure nutrients management plan (Sims et al. 2000). The use of manure as fertilizer in a proper way not only helps in the buildup of fertile soil but also plays a substantial role in the minimization of nutrient pollution in groundwater and running water. In addition to its use as a fertilizer for crops, manure can also be used to grow insect larvae, worms, algae, and also some other organisms. These organisms transfer the nutrients present in manure to biomass, these nutrients from their biomass are then harvested and used for soil amendments, as fertilizer, and as animal feed (Huis and Oonincx 2017).

9.3.1 Organic Matter

Organic matter present in manure is the undigested animal feed and bacteria that are part of the gastrointestinal tract of animals. An outstanding way to increase the organic matter in the soil is the addition of manure to it. Organic matters present in soil help in the overall health of soil including its sustainability and ability to act

Table 9.1 Macro-nutrients composition in different manures

Composition macro-elements	Form	Role	Nutritive element percentage in the case of cow dung manure (%)
Nitrogen	Present both in organic and inorganic form	It is one of the essential macro-nutrients. It plays an important role in plant function. Nitrogen is a component of nitrogenous bases and amino acids and both are essential for life survival.	2.5
Phosphorus	Inorganic form	It is easily accessible by plants. It is important for cell division and the development of the growing tip of the plants.	0.2
Potassium	Inorganic form	It has a role in the movement of water and nutrients in plant tissues.	0.5

as a living ecosystem (Liang et al. 2019). Moreover, organic matters also play an important role in the improvement of soil structure as well as its capacity to hold water, enhance the retention of essential nutrients, and promote the growth of plants and growth-promoting microorganisms in the soil. Organic content also prevents soil from both water and wind erosion. It can also protect the quality of water by lowering the contaminated runoff (Baumhardt et al. 2015). The quantity of organic content in manure is hard to quantify but the higher yield and economic return of crops are associated with high organic content.

9.3.2 Fibers

A huge number of fibers are present in manure. Some of these fibers are mixed with manure from straws, undigested animal feed, sawdust, and some other beddings. Various products are produced from the fibers of manure, such as plant pots, plant growth mediums, paper, building materials, and fertilizer garden sculptures (Miller et al. 2015).

9.3.3 Nitrogen Content

Nitrogen in animal manure is present in both organic and inorganic forms. The organic form of nitrogen releases slowly compared to the inorganic form. The organic form of nitrogen is not easily accessible to plants, but it is vulnerable to ammonia volatilization that occurs during the storage and application of manure as fertilizer. Utilization of manure as fertilizer reduces the nitrogen loss in soil because inaccessible nitrogen become available to the next crop. Fertilizers' nutrients in animal manure is the guide that provides information about the amount of nitrogen accessible to plants in 1st year and subsequent years (Rashmi et al. 2020).

9.3.4 *Phosphorus and Potassium Contents*

The availability of phosphorous and potassium in manure is mostly in inorganic form and is easily accessible to plants. Most nutrient management plans limit the utilization of animal manure as fertilizer in the field without determining the nutrients required by the crops by the testing of soil to achieve yield goals.

9.3.5 *Micronutrients Content*

Micronutrients (calcium, magnesium, and sulfur) are important for the promotion of plant growth, these nutrients are found in manure in sufficient form (Möller and Müller 2012).

9.3.6 *Energy*

A huge amount of carbon and other elements are present in manure that are used for the generation of various types and forms of biofuels (Fig. 9.2). A technology known as anaerobic digestion is used in which microbes process manure and convert it into biogas. And this biogas is further used to generate electricity and heat (Zhou et al. 2016). We can also obtain biodiesel from manure which has properties like petroleum diesel. Manure has also an application in gasification in which manure is converted into syngas, and this synthetic gas can be used to run power engines, fuel cells, and turbines. Using manure to generate biogas and biofuel can reduce our dependency on non-renewable fossil fuels, so farmers can save money (Ghaffarpour et al. 2018).

9.4 Antibiotic Use in Animal Farms

Antibiotics in animal feed are either used for therapeutic purposes in higher doses for specific diseases or used sub-therapeutically in small doses for increased feed

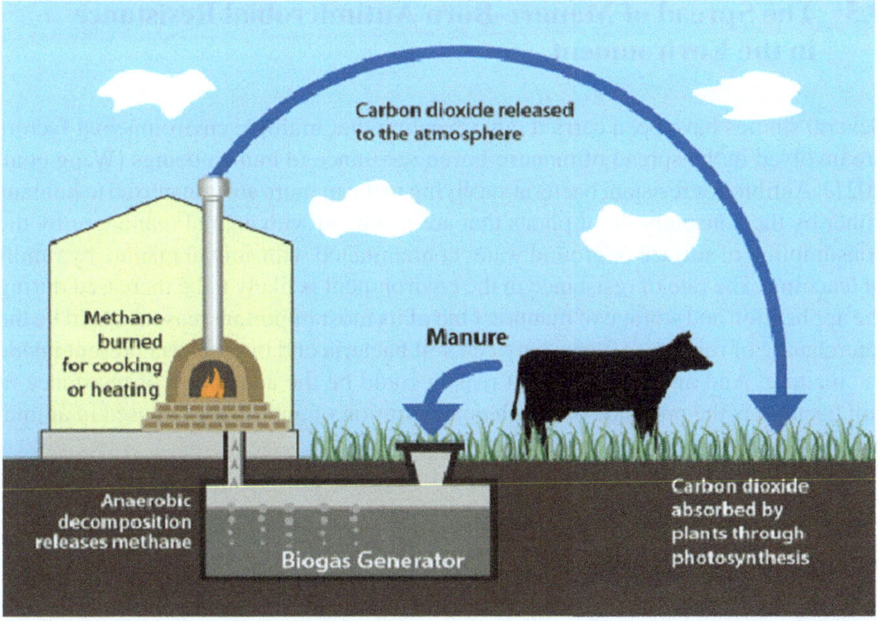

Fig. 9.2 Recycling process of manure and methane production (Reproduced from Benali 2019)

efficiency rate and the prevention of disease (Diaz-Sanchez et al. 2015). However, some antibiotics are not fully degraded and absorbed in the animal gut and sufficient quantities of the antibiotics and their residual metabolites are passed with manure and urine into the soil that leads to the emergence of antibiotics resistance in environmental microbes and dissemination of resistance to human pathogens (Kim et al. 2011). About 0.7 million deaths occur worldwide per annum due to antimicrobial resistant bacteria (Gaur 2017). In 1928, the discovery of penicillin became a source of cure for many lethal and life-threatening diseases that were more beneficial for animal breeders and veterinarians (Kirchhelle 2020). But since 1960, these antibiotics are extensively used in sub-lethal doses in animal feed as growth promoters. The actual mode of action is still unclear but it has been assumed that when antibiotics are used in a sub-lethal dose, they stimulate the intestinal synthesis of vitamins, they also reduce the count of intestinal tract bacteria by lowering competition between host and microorganisms for nutrients, help in modification of the microbial metabolic system of the rumen (Capita and Alonso-Calleja 2013), also involve in the growth inhibition of pathogenic bacteria, but the extensive and uncontrolled use of sub-lethal doses aid the selection of antibiotic-resistant bacteria by introducing novel mutation or also by promoting their growth. The presence of antibiotic-resistant bacteria can help in the transfer of ARGs (antibiotic-resistant genes) between the intestinal tract's enteric bacteria of the host organism, may affect their quorum sensing (Gilmore et al. 2013), and can also stimulate the presence of biofilm.

9.5 The Spread of Manure-Born Antimicrobial Resistance in the Environment

Several studies have been carried out, revealing that multiple environmental factors are involved in the spread of manure-borne resistance in human beings (Wang et al. 2021). Antibiotics resistant bacteria surviving in the manure are transferred to humans either by the consumption of plants that are fertilized with animal manure or by the consumption of surface or ground water contaminated with animal manure by runoff of leaching. The rate of resistance in the environment is likely to be increased during the application and storage of manure. One of its most important reasons could be the interchange of resistance genes between soil bacteria and the bacteria present inside the manure. And another important reason could be the adaptation of resistance in soil bacteria is the presence of a high concentration of antibiotics released in animal manure in un-metabolized form (Jechalke et al. 2014).

9.6 Issues Associated with Manure

9.6.1 Storage

Many people believe that not only the smell of manure is unpleasant, but also it is dangerous too, this nasty smell is due to the production of some toxic gases that can affect both human beings and animals. Overtime, the gases are generated by manure and as long as the manure is stored its smell becomes more intoxicating. This unpleasant smell becomes nastier if the manure is stored without ventilation in a closed location (Nimmermark 2004).

9.6.2 Gases Produced by Manure and Their Effects

When a large quantity of any type of animal waste is being stored, multiple hazards must be present there for both human beings and animals. One of the most hazardous dangers is the production of toxic gases. These gases are produced during the decomposition of manure by bacteria (Driehuis et al. 2018). Before dealing with manure, we should know the risk factors of the gases produced by manure.

9.6.2.1 Hydrogen Sulfide

According to the United States Department of Agriculture, hydrogen sulfide is the most dangerous gas produced in manure that smells like rotten eggs (Blunden et al. 2008). Humans and animals suffer from nausea, dizziness, and headache when they breathe this gas. Hazardous effects increase when hydrogen sulfide is produced in higher concentrations, because, at a higher level, the gas loses its smell, and its inhalation causes the most terrible effects including comma, sudden failure of the respiratory system, and even death.

9.6.2.2 Ammonia

Ammonia has a very sharp, penetrating, and distinct odor that can be very easily detected even at lower concentrations. Its medium concentration can irritate the respiratory tract and the eyes. Its high concentration can cause severe respiratory tract irritation and ulceration to the eyes. If someone got exposed to ammonia the best first aid is to flush the irritated eyes or skin (Timmer et al. 2005).

9.6.2.3 Carbon Dioxide

It is difficult to detect carbon dioxide because it is heavier than air. In the air, it replaces O_2 and acts asphyxiating. At medium concentration, it is responsible to cause dizziness and shortness of breath. One of the major causes of the death of animals is asphyxiation caused by carbon dioxide in a building with an improper ventilation system. On the other hand, in manure decomposition, CO_2 is also a livestock respiration byproduct (Sonwani and Saxena 2016).

9.6.2.4 Methane

The anaerobic digestion of animal manure in many countries is being used for the production of valuable products like biogas under the waste-to-energy production system. Besides, biogas is the ultimate digestate that can be used as a biofertilizer for increased crop production technology (Karim et al. 2005). Methane is an odorless gas and lighter gas and its concentration is high at the top of manure dents. At extremely high concentrations, it is responsible for asphyxiation. Its major hazard is that it is explosive and flammable. As methane is odorless so it is very difficult to detect it without any proper instrumentation (Carson 2002).

9.7 Introduction of Heavy Metals in Soil

Manure of different animals is added to agricultural land to increase organic contents as well as to improve the fertility of the soil, but this practice is also becoming a cause of some severe environmental problems such as surface water contamination due to phosphate and nitrate. Another serious problem persuaded by manure is metal pollution. A high concentration of metals including Zinc, Cadmium, Arsenic, and

Copper is present in animal manure (Dhaliwal et al. 2020). Heavy metal residues can be accumulated on the surface of soil due to the long-term use of manure on agricultural land. Heavy metal accumulation in soil not only affects the fertility of the soil but also contaminates surface and ground water through runoff and leaching (Yan et al. 2018).

9.8 Antibiotic Resistance Bacteria in Manure

One of the public health problems which is of growing concern is antibiotic resistance. The application of animal manure to the soil is known to be the major cause of dissemination and propagation of ARB (Antibiotic-resistant bacteria), antibiotic residues, and ARGs (antibiotic resistance genes) in soil and water system (Checcucci et al. 2020). In the recent era, studies have been increased on the effect on soil microbiomes by antibiotics-contaminated manure. The antibiotic resistance genes are mostly present as MGEs (Mobile genetic elements). Horizontal gene transfer of mobile genetic elements has been recognized as a leading cause dissemination and persistence of antibiotics resistance. Bio-sanitizing and chemical treatment can reduce the ARB and antibiotic load (Zhao et al. 2019).

In recent eras, the misuse and overuse of antibiotics in veterinary practices have become one of the most serious public health problems (Ferri et al. 2017). The increased number of commensal and pathogenic resistant bacteria has been linked with the environmental blowout of antibiotics and the proliferation of ARGs (Antibiotic-Resistant genes). In soil, microbial diversity can be lost and changed due to antibiotic environmental diffusion. Worldwide antibiotics are used in the production of livestock which is becoming the leading cause of the spread of antimicrobial resistance. Selective pressure of antibiotics directly increases when they are used for prophylactic purposes, which favors the antibiotic resistance bacteria generation. In order to minimize antibiotics resistance during animal husbandry, improved waste management, and livestock strategies should be adopted, that includes diet, waste treatment, and proximity between animals, operating conditions, and use of additives (Tian et al. 2021).

There are multiple pathways through which ARGs (Antibiotic-resistant genes) can persist and enter the ecosystem. They proliferate across crops, soil, and the gut microbiota of livestock and wild animals as well as human. The ARGs proliferate through HGT (horizontal gene transfer) of mobile genetic elements (MGEs) in the form of plasmids, phages, integrons gene cassettes, or transposons. The acquirement of antimicrobial resistance may be due to antimicrobial resistance genes from other bacteria through horizontal gene transfer or due to spontaneous mutation (Sobecky and Hazen 2009).

Generally, antibiotics are given to livestock utilizing drinking water or feed for their treatment and protection from different types of diseases. After administration, these antibiotics are not fully absorbed, and 30–90% is excreted in the urine and feces of animals after excretion these antibiotics can make a favorable condition for the development and spread of antibiotic-resistant bacteria in the environment (Manyi-Loh et al. 2018). On the other hand, antimicrobial resistance genes and bacteria are also naturally present in the GIT of animals that can be excreted in urine and feces. So, manure can act as a means for the spread of antimicrobial resistance in the environment (Chee-Sanford et al. 2009).

9.9 Pathogens Present in Manure

With the increasing population of humans, livestock industries are also terrifically increasing which are becoming the rising source of organic residues and production that can become the cause of a huge number of hazards if proper strategies are not set for their management and disposal (Karim et al. 2005). Agricultural animals are producing abundant quantities of animals' manure which consist of animals' urine and feces along with microorganism, bedding, spilled and undigested feed, process-generated wastewater, fur, antibiotics, nutrients, and secretions of the nose, throat, mammary glands, blood, placenta, and skin (Sakar et al. 2009). Moreover, all these animals' manure is also full of a huge variety of both pathogenic and non-pathogenic microorganisms that can be harmful to both humans and animals. The types and levels of pathogens present in manures are different, depending on animal species, health status, dietary sources, age of animals, and chemical and physical characteristics of manure as well as manure storage facilities (Hutchison et al. 2005; Spiehs et al. 2007).

The pathogenic microorganism found in animal manure and excreta consists of different types of bacteria, protozoa, and viruses. This pathogenic and microbial diversity present in the manure mainly depends on the characteristics of organisms, its chemical composition, and the source of manure such as ammonium content, temperature, pH, oxygen level, dry matter, moisture, and their competition for nutrients among the microbial community. Therefore, it is recommended to manage the proper handling, storage, and disposal of manure properly, otherwise, it can cause severe infection in both humans and animals either by direct contact with animals or their waste or utilizing contaminated water or food (Martin et al. 2005).

The potential risk should be minimized in animal manure by reduction of the pathogen during collection, storage, and finally during application of manure to the agricultural land. On the other hand, inadequately treated and raw manure can act as a means of pathogenic contamination, responsible to cause water, air, and soil pollution which will cause critical and severe public health issues (Manyi-Loh et al. 2013).

Various physicochemical and biological environments are provided to microorganisms by manure. Different types of manure-based pathogenic microbes have been found the major ones are zoonotic bacteria that include *Campylobacter* spp., *Salmonella* spp., *Yersinia enterocolitica*, *Escherichia coli*, and *Listeria monocytogenes* and protozoa viz. *Giardia lamblia* and *Cryptosporidium parvum* (Table 9.2). Another group of pathogens present in manure especially in cattle manure is viruses;

Table 9.2 Various types of bacteria present in manure

Bacteria	Manure source	Diseases associated with pathogens
Escherichia coli	Cattle	Enterohemorrhagic, verocytotoxic, or Shigatoxin-producing *Escherichia coli*
*Salmonella*spp.	Cattle, poultry, swine and birds	Salmonellosis
*Campylobacter*spp.	Poultry, cattle and birds	Stillbirths, abortion, the birth of weak sheep's lambs during some late pregnancy stages
Yersinia enterocolitica	Swine	Acute-enterocolitis, lymphadenitis, septicemia, polyarthritis, nodosum erythema, and even death
Listeria monocytogenes	Cattle, poultry	Localized-cutaneous infections

primarily these pathogens are inhibited in the animals' intestinal tract and then asymptomatically shed in the environment (Abdalla 2021).

The chances of infection depend on infective doses, types of bacteria, and the immune status of the individual. Due to differences in virulence factors and pathogenicity across the domains of bacteria, the numbers of bacteria that are responsible for infection differ from one bacterium to another (Sarowska et al. 2019). It has been found that in the case of some bacteria only a few that are around 10 cells are enough to cause infection while in some cases larger numbers of bacterial cells are required for the same purpose. However, in some cases, a small numbers of bacteria are shed in manure but after finding favorable conditions these bacteria multiply and result in a higher risk and greater level of contamination in soil, food, and water (Newell et al. 2010).

9.10 Some Important Bacteria Present in Manure and Their Pathogenicity

9.10.1 Escherichia coli

Escherichia coli is a rod-shaped, gram-negative, motile, and facultative anaerobic bacteria that is found as normal flora in the intestinal tract of healthy humans and animals. It acts as an indicator organism of fecal contamination (Gillen and Augusta 2018). *Escherichia coli* O157:H7 is a bacterial strain found in the manure of cattle which has been reported as the most hazardous pathogen and produces an intoxicating toxin that can cause severe infections in humans. This strain of *E. coli* can also be called Enterohemorrhagic, Verocytotoxic, or Shigatoxin-producing *E. coli.* Human beings can be infected by this pathogenic *E. coli,* either by consuming contaminated water or food or by direct contact with the feces of livestock. It can also be transferred by person-to-person contact (Mohawk and O'Brien 2011).

9.10.2 *Salmonella* spp.

Salmonella spp. belongs to the family of Enterobacteriaceae, rod-shaped, gram-negative, non-spore-forming, non-capsulated, motile, and facultatively anaerobic bacteria. These pathogenic bacteria have been observed in humans and a wide range of animals, i.e., cats, dogs, birds, pigs, and cattle (Khan et al. 2019) and cause salmonellosis. The mode of transmission of salmonella is the consumption of water or food that has been contaminated with the feces of different types of animals and it can also be transmitted by direct contact with animals' feces. This type of infection is characterized by 3 major symptoms, i.e., acute enteritis, septicemia, and chronic enteritis. The severity of salmonellosis varies from species to species and from host to host. The multi-drug resistance strains of *Salmonella* are very hazardous due to their resistance to the available classes of antibiotics. This high resistance rate can lead to high rates of death and can present a path toward epidemic outbreaks (Wu and Hulme 2021).

9.10.3 *Campylobacter* spp.

Most commonly *Campylobacter* spp. is found in the intestinal tracts such as pigs, cattle, wild-living mammals, chickens, and birds. This pathogenic bacterium is the main cause of a wide range of infections in sheep, pigs, and cattle (Pao et al. 2014). The *Campylobacter fetus* and *Campylobacter jejuni* cause stillbirths, abortion, and the birth of weak sheep's lambs during some late pregnancy stages (Givens and Marley 2008). In the same way, an infection called Campylobacteriosis is caused in humans by this pathogenic bacterium by consuming contaminated water and uncooked food products. *Campylobacter coli* and *C. jejuni* are the two most important species that are bacterial human gastroenteritis.

9.10.4 *Yersinia enterocolitica*

It is a gram-negative coco-bacillus, a member of the genus *Yersinia,* and belongs to the Enterobacteriaceae family. It is a non-capsulated, non-spore-forming, that is found in the intestinal flora of humans as well as of many domestic and wild animals, such as cattle. Yersiniosis is an infection caused by this bacterium in both humans and animals manifested as acute enterocolitis, lymphadenitis, septicemia, Polyarthritis, Nodosum erythema, and even death (Aziz and Yelamanchili 2021).

9.10.5 *Listeria monocytogenes*

It is a Gram-positive, rod-shaped, facultative, and intracellular bacteria associated with many food-borne illnesses of humans and transmission is by oral-fecal route. The main reservoirs of this pathogen are dairy cows (Dhama et al. 2015).

9.10.6 *Enterococcus Species*

Enterococcus spp. belongs to the subgroup of Group-D fecal *Streptococcus,* is Gram-positive, spherical, and found either singly, in form of short chains or pairs. They are facultative anaerobic and lactic acid producers that live in the gastrointestinal tract (GIT) of animals and humans as commensal bacteria. Members of this subgroup are significantly different from each other based on their virulence factors, their distribution in dry and fresh cattle manure is based on their antibiotic resistance genes (Herrera et al. 2009).

9.10.7 *Mycobacterium Species*

The acid-fast Mycobacterium species have been reported in the manure of cattle. These bacteria can survive in the stored manure for a long period even in harsh conditions such as fluctuations in pH and temperature, exposure to sunlight, and dehydration (Manyi-Loh et al. 2016).

9.11 Control and Prevention of Pathogens in Animal Manure

As animal manure is the source of pathogenic microbes which can harm the environment and have a potential risk to the human population, there are some methods by which the pathogens can be minimized (Table 9.3; Fig. 9.3). These methods include physical, chemical, and biological or combinations of these methods (Gerba and Smith 2005).

Table 9.3 Methods to control pathogens in manure

Physical methods	Drying
	Heating Irradiation
Chemical methods	Use of lime substances Hydrogen peroxide
Biological methods	Anaerobic storage of manure Composting Anaerobic digestion

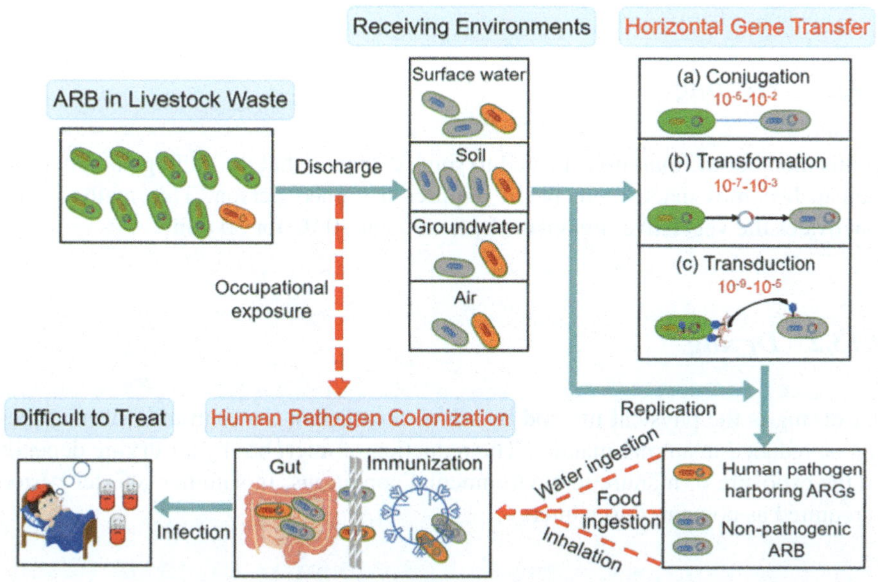

Fig. 9.3 Pathogen control using different techniques (Reproduced from He et al. 2020)

9.12 Chemical Methods

9.12.1 Use of Lime Substances

Chemicals like Calcium oxide (CaO) or Calcium Hydroxide (CaOH) are used to inactivate the microbes in animal manure. For this purpose, some amount of lime is homogenized with manure to attain a pH of 12 for two to three hours of contact time. At that pH, ammonia production occurs which ultimately inactivates the pathogens (Bean et al. 2007).

9.12.2 Hydrogen Peroxide

Hydrogen peroxide, an oxidizing agent is used for the inactivation of pathogens and it oxidizes the cellular component of most microbes (Guo et al. 2021).

9.13 Physical Methods

Physical methods include heating, drying, and irradiation.

9.13.1 Heating

In this method, the manure is heated to some extent so that the pathogenic microorganism becomes inactivated, but it requires a trained person. This method only inactivates the vegetative organisms by heating at 70 °C for 30 min.

9.13.2 Drying

Air drying is the physical method by which a sufficient concentration of microbes can be reduced in animal manure. The reduction of microbes by air drying depends on the exposure of manure to environmental conditions. In summer, shorter period is required as compared to winter.

9.13.3 Irradiation

The inactivation of pathogens in manure by irradiation like ultraviolet radiation (UV Light) depends on the type of pathogen, the intensity of the light, the quantity of the solid materials in the manure, and time and duration of exposure to ultraviolet light (Bilotta and Kunz 2013). This method has an advantage as it only destroys the pathogenic microbes without any change in the consistency of the manure, i.e., physiochemical and nutritional characteristics of the manure (Spiehs et al. 2007).

9.14 Biological Methods

Biological methods include anaerobic storage of manure, composting, and anaerobic digestion.

9.14.1 Anaerobic Storage of Manure

For this purpose, lagoons or tanks are constructed away from the animal farm on elevated ground with a well-drained area and a concrete floor covered with a roof. The animal manure in solid, semisolid, and liquid forms is placed in these lagoons for a longer period for optimum reduction of the pathogen level (Nicholson et al. 2005).

9.14.2 Composting

Composting is the degradation of organic material in animal manure under aerobic conditions. It is an effective biological method during which the living aerobic microbes degrade or decompose the organic material in manure into carbon dioxide, water, and heat. This compost (manure) is then used as a biofertilizer for enhanced crop production (Spiehs et al. 2007).

9.14.3 Anaerobic Digestion

It is also one of the most important biological methods by which microorganisms can be inactivated to a level acceptable for public health concerns (Manyi-Loh et al. 2013). The production of methane and carbon dioxide gas (greenhouse gases) which are produced during the anaerobic digestion of manure treatment has no adverse or detrimental effect on public health because these gases are under the controlled environment of the digester. These gases have a challenging effect on global warming and climatic change when released into the environment when the animal manure is not treated properly (Rao et al. 2010).

However, with the reduction in pathogenic microbes, anaerobic digestion also synthesizes biogas. This gas leaves no carbon soot (Arthur et al. 2011). Additionally, anaerobic digestate has high nutritive value and can be used as soil fertilizer (biofertilizer).

9.15 Conclusion

Animal manure is an important source of biogas and nutrient-rich fertilizer for agricultural crop production. It is inappropriate and harmful to use as fertilizer without treatment and has very serious consequences to other livestock, the environment, and the human population in terms of pathogenic and antibiotic resistance genes and antibiotic-resistant microorganisms. Before the development of technology, the manure after collection from the animal was directly applied to agricultural land as a fertilizer. Such type of practices leads to the emergence of resistant microbes and the development of new challenges to the environment and human health. With the advancement of technology such as aerobic composting and anaerobic digestion, animal manure can be efficiently used to produce biogas, bioenergy, and biofertilizer.

It is concluded that animal manure can be an alternative source of energy, nonrenewable sources are decreasing with time and an alternative source for energy production should be investigated to fulfill the energy demand of the increasing global human population in the future. Research needed other efficient and advanced technology for treating animal manure for enhanced biogas production and pathogen-free digestate as biofertilizer for increased crop production.

9.16 Future Recommendations

Animal manure is used in the production of energy and biogas as an alternative to sustainable energy technology globally nowadays. Livestock rearing produces large quantities of livestock manure worldwide, and it should be properly operated, handled, managed, and treated. Traditionally, animal manure was directly applied to agricultural land as a fertilizer, which should be avoided, and it is recommended that the manure must be properly collected and stored before treatment. Therefore, proper procedures for handling, management, and treatment of animal manure must be taken in safeguarding the land, water, environment, and public health. Also, all the methods and procedures adopted must be cost-effective and environment-friendly (Tomasch et al. 2018).

As animal gastrointestinal tract contains a diversified community of microbes' containing both pathogenic and commensals and these microbes passed out with feces. Therefore, the pathogenic microbes of animal manure after land application may pose a serious threat to the environment. Since animal manure also contains antibiotic residue and antibiotic resistance genes (ARG), which could lead to antibiotic resistance bacteria (ARB). So, every management practice should be evaluated on case-to-case bases for the complete removal of antibiotic residues, antibiotic resistance genes, and pathogenic microbes (Sun et al. 2019). In storage facilities, the manure should be covered properly to prevent any leakage or runoff into the environment (Zhang et al. 2015). It is evident from various studies that anaerobic digestion of batch methods at various temperatures could efficiently lower the antibiotic residues as well as the antibiotic resistance genes and pathogenic microbes. Also, it is of prime importance that inappropriate antibiotic administration in veterinary practices at the sub-therapeutic level for disease prevention and growth promotion of animals should be avoided. For this purpose, strict rules and regulations by the drug regulatory authority should be initiated in true spirit to stop the inappropriate usage of veterinary antibiotics. The animal manure after various treatments can be used as a biofertilizer for agricultural land for enhanced crop production. It is also recommended that public campaigns and awareness programs should be arranged to transfer knowledge to the farmer community regarding safe handling, storage, and treatment of animal manure before land application. At last, further new research is suggested to carry out improved management and technology which have low cost and that would minimize the transfer of the pathogen from animal to the environment.

References

Abdalla MAA (2021) A study on diary management practices in Khartoum State. Sudan University of Science & Technology

Agga GE, Couch M, Parekh RR, Mahmoudi F, Appala K, Kasumba J, Loughrin JH, Conte ED (2022) Lagoon, anaerobic digestion, and composting of animal manure treatments impact on tetracycline resistance genes. Antibiotics 11(3):391

Arthur R, Baidoo MF, Antwi E (2011) Biogas as a potential renewable energy source: a Ghanaian case study. Renew Energy 36(5):1510–1516

Ayilara MS, Olanrewaju OS, Babalola OO, Odeyemi O (2020) Waste management through composting: challenges and potentials. Sustainability 12(11):4456

Aziz M, Yelamanchili VS (2021) *Yersinia enterocolitica* StatPearls [Internet]. StatPearls Publishing

Baumhardt RL, Stewart BA, Sainju UM (2015) North American soil degradation: processes, practices, and mitigating strategies. Sustainability 7(3):2936–2960

Bean CL, Hansen JJ, Margolin AB, Balkin H, Batzer G, Widmer G (2007) Class B alkaline stabilization to achieve pathogen inactivation. Int J Environ Res Publ Health 4(1):53–60

Benali M (2019) Experimental investigation of biogas production from cow dung in an anaerobic batch digester at mesophilic conditions. Iranian (Iranica) J Energy Environ 10(2):121–125

Bilotta P, Kunz A (2013) Swine manure post-treatment technologies for pathogenic organism inactivation. Engenharia Agrícola 33:422–431

Blunden J, Aneja VP, Overton JH (2008) Modeling hydrogen sulfide emissions across the gas–liquid interface of an anaerobic swine waste treatment storage system. Atmos Environ 42(22):5602–5611

Capita R, Alonso-Calleja C (2013) Antibiotic-resistant bacteria: a challenge for the food industry. Crit Rev Food Sci Nutr 53(1):11–48

Carson PA (2002) Hazardous chemicals handbook. Elsevier

Checcucci A, Trevisi P, Luise D, Modesto M, Blasioli S, Braschi I, Mattarelli P (2020) Exploring the animal waste resistome: the spread of antimicrobial resistance genes through the use of livestock manure. Front Microbiol 11:1416

Chee-Sanford JC, Mackie RI, Koike S, Krapac IG, Lin Y-F, Yannarell AC, Maxwell S, Aminov RI (2009) Fate and transport of antibiotic residues and antibiotic resistance genes following land application of manure waste. J Environ Quality 38(3):1086–1108

Chiumenti A, Da Borso F, Rodar T, Chiumenti R (2007) Swine manure composting by means of experimental turning equipment. Waste Manage 27(12):1774–1782

Dhaliwal SS, Singh J, Taneja PK, Mandal A (2020) Remediation techniques for removal of heavy metals from the soil contaminated through different sources: a review. Environ Sci Pollut Res 27(2):1319–1333
Dhama K, Karthik K, Tiwari R, Shabbir MZ, Barbuddhe S, Malik SVS, Singh RK (2015) Listeriosis in animals, its public health significance (food-borne zoonosis) and advances in diagnosis and control: a comprehensive review. Veterinary Q 35(4):211–235
Diaz-Sanchez S, Moscoso S, Solis de los Santos F, Andino A, Hanning I (2015) Antibiotic use in poultry: a driving force for organic poultry production. Food Prot Trends 35(6):440–447
Driehuis F, Wilkinson JM, Jiang Y, Ogunade I, Adesogan AT (2018) Silage review: animal and human health risks from silage. J Dairy Sci 101(5):4093–4110
Eghball B, Wienhold BJ, Gilley JE, Eigenberg RA (2002) Mineralization of manure nutrients. J Soil Water Conserv 57(6):470–473
Ferri M, Ranucci E, Romagnoli P, Giaccone V (2017) Antimicrobial resistance: a global emerging threat to public health systems. Crit Rev Food Sci Nutr 57(13):2857–2876
Gaur RK (2017) Antibiotic resistance: alternative approaches. Ind J Pharmacol 49(2):208
Gerba CP, Smith JE (2005) Sources of pathogenic microorganisms and their fate during land application of wastes. J Environ Quality 34(1):42–48
Ghaffarpour Z, Mahmoudi M, Mosaffa AH, Farshi LG (2018) Thermoeconomic assessment of a novel integrated biomass based power generation system including gas turbine cycle, solid oxide fuel cell and Rankine cycle. Energy Convers Manage 161:1–12
Gillen AL, Augusta M (2018) The coliform kind: *E. coli* and its "cousins" the good, the bad, and the deadly
Gilmore MS, Lebreton F, Van Schaik W (2013) Genomic transition of enterococci from gut commensals to leading causes of multidrug-resistant hospital infection in the antibiotic era. Curr Opin Microbiol 16(1):10–16
Givens MD, Marley MSD (2008) Infectious causes of embryonic and fetal mortality. Theriogenology 70(3):270–285
Goudie AS (2018) Human impact on the natural environment. Wiley
Guo Y, Dundas CM, Zhou X, Johnston KP, Yu G (2021) Molecular engineering of hydrogels for rapid water disinfection and sustainable solar vapor generation. Adv Mater 33(35):2102994
He Y, Yuan Q, Mathieu J, Stadler L, Senehi N, Sun R, Alvarez PJ (2020) Antibiotic resistance genes from livestock waste: Occurrence, dissemination, and treatment. NPJ Clean Water 3(1):1–11
Herrera P, Kwon YM, Ricke SC (2009) Ecology and pathogenicity of gastrointestinal *Streptococcus bovis*. Anaerobe 15(1–2):44–54
Huang PCC (2002) Development or involution in eighteenth-century Britain and China? A review of Kenneth Pomeranz's the great divergence: China, Europe, and the making of the modern world economy. J Asian Stud 61(2):501–538
Hutchison ML, Walters LD, Avery SM, Munro F, Moore A (2005) Analyses of livestock production, waste storage, and pathogen levels and prevalences in farm manures. Appl Environ Microbiol 71(3):1231–1236
Jechalke S, Heuer H, Siemens J, Amelung W, Smalla K (2014) Fate and effects of veterinary antibiotics in soil. Trends Microbiol 22(9):536–545
Karim K, Hoffmann R, Klasson T, Al-Dahhan MH (2005) Anaerobic digestion of animal waste: waste strength versus impact of mixing. Biores Technol 96(16):1771–1781
Khan MA, Cao L, Riaz A, Li Y, Jiao Q, Liu Z, Wang H, Meng F, Ma Z (2019) The harm of Salmonella to pig industry and its control measures. Int J Appl Agric Sci 5:24
Kim K-R, Owens G, Kwon S-I, So K-H, Lee D-B, Ok YS (2011) Occurrence and environmental fate of veterinary antibiotics in the terrestrial environment. Water Air Soil Pollut 214(1):163–174
Kirchhelle C (2020) Pyrrhic progress: the history of antibiotics in Anglo-American food production
Kumar R, Kumar R, Prakash O (2019) Chapter-5 the impact of chemical fertilizers on our environment and ecosystem. Chief Ed 35:69
Liang C, Amelung W, Lehmann J, Kästner M (2019) Quantitative assessment of microbial necromass contribution to soil organic matter. Glob Change Biol 25(11):3578–3590

Liu C, Liu Y, Feng C, Wang P, Yu L, Liu D, Sun S, Wang F (2021) Distribution characteristics and potential risks of heavy metals and antimicrobial resistant *Escherichia coli* in dairy farm wastewater in Tai'an, China. Chemosphere 262:127768

Logan M, Visvanathan C (2019) Management strategies for anaerobic digestate of organic fraction of municipal solid waste: current status and future prospects. Waste Manage Res 37(1_suppl):27–39

Manyi-Loh C, Mamphweli S, Meyer E, Okoh A (2018) Antibiotic use in agriculture and its consequential resistance in environmental sources: potential public health implications. Molecules 23(4):795

Manyi-Loh CE, Mamphweli SN, Meyer EL, Makaka G, Simon M, Okoh AI (2016) An overview of the control of bacterial pathogens in cattle manure. Int J Environ Res Publ Health 13(9):843

Manyi-Loh CE, Mamphweli SN, Meyer EL, Okoh AI, Makaka G, Simon M (2013) Microbial anaerobic digestion (bio-digesters) as an approach to the decontamination of animal wastes in pollution control and the generation of renewable energy. Int J Environ Res Publ Health 10(9):4390–4417

Martin H (2005) Manure composting as a pathogen reduction strategy. Ministry of Agriculture & Food

Miller JJ, Beasley BW, Drury CF, Larney FJ, Hao X (2015) Influence of long-term application of manure type and bedding on yield, protein, fiber, and energy value of irrigated feed barley. Agron J 107(1):121–128

Mohawk KL, O'Brien, Alison D (2011) Mouse models of *Escherichia coli* O157: H7 infection and Shiga toxin injection. J Biomed Biotechnol

Möller K, Müller T (2012) Effects of anaerobic digestion on digestate nutrient availability and crop growth: a review. Eng Life Sci 12(3):242–257

Newell DG, Koopmans M, Verhoef L, Duizer E, Aidara-Kane A, Sprong H, Opsteegh M, Langelaar M, Threfall J, Scheutz F (2010) Food-borne diseases—the challenges of 20 years ago still persist while new ones continue to emerge. Int J Food Microbiol 139:S3–S15

Nicholson FA, Groves SJ, Chambers BJ (2005) Pathogen survival during livestock manure storage and following land application. Biores Technol 96(2):135–143

Nimmermark S (2004) Odour influence on well-being and health with specific focus on animal production emissions. Ann Agric Environ Med 11(2)

Pao S, Hagens BE, Kim C, Wildeus S, Ettinger MR, Wilson MD, Watts BD, Whitley NC, Porto-Fett AC, Schwarz JG (2014). Prevalence and molecular analyses of *Campylobacter jejuni* and *Salmonella* spp. in co-grazing small ruminants and wild-living birds. Livestock Sci 160:163–171

Rao PV, Baral SS, Dey R, Mutnuri S (2010) Biogas generation potential by anaerobic digestion for sustainable energy development in India. Renew Sustain Energy Rev 14(7):2086–2094

Rashmi I, Roy T, Kartika KS, Pal R, Coumar V, Kala S, Shinoji KC (2020) Organic and inorganic fertilizer contaminants in agriculture: impact on soil and water resources. Contaminants Agric. Springer, pp 3–41

Rawal N, Pande KR, Shrestha R, Vista SP (2022) Soil nutrient balance and soil fertility status under the influence of fertilization in maize-wheat cropping system in Nepal. Appl Environ Soil Sci

Reed K (2015) From the field to the hearth: plant remains from Neolithic Croatia (ca. 6000–4000 cal BC). Vegetation History Archaeobotany 24(5):601–619

Ribaudo M, Kaplan JD, Christensen LA, Gollehon N, Johansson R, Breneman VE, Aillery M, Agapoff J, Peters M (2003) Manure management for water quality costs to animal feeding operations of applying manure nutrients to land. USDA-ERS Agric Econ Rep 824

Sakar S, Yetilmezsoy K, Kocak E (2009) Anaerobic digestion technology in poultry and livestock waste treatment—a literature review. Waste Manage Res 27(1):3–18

Sarowska J, Futoma-Koloch B, Jama-Kmiecik A, Frej-Madrzak M, Ksiazczyk M, Bugla-Ploskonska G, Choroszy-Krol I (2019) Virulence factors, prevalence and potential transmission of extraintestinal pathogenic *Escherichia coli* isolated from different sources: recent reports. Gut Pathogens 11(1):1–16

Sharma M, Shilpa K, Manpreet S, Kumar A, Sharma P (2022) Influence of different organic manures, biofertilizers and inorganic nutrients on performance of pea (*Pisum sativum* L.) in North Western Himalayas. J Plant Nutrit:1–18

Sims JT, Edwards AC, Schoumans OF, Simard RR (2000) Integrating soil phosphorus testing into environmentally based agricultural management practices. J Environ Qual 29(1):60–71

Sobecky PA, Hazen TH (2009) Horizontal gene transfer and mobile genetic elements in marine systems. Horizontal Gene Transf:435–453

Sonwani S, Saxena P (2016) Identifying the sources of primary air pollutants and their impact on environmental health: a review. IJETR 6(2):111–130

Spiehs MJ, Goyal SM (2007) Best management practices for pathogen control in manure management systems. Minnesota Extension

Sun W, Gu J, Wang X, Qian X, Peng H (2019) Solid-state anaerobic digestion facilitates the removal of antibiotic resistance genes and mobile genetic elements from cattle manure. Biores Technol 274:287–295

Szpak P, Millaire J-F, White CD, Longstaffe FJ (2012) Influence of seabird guano and camelid dung fertilization on the nitrogen isotopic composition of field-grown maize (Zea mays). J Archaeol Sci 39(12):3721–3740

Tabitha K, Wilson T, Joseph P-O (2018) Influence of Organic and inorganic manures on macronutrients, micro-nutrients, and anti-nutrients in two *Amaranth* spp. in Kiambu County, Kenya. Asian J Res Crop Sci 1(1):1–17

Tian M, He X, Feng Y, Wang W, Chen H, Gong M, Liu D, Clarke JL, van Eerde A (2021) Pollution by antibiotics and antimicrobial resistance in livestock and poultry manure in China, and countermeasures. Antibiotics 10(5):539

Timmer B, Olthuis W, Van Den Berg A (2005) Ammonia sensors and their applications—a review. Sensors Act B Chem 107(2):666–677

Tomasch J, Wang H, Hall ATK, Patzelt D, Preusse M, Petersen J, Brinkmann H, Bunk B, Bhuju S, Jarek M (2018) Packaging of *Dinoroseobacter shibae* DNA into gene transfer agent particles is not random. Genome Biol Evol 10(1):359–369

Van Huis A, Oonincx D GAB (2017) The environmental sustainability of insects as food and feed. A review. Agron Sustain Dev 37(5):1–14

Wang F, Fu Y-H, Sheng H-J, Topp E, Jiang X, Zhu Y-G, Tiedje JM (2021) Antibiotic resistance in the soil ecosystem: a one health perspective. Curr Opin Environ Sci Health 20:100230

Wu S, Hulme JP (2021) Recent advances in the detection of antibiotic and multi-drug resistant salmonella: an update. Int J Mol Sci 22(7):3499

Xiong X, Yanxia L, Wei L, Chunye L, Wei H, Ming Y (2010) Copper content in animal manures and potential risk of soil copper pollution with animal manure use in agriculture. Resour Conserv Recycl 54(11):985–990

Yan X, Liu M, Zhong J, Guo J, Wu W (2018) How human activities affect heavy metal contamination of soil and sediment in a long-term reclaimed area of the Liaohe River Delta, North China. Sustainability 10(2):338

Zhang H, Shi J, Liu X, Zhan X, Chen Q (2014) Occurrence and removal of free estrogens, conjugated estrogens, and bisphenol A in manure treatment facilities in East China. Water Res 58:248–257

Zhang Q-Q, Ying G-G, Pan C-G, Liu Y-S, Zhao J-L (2015) Comprehensive evaluation of antibiotics emission and fate in the river basins of China: source analysis, multimedia modeling, and linkage to bacterial resistance. Environ Sci Technol 49(11):6772–6782

Zhao Y, Cocerva T, Cox S, Tardif S, Su J-Q, Zhu Y-G, Brandt KK (2019) Evidence for co-selection of antibiotic resistance genes and mobile genetic elements in metal polluted urban soils. Sci Total Environ 656:512–520

Zhou J, Zhang R, Liu F, Yong X, Wu X, Zheng T, Jiang M, Jia H (2016) Biogas production and microbial community shift through neutral pH control during the anaerobic digestion of pig manure. Bioresour Technol 217:44–49

Chapter 10
Utilization of Animal Wastes to Mitigate the Climate Changes

Sadia Javed, Sher Zaman Safi, Saboor Gul, Nazima Anwaar, Amreen Aftab, and Muhammad Arfan Zaman

Abstract Over the past ten years, consumption of animal by-products has grown significantly. This is because it has the potential to help many nations' battle protein deficiency and food insecurity. Animal by-products are divided into edible and inedible portions shortly after slaughter. 55% of the production is considered to be edible, with the remaining 45% being inedible by-products (IEBPs). These IEBPs can be processed again to create sustainable goods for use in industry and agriculture. The cost and scarcity of feed sources, which are now high due to fierce competition from both the humans and the animals, can be decreased through the efficient utilization of animal wastes. Additionally, this will help the society's environmental pollution be reduced. In this sense, effective use of animal by-products, like rumen digesta, can lead to cheaper feed, less competition, and lower production costs. Animal by-products like rumen digesta have been successfully used in livestock feed over the years without having any negative effects on the animals. However, there are growing gaps in the items' sustainability and food security that need to be further addressed. The goal of this review is to emphasize the efficiency and usefulness of employing animal waste and by-products as substitute sources for feed components in the pharmaceutical and leather industries, among other industries.

S. Javed (✉) · S. Gul · N. Anwaar · A. Aftab
Department of Biochemistry, Government College University, Faisalabad, Pakistan
e-mail: sadiajaved@gcuf.edu.pk

S. Z. Safi
Faculty of Medicine, Bioscience and Nursing, MAHSA University, Jenjarom, Selangor, Malaysia

M. A. Zaman
Department of Pathobiology, University of Veterinary and Animal Sciences Lahore, Jhang-Campus, Lahore, Pakistan

M. Arshad (ed.), *Climate Changes Mitigation and Sustainable Bioenergy Harvest Through Animal Waste*, https://doi.org/10.1007/978-3-031-26224-1_10

Graphical Abstract

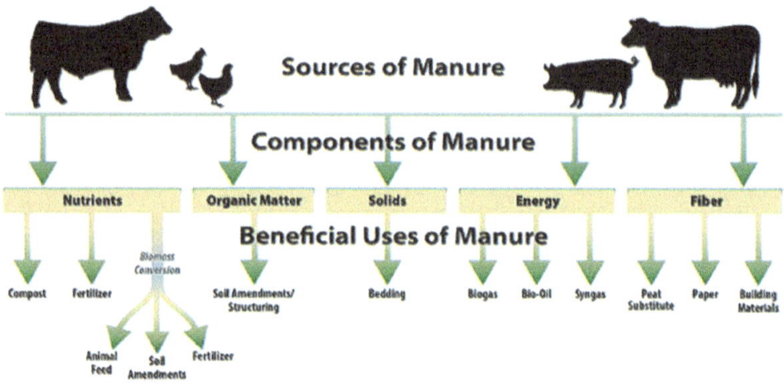

Keywords Animal manure · Environmental pollution · Climate changes · Novel technologies · Anaerobic digestion

10.1 Introduction

The high proportion of waste related to a product that characterizes waste in the food industry means that not only is its generation inevitable, but it also means that it is almost impossible to change the quantity and type of waste product which primarily contains the organic waste left behind after processing raw materials while maintaining the consistency of the finished product's quality. Product-specific waste is challenging to use because of its low biological stability, potential pathogenicity, increased water content, likelihood of quick auto oxidation, and high level of enzymatic activity. The numerous waste types created by the various branches of the food industry can be measured based on the equivalent amount of production.

The human diet includes a significant portion of meat and animal products because they offer necessary nutrients that are difficult to obtain from the vegetables and the by-products of vegetables (Byers et al. 2002). They offer a way to improve household food security and lessen malnutrition (Chikwanha et al. 2018). The demand for flesh and the meat products has expanded over the past 20 years in many regions of the world, which has caused a rapid expansion in livestock production for long-term food security (Sans et al. 2015). Abattoirs typically produce a large amount of by-products during the process of turning livestock into meat, which can then be used by people as food or processed again as secondary by-products for use in both industrial and agricultural applications (Liu et al. 2002). Although animal by-products make up around two-thirds of the animal after slaughter, it has been stated that the yield of these by-products accounts for 10–15% of the value of the live animal in developed countries (Irshad et al. 2015). For usage by humans and livestock, bones, hides and skin, feathers, hooves, horns, hair, bristles, and rumen digesta can be made into beneficial and valuable items. It is generally believed that while skin/hide and feathers can be processed and used in the upholstery, leather, and textile industries, bone can be recycled and used as livestock feed (a source of minerals). However, since most developing nations are experiencing a lack of feed components due to extreme weather conditions, rising feed costs, and rivalry for cereal crops between humans and livestock, the use of rumen digesta could serve as an alternate feed source for the livestock business (Elfaki et al. 2015). As a result, this analysis emphasizes the use of animal by-products in the food system as sources of nutrients for people and as substitute ingredients in animal feed. The recycling of organic waste from on-farm like livestock manure or off-farm like sewage sludge, industrial wastes sources, is a crucial aspect of the circular economy to develop more sustainable food production systems. For the safe, effective, and efficient use of organic "waste" streams as resources for fertilizer provision and soil enhancement in agricultural systems, Bernal (2017) identifies a number of obstacles that must be solved:

i. to improve the availability of nutrients and land cycling.
ii. to invent the technologies for re-use of the nutrients.
iii. to lessen contaminations and enhance the safety of food.
iv. to reduce the environmental emissions.
v. to improve soil or land health and functioning.

Although livestock manure is an important natural resource for agricultural re-use, it is frequently treated as waste due to the growing specialization and geographic separation between the production of livestock and crops in many regions of the world, as well as the accessibility and affordability of synthetic fertilizers. Manure-based inputs are widely used by both conventional and organic growers, not only for their usefulness as fertilizer but also for improving and preserving soil quality. A number of safety requirements must be followed in order for organic food production and distribution to continue to grow globally (Cooper et al. 2007).

A typical tons of cow manure will likely contain 10 pounds of N (nitrogen), 5 pounds of P (phosphate), and 10 pounds of potash. Beneficial bacteria found in manure can help improve the biological activity and soil structure. But over the past 50 years, the use of manure has gradually decreased on many farms for a variety of reasons, such as:

- Specialized farms with rising separation of reproduction and cultivation (livestock and crop production).
- Cost factors for transporting manure.
- Greater availability of synthetic fertilizer with preferable compositions and concentrations at less expensive prices.

A stunning amount of manure is produced nowadays by mass producing animals in industrial farms to meet the world's food needs—nearly 130 times as much as human waste.

The main method for utilizing the animal by-products produced in slaughterhouses is to process them through renderers to produce pet food and animal feed. However, this approach frequently yields low-value goods including poultry waste meal, raw meat, feather meal, and flesh and bone meal. The annual average cost of such manufactured products has been falling in recent years. For instance, compared to their pricing in 2015, the estimated yearly price of pig meat, bone, and raw meat declined by 17% on average, while the cost of feather meal declined by 25%. Recent developments on using slaughterhouse leftovers in variety of applications include using them as feedstock for anaerobic digestion as a reservoir of protein hydrolysate, catalysts, and lipids for making processed food for consumption by humans as well as for recovering bioactive peptides (Bah et al. 2014). Therefore, establishing the appropriate processing conditions and protein enriching techniques to transform them into more homogeneous, soluble, easily processable, and safe protein-enriched feedstock is a crucial step in creating technical applications of such wastes.

Among the various available alternative sources of energy, biogas is an environmentally responsible and practical option for a sustainable power source. Anaerobic decomposition of biodegradable waste and materials high in nutrients results in the creation of a significant volume of biogas in the biogas production process. The degradation of natural materials by bacteria devoid of oxygen is known as biomethanation. Natural carbon is converted generally into carbon dioxide and methane through a multi-step process (Angelidaki et al. 2003). A variety of substrates, such as animal compost, energy crops, waste from specific sectors, and so on, can be used to make biogas. Anaerobic reactions typically take place in biogas setups where methanogenic reactions promote the use of poultry and animal waste as the waste substrate. According to the energy report development series, methane production ranges from 50 to 70%, while carbon dioxide production ranges from 30 to 40%, hydrogen production ranges from 5 to 10%, nitrogen production ranges from 1 to 2%, and water vapor and hydrogen sulfide (H_2S) production are minimal (Chatterjee et al. 2015). The common anaerobic process that occurs when bio-waste decomposes (Heng 2017).

To meet the population's demand for protein, more than 25,000 poultry farms are currently in operation in Pakistan. These farms include of laying hens (for eggs), broilers (for meat), and hatcheries (for breeding) forms. This figure is rising because of the nation's growing population. Large amounts of trash are created as a result of processing chickens (Hussain et al. 2015). Pure organic matter makes up all of the poultry waste, which is now produced as waste because it has no use in any industry (Salminen et al. 2002). Feathers, blood, and manure—all organic wastes that can be utilized to create biogas—are created during the processing of poultry. The generated biogas can be used to generate electricity (León and Martín 2016). By using the anaerobic digestion process, it is possible to create biogas from poultry manure. The biodigestate created during the anaerobic digestion process is a nutrient-rich by-product that is used as fertilizer (Balsam 2022). Along with power generation, the generated biogas is a useful energy source with many other uses. Therefore, using chicken manure to produce biogas for energy production is a good idea (Mandeno et al. 2005).

10.2 Classification of Animal By-Products

Animal by-products are described and categorized differently from one country to the next and in accordance with various uses in the meat business. They can be categorized as organs and non-organs by-products, or EBPs and IEBPs. They can also be divided into groups according to their muscle composition, appearance, and color (Pérez-Alvarez 2011).

10.2.1 List of Animal By-Products

Includes the fat, skin or flesh, blood, milk, whey, eggs, and gelatin. Animal organs such as the pancreas, blood, bones, pituitary gland, and kidney and liver.

Animal by-products are broken down into 3 categories, with Category 1 including the materials with the highest danger. Category 3 by-products have the greatest potential for future usage because it carries the lowest amount of risk. Every by-product should always be treated independently in each of its categories.

10.2.1.1 Category 1

All healthy ruminants that have passed away naturally on farms fall into Category 1, along with particular ruminant organs like the brain and spinal cord. The category also includes animal carcasses and organs that may have been affected by transmittable spongiform encephalopathy (TSE) or another illness that can affect both humans and animals. Additionally, animals that have been shown to have remnants of drugs that are unlawful or detrimental to public health are under Category 1.

10.2.1.2 Category 2

Comprises healthy pigs and chicken that have died naturally on farms, which lowers the risk level. Additionally, animal by-products that exceed the allowable limit for residues of authorized pollutants or chemicals are listed in Category 2.

10.2.1.3 Category 3

The broadest range of by-product categories and the lowest amount of risk. It includes all of the healthy animals' carcasses, organs, and other components that have been slain for the production of meat.

All animal by-products, as well as waste from the meat and food sectors, are classified as by-products in their entirety. It specifies, for instance, how to manage meals containing by-products from non-farm animals, overseas animals, and products of animals of animal origin.

10.2.2 By-Products to Expensive Industrial Products

10.2.2.1 Category 1

Category 1 by-products may only be used as raw materials in products that are not disseminated on the ground as fertilizers, used as animal feed, or in contact with humans. Category 1 fats are processed in Honkajoki to be used as the base for biodiesel. In huge industrial co-incineration plants or industrial boilers, dry goods are refined to create fuel for electricity generation.

10.2.2.2 Category 2

By heating and drying leftovers, protein products are made from them. The materials are employed when making animal feed for furry creatures, and because they are abundant in nitrogen and phosphorus, they are also suitable for the creation of organic fertilizers. Organic fertilizers are easier to incorporate into the soil, persist longer, and enhance soil quality over time. The majority of the technical uses for refined Category 2 fats are in the manufacturing of biodiesel.

10.2.2.3 Category 3

For further industrial application, by-products are processed into PAP products and triglycerides. The PAP products are an excellent source of nutrients for making foods for animals, such as pet food, which can also be made using Category 3 fats. Additionally, fats are widely used in a variety of industries, including biofuels, cosmetics, and pharmaceuticals. It is also possible to deliver fresh Category 3 by-products to fur feed kitchens.

10.3 Utilization of Poultry Wastes

As chicken farms increase, huge amounts of excrement are produced. The European Union produces more than 107 tons of poultry dung, a significant agricultural fertilizer (Regueiro et al. 2018). However, poorly balanced use of chicken manure in agriculture can result in serious environmental issues such as air pollution, nitrogen leaching into aquifers, and eutrophication, emissions of greenhouse gases and disease spread, despite of the fact that it is nutrient-rich in nitrogen, phosphorus, and potassium natural wastes. Given that poultry manure is stabilized and energy is produced, anaerobic digestion (AD) of poultry litter is appealing for the management of bio-waste.

10.3.1 Biogas Production from Poultry Waste

Among the various available alternative sources of energy, biogas is an environmentally responsible and practical option for a sustainable power source. Anaerobic decomposition of biodegradable waste and materials high in nutrients results in the creation of a significant volume of biogas in the biogas production process. The degradation of natural materials by bacteria devoid of oxygen is known as biomethanation. Natural carbon is converted generally into carbon dioxide and methane through a multi-step process (Angelidaki et al. 2003). A variety of substrates, such as animal compost (including cow manure, poultry manure, and horse manure), energy crops (rice hulls, wheat hulls), waste from specific sectors, and so on, can be used to produce biogas. Anaerobic processes typically take place in biogas setups where methanogenic reactions promote the use of poultry and animal manure as the waste substrate.

The organic biodegradable components of the poultry waste include lipids, proteins, and carbs. Through an acid hydrolysis, the biomaterial is broken down into glucose, amino acids, and fatty acids (Sowunmi et al. 2016). Overall, there were four stages to the biogas manufacturing process: hydrolysis, acylation, acetogenesis, and methanogenesis (Mani et al. 2016). Acidogenic bacteria have the ability to generate biogas by hydrolyzing breakdown products into CO_2, H_2, acetates, and flammable fatty acids. Acetate and H_2 are formed from the breakdown of the volatile fatty acids. Lastly, methanogenic bacteria create CH_4 (Rasheed et al. 2016). The dead bird parts and the enormous amounts of manure that are also produced from laying birds are used to make manure. It has already been proposed that rice husk and poultry manure may be used together to create electricity, and the method was seen as both cost-effective and environmentally benign. The process of anaerobic digestion is a practical way to turn manure into fertilizer since it can produce biogas from it and manage trash in an environmentally acceptable way. The leftovers from anaerobic digestion contain essential nutrients (Smith et al. 2014). The production of biogas from the poultry waste has shown in the Fig. 10.1.

10.3.2 Birds Feathers and Their Utilization

An estimated 40–109 kg of feather by-products are produced annually by the processing of poultry meat globally. Despite the fact that some feathers are frequently converted into useful goods like feather meal and fertilizers, feathers are still seen as waste. A detailed examination of the structure and makeup of feathers reveals that

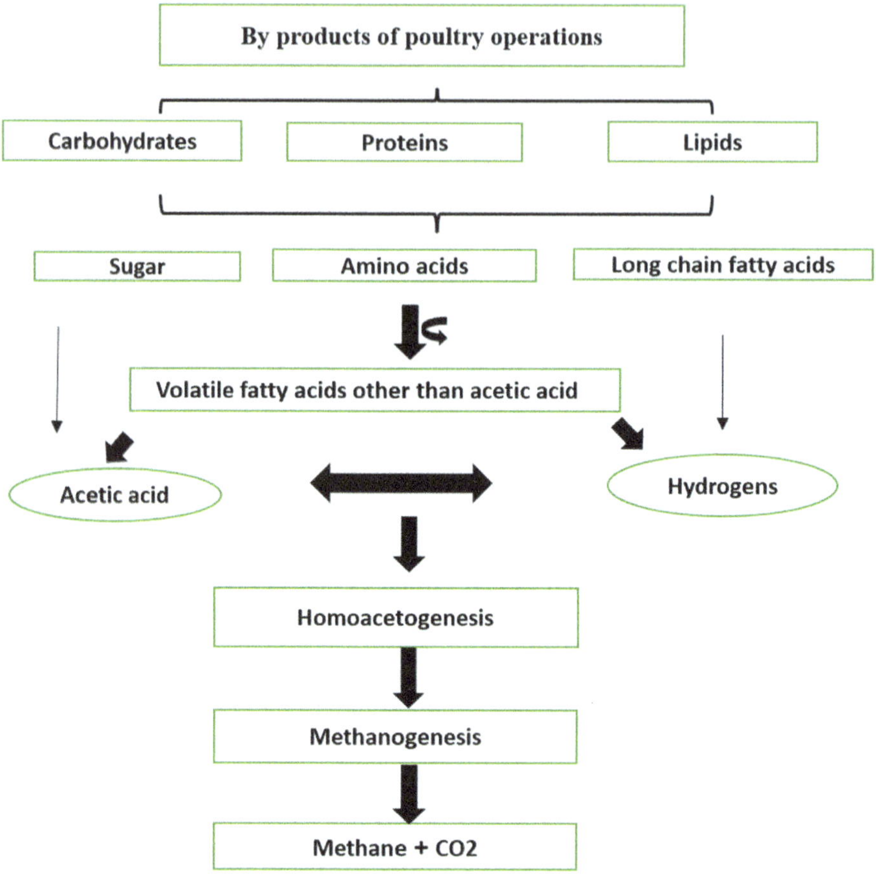

Fig. 10.1 Biogas production from poultry wastes

the entire chicken feather can be used as a source of keratin, a pure structural protein that can be used to create a variety of high-value bio products. Additional biological feather components can be transformed into high-value goods using a variety of technologies. Feathers may thus be a desirable raw material for the development of bio products due to the conversion of waste into useful products. Therefore, turning leftover feathers into a useful resource can aid the poultry business in getting rid of leftover feathers in a way that is both ecologically friendly and brings in additional revenue.

10.3.2.1 Feathers Used for Decoration

Large bird feathers have been used to create artificial flowers. The shading, shape, size, and plumage patterns of feathers must be taken into consideration while

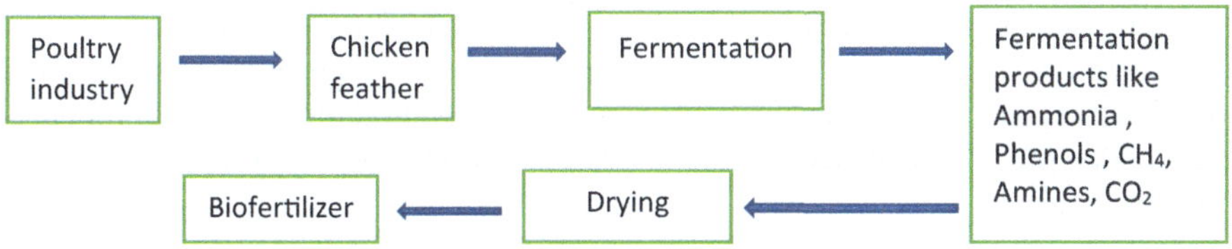

Fig. 10.2 Utilization of feathers into bio fertilizers

choosing them for decorative purposes. Due to their exquisite shading, cock pheasant feathers are in high demand for decorative uses (Levine 1991).

10.3.2.2 Feathers Used for Medical Purposes

Traditional medicines often include the use of chicken feathers. For example, in India, Indian peacock feathers have been used in traditional medicine for barrenness, hacking, and snakebites. In South America, condor feather mixes are used in traditional pharmaceutical products (Murari et al. 2005).

10.3.2.3 Feathers Used as Fertilizer

Feathers are amazing for composting since they have a nitrogen concentration of more than 13% (Tesfaye et al. 2017), which is better than the blood feeding used for these applications (Choi and Nelson 1996). As a result, feathers are utilized in plant-growing processes that ask for dressings high in nitrogen. However, feathers are difficult to disintegrate due to their high cysteine crosslinking (Park et al. 2000). As a result, there is very little nitrogen from the feathers that can be used as fertilizer. Feathers are a good mulching material as well. This is because they steadily decay and continuously release nitrogen. Most proteins-degrading substances cannot break down their stiff, fibrous structure, but when combined with manure, they disintegrate well (Gurav and Jadhav 2013). When feathers break down, their metabolites reverse direction and are incorporated into the soil as organic matter, increasing the soil's fertility. Given that they supply nitrogen, a vital component of fertilizer, they create an important poultry composites blend (Veerabadran et al. 2012). The feathers obtained as poultry waste are utilized to produce organic fertilizers and the process has been shown in Fig. 10.2.

10.3.2.4 Feathers to Create Leather Composites

There is a need to switch to eco-friendly materials because the various treatment techniques employed in tanning leather may result in skin and respiratory diseases as well as cancer. In this regard, wool and colleagues created bio-composites by processing scraped, powdery chicken feather fibers into synthetic leather using methods created by aeronautical engineers. Under pressure and heat, wool consolidates natural fiber and plant oil polymers to create a Wool consolidates natural fiber and plant oil polymers under pressure and heat to produce a nanocomposite that resembles leather (Sydney 2015).

10.3.2.5 Feather Meal as a Feedstock

Due of their harmfulness (microbiological infections present) and poor digestion when consumed on land, the majority of feathers are not suited for the applications especially. The conversion of feather waste into feather meal for use as stock fodder is thus the main method of managing feather waste. The rhizome must undergo hydrolysis to become digestible in order to produce feather meal. A typical method is as following:

- After being collected from industries, feathers are cleaned in water and then de-watered and heat is preferred above mechanical pressure.
- After the water was removed, they would have been heated and wet-cooked for one to two hours under pressure to facilitate hydrolysis.
- The feathers are subsequently dried, chilled, and crushed.
- To remove the large metal particles, the powdered food is next placed through metal detectors (El Boushy et al. 1990).

The quantity of hydrolysis (cooking time and pressure) has a direct impact on how digestible feather meal (McCasland and Richardson 1966). However, the protein edibility is very low due to the presence of disulfide bonds, resistant to the digestive enzymes found in chickens. Feather feed contains about 92% crude protein (ranges from 70 to 80% as digestive protein). Methionine, histidine, lysine, and tryptophan, the four essential amino acids, are insufficient in feather meal, whereas threonine, cysteine, and arginine are abundant (El Boushy et al. 1990). About 0.5–1.5% is a reasonable percentage to use feather meal as a feedstock (Park et al. 2000). The production of animal feed from the feathers waste of poultry has been presented in Fig. 10.3.

10.4 Utilization of Wastes of Slaughter Animals

10.4.1 Utilization of Animal Blood

Animal is a significant food by-product that is rich in myoglobin and protein (Wan et al. 2002). Blood dessert, blood meatballs, biscuits, and bread have all been made with animal blood throughout Europe for a very long time. In Asia, it is used to make blood yogurt, blood pudding, and blood sausage (Ghosh 2001). Additionally, it is

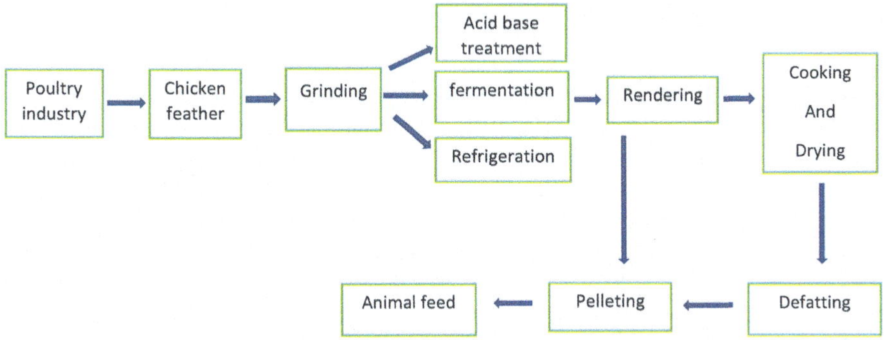

Fig. 10.3 Animal feed preparation using chicken feathers is shown schematically

used for non-food items including binders, feeds, and fertilizers. According to the United States' Meat Inspection Act, blood that has been drawn from an animal that has undergone inspection and been given the all-clear for its use in meat items is permitted.

Normally, an animal in good health has sterile blood. It contains a high protein content of 17.0 and a decently balanced amino acid profile. (2.4–10.0% of the animal weight) The animal's blood makes up a sizeable portion of its body mass. Pigs, cattle, and lambs can each recover an average of 3.0–4.0, 3.0–4.0, and 3.5–4.0% of their blood, respectively. However, using blood during the processing of meat could result in a product that is unappealingly black in color. Due to its beneficial properties and absence of color, plasma is the component of blood that attracts the most interest.

10.4.2 Blood Usage in Medicine and Pharmaceuticals

Several therapeutically useful fractions of blood can be extracted from whole blood. The biggest fraction (63.0%) is liquid plasma. 3.5% albumin, 4.0% globulin, and 4.0% fibrinogen make up its composition. Numerous blood products are utilized in the laboratory as nutrients for tissue culturing media, an essential component of agar media, and also as peptones for microbiological usage (Kurbanoglu 2004). For biological assays, glycerophosphates, albumins, globulins, sphingomyelins, and catalase are also employed. Numerous blood components, including plasminogen, fibrinogen, serotonin, and immunoglobulins, are separated for use in drugs or chemicals. Animals who have lost blood or fluids can be helped by administering purified bovine albumin. It serves as a stabilizer for vaccines and is used to screen for the Rh factor in human beings. Antibiotic sensitivity testing also employs it.

10.4.3 Utilization of Hides and Skins

Humans have utilized animal hides as storage, garments, and shelter since the beginning of time. Between 4 and 11% of the living animal's weight is made up of the hides. One of the most important by-products from animals is typically hides and skins. Cattle, pig, and sheep hides can be used to make completed goods including leather shoes and purses, rawhide, sports equipment, reformation of sausage casing, cosmetics, palatable gelatin, and adhesives (Benjakul et al. 2009).

10.4.3.1 Gelatin from Hides and Skins

A water-insoluble collagen formed from protein is controlled hydrolyzed to create gelatin. It is created using recent, consumable raw materials like edible skins or bone. Collagen is present in great abundance in both skins and bones. Three main processes are involved in the production of gelatin from hides. The first step is the separation of non-collagenous components from the raw material. Controlled collagen hydrolysis into gelatin occurs next. Recovery and drying of the finished product are the last two steps.

10.4.3.2 Uses of Gelatin in the Food and Pharmaceutical Industries

Gelatin made from animal hides and skins can be consumed (Choa et al. 2005). Fat can also be produced by rendering the source material. Pig skin is dipped, boiled, dried, and then fried to create pork peels, also known as "pork scratching," in the United States, Latin America, Europe, and some Asian nations. Since it can bind a lot of fat, collagen from hides and skins also serves as an emulsifier in meat products. It can, therefore, act as a filler or ingredient for meat food. The meat's utilization of collagen sausage can also be made by extracting collagen from cow skins.

In addition to being a key component in jellies and aspic, gelatin is added to a variety of meals (Jamilah and Harvinder 2002). Due to its "melt in the mouth" qualities, it is primarily used to create flavored desserts, but it is also added to a variety of meat items, most notably meat pies. A common stabilizer for desserts that are frozen, such as ice cream is gelatin. For further protection, high-bloom gelatin is added to pies with cream, curd, and ice cream. It is believed that the gelatin prevents lactose from recrystallizing and ice crystals from forming during storage. In the pharmaceutical sector, gelatin is used for about 6.5% of total output (Hidaka and Liu 2003). The majority of it is employed in the creation of capsules' exterior shell. In addition, gelatin can be utilized as a binding and compounding agent in the production of pharmaceutical pastilles and tablets. When treating ulcerated varicose veins, it is a key component of preventive ointments such as zinc gelatin. By beating gelatin into foam, applying formaldehyde, and then drying it, gelatin could be transformed into a sterile sponge. These sponges are employed during surgery as well as for the direct implantation of medications such as antibiotics. Gelatin, a protein, is used to expand blood plasma in cases of really severe shock and injury. For many emulsions and foams, gelatin works great as an emulsifier and stabilizer. It is utilized in cosmetics as well as printing processes including silk screen printing and photogravure.

Although collagen casing goods were created in Germany in the 1920s, American consumers didn't start using them until the 1960s. Like with gelatin, the procedure does not make the collagen into a soluble substance. Instead, it creates a product that is robust enough to be utilized as a casing for sausages and other items while yet retaining a significant amount of the native collagen fiber. The purified collagen is combined with water to create a dough that may be extruded either wet or dry. The collagen is subsequently precipitated by passing the collagen extrusion tube through a chamber of ammonia and a saturated salt solution. Gel that has swelled contracts to form a film with respectable strength. Glycerin can be added to it to make it better and more flexible. The tube is subsequently dried until it contains 10–15% of water.

10.4.3.3 Medicinal Uses of Hides and Skins

During surgery, a collagen-derived substance can encourage blood clotting. Pork skin can be used as a bandage for wounds or skin ulcers since it resembles human skin. Pork skin that will be used as a dressing needs to be broken into 0.2–0.5 mm thick strips or patches, cleaned, sanitized, and packaged. It is appropriate for skin transplantation.

10.4.4 Utilization of Waste as Biofuel

The accessibility of wet biomass as a by-product of industrial processes and the requirement to adhere to environmental requirements serve as the primary impetuses for exploring various disposal alternatives for this waste. Operators of power plants are becoming increasingly interested in the thermal recycling of leftovers as secondary fuel (Arvanitoyannis and Ladas 2008). Poultry litter has been used as a substitute for the production of natural fuel sources, according to studies. Notably, chicken litter with a water content of less than 9% can burn on its own without additional fuel. As a result, these samples could be employed as fuel for the production of electrical power. To improve the effectiveness of removing organic matter, wastewater from the meat sector is subjected to physicochemical treatment, which produces a significant volume of sludge. This particular wastewater was treated using commercial ferric sulfate as a coagulant, which led to high organic material removals, significantly lowering the volume of waste that needed to be handled in natural systems and enabling the production of 0.83–0.87 kg of biomass fuel for every m^3 of treated water (De Sena et al. 2008). Due to hygienic, environmental, and operational concerns related to the discharge, land disposal, and re-use of wastes, the use of biofuels for steam production has shown to be a realistic choice. This fuel type is a source of renewable energy and has a high heating value. The 4:1 biofuel to sawdust ratio used in the combustion test met the technical criteria for characterizing this prospective fuel, but optimal operating conditions are still needed to keep NO_2 and SO_2 emissions within regional and/or global restrictions.

Diesel fuel made from petroleum can be substituted for or combined with fuel made from animal and fish fats and oils called biodiesel. There is a wealth of information available on the manufacture of biogas from cow manure, pig waste, and fishery leftovers (Arvanitoyannis and Kassaveti 2008).

10.4.5 Organ and Gland Applications in Medicine and Pharmaceuticals

Traditional medical practices in numerous nations, including China, India, and Japan, include the use of animal glands and organs. Hormones, which are actually enzymes that control the body's metabolism, are secreted by the endocrine glands. These include the kidney, corpus luteum, ovary, follicle, pituitary, thyroid, pancreas, stomach, parathyroid, and adrenal glands. Only healthy animals are used in the collection of the glands. The glands require considerable experience to locate. They frequently contain other tissue and are tiny.

Distinct glands have different roles in various species. The species, gender, and age of the animal all affect how the glands work. Rapid freezing is the most effective way to preserve most glands in order to halt bacterial growth and tissue destruction. The glands must be cleansed and the fat and connective tissue around them removed before freezing. After that, the glands are put on waxed paper and maintained at or below 18 °C. When the glands arrive at the pharmaceutical facility, they are examined before being cut up and combined with various extraction solutions or put in a vacuum dryer. Solutions like acetone, ethylene, light petroleum, or gasoline are used to dissolve excess fat from desiccated glands if there is too much of it.

The raw materials for vitamin D3 synthesis are found in the brains, neurological systems, and spinal cords. Additionally, cholesterol serves as cosmetics emulsifier (Ejike and Emmanuel 2009). For the same objective, other components of the brain's hypothalamus can be removed. Mental retardation, sleeplessness, and other issues are being studied as potential treatments for with the hormone melatonin, which is derived from the pineal gland.

Bile can be obtained from the gall bladder and is made up of acids, pigments, proteins, cholesterol, and other substances. It is used to treat biliary tract problems, constipation, and indigestion. It is also employed to boost the liver's secretory activity. It is possible to buy liquid or dry extracts of bile from cows or pigs. Prednisone and cortisone are two examples of bile components that can be isolated and utilized as medications. Gallstones can be purchased for a high price and are said to have aphrodisiac characteristics. Typically, they are incorporated into necklaces and brooches as ornaments.

The largest gland in an animal is the liver. A mature cow's liver typically weighs 5 kg, but a pig's liver weighs roughly 1.5 kg. Raw powdered liver is combined with heated, just little acidic water for the extraction of liver. The pharmaceutical sector uses the stock as a raw material once it is condensed into a paste in a low-temperature vacuum environment. Liver extract, which may be derived from cattle, has long been used as a dietary supplement and as a source of the vitamin B12 for treating various types of anemia (Devatkal et al. 2004a, b; Colmenero and Cassens 1987). The liver, lungs, and small intestine lining are among the organs from which heparin can be obtained. It serves as an anticoagulant to postpone the time at which blood clots. Additionally, it is employed in organ transplants to thin the blood and stop blood clots from forming during surgery.

Pig ovaries can be used to extract progestin and estrogen. It can be used to treat female patients with reproductive issues. The hormone relaxin, which is extracted from the ovary of the pregnant sows, is frequently utilized during birthing.

Insulin, which controls how sugar is metabolized and is used to treat diabetes, is produced by the pancreas. Blood sugar is raised using pancreatic tissue-derived glucagon, which is also used to cure insulin overdose and low blood sugar brought on by alcoholism. Trypsin and chymotrypsin are utilized to speed up the recovery process following surgery or injury.

For the production of catgut, which is utilized to create internal surgical sutures, sheep and calf intestines are employed. While the intestines are being prepared for use as casings, the cattle's small intestinal lining can be extracted. For distribution to heparin producers, it is either stored in a raw form or turned into a dry powder. The applications of animal waste are presented in Table 10.1.

Table 10.1 Types of animal wastes and their uses

Types of animal wastes	Uses
Manure and litter for poultry	Recycled feed and agricultural land surface dressing
Hatchery products	Egg shell meal as high calcium diet
Feathers	Decorative purpose, feather meal
Heads	Poultry meal
Blood	Blood meal
Skin	Leather industry
Wool	Dress making

10.5 Utilization of Livestock Animal Wastes

10.5.1 Manure Utilization on Crops

Around the world, food production area has been fertilized with animal excrement for millennia. Although this method has generally met the needs of the farming community and is a respectable way to conserve resources on a variety of grain, bean, and cotton crops, it disperses surviving viruses across a wide area (Bicudo and Goyal 2003). It is obvious that using animal manure in the initial stages of growing fruit and vegetable crops has risks and raises the potential for enteric pathogens, which can survive in animal waste inputs, to cause contamination. In the United States between 1990 and 1998, outbreaks of foodborne disease connected with tainted produce (24%) were calculated to be virtually equivalent to those linked with meats (29%).

Several outbreaks involving produce were documented during this time period from tiny, organic gardens where raw manure had just been administered (Cieslak et al. 1993; Guan and Holley 2003; Nelson 1997). Organic farming uses animal manures, crop rotation and residues, ammonia legumes, soil additives, and mineral limestone granules to preserve soil quality and feed plant nutrients. Insects, weeds, and other pests are controlled via cultivation, cultural treatments, and biocontrol. The current USDA certified organic requirements mandate that growers' compost manure under thermophilic conditions, or if they utilize raw manure, harvest cannot start before 90–120 days after application. The USDA National Organics Program (NOP) and National Agricultural Statistics Service do not keep detailed records of the number of farms, certified natural or conventional, that use manure or dung-based products to produce fresh food. The NOP mandates certification for businesses with organic sales of more than $5000 and compels organic growers and food handlers to adhere to a unified organic standard. Through independent certifiers that it audits, the NOP puts the regulations into effect. The NOP has accredited approximately 40 overseas programs and about 50 states and the private certification programs in the United States.

Manure-based inputs are widely used by both conventional and organic growers, not only for their usefulness as fertilizer but also for improving and preserving soil quality. A number of safety requirements must be followed in order for organic food production and distribution to continue to grow globally (Cooper et al. 2007). According to records, both commercially and organically grown produce has had epidemics. It is evident that both production processes use similar inputs. The non-preferential contamination of fresh fruit outbreaks across conventional and organic sources suggests that actual on-site circumstances and practices, rather than marketing-based designations (such as "organic"), are the key determinants of the hygienic state of fresh produce. However, in order to employ their own composts, small farmers and backyard gardeners might need to be continually informed on proper agricultural methods.

Along with applying manure directly to the soil, runoff from key fresh market crop producing locations such as enclosed lots where animals graze might raise the threat of disease infection of crops for fresh produce crops. In cattle and poultry production facilities both domestically and overseas, the economics and efficiency of animal husbandry have resulted in an increase in the animal density per unit area over the past few decades. Within very small landscape areas, these intensive production practices now employed for broilers, layers, turkeys, swine, beef, and dairy animals produce significant amounts of excrement. In order to use these manures as fertilizers properly and prevent the contamination of surface and groundwater by pathogens, organic matter, sediments, nutrients (nitrate, phosphorous), and other substances, it is necessary to calculate the nutrient content, especially nitrogen and phosphorus, relative to crop needs and existing soil test values (Al-Kaisi et al. 1998a, b). Due to the high volumes of manure produced by intensive livestock and poultry operations, numerous complex manure management methods have been created (Vanotti et al. 2003, 2005b). Development has concentrated on pathogen disinfection in both liquid and solid-phase materials in addition to nutrients (Vanotti et al. 2005a). Although these innovations were aimed at swine manure, they can be used in dairy systems that employ liquid collection techniques.

10.5.2 Transesterification of a Lipid Component in Livestock Waste to Produce Biodiesel

In majority of nations that use biodiesel, the fuel is created by trans esterifying the lipids found in palatable crops, animal fat oils, and used oils for cooking (Khounani et al. 2019). Triglycerides (lipids) are initially extracted from the oil containing biomass feed stocks including chicken fat, rapeseed, sunflower, rapeseed oil, soybean, palm tree, and used fish oils, etc., in the biodiesel production process. There are very few fatty acids present in lipids, and the triglycerides that are present have three distinct or partially distinct fatty acids bound to the backbone of glycerol. Triglycerides and free fatty acids are then (trans)esterified with the addition of methanol or ethanol to produce fatty acid methyl esters (FAMEs) or fatty acid ethyl esters (FAEEs). In order to separate the lipids in the unprocessed biomass materials, lipid extraction is a pretreatment procedure. Currently, a number of strategies have been developed, including solvent extraction using a Supercritical CO_2 expunction, the Soxhlet device, and ultrasonic-assisted extraction (Siddiquee and Rohani 2011).

The process of turning lipid into biodiesel requires both the free fatty acid esterification and triglycerides transesterification when alcohol is present. When utilizing acidic or alkaline catalysts to transesterify lipids, methanol is frequently used as an alcohol. The type of catalyst, reaction temperature (30–120 °C for homogeneous catalysts), reaction time (0.1 h–1 day), and molar ratio of lipids to methanol have all had a significant impact regarding the yield and manufacturing rate of biodiesel (1:1–40:1).

It is an undeniable truth that the cost of the feedstock (> 70%) predominates the entire production cost of biodiesel, even if many oil-bearing feed stocks can provide high yields of biodiesel. As a result, the cost of producing biodiesel rises, necessitating the search for low-priced biomass waste that contains oil. Gomaa and Abed transesterified livestock manures due to the significant lipid content (11–14 wt%) of chicken, goat, and cow manures. Before using the samples during 24 h at 50 °C for concurrent lipid extraction and transesterification using a co-solvent, cattle dung was dried and homogenized. Based on the weight of the dried dung, the biodiesel yields from chicken, goat, and cattle manures were 4, 6, and 6.5 wt%, respectively. The biodiesel yield ranged from 35.7 to 54.1 wt% based on the lipid content (Gomaa et al. 2017).

Kim et al. explored the production of biodiesel from swine dung, which has a dry basis lipid content of 12 wt% (Kim et al. 2020). In order to separate the solvent from the solvent, lipid was extracted using a Soxhlet extractor in the presence of n-Hexane, and the acid value of the lipid extractive was 72.25 mg KOH g1 lipid. The considerable concentration of free fatty acids is indicated by the high acid value. In order to produce biodiesel, transesterification of lipid with an alkaline catalyst is kinetically quicker and more effective than transesterification with an acidic catalyst. But when there are more than 5 free fatty acids, an alkaline catalyst reacts with them to form soap (soap production). The transesterification of triglycerides and the esterification of free fatty acids should be carried out separately utilizing two procedures or in addition to the acidic catalyst in order to prevent the saponification reaction.

Using an acid catalyst, synchronous (trans)esterification of swine dung lipid extract (H_2SO_4) was carried out by Kim et al. for 26 h at 60 °C. Based on lipid extraction, the production of biodiesel was 14.2 wt%. The amount of biodiesel produced from swine excrement was minimal because (trans)esterification is very vulnerable to the presence of contaminants. They devised a thermochemical non-catalytic transesterification to address the technological difficulty. In this procedure, silica, a porous material, and methanol were used to quickly transform lipid extractive into biodiesel at high temperature (200–400 °C). When swine dung lipid was transesterified at a temperature of 360 °C, the yield of biodiesel increased to more than 90 wt% (lipid basis). Without lipid extraction, pig excrement might likewise be converted directly into biodiesel using this method. At 400 °C, the maximum biodiesel yield produced by transesterification was greater than 95 wt% (Kim et al. 2020). This technique also had the advantage of not requiring the steps for separating and extracting lipids, which can improve reaction efficiency overall and lower process costs. With the concept of valorizing manure leftovers following the generation of biodiesel, Fig. 10.4 compares thermochemical non-catalytic transesterification procedures with traditional transesterification.

10.6 Conclusion

One of the waste materials produced during the production of cattle is animal waste. Animal waste has significant negative effects on both people and the environment,

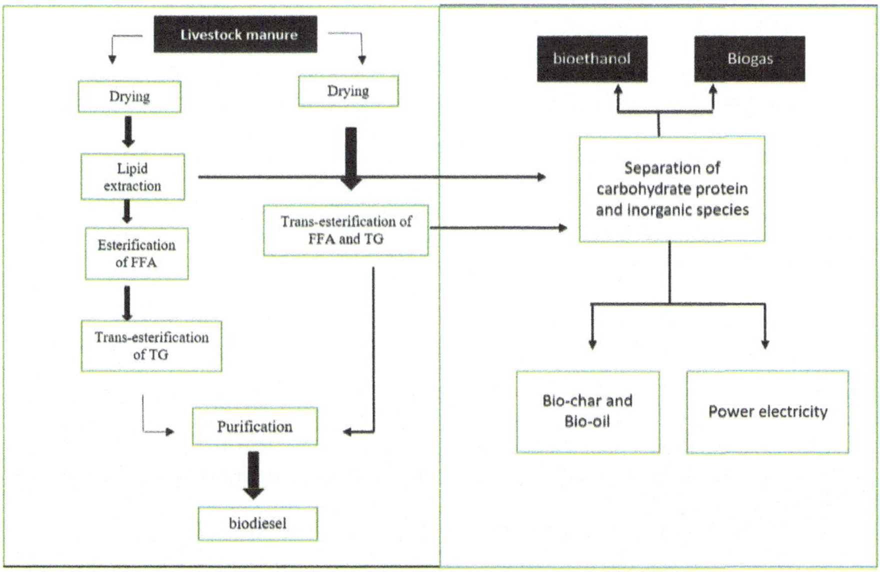

Fig. 10.4 Transesterification of a lipid component in cattle waste to produce biodiesel

but it also has extremely significant positive effects. Animal wastes can be used as a growing medium for earthworms, a source of nourishment for plants, and an energy source in the form of methane gas. Environmental pollution can result from improperly treated animal waste. In an animal, around one-third of its body—the meat portion—is consumed as food, and the other two-thirds are non-meat. In fact, if technology is added to the non-meat element, its economic potential can be maximized. Applications for livestock by-products can be found in a wide range of large- and small-scale industries. The government's initiatives to raise the cattle population will undoubtedly enhance the potential for by-products as well. One exportable livestock product is animal skin. The method of maintaining cattle during the cultivation phase has a significant impact on the quality of leather export. The quality of the finished skin is significantly influenced by the nutrition, management, maintenance, and treatment of cattle in slaughterhouses.

References

Al-Kaisi MM, Waskom RM, Davis JG (1998a) Liquid manure application to cropland (Doctoral dissertation, Colorado State University. Libraries)

Al-Kaisi MM, Davis JG, Waskom RM, More OL (1998b) Liquid manure application methods no. 1.223

Angelidaki I, Ellegaard L, Ahring BK (2003) Applications of the anaerobic digestion process. Biomethanation II:1–33

Arvanitoyannis IS, Kassaveti A (2008) Fish industry waste: treatments, environmental impacts, current and potential uses. Int J Food Sci Technol 43(4):726–745

Bah H, Zhang W, Wu S, Qi D, Kizito S, Dong R (2014) Evaluation of batch anaerobic co-digestion of palm pressed fiber and cattle manure under mesophilic conditions. Waste Manage 34(11):1984–1991

Balsam J, Ryan D (2006) Anaerobic digestion of animal wastes: factors to consider. Nat Sustain Agric Inf Service:2–3

Benjakul S, Oungbho K, Visessanguan W, Thiansilakul Y, Roytrakul S (2009) Characteristics of gelatin from the skins of bigeye snapper, *Priacanthus tayenus* and *Priacanthus macracanthus*. Food Chem 116(2):445–451

Bernal MP (2017) Grand challenges in waste management in agroecosystems. Front Sustain Food Sys 1:1

Bicudo JR, Goyal SM (2003) Pathogens and manure management systems: a review. Environ Technol 24(1):115–130

Byers T, Nestle M, McTiernan A, Doyle C, Currie-Williams A, Gansler T, Thun M, American Cancer Society 2001 Nutrition and Physical Activity Guidelines Advisory Committee (2002) American Cancer Society guidelines on nutrition and physical activity for cancer prevention: reducing the risk of cancer with healthy food choices and physical activity. CA Cancer J Clin 52(2):92–119

Chatterjee RN, Rajkumar U (2015) An overview of poultry production in India. Ind J Animal Health 54(2):89–108

Chikwanha OC, Vahmani P, Muchenje V, Dugan ME, Mapiye C (2018) Nutritional enhancement of sheep meat fatty acid profile for human health and wellbeing. Food Res Int 104:25–38

Cho SM, Gu YS, Kim SB (2005) Extracting optimization and physical properties of yellowfin tuna (*Thunnus albacares*) skin gelatin compared to mammalian gelatins. Food Hydrocolloids 19(2):221–229

Choi JM, Nelson PV (1996) Developing a slow-release nitrogen fertilizer from organic sources: III. Isolation and action of a feather-degrading actinomycete. J Am Society Horticultural Sci 121(4):639–643

Cieslak P, Barrett T, Griffin P, Gensheimer K, Beckett G, Buffington J, Smith MG (1993) *Escherichia coli* 0157: H7 infection from a manured garden. Lancet 342(8867):367

Cooper J, Leifert C, Niggli U (eds) (2007) Handbook of organic food safety and quality

De Sena RF, Claudino A, Moretti K, Bonfanti ÍC, Moreira RF, José HJ (2008) Biofuel application of biomass obtained from a meat industry wastewater plant through the flotation process—a case study. Resour Conserv Recycl 52(3):557–569

Devatkal S, Mendiratta SK, Kondaiah N (2004a) Quality characteristics of loaves from buffalo meat, liver and vegetables. Meat Sci 67(3):377–383

Devatkal S, Mendiratta SK, Kondaiah N, Sharma MC, Anjaneyulu ASR (2004b) Physicochemical, functional and microbiological quality of buffalo liver. Meat Sci 68(1):79–86

Ejike CE, Emmanuel TN (2009) Cholesterol concentration in different parts of bovine meat sold in Nsukka, Nigeria: implications for cardiovascular disease risk. Afr J Biochem Res 3(4):095–097

El Boushy AR, Van der Poel AFB, Walraven OED (1990) Feather meal—a biological waste: its processing and utilization as a feedstuff for poultry. Biol Wastes 32(1):39–74

Elfaki MOA, Abdelatti KA, Malik HEE (2015) Effect of dietary dried rumen content on broiler performance, plasma constituents and carcass characteristics. Glob J Animal Sci Res 3(1):264–270

Ghosh R (2001) Fractionation of biological macromolecules using carrier phase ultrafiltration. Biotechnol Bioeng 74(1):1–11

Gomaa MA, Abed RM (2017) Potential of fecal waste for the production of biomethane, bioethanol and biodiesel. J Biotechnol 253:14–22

Guan TT, Holley RA (2003) pathogen survival in swine manure environments and transmission of human enteric illness—a review a. In: Hog manure management, the environment and human health, pp 51–71

Gurav RG, Jadhav JP (2013) A novel source of biofertilizer from feather biomass for banana cultivation. Environ Sci Pollut Res 20(7):4532–4539

Heng LK (2017) Bio gas plant green energy from poultry wastes in Singapore. Energy Proc 143:436–441

Hidaka S, Liu SY (2003) Effects of gelatins on calcium phosphate precipitation: a possible application for distinguishing bovine bone gelatin from porcine skin gelatin. J Food Compos Anal 16(4):477–483

Hussain J, Rabbani I, Aslam S, Ahmad HA (2015) An overview of poultry industry in Pakistan. Worlds Poult Sci J 71(4):689–700

Irshad A, Sharma BD (2015) Abattoir by-product utilization for sustainable meat industry: a review. J Animal Prod Adv 5(6):681–696

Jamilah B, Harvinder KG (2002) Properties of gelatins from skins of fish—black tilapia (*Oreochromis mossambicus*) and red tilapia (*Oreochromis nilotica*). Food Chem 77(1):81–84

Jiménez-Colmenero F, Cassens RG (1987) Influence of an extract of liver on colour and shelf stability of sliced bologna. Meat Sci 21(3):219–230

Khounani Z, Nazemi F, Shafiei M, Aghbashlo M, Tabatabaei M (2019) Techno-economic aspects of a safflower-based biorefinery plant co-producing bioethanol and biodiesel. Energy Convers Manage 201:112184

Kim M, Jung S, Lee DJ, Lin KYA, Jeon YJ, Rinklebe J, Klinghoffer NB, Kwon EE (2020) Biodiesel synthesis from swine manure. Biores Technol 317:124032

Kurbanoglu EB, Kurbanoglu NI (2004) Utilization as peptone for glycerol production of ram horn waste with a new process. Energy Convers Manage 45(2):225–234

León E, Martín M (2016) Optimal production of power in a combined cycle from manure based biogas. Energy Convers Manage 114:89–99

Levine VL (1991) Feathers in southeast American Indian ceremonialism. Expedition 33(2):3

Liu DC (2002) Better utilization of by-products from the meat industry. Food and Fertilizer Technology Center

Mandeno G, Craggs R, Tanner C, Sukias J, Webster-Brown J (2005) Potential biogas scrubbing using a high rate pond. Water Sci Technol 51(12):253–256

Mani S, Sundaram J, Das KC (2016) Process simulation and modeling: anaerobic digestion of complex organic matter. Biomass Bioenerg 93:158–167

McCasland WE, Richardson LR (1966) Methods for determining the nutritive value of feather meals. Poult Sci 45(6):1231–1236

Murari SK, Frey FJ, Frey BM, Gowda TV, Vishwanath BS (2005) Use of *Pavo cristatus* feather extract for the better management of snakebites: neutralization of inflammatory reactions. J Ethnopharmacol 99(2):229–237

Nelson H (1997) The contamination of organic produce by human pathogens in animal manures. In: Ecological agriculture projects, Faculty of Agricultural and Environmental Science, McGill University (Macdonald Campus), Ste-Annede-Bellevue, QC, Canada

Park SK, Bae DH, Hettiarachchy NS (2000) Protein concentrate and adhesives from meat and bone meal. J Am Oil Chem Soc 77(11):1223–1227

Pérez Álvarez F (2011) Temas actuales de investigación en ciencias penales: I Congreso internacional de jóvenes investigadores en ciencias penales, 26, 27 y 28 de octubre de 2009. Temas actuales de investigación en ciencias penales, pp 1–452

Rasheed R, Khan N, Yasar A, Su Y, Tabinda AB (2016) Design and cost-benefit analysis of a novel anaerobic industrial bioenergy plant in Pakistan. Renew Energy 90:242–247

Rodriguez-Verde I, Regueiro L, Lema JM, Carballa M (2018) Blending based optimisation and pretreatment strategies to enhance anaerobic digestion of poultry manure. Waste Manage 71:521–531

Salminen E, Rintala J (2002) Anaerobic digestion of organic solid poultry slaughterhouse waste—a review. Biores Technol 83(1):13–26

Sans P, Combris P (2015) World meat consumption patterns: an overview of the last fifty years (1961–2011). Meat Sci 109:106–111

Siddiquee MN, Rohani S (2011) Lipid extraction and biodiesel production from municipal sewage sludges: a review. Renew Sustain Energy Rev 15(2):1067–1072

Smith J, Abegaz A, Matthews RB, Subedi M, Orskov ER, Tumwesige V, Smith P (2014) What is the potential for biogas digesters to improve soil fertility and crop production in Sub-Saharan Africa? Biomass Bioenerg 70:58–72

Sowunmi A, Mamone RM, Bastidas-Oyanedel JR, Schmidt JE (2016) Biogas potential for electricity generation in the Emirate of Abu Dhabi. Biomass Convers Biorefinery 6(1):39–47

Sydney B (2015) Green chemists use rocket science and chicken feathers to create cow-less leather

Tesfaye T, Sithole B, Ramjugernath D, Chunilall V (2017) Valorisation of chicken feathers: characterisation of chemical properties. Waste Manage 68:626–635

Vanotti MB, Millner PD, Hunt PG, Ellison AQ (2005a) Removal of pathogen and indicator microorganisms from liquid swine manure in multi-step biological and chemical treatment. Biores Technol 96(2):209–214

Vanotti MB, Rice JM, Ellison AQ, Hunt PG, Humenik FJ, Baird CL (2005b) Solid-liquid separation of swine manure with polymer treatment and sand filtration. Trans ASAE 48(4):1567–1574

Vanotti MB, Szogi AA, Hunt PG, Millner PD, Humenik FJ (2007) Development of environmentally superior treatment system to replace anaerobic swine lagoons in the USA. Biores Technol 98(17):3184–3194

Vanotti MB, Hunt PG, Szogi A, Humenik FRANK, Millner P, Ellison A (2003, October) Solids separation, nitrification-denitrification, soluble phosphorus removal, solids processing system. In: Proceedings of the North Carolina animal waste management workshop, pp 30–35

Veerabadran V, Balasundari SN, Devi DM, Kumar DM (2012) Optimization and production of proteinacious chicken feather fertilizer by proteolytic activity of Bacillus sp. MPTK 6. Ind J Innovations Dev 1(3):193–198

Wan Y, Ghosh R, Cui Z (2002) High-resolution plasma protein fractionation using ultrafiltration. Desalination 144(1–3):301–306

Chapter 11
Circular Bioeconomy of Animal Wastes

Nasib Zaman, Sher Zaman Safi, Shahid Ali, Ghulam Mustafa, Raja Tahir Mahmood, Dawood Ahmad, Muhammad Nazir Uddin, Aziz ur Rehman, Abdur Rahman Ansari, Aqsa Mumtaz, and Muhammad Arshad

Abstract Animals and poultry produce a significant amount of waste or by-products on a global scale. They are currently underutilized in high-value applications or used to produce relatively low-value goods like animal feed and pet food. The disposal of some animal by-products poses a significant environmental risk since they cannot be used to produce food or feed. This chapter aims to approach possible solutions to environmental pollution by highlighting the waste biorefinery as a sustainable bio-based circular economy. The circular bioeconomy is also made possible by incorporating trash into bioprocesses to create valuable goods and metabolites. The concept of a circular bioeconomy which manufactures valuable bioproducts from biological

N. Zaman · M. N. Uddin
Centre for Biotechnology and Microbiology, University of Swat, Khyber Pakhtunkhwa, Swat, Pakistan

S. Z. Safi
Faculty of Medicine, Bioscience and Nursing, MAHSA University, Jenjarom, Selangor, Malaysia

S. Ali
School of Economics and Management, North China Electric Power University, Beijing, China

G. Mustafa
Department of Biochemistry, Government College University Faisalabad, Faisalabad 38000, Pakistan

R. T. Mahmood
Department of Biotechnology, Mirpur University of Science and Technology, AJK, Pakistan

D. Ahmad
Department of Medical Lab Technology, University of Haripur, Khyber Pakhtunkhwa, Haripur, Pakistan

A. Rehman
Department of Pathobiology, Jhang-Campus, University of Veterinary and Animal Sciences Lahore, Lahore, Pakistan

A. R. Ansari · A. Mumtaz · M. Arshad (✉)
Department of Basic Sciences, Jhang-Campus, University of Veterinary and Animal Sciences Lahore, Lahore, Pakistan
e-mail: muhammad.arshad@uvas.edu.pk

M. Arshad (ed.), *Climate Changes Mitigation and Sustainable Bioenergy Harvest Through Animal Waste*, https://doi.org/10.1007/978-3-031-26224-1_11

sources aims to reduce environmental pollution. Animals are an essential component of the modern world economy, and new technologies are needed to upgrade wastes and coproducts and produce high-value products. The successful application of these technologies will address the environmental and productivity challenges that are increasingly important to producers and consumers while also attempting to bring commercial value.

Graphical Abstract

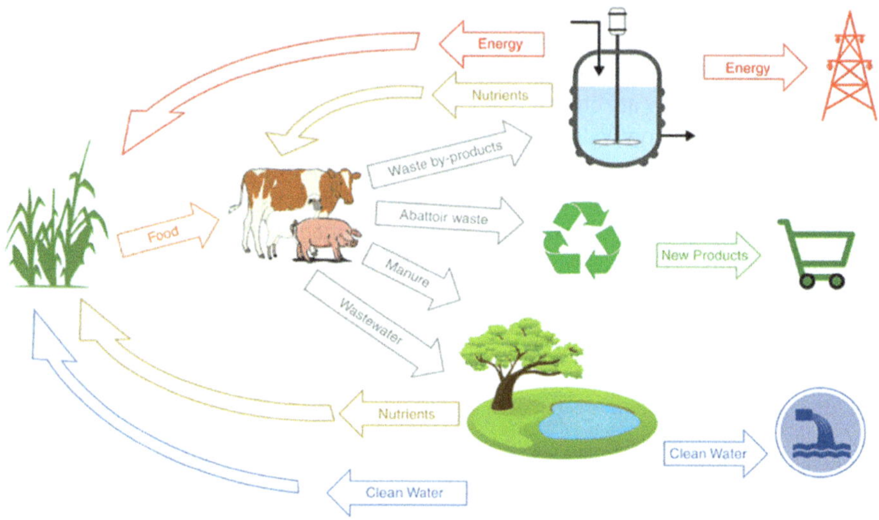

Keywords Bioeconomy · Circular economy · Animals' wastes

11.1 Introduction

Along with the global expansion in farm livestock units, animal waste is also increasing. The advantages of a circular bioeconomy include: (1) increased resource and ecofriendly; (2) decreased greenhouse effect; (3) a decreased dependence on fossil fuels; and (4) the restraint of waste and side products from a variety of sources as it create a sustainable and environmental-friendly environment, and it can thus be viewed as a low-carbon economy (Astals et al. 2014; Australian Meat Processor Corporation 2016).

Biological resources are collected and reused to the greatest extent possible in a circular bioeconomy. This economic strategy is gaining ground as a means of addressing sustainability concerns while meeting societal needs. In the circular economy, both biological and technological cycles are taken into account. Only the biological cycle, in which products generated from biological nutrients (bio-based products) are used again in the biosphere, is connected to the bioeconomy. The World Meter estimates that in September 2022, 7.9 billion people were living in the world; by 2050, that number is expected to rise to 9.7 billion (Bruinsma 2003). The creation of biomaterials and energy can support the energy-environment nexus and replace petroleum as the production feedstock, resulting in lower carbon emissions and a cleaner environment. The goal of this assessment is to promote a cleaner environment (Carrez et al. 2017). A large portion of the global bioeconomy and a crucial component of society are animal-based products, including food, leather, and wool. The use of animal products is pervasive in modern society, and as the world's population has grown, so has consumer demand for these items (Carus and Dammer 2018). The livestock industry is reliant on a variety of predictable factors. Maintaining agricultural profitability is made more difficult by the rising prices and unavailability of essential farm inputs, including electricity, water, capital, and labor (Chan et al. 2018). Agriculture-related sectors have been working to meet market expectations for high-quality and safe products while improving yields, feed conversion efficiency, and lowering production costs to improve economic sustainability. The sustainability of production systems is necessary to fulfill the increased global demand for food, energy, or other consumer goods. Worldwide, livestock sectors are developing plans to improve and show their environmental sustainability. (Commission 2005b; Cromwell 1980). However, a more recent estimate puts the number at 50% of farmers worldwide (Djissou et al. 2016). Over 50% of the world's croplands are fertilized organically by livestock dung, turning waste into ingredients for creating high-value foods (Directive 2006; Drummond et al. 2019). Manure contributes significantly to replenish soil organic matter, which is essential for preserving the soil's health and quality, supporting crop productivity, and rebuilding degraded soils (Drummond et al. 2019).

11.2 Bioeconomy, Bio-Based Economy, Green Economy, and Circular Economy

Various terminologies are linked to the term "bioeconomy," including "bio-based economy," "green economy," and "circular economy." Fig. 11.1 depicts how the phrases are related and where they overlap. It is believed to "increase human well-being and social equity while substantially lowering environmental dangers and ecological scarcities." A green economy can be defined as low carbon, resource-efficient, and socially inclusive in its basic form (Fagbenro and Jauncey 1995). The bioeconomy is typically regarded as a component of the green economy (Fig. 11.2). Early on, the terms "bioeconomy" and "circular economy" were used to describe the same. The principles used are summarized in Fig. 11.2. The processing of agricultural products is typically a part of the feeding; hence, it fits into the bio-based economy. The bioeconomy can complement the circular economy, which is becoming more popular (Fagbenro and Jauncey 1995; FAO 2018). The idea of circularity is not new, and it has served as the foundation for economic modeling since the writings of François Quesnay and the Physiocratic school of thought in the eighteenth century in France (Gasco et al. 2020).

Fig. 11.1 Main sources of keratin (Seerley 1991)

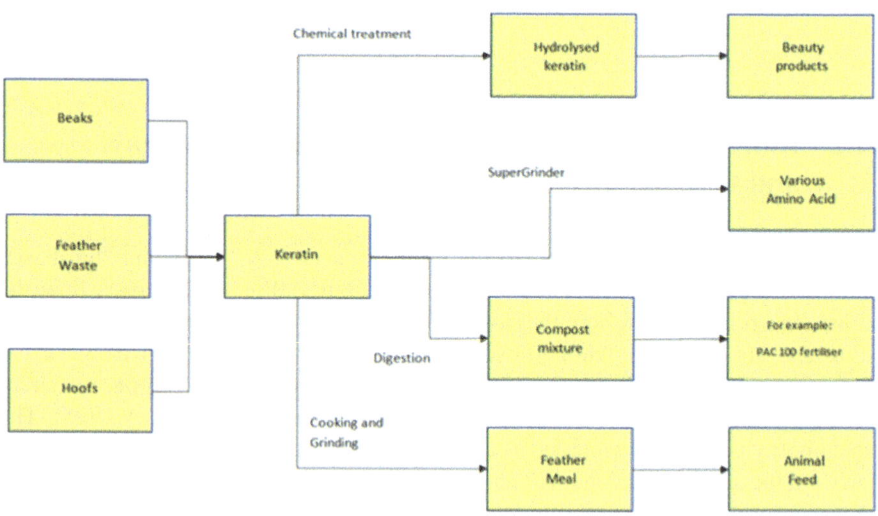

Fig. 11.2 Flowchart demonstrating the keratin supply chain and potential end applications (Staroń et al. 2014)

11.3 Approaches to Waste Management and Value-Addition in Livestock Industries

The wide spectrum of organic wastes produced during animal processing is normally categorized as liquid waste (wastewater) or solid waste, depending on the water content and material handling characteristics (Godfray 2011; Gollnow et al. 2014). Rumen waste rejected meat, and offal from screens, settling tanks, and dissolved air flotation are the main components of meat processing wastes. Many processing facilities use rendering to decrease waste further and produce by-products by using offal and waste materials from the boning room and slaughter floor. Rendering and other coproduct processing steps (like tanning) also produce organic wastes, including hiding trimmings, hair, and hooves. Whey is one of the coproducts that dairy processing has created to reduce waste. However, minor amounts from spills and rejected products still end up in waste streams. Large amounts of organic (and, occasionally, nutrient-rich) effluent are produced when animals, pens, processing areas, and equipment are washed. These wastewaters are typically sent to wastewater treatment plants. The solid waste produced by these treatment facilities may include aerobic sludge. The livestock industry produces single-use packaging and other plastic trash on and off farms. About 5.9 kg/hot standard carcass weight is the average amount of meat processing waste in landfills. Although materials that come into contact with meat, blood, or feces are deemed polluted and are landfilled, packaging wastes like cardboard and plastic are recycled at high rates (79% in 2009). Regarding overall waste generation, animal manure also makes up a sizeable portion of organic waste. Table 11.1 displays the various wastes from animal production (Cromwell 1980; Guerrero and Cremades 2012).

11.4 Waste Valorization Technologies and Markets

The creation of new technologies benefits the economy, the environment, and society. Figure 11.2 illustrates a circular economy strategy for the livestock industry, with multiple opportunities to turn waste materials into high-value goods. What was formerly referred to as "trash" has changed as a result of the development of the potential to upgrade low-value by-products with the advent of modern technologies (Hatfield and Stewart 1997). Recycling and waste recovery result in significant sustainability gains if carried out along the value chain of animal production. Closing nutrient loops in agriculture, lowering carbon dioxide emissions, and boosting soil organic matter composition can all be accomplished using organic wastes (Ellen). This is seen on farms through the usage of manure and wastewater. Since nutrients from leftover fractions are reintroduced into the system and new inputs from elsewhere are reduced, the total nutrient usage efficiency may be higher, e.g., mineral fertilizers (Hezron et al. 2019). The carbon dioxide and methane emissions caused by burning fossil fuels are also decreased by biogenic energy

Table 11.1 Summary of general and current applications of fisheries' waste and by-products in Bangladesh

Fishery waste product	General use	Present situation/use in Bangladesh	Recommendations
Air bladder	Use for the production of surgical yarn and capsule cover caps [80]	Smaller size is utilized as chicken feed, while larger sizes are exported to China, Thailand, and Malaysia	To properly use them, build the pharmaceutical industry
Viscera	Used as feed for the poultry industry and several fish species, including catfish (Leong et al. 2021)	Widely employed in the fish and poultry feed industries	Feeding materials used as a protein alternative in fish and animal diets, such as fish silage and fishmeal
Skin	To recover burned skin to human beings [80]	These are exported to China, Vietnam, and Thailand	To get it extensively used in Bangladesh's medical community, empirical study is required
Oil	Used as a raw material in the production of feed for livestock, poultry, and fishes [81]	Utilized in exports for the animal as well as fish feed	Used in biodiesel after being refined as edible oil
Powder/silage	Used mostly in moistened feed pellets and fish diets [82]	Used as fish and animal feed	It can be used in animal feed as hydrolysate and in poultry feed

from trash. Advanced technologies recycle and recover wastes, such as anaerobic digestion and composting (Hezron et al. 2019b). Although conversion rates and cost-effectiveness are still be enhanced, the advantages of these techniques are well understood. Creating economically viable applications might be challenging, given these items' operating and shipping costs (https://www.keenanrecycling.co.uk/recycling-services/organic-recycling/; https://www.statista.com/statistics/263962/number-of-chickens-worldwide-since-1990/).

11.5 Utilization of Waste Materials from Meat

11.5.1 Meat Industry

Utilizing the proper meat by-products is essential for the meat industry to be profitable. According to estimations, the income from pigs and beef is made up of

7.5% and 11.4%, respectively, of the by-products. As a result of low costs and hygienic concerns, traditional markets for edible beef by-products have been gradually vanishing. In order to address these issues, meat producers have focused their marketing and development efforts on non-food applications. According to the available research, the live weight of cattle, pigs, and lambs, respectively, is made up of 66.0, 52.0, and 68.0% of their by-products. Because of this, a significant earnings' stream is vanished, along with the price of getting rid of things of this type which is rising (Leiva-Candia et al. 2014). According to data from the US Department of Agriculture's Economic Research Service, by-products account for 11.4% of the beef industry's gross revenue. Pork makes up 7.5% of the total. Unused meat products harm the environment severely in addition to causing financial losses.

11.6 Recovery of Nutrients from Animal Wastes

In large amounts, ammonia produced from the breakdown of proteins, urea, or nucleic acids might stifle the digestive process (McCabe et al. 2016, 2020). Proteins, urea, and fat concentrations in animal processing wastes can be significant (MacArthur 2013). The rates at which biogas is produced in lagoons, for instance, can vary depending on the seasonal temperature (Meadowcroft 2009). All the wastes are not acceptable for lagoon-style digesters. Animal farming businesses are better suited for more sophisticated and pricey in-vessel reactors. Unfortunately, these wastes sometimes necessitate lengthy processing durations and provide modest amounts of biogas; as a result, anaerobic digestion may become less profitable.

11.7 Producing Nutritionally Advanced Feeds

Anaerobic digestion by microorganisms offers a practical method for treating wastes and producing fertilizer and other higher-value bio-based products. Microalgae like Chlorella sp. and Scenedesmus sp., as well as macroalgae like Lemna minor, have the potential function in wastewater systems (Mozumder et al. 2022; Musyoka et al. 2020; Navone and Speight 2018). These microbes are photosynthetic, and despite the nutrient-rich environment of wastewater, these species require additional carbon to support growth. These products also contain essential amino acids, vitamins, and probiotics, which can improve animals' productivity and digestive health. Microalgae have recently been developed and used in wastewater treatment applications and have been touted as viable biofuel. Fishmeal is one of the high-priced and unsustainably obtained commodities that must be replaced when developing alternative protein sources for animal and aquaculture feed (Rajagopal et al. 2013).

11.8 Waste to New Materials

Bio-based polymers and biocomposites can be made from both liquid and solid wastes. Blood is converted into biodegradable bioplastic rather than blood meal for use as animal feed or fertilizer (Ramirez et al. 2021). Plastic waste is typically disposed of in landfills since it cannot be recycled or reused due to contamination. High-performance plastics are necessary for packaging applications to preserve freshness and avoid product contamination. If certain physical requirements are met, newly developed biodegradable or compostable polymers might be appropriate for various uses in the preparation of meat. Some varieties of polyhydroxyalkanoates, for instance, can be utilized in food packaging (Rivera et al. 2000). The biocomposites' strength can be increased and prices can be reduced by adding additives like natural fibers (Schenk 2016). In order to reduce plastic contamination and overall landfilled trash, it is essential to continue developing food-grade and biodegradable plastic substitutes.

11.8.1 Processing of Keratin Wastes

It includes keratin present in feathers. The richest sources of keratin are stratum corneum, feathers, wool, hair, and hooves (Schmidt et al. 2019; Seerley 1991). Feather waste is pervasive (Shepherd and Jackson 2013). The Super Grinder provides an alternate method to eliminate a problematic waste stream that, in some parts of the world, is even thrown on the streets. Other significant sources of keratin exist as well (Fig. 11.2) although they are either less common or are not widely collected. Waste keratin is chemically hydrolyzed using a technique called chemical hydrolysis (acid, base, catalyst). Chemical hydrolysis presents a bigger environmental danger and calls for more aggressive reaction conditions (high temperature and pressure) (Sogbesan 2006). The supply chain has shown in flowchart of Fig. 11.2 for explaining the supply chain of keratin (Staroń et al. 2014).

Keratin recycling already uses a variety of various procedures and products. Keratin is hydrolyzed for use in a variety of hair products. Feather waste can be composted into fertilizer by mixing wood chips and bacteria with it; this is an easy and reasonably priced way to get rid of feathers. Feather meal is a product made from cooked, dried, and ground feathers that is frequently used as an animal feed supplement due to its high nitrogen and protein contents.

Globally, seafood waste has been a major issue that results in annual losses of billions of dollars. The sector produces about 100 tons of garbage annually, the majority of which is dumped or transformed into low-value items. The practice of recycling seafood trash is not well-established, and a significant portion of this material is dumped directly into the environment. Some of these raw resources can, however, be used to create products with significant value. Chitosan, a polymer produced by deacetylating chitin and present in the exoskeletons of insects, arthropods, and mollusks, is one example. Chitosan possesses antibacterial and antifungal properties that could be employed in the textile, culinary, cosmetic, and agricultural industries. The majority of the current extraction techniques for chitin rely on chemical processes, which may be costly and ineffective from an energy standpoint. New processing and management strategies are required in order to process and produce such goods from seafood trash. We think that by utilizing the Super Grinder, we might enhance the present extraction techniques and offer a more environmental-friendly, long-lasting substitute.

11.9 Bioprocesses with Waste For Bio-Lipids' Synthesis

The bioprocessing of waste contributes to the accumulation of bio-lipids as well as the creation of green biopolymers. As a substitute feedstock for the manufacturing of biofuels, health, food supplements, and oleo-chemicals, the generation of microbial lipids utilizing inexpensive substrates from waste materials has drawn significant interest from both the industry and research fields. Oleaginous microorganisms, including yeast, cyanobacteria, algae, certain bacteria, and fungi, can collect significant lipids up to 80% of their body weight (Stevens et al. 2018). The microbial oils are non-toxic, biodegradable, and safe to use so that they can be used in industrial settings without the need for chemicals derived from petroleum. These characteristics support a greener society and can assist in resolving many global challenges (Tait et al. 2009; Thornton 2010; Vadiveloo et al. 2019). Acid at a concentration of 4–20 g/L is favourable, and the previous study used acetic acid as the only carbon source (5).

11.10 Reutilization of Aquaculture Wastes

Over the past few years, the European Union has promoted sustainable resource usage, proper waste management, raw materials like biomass and biological resources, and environmental media like water (Wiedemann et al. 2017; Worldometer 2022; Yong et al. 2021). Aquaculture has started to develop sustainable production methods that are both profitable and environmentally friendly. It will play a crucial part in supplying animal protein in the future. Aquaculture wastes, such as metabolic wastes or uneaten food, must be considered as possible sources of minerals, vitamins, proteins, and lipids for further usage to increase the productivity of aquaculture systems and lessen their environmental impact (Sabir et al. 2021). According to this production design, a small number of species from various trophic levels are cultivated; like in natural ecosystems, wastes produced by species at higher trophic levels are used as nutrients by species at lower trophic levels (Arshad et al. 2021). Fish, crustaceans, and cephalopods are frequently found on the first level.

The second type involves filtering and suspensivorous invertebrates (such as filter molluscs, anemones, sea cucumbers) that consume the organic materials produced by the first level, such as feed residues or by-products. Third-level marine macroalgae utilize inorganic substances, such as those from excretory products generated by earlier stages. Monoculture farms release a lot of pollutants that are rich in dissolved nutrients and solid waste (such as uneaten food and feces) that can cause water eutrophication and impact the ecosystems around (Arshad et al. 2022).

11.11 Fishery Waste and Fishery By-Products in Bangladesh

In Bangladesh, a large amount of work has been done on researching the usage of fish scales from seafood industrial wastes. Bangladesh produces around 93,000 metric tons of waste products from fish processing annually. However, just 900 tons are employed to manufacture various items. The seafood sector in Bangladesh generates approximately 43,320.88 tons of seafood waste each year, with a value of USD 13.73 per tons totaling USD 44.09 million. Khulna produces the most fish and shrimp waste, followed by Chittagong, Cox's Bazar, Dhaka, and Sylhet. Based on primary and secondary data, the authors outlined the general applications of fishing waste and by-products and their current usage in Bangladesh, followed by recommendations for the sustainable use of such resources (Table 11.1) (Arshad et al. 2014).

11.12 Fishery Waste Utilization: Opportunities and Challenges to Sustainable Use

This portion of the book chapter summarizes the respondents' perception of and knowledge about the potential and challenges of utilizing fishery, the idea of a circular bioeconomy, fish feeds generated from fishing waste, nutrition gained from it, recycling, and lowering pollution. A lot of things will be feasible in this situation. First and foremost, we receive high-quality goods. Additionally, value can be added. Third, efforts are being made to reduce pollution. The waste produced by the fishing industry offers innovation potential. There are numerous products on the market; however, we are unable to perform extraction operations of that nature in Bangladesh, including fish scales, cosmetics, air bladders, and medications (surgical yarn). Shark soup, a popular meal in China, can be produced with shark fin. Additionally, numerous kinds of ornaments and jewelry can be manufactured from bones or teeth. Leather items made of leather can be made from fish scales and skin; shells can be used to produce chitin and chaetocin (Arshad et al. 2014).

11.13 Fisheries and Aquaculture

A Latvian case study including round-goby fish wastes from the Baltic Sea conducted laboratory-scale experiments to investigate the biochemical methane potential of such wastes when co-digested effectively and sustainably with sewage sludge. Indeed, as the authors point out, what applies to goby fish wastes also applies to other species across the planet. Shells from gastropods, oysters, and crabs have all been used as bio-calcite phosphorus adsorbents in wastewater treatment and for other value-added bioproducts. As bio-extractors of nitrogen, phosphorus, CO_2, and toxic heavy metals from wastewater (Commission 2005a), sources of third-generation biofuels, and effective remedies for the food-fuel-fiber-feed impasse (Thornton 2010). Aquaponics is the combined practice of growing food plants, such as vegetables and raising fish. According to Pous, the most effective way to recycle the wastewater from the aquaculture sub-system to the hydroponics sub-system is by employing bio-trickling filters. The extra algal biomass can be utilized as a substrate in microbial fuel cells to produce power when eutrophic lakes need to be cleaned up. Microcystis aeruginosa exhibits increased electrochemical performance compared to traditional substrates, according to experiments by Thornton (2010).

11.14 Animal Manures

In all parts of the world, soil biology is mostly supported by organic compounds from animal manure. Manures are valuable fertilizers for agricultural soils because of their significant nutrient concentrations. Most of the manure on animal farms is sold to neighbors or spread back onto nearby fields. Processing and properly utilizing such vast amounts of animal manure have become particularly challenging as massive concentrated livestock feeding facilities have grown. By driving off C through decomposition, concentrating nutrients, and extracting value from these vast amounts of manure, composting and biogas digestion facilities have been used to minimize volume.

11.14.1 Nutrients in Manure (Animal Wastes)

Nitrogen corn and soybean meal make up more than 80% of the grain and protein supplements given to pigs and poultry in the USA. When employed as the main source of energy in swine diets, corn falls short of meeting the needs of the growing pig for some essential amino acids. Corn's primary and secondary limiting amino acids are lysine and tryptophan, with threonine and isoleucine coming in third and fourth. These essential amino acids must be coupled with maize in sufficient concentrations to give a diet enough levels of each. Soybean meal is included in swine diets made of maize and soybean meal in an amount sufficient to meet the lysine requirements of the pigs. When prepared in this way, all additional amino acids required to meet the nutritional needs of a growing pig are included in the diet (Arshad et al. 2022).

11.15 Animal-Based Ingredients and By-Products

Food and animal's wastes can be used to produce proteins and other useful substances; for instance, the manures from chickens, pigs, and cattle are used to grow houseflies (*Musca domestica*) and maggots, which are then utilized as fish meal supplements or as a feed ingredient for fish. Production of biofuel from animal waste has been shown in Fig. 11.3. Maggots can be collected and processed to provide a meal that can be used in replacement for fish meals. Cattle blood and wheat bran were used to develop the maggots, which had 92.7% dry matter, 47.6% crude protein, 25.3% fat, 7.5% crude fiber, and 6.25% ash, as well as an amino acid profile similar to fish meal. Maggots were used to replace 25% of the fishmeal in catfish feed, culture, and diet, and the results were higher growth rates and greater profitability. Many studies have used red worms, black soldier flies, common houseflies, and yellow mealworms as protein sources in place of fishmeal (Arshad et al. 2014).

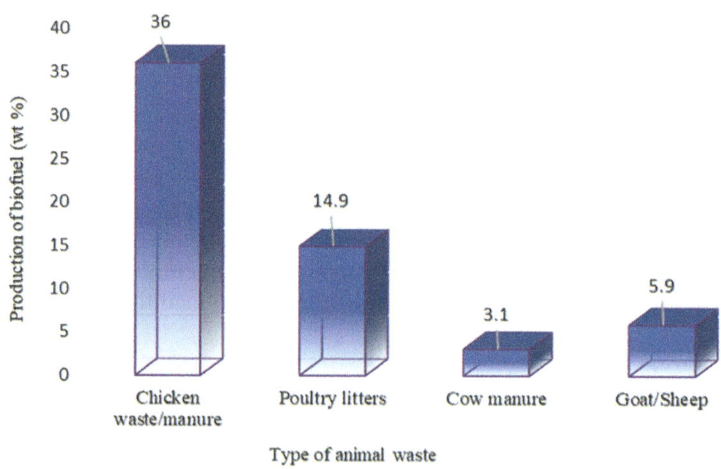

Fig. 11.3 Production of biofuel from wastes collected from different sources of animals

11.16 Conclusion

The development of an industry based on converting animal's wastes to high-value products addresses complex issues facing the meat and livestock industries. By-product underuse or non-use results in a loss of prospective income as well as increased disposal costs. The improper use of animal by-products have detrimental consequences on human's health and appearance. Along with environmental and safety concerns, meat, poultry, and fish processing wastes frequently have the ability to be recycled into raw materials or turned into useful goods with higher value.

References

Arshad M, Adil M, Sikandar A, Hussain T (2014) Exploitation of meat industry by-products for biodiesel production: Pakistan's perspective. Pak J Life Soc Sci 12: 120–125

Arshad M, Javed S, Ansari AR, Fatima A, Shahzad MI (2021) Biogas: a promising clean energy technology. In: Bioenergy resources and technologies, pp 91–120

Arshad M, Ansari AR, Qadir R, Tahir MH, Nadeem A, Mehmood T, Alhumade H, Khan N (2022) Green electricity generation from biogas of cattle manure: an assessment of potential and feasibility in Pakistan. Front Energy Res 10

Astals S, Batstone D, Mata-Alvarez J, Jensen P (2014) Identification of synergistic impacts during anaerobic co-digestion of organic wastes. Biores Technol 169:421–427

Australian Meat Processor Corporation (2016) Inorganic waste management at abattoirs inorganic waste management at abattoirs (2016)

Bruinsma J (2003) World agriculture: towards 2015/2030: an FAO perspective. Earthscan. Food and Agriculture Organization, London/Rome

Carrez D, Carus M, Griffon M, Jilkova J, Juhász A, Lange L, Consuelo Varela O (2017) Expert group report-review of the EU bioeconomy strategy and its action plan

Carus M, Dammer L (2018) The circular bioeconomy—concepts, opportunities, and limitations. Ind Biotechnol 14(2):83–91

Chan CM, Vandi L-J, Pratt S, Halley P, Richardson D, Werker A, Laycock B (2018) Composites of wood and biodegradable thermoplastics: a review. Polym Rev 58(3):444–494

Commission E (2005a) Taking sustainable use of resources forward: a thematic strategy on the prevention and recycling of waste. Communication from the commission of the European communities, Brussels, Belgium. COM (2005a), p. 666

Commission, E. (2005b). Thematic strategy on the sustainable use of natural resources. In: European commission Brussels, Belgium

Cromwell G (1980) Biological availability of phosphorus for pigs. Feedstuffs (USA)

Directive E (2006) Directive 2006/12/EC of the European parliament and of the council of 5 April 2006 on waste. Off J Eur Union 50:114

Djissou AS, Adjahouinou DC, Koshio S, Fiogbe ED (2016) Complete replacement of fish meal by other animal protein sources on growth performance of Clarias gariepinus fingerlings. Int Aquat Res 8(4):333–341

Drummond L, Álvarez C, Mullen AM (2019) Proteins recovery from meat processing coproducts. Sustain Meat Prod Proc 69–83

Fagbenro O, Jauncey K (1995) Water stability, nutrient leaching and nutritional properties of moist fermented fish silage diets. Aquacult Eng 14(2):143–153

FAO (2018) Nitrogen inputs to agricultural soils from livestock manure: new statistics. Food and agriculture organization of the United Nations

Gasco L, Acuti G, Bani P, Dalle Zotte A, Danieli PP, De Angelis A, Piccolo G (2020) Insect and fish by-products as sustainable alternatives to conventional animal proteins in animal nutrition. Ital J Anim Sci 19(1):360–372

Godfray HCJ (2011) Food for thought. Proc Natl Acad Sci USA 108:19845. https://doi.org/10.1073/pnas.1118568109

Gollnow S, Lundie S, Moore AD, McLaren J, van Buuren N, Stahle P, Rehl T (2014) Carbon footprint of milk production from dairy cows in Australia. Int Dairy J 37(1):31–38

Guerrero S, Cremades J (2012) Integrated multi-trophic aquaculture (IMTA): a sustainable, pioneering alternative for marine cultures in Galicia. In: Regional Government of Galicia (Spain)

Hatfield JL, Stewart BA (1997) Animal waste utilization: effective use of manure as a soil resource. CRC Press

Hezron L, Madalla N, Chenyambuga SW (2019) Alternate daily ration as a feeding strategy for optimum growth, nutrient utilization and reducing feed cost in Nile tilapia production. Livest Res Rural Dev 31(7):113

Hezron L, Madalla N, Chenyambuga SW (2019b) Mass production of maggots for fish feed using naturally occurring adult houseflies (Musca domestica). Livestock Res Rural Dev 31(4)

https://www.keenanrecycling.co.uk/recycling-services/organic-recycling/

https://www.statista.com/statistics/263962/number-of-chickens-worldwide-since-1990/.

Leiva-Candia D, Pinzi S, Redel-Macías M, Koutinas A, Webb C, Dorado M (2014) The potential for agro-industrial waste utilization using oleaginous yeast for the production of biodiesel. Fuel 123:33–42

Leong HY, Chang C-K, Khoo KS, Chew KW, Chia SR, Lim JW, Show PL (2021) Waste biorefinery towards a sustainable circular bioeconomy: a solution to global issues. Biotechnol Biofuels 14(1):1–15

MacArthur E (2013) Towards the circular economy, economic and business rationale for an accelerated transition. Ellen MacArthur Found: Cowes, UK 21–34

McCabe BK, Harris P, Antille DL, Schmidt T, Lee S, Hill A, Baillie C (2020) Toward profitable and sustainable bioresource management in the Australian red meat processing industry: a critical review and illustrative case study. Crit Rev Environ Sci Technol 50(22):2415–2439

McCabe BK, Antille DL, Birt HW, Spence JE, Fernana JM, van der Spek W, Baillie CP (2016) An investigation into the fertilizer potential of slaughterhouse cattle paunch. Paper presented at the 2016 ASABE Annual International Meeting

Meadowcroft J (2009) Minding the stock: bringing public policy to bear on livestock sector development

Mozumder MMH, Uddin MM, Schneider P, Raiyan MHI, Trisha MGA, Tahsin TH, Newase S (2022) Sustainable utilization of fishery waste in Bangladesh—a qualitative study for a circular bioeconomy initiative. Fishes 7(2):84

Musyoka SN, Liti D, Ogello EO, Meulenbroek P, Waidbacher H (2020) Earthworm, *Eisenia fetida*, bedding meal as potential cheap fishmeal replacement ingredient for semi-intensive farming of Nile Tilapia, *Oreochromis niloticus*. Aquac Res 51(6):2359–2368

Navone L, Speight R (2018) Understanding the dynamics of keratin weakening and hydrolysis by proteases. PLoS ONE 13(8):e0202608

Rajagopal R, Massé DI, Singh G (2013) A critical review on inhibition of anaerobic digestion process by excess ammonia. Biores Technol 143:632–641

Ramirez J, McCabe B, Jensen PD, Speight R, Harrison M, Van Den Berg L, O'Hara I (2021) Wastes to profit: a circular economy approach to value-addition in livestock industries. Anim Prod Sci 61(6):541–550

Reducing enteric methane for improving food security and livelihoods. Project highlights 2015–2017. Rome, Italy, FAO

Rivera JA, Sebranek JG, Rust RE, Tabatabai LB (2000) Composition and protein fractions of different meat by-products used for petfood compared with mechanically separated chicken (MSC). Meat Sci 55(1):53–59

Sabir A, Idrees H, Shafiq M, Butt MT, Jacob KI, Arshad M (2021) Impact of CO_2 discharge from distilleries on climate changes: key facts. In: Sustainable ethanol and climate change: sustainability assessment for ethanol distilleries, pp 113–140

Schenk P (2016) On-farm algal ponds to provide protein for northern cattle. MLA: Sydney, NSW, Australia

Schmidt T, Harris P, Lee S, McCabe BK (2019) Investigating the impact of seasonal temperature variation on biogas production from covered anaerobic lagoons treating slaughterhouse wastewater using lab scale studies. J Environ Chem Eng 7(3):103077

Seerley R (1991) Major feedstuffs used in swine diets. In: Swine nutrition. Elsevier, pp 451–481

Shepherd C, Jackson A (2013) Global fishmeal and fish-oil supply: inputs, outputs and markets. Wiley Online Library 83:1046–1066

Sogbesan O (2006) Nutritive potentials and utilization of garden snail (Limicolaria aurora) meat meal in the diet of Clarias gariepinus fingerlings. Afr J Biotechnol 5(20)

Staroń P, Banach M, Kowalski Z, Staroń A (2014) Hydrolysis of keratin materials derived from poultry industry. Proc ECOpole 8(2):443–448

Stevens JR, Newton RW, Tlusty M, Little DC (2018) The rise of aquaculture by-products: increasing food production, value, and sustainability through strategic utilisation. Mar Policy 90:115–124

Tait S, Tamis J, Edgerton B, Batstone DJ (2009) Anaerobic digestion of spent bedding from deep litter piggery housing. Biores Technol 100(7):2210–2218

Thornton PK (2010) Livestock production: recent trends, future prospects. Philos Trans R Soc B: Biol Sci 365(1554):2853–2867

Vadiveloo A, Nwoba EG, Moheimani NR (2019) Viability of combining microalgae and macroalgae cultures for treating anaerobically digested piggery effluent. J Environ Sci 82:132–144

Wiedemann SG, McGahan EJ, Murphy CM (2017) Environmental impacts and resource use from Australian pork production determined using life cycle assessment. 2. Energy, water and land occupation. Anim Prod Sci 58(6):1153–1163

Worldometer (2022) Current world population. 7.9 billion people

Yong JJY, Chew KW, Khoo KS, Show PL, Chang J-S (2021) Prospects and development of algal-bacterial biotechnology in environmental management and protection. Biotechnol Adv 47:107684

Chapter 12
Sustainable Solutions to Animal Waste: Climate Change Mitigation and Bioproduct Harvest

Asha Sohil and Muzaffar A. Kichloo

Abstract Livestock waste management is one of the global environmental problems with significant impact on air, water, soil, and biodiversity. Global livestock output has increased drastically to keep up the pace with the growing demand for meat, milk, egg, fiber, leather, and other animal products. However, the issue is the growing harm to the environment posed by animal waste production. In addition to nitrogen and phosphorus emissions into soil and water, animal manure is also responsible for the emission of gases like ammonia, nitrous oxide, and methane. The emissions from animal waste are directly linked to environmental problems including eutrophication, acidification, and climate change. The usage of hazardous chemicals in the livestock business and the prevalence of zoonotic viruses in animal manure have created concern for both human and environmental health. Animal waste management techniques like bioenergy production, composting, vermicomposting, use of animal waste as nutrient media are some of the environmental sound practices to be taken so as to lessen the negative impacts on environment and human health. The chapter aims to provide the best animal waste treatment methods so as to recover valuable products such as renewable energy, nutrients and prepare products like compost, vermicompost utilized as biofertilizer. Moreover, animal waste can also be used as a nutrient media for the production of variety of biochemicals used in pharmaceutical industry. In addition to improving human health and minimizing environmental damage, appropriate animal waste management procedures boosts regional and local economy.

A. Sohil (✉)
Government Degree College, Udhampur, Jammu & Kashmir, India
e-mail: ashasohil04@gmail.com

M. A. Kichloo
Government Degree College, Banihal, Jammu & Kashmir, India

M. Arshad (ed.), *Climate Changes Mitigation and Sustainable Bioenergy Harvest Through Animal Waste*, https://doi.org/10.1007/978-3-031-26224-1_12

Graphical Abstract

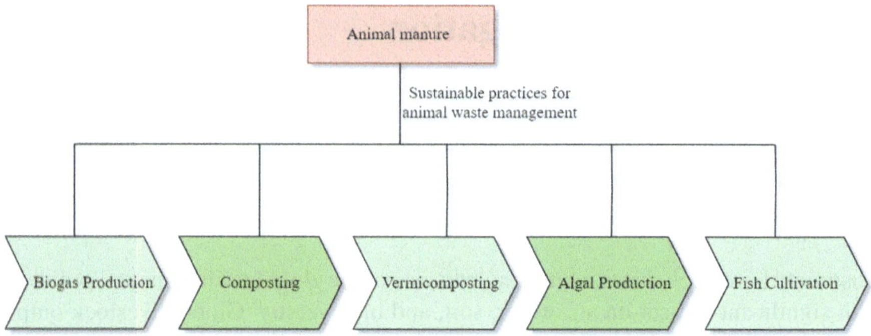

Keywords Livestock · Management · Sustainable · Hazardous · Zoonotic · Impacts

12.1 Introduction

12.1.1 Concept of Sustainability

The definition of term sustainable is "which can be sustained throughout time" (Heinberg and Lerch 2010). To start with the term's etymology, the word *sustainable* or *sustainability* has its origin in the Latin word *sustinēre* formed by combination of words *sub* and *tenēre* that means to "maintain," "support," "endure," or "to restrain" (Caradonna 2014). Future preservation of natural resources has always been a concern. Many indigenous communities across the world have embraced the idea of sustainability. For instance, the leader of the Iroquois Confederacy's Gayanashagowa, or Great Law of Peace, tribe took into account how his judgments will affect the seventh generation (Heinberg 2007). Hans Carl von Carlowitz, a German scholar and forester, used the term "sustainability" (German: Nachhaltigkeit) to describe sustained-yield forestry in his work *Sylvicultura Oeconomica* in 1713. The Brundtland Report on the United Nations' *World Commission of Environment and Development* provided a comprehensive overview of sustainable development that means the development that "meets the needs of the present generation without compromising the ability of future generations to meet their own needs," which led to rise in usage of the term after 1987 (WCED 1987). Despite its widespread use, the notion of sustainability has come under fire for failing to take into account the problem of population increase and the unsustainable use of non-renewable resources (Barlett 1998).

According to the economist Jeffrey D. Sachs, we are living in the age of the Anthropocene, where humans have caused havoc to natural environment. The ecological assault on the planet was triggered by industrialization, and therefore, sustainability is the sole means of reversing and replenishing the imbalance. In 2005, UN World Summit endorsed a sustainability model that demonstrated the interconnectedness of the "three Es:" environment, economy, and equity or social equality. A more recent sustainability model depicted society and the economy as nested subsystems inside the larger biophysical system of the environment in the form of a series of concentric circles (Daly 1973; Victor 2008). In 1989, Swedish oncologist Dr. Karl-Henrik Robèrt founded an organization named *the Natural Step* with the goal to promote sustainable society. He formulated four conditions for sustainability which are: In a sustainable society, nature is not subjected to steadily rising: 1. the volume of materials removed from the earth's surface, 2. the volume of materials produced by people, 3. deterioration by physical means, and 4. they do not live in circumstances that make it difficult for them to achieve their fundamental requirements (www.thenaturalstep.org). According to Heinberg and Lerch (2010), any society that adopts unsustainable lifestyles fails to persist for a longer period of time. Richard Heinberg, world's leading expert on the challenge of sustainability in the energy and environmental sectors and a senior scientist at the Post Carbon Institute, has developed five axioms (or "self-evident truths") of sustainability suggesting sustainable use of earth's natural deposit, judicious use of non-renewable resources, and minimizing release of harmful substance that impair biosphere functions.

Livestock plays a substantial role in livelihood and economy of both developing and developed countries (Singh and Rashid 2017). Livestock population is one of the country's valuable resource and key source of income, employment, and nourishment for rural as well as urban households. These socio-economic roles contribute to its fast global growth (Herrero et al. 2013). According to the United Nations' Food and Agriculture Organization (FAO 2004), livestock industry is one of the quickest developing sectors. To offer these benefits, the livestock industry uses significant amount of land, water, biomass, and other resources and releases considerable quantity of unwanted substances into the environment chiefly in the form of animal manure posing a serious threat to the environment. With increasing human population and changing lifestyle, the demand of animal products is increasing; therefore, livestock population is also growing at a faster rate. As a result, the quantity of manure generated by cattle particularly in confined animal feeding operations (CAFOs) has grown tremendously. Animal agriculture industry is one of the most polluting industries on earth. The unsustainable approaches of livestock production and waste disposable have led to habitat degradation, environmental pollution, climate change, and spread of deadly diseases (Gerber et al. 2005). Therefore, disposal of animal waste is a challenge in terms of cost, environmental safety, and biosecurity. Now, the concern arises how would these livestock wastes be handled without having any negative impact on human health, natural resources, or food security. The implementation of economically viable, sustainable, and eco-friendly animal waste management systems through appropriate research and development is required so as to develop more sustainable solutions. This review was necessitated as an attempt to answer these questions.

12.1.2 Sustainable Utilization of Livestock Waste

Objectives

1. Biological stabilization prior land application.
2. Energy recovery from animal waste.
3. Sustainable animal waste management strategies.
4. Identify environmental impacts of animal waste.
5. Treatment protocols of animal waste.
6. Climate change adaptation and mitigation.

Benefits

1. Animal waste management reduces possible environmental impacts on air, water, soil, and wildlife.
2. Minimizes the risks associated with human health posed by animal waste.
3. Animal manure can be used as a soil fertilizer. Therefore, it aids in enhancing soil's nutritional quality.
4. Organic matter in animal manure enhances soil structure and water-holding capacity.
5. Solids present in animal manure can be used as substitute of bedding material.
6. Animal manure can be used as an alternative source to meet energy demand.
7. Reduced greenhouse gas emissions from animal waste will aid in climate change adaptation and mitigation strategies.
8. Creates employment and business opportunities for the locals.

Negative impacts of unsustainable animal waste disposable practices

1. **Source of some major pollutants**: Plants absorb various nutrients from the soil to support healthy growth and development. Animal manure serves as an effective fertilizer, but if it is applied at a pace faster than the plants can absorb it, a buildup of nutrients occurs (Tamminga 1992), especially nitrogen and phosphorus, which are present in constant amounts causing significant environmental contamination (Headon and Walsh 1994). Excessive volume of dissolved carbon from animal waste in addition to soluble phosphorus rapidly reduces O_2 content in water and increases the need for biochemical oxygen demand (BOD).
2. **Water pollution by phosphorus**: Phosphorus often remains in association with soil's top layer. When huge amount of animal manure is spread on cropland, excess amount of phosphorus in the manure flows off from the soil surface layer into nearby water bodies increasing the amount of phosphorus in water and triggers the process of eutrophication.
3. **Environmental pollution by nitrogen**: The excess amount of nitrogen is present as an inorganic ammonium ion (NH_4^+) in the soil. Some of it is lost as an ammonia gas (NH_3) into the atmosphere (MAFF 1996) which is one of the toxic gases associated with animal waste (Tamminga and Verstegen 1992). The unassimilated ammonium ion is converted into nitrate. Some portion of nitrate is converted into nitrogen gas (N_2), and much of it finds its way into groundwater causing

ground water pollution (Headon and Walsh 1994). The significant levels of phosphates and nitrogen in animal manure runoff create anaerobic conditions which are extremely deleterious to stagnant water bodies. Phosphorus and nitrogen promote excessive microbial growth lowering oxygen concentration necessary to support other life forms in water bodies. In addition to soluble nitrogen and phosphorus, Na, Cl, Ca, Mg, K, and total dissolved solids (TDSs) are the contaminants introduced in substantial amounts into the environment posing threat to environment. Most animal excrement has significant levels of TDS ($\gg$ 10,000 mgL^{-1}) (Brusseau and Artiola 2019).

4. **Heavy metal contamination**: Unregulated animal waste disposal practices cause heavy metal contamination like copper, zinc, and lead from animal diets, use of pesticides, antibiotics, and growth regulators into the soil and water bodies (Brusseau and Artiola 2019; Li et al 2005; Rodríguez-Eugenio et al. 2018).
5. **Greenhouse gas emission**: In addition to major greenhouse gases like methane and nitrous oxide, animal waste also emits harmful gases like ammonia, hydrogen sulfide, and other unpleasant gases. Large volume of particulate matter is released by animal dung, especially in arid areas where the manure easily transforms into dust and spreads out quickly. According to the United Nations *Food and Agriculture Organization (FAO)*, the animal agriculture sector is responsible for approximately 18% of greenhouse gas (GHG) emissions.
6. **Effects quality of living**: Untreated animal waste releases air pollutants like H_2S, mercaptans, indoles, org-sulfides that creates health problems, causing health issues, especially for adjacent community residents (Brusseau and Artiola 2019).
7. **Health hazard**: Animal waste contaminates meat and farm fields and therefore spreads bacteria and viruses that could be health hazard for thousands of people. Antibiotic overuse might result in a global public health emergency. Studies conducted by Nadimpalli et al. (2018) have discovered high concentrations of drugs and antibiotic-resistant genes in feedlot air samples.

12.2 Biogas Production from Animal Waste

Biogas is a clean, economical, and renewable energy utilized as a substitute for other non-renewable fuels for household application in rural areas most often in developing nations in Asia and Africa (Sorathiya et al. 2014). Biogas, commonly referred to as "marsh gas," is a byproduct of the anaerobic breakdown of organic waste depicted as follows:

$$\text{Organic matter} + \text{anaerobic bacteria} \rightarrow CO_2 + H_2S + NH_3 + CH_4 + \text{other end products} + \text{energy}.$$

12.2.1 Composition of Biogas

Most often, "waste materials" like human excreta, animal waste like livestock manure, sewage sludge, and crop residues are most frequently used as a raw material for biogas production. According to Polprasert (2007), biogas composition is determined by the type of organic raw materials used and the time and temperature of anaerobic decomposition which is approximately given as follows:

1. Methane (CH_4): 55–65%.
2. Carbon dioxide (CO_2): 35–45%.
3. Nitrogen (N_2): 0.3%.
4. Hydrogen (H_2): 0–1%.
5. Hydrogen sulfide (H_2S): 0–1%.

Of all the gases produced, CH_4 is a preferred gas because of high calorific value of $\approx$ 9000 kcal/m^3 at standard temperature and pressure.

12.2.2 Raw Material

According to Polprasert (2007), the three primary categories of suitable organic source materials for ethanol production are:

(a) Raw materials containing sugar that include sugarcanes, molasses, sweet sorghum, etc.
(b) Raw materials containing starch such as corn, cassava, potato.
(c) Cellulose containing substances like wood, agricultural residues, etc.

Starch and cellulose-based materials undergo biochemical conversion to generate sugars from carbohydrates before fermentation. In contrast, sugar-rich materials simply undergo fermentation to produce ethanol. Fermented ethanol is then separated by the process of distillation.

12.2.3 Biochemical Reactions and Microbiology of Anaerobic Digestion Involved in Biogas Production

Anaerobic digestion of animal waste involves following three stages:

(1) Hydrolysis or polymer breakdown

Animal waste consists of complex organic compounds in the form of proteins, fats, carbohydrates, cellulose, lignin, etc. During the process of hydrolysis, these complex organic polymers are broken down by extracellular enzymes produced by the hydrolytic bacteria into simple, soluble monomers. The hydrolysis reactions occurring in this stage will convert protein into amino acids, carbohydrate into simple sugars, and fat into long-chain fatty acids (NAS 1977). Rate of hydrolysis depends on type of substrate, bacterial concentrations, and environmental factors such as pH and temperature.

(2) Acidogenesis

During the process of acidogenesis, the monomeric compounds formed by the hydrolytic breakdown are converted into acetic acid and H_2/CO_2. Various fatty acids are produced as the end products of bacterial metabolism of protein, fat, and carbohydrate like acetic acid, propionic acid, butyric acid, succinic acid, lactic acid, etc., of which acetic, propionic, and lactic acids are the major products. Carbon dioxide, hydrogen, and ammonia (gases) and methanol and other forms of alcohols (liquids) are the other possible by-products produced during waste catabolism. The diversity of the microbial consortium involved in fermentation includes microbes involved in hydrolysis step along with microbes belonging to the genera Enterobacterium, Acetobacterium, and Eubacterium (Schnurer and Jarvis 2010). The type and number of by-products produced depend on the type of waste (substrate) and microbes present and set of environmental conditions (Schnurer and Jarvis 2010).

(3) Methanogenesis

The products like acetic acid (CH_3COOH), methanol (CH_3OH), carbon dioxide (CO_2), and hydrogen gas (H_2) produced during acidogenesis are modified into methane (CH_4) and other by-products by methanogenic bacteria. Methanogenesis comprises two pathways: acetoclastic, i.e., the cleavage of acetic acid in methane and carbon dioxide and hydrogenotrophic, i.e., the formation of methane from hydrogen and carbon dioxide (Schattauer et al. 2011). Acetic acid is the most important substrate which approximately produces 70% of the methane, while the remaining is produced from CO_2 and H_2 gas. Methane (CH_4) can also be produced using other substrates such as formic acid.

The efficiency of the overall process is influenced by the interactions between the hydrolysis, acidogenesis, and methanogenesis. For instance, during acidogenesis, nitrogen containing organic compound must be converted into ammonia to ensure effective usage of nitrogen by methanogenic bacteria. Likewise, methanogens convert volatile fatty acids like H_2 into CH_4 and other gases that control and balance the pH of the digester slurry, lowering the partial pressure of H_2 in the slurry and promoting the activity of the acetogenic bacteria. If the methanogenic bacteria are unable to create CH_4 properly, organic molecules will be transformed into volatile fatty acids, which can pollute water bodies and land surfaces if discharged into them.

The four main groups of bacteria that participate in the processes of biogas formation are (Brown and Tata 1985):

(1) Bacteria that produce acid through hydrolysis and fermentation (acid-forming bacteria).
(2) Bacteria that produce acetate and hydrogen gas (acetogenic bacteria).
(3) Bacteria that produce methane (acetoclastic bacteria).
(4) Hydrogen exploiting methane bacteria.

Acid, acetate, and hydrogen gas-forming bacteria are collectively known as non-methanogenic bacteria, whereas methane-producing and hydrogen-exploiting bacteria are together known as methanogenic bacteria. According to Gunnerson and Stuckey (1986), there are two primary routes through which acid-forming and acetogenic bacteria assist the process of hydrolysis and breakdown complex organic substances into simple products like such as CO_2, H_2, and other volatile fatty acids as shown in reaction (12.1).

McInerney and Bryant (1981) stated that at low H_2 partial pressure, formation of $CO_2 + H_2$ will predominate, whereas at high H_2 partial pressure, production of volatile fatty acids like propionate ($C_3H_6O_2$), butyrate ($C_4H_8O_2$), and ethanol is favored as shown in reaction (12.2). These products are further converted into CH_3COOH, H_2, and CO_2 through acetogenic dehydrogenation reaction by employing acetogenic bacteria.

$$\text{Substrate} \rightarrow CO_2 + H_2 + CH_3COOH \tag{12.1}$$

$$\text{Substrate} \rightarrow C_3H_6O_2 + C_4H_8O_2 + C_2H_6O \tag{12.2}$$

The products thus formed by non-methanogenic bacteria are utilized by acetoclastic and hydrogen exploiting methane bacteria to produce CH_4 as shown in the following reaction (12.3). Brown and Tata (1985) found that the methane-forming acetoclastic bacteria have a longer generation period (2 to 3 days at 35 °C under optimal environmental conditions) than the acid-forming bacteria (2 to 3 h at 35 °C under optimal environmental conditions). Therefore, biomass loadings into the anaerobic digesters should be avoided as the acid-forming bacteria will generate more volatile fatty acids at the rate faster than the acetoclastic bacteria can utilize them. The growth of acetogenic bacteria in the anaerobic digester is sensitive to the H_2 partial pressure (McInerney and Bryant 1981). If the H_2 partial pressure increases, Eq. (12.3) reaction is favored and the production of acetate will be less. Since about 70% of CH_4 is formed from acetate, therefore rate of biogas production is reduced. In order to maintain the low H_2 partial pressure and eliminate H_2 gas from the system, Eq. (12.4) is crucial to the anaerobic digestion process. Figure 12.1 depicts the entire process of methane generation from animal feces.

$$CH_3COO - +H_2O \rightarrow CH_4 + HCO_3^- + \text{energy} \tag{12.3}$$

$$4H_2 + HCO_3^- + H^+ \rightarrow CH_4 + CO_2 + \text{energy} \tag{12.4}$$

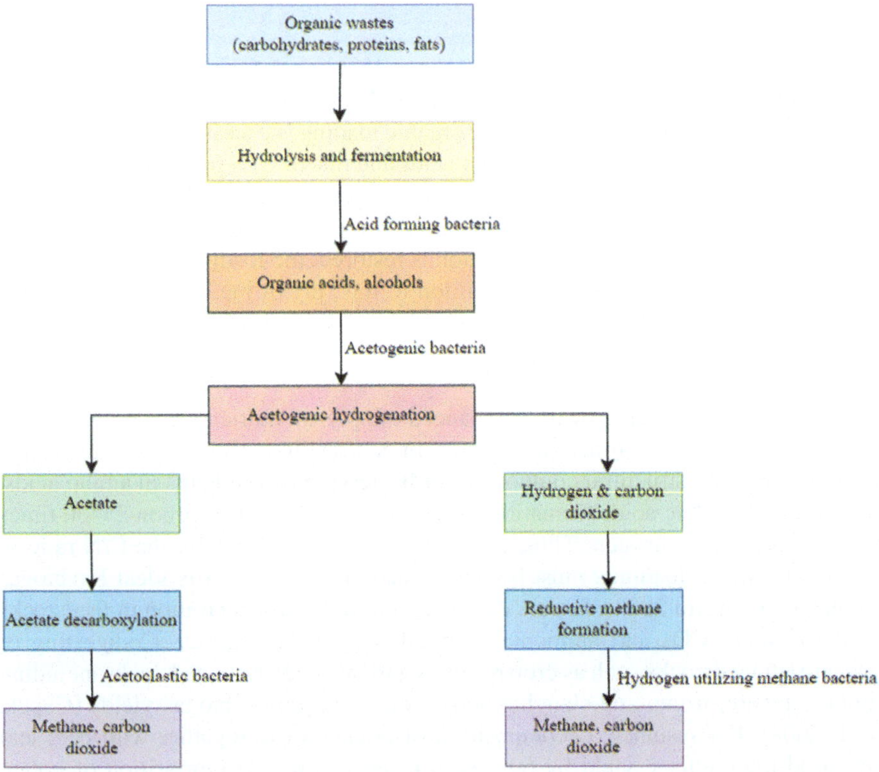

Fig. 12.1 Methane formation of organic wastes by microorganisms (Brown and Tata 1985)

12.2.4 Environmental Requirements for Anaerobic Digestion

1. **Temperature**

The rate of gas generation is significantly impacted by daily and seasonal temperature variations. Generally, two temperature ranges are taken into account for methane production, mesophilic (25–40 °C) and thermophilic (50–65 °C). As the temperature increases, simultaneously the rate of methane production also increases. But, at 45 °C, methane production stops as at this temperature the mesophilic as well as the thermophilic bacteria fail to thrive. During winters, growth of anaerobic bacteria ceases; therefore, heating of biogas digesters is required. The temperature in the digester can be maintained 5–10 degrees Celsius higher than the surrounding air simply by insulating it with a thick, clear plastic sheet or by placing compostable materials like leaves (Brown and Tata 1985). Polprasert (2007) suggest that the digester can also be heated by recirculating hot water through pipe coils that are put inside of it or by heating the influent feeding materials.

2. **pH**

pH level in anaerobic digesters should range between 6.6 and 7.6 for hydrolysis and acidogenesis, whereas methanogens favor pH in the range of 7.8–8.2 (Kim et al. 2003a, b; Yu and Fang 2002). When the organic loading is excessive, pH may drop to below 6.6 due to buildup of volatile fatty acids and rate of CH_4 production decreases. Therefore, suitable measures should be taken quickly, such as stopping feeding of the digester so that the methanogens can use the accumulated volatile fatty acids. After the optimum rates of gas production is restored, the loading of organic matter in the digester can be resumed. In addition to this, pH of the digester can also be neutralized by adding lime

3. **Carbon-to-Nitrogen (C/N) Ratio**

For optimum microbial growth, a balanced supply of nutrients for an anaerobic digester is necessary to function properly. In contrast to carbon, which acts as the energy source and structural component for bacteria, nitrogen helps in amino acids, proteins, and nucleic acids formation. Anaerobic microbes use carbon 25–30 times more quickly than nitrogen. Thus, for efficient biogas production, the C/N ratio in the organic waste loadings must be maintained at 25:1, which is ideal for biogas production (Gerardi 2003). Excess carbon and nitrogen concentration in feedstocks results in more CO_2 and ammonia accumulation in the digester. Codigesting of carbon-rich feedstocks such as crop residues with nitrogen rich feedstocks including animal manure, urine, and slaughterhouse wastes improves biogas yields (Capela et al. 2008). For instance, decomposition of potato waste together with beet leaf increased the methane yield by 60% with respect to the decomposition of potato waste alone (Parawira et al. 2004).

4. **Organic Loading Rate**

The amount of organic matter to be feed in the digester depends upon the reactor configuration, quantity of organic waste contained in the bioreactor, substrate amount and biodegradability, environmental factors like temperature and pH (Khanal et al. 2019). Some key indicators of overloading of organic matter include reduction in methane production, biogas formation, a drop in pH, and a sharp rise in amount of volatile fatty acids (VFAs). The organic matter should be loaded at a rate that keeps the pH in the neutral range, the VFAs between 1500 and 4500 mg as acetic acid equivalent (HAc)/L, and the ammonium nitrogen concentration below 4500 mg/L (Braun 2007).

5. **Toxicity**

Anaerobic microorganisms are sensitive to certain compounds like heavy metals, halogenated compounds, cyanides, antibiotics (present in feedstocks) as well as some metabolic by-products like ammonia, sulfide, and VFAs (Khanal et al. 2019). These compounds are associated with anaerobic digester failure. The presence of molecular oxygen is also inhibitory to the methanogenic bacteria. The inhibition of anaerobic digestion depends on the ionic strength of the toxic substance, ratio of the concentration of toxic substances to microbial biomass, pH, organic loading rate, temperature, and other materials (Kugelman and McCarty 1965). Inhibition caused by excess concentrations of certain ions can be counterbalanced by some antagonistic ions and exacerbated by synergistic ions (Polprasert 2007). According to Price and Cheremisinoff (1981), the toxicity of heavy metals can be lowered either by adding precipitating anions such as sulfide or by operating the digester at a maximum allowable pH, as most of the heavy metal hydroxides have less solubility at higher pH. Similarly, VFA toxicity can be reduced by limiting substrate feeding into the digester until VFA levels drop to an acceptable limit. Toxicity caused by ammonia can be corrected by diluting the digester substrate with water. Also, ammonium nitrogen toxicity can overcome by maintaining proper C/N ratio by feeding carbon-rich feedstocks into the digester.

6. **Nutrients and trace elements**

A balanced supply of nutrients and trace elements is necessary for proper working of an anaerobic digester. Nutrients and trace elements provide a suitable physico-chemical condition for optimum growth of microorganism (Khanal et al. 2019). Few strains of microorganisms, especially *Archaea* involved in the hydrogenotrophic pathway of methanogenesis (Nettmann et al. 2010), need several trace elements as micronutrients of which cobalt, molybdenum, nickel, and selenium are the most important (Gronauer et al. 2009; Lebuhn, et al. 2008). Deficiency of these trace elements leads to reduced anaerobic digestion performance and is usually supplied to biogas digesters with feedstock. Selection of feedstock directly influences the abundance of trace elements. The amount of trace elements suggested by different authors for proper functioning of anaerobic digester is given in Table 12.1.

Table 12.1 Recommended concentration of trace elements is anaerobic digester by different authors

	Bischofsberger et al. (2005)	Weiland (2006)	Seyfried et al. (1990)	Sahm (1981)	Pobeheim et al. (2010)
Boron				0.001–11	
Cobalt	0.06	0.003–0.06	0.003–0.06	0.06	0.024–10
Chromium	0.05–50			0.005–52	
Copper				0.06–64	
Iron		1–10	1–10		
Manganese	0.005–50			0.005–65	
Molybdenum	0.05	0.005–0.05	0.005–0.05	0.05	0.16–50
Nickel	0.006	0.005–0.5	0.005–0.5	0.006	0.024–0.62
Lead	0.02–200				
Selenium	0.008		0.008	0.008	
Tungsten			0.1–0.4		

7. **Mixing**

Maintaining homogeneity inside the digester requires mixing. Effective substrate usage, prevention of temperature stratification, reduction in settling of digested solids at the bottom, lowering of scum formation at the slurry surface are ensured only by proper mixing so as to enhance biogas production (Polprasert 2007; Khanal et al. 2019). In commercial-scale digesters, propeller and screw mixers are installed which are quite costly and energy intensive. However, for small-scale digester such as household digester, recirculating slurry and gas or mixing the slurry manually could be sufficient to mix the digester contents (Khanal et al. 2019).

12.2.5 Advantages of Biogas Technology

1. **Energy service**: The most substantial benefit of biogas technology is the production of biogas from organic wastes. To minimize domestic energy demand, many countries have launched domestic biogas programs (Feng et al. 2009; Kamp and Forn 2016; Katuwal and Bohara 2009; Yu et al. 2008). Bioenergy has become the major domestic energy for rural households, especially in developing countries, where biogas can be directly used for cooking purposes, electricity generation, etc. Rural and regional energy reliability and security are enhanced by production and consumption of domestic bioenergy. According to Feng et al. (2009), proper use of biogas not only reduces pressure on non-renewable energy resources but also improves health, economy, and environment. In India, a novel system of biogas purification and bottling was recently developed at IIT, New Delhi (Vijay 2011). The initial biogas bottling projects started in states of India like Maharashtra and Punjab have successfully produced pure biogas with 98% methane content compressed to 150 bar pressure for filling in cylinders. Pure biogas stored in cylinders is a marketable product and can be easily used any time anywhere just like LPG cylinders. Further, the stored biogas is also used to run petrol-based auto rickshaws (Kapdi et al. 2006) and diesel engines (Ilyas 2006).
2. **Soil fertility and waste stabilization**: The organic waste materials are reduced by 30–60% in biogas digester, and a stabilized sludge is produced. In addition to being used as a fertilizer, the bio-slurry also embraces important substances like cellulose, protein, and lignin which acts as a soil conditioner and helps in improving the physical properties of the soil. The application of digester slurry to unproductive soils improves soil quality and helps in waste lands' reclamation. Furthermore, it generates favorable condition for soil's microorganisms for transformation of soil nutrients into plant usable forms.
3. **Nutrient reclamation**: Nitrogen, phosphorus, potassium, and other nutrients present in organic wastes are usually present in complex organic forms which are not readily available for the crops. Anaerobic digestion does not destroy any of the nutrients from domestic and farm wastes, but makes them easily available

to plants. Bio-slurry comprises both major and micronutrients valuable for the plant's growth (Barbosa et al. 2014). After digestion, at least 50% of the nitrogen present in complex form is converted into inorganic form. Digestion increases the availability of nitrogen up to a range of about 30–60%. Nitrogen present in bio-slurry has fast and immediate fertilizer effect than the one in the fresh manure (Bonten et al. 2014; ESCAP 2007). The phosphate and potash contents are not decreased, and their availability of around 50% and 80%, respectively, is not changed during digestion and is converted into plant usable forms.

4. **Reduction in use of chemical fertilizer**: Among biogas-user households, use of bio-slurry decreased their dependency on the use of chemical fertilizers. Chemical fertilizer production is an energy intensive process; therefore, the amount of GHG emission from chemical fertilizer synthesis has reduced.
5. **Pathogen inactivation**: During anaerobic decomposition, the organic waste is held in oxygen-free environment for a considerable period of time (approximately for 15–50 days), at a temperature of about 35 °C. These conditions are sufficient to inactivate some of the pathogenic bacteria, viruses, protozoa, and helminth ova (Polprasert 2007)
6. **Reduction in greenhouse gas (GHG) emission**: The use of biogas has enabled the biogas-user households to reduce the consumptions of various traditional biomass fuels like kerosene, cow dung, and in turn, emissions of GHGs (Mengistu et al 2016). Studies conducted by Katuwal and Bohara (2009); Laramee and Davis (2013); Pathak et al. 2009; Zhang et al. (2013) concluded that GHG emission reductions ranges from 1.3 t to 9.7 t of CO_2 equivalents per unit domestic biogas plant.
7. **Minimize pressure on forests and other energy resources**: The use of biogas energy has somewhat replaced the consumptions of wood fuel and other household energy sources, therefore minimizing pressure on traditional biomass fuels and other non-renewable energy resources (Feng et al. 2009). Biogas generation has reduced dependency on forests, therefore saved forests also from felling (Subedi et al. 2014).
8. **Economic benefit**: Switching from non-renewable energy resources to biogas is a financial benefit to rural households as it is quite cheap and needs little investment. Biofuel production and consumption also lower fossil fuel use and generate economic benefits to consumers worldwide (Huang et al. 2013). Many countries worldwide like China generate huge revenue out of biogas power generation (Zhang and Xu 2020). In addition to monetary benefits of saving fuel cost, biogas provides several other indirect benefits. The slurry or residue produced from biogas plant can be used as biofertilizer saving the cost of chemical fertilizers. Biofertilizers produced from biogas plants generate a source of income for people because of its good selling price (Zhang and Xu 2020). It also provides economic benefits due to employment generation.
9. **Ecological benefit**: Domestic animals such as poultry, cattle, sheep, pigs produce more than 85% of the world's fecal waste. The animal waste harbors zoonotic pathogens that pose high risk to human health. Surface waters are more susceptible to pathogen contamination by animal feces transported through

runoff (Daniels et al. 2015; Odagiri et al. 2016; Schriewer et al. 2015). Indiscriminate defecation from cattle contaminates fields resulting into soil pollution (Lupindu et al. 2014). This causes the pathogen to spread from animal waste into soil and water bodies, causing environmental pollution and human infection (Manyi-Loh et al. 2016). Anaerobic digestion is one of the potent techniques employed to convert organic wastes into useful products. Besides economic, social, and environmental benefits (Yu et al. 2020), biogas is a well-developed technology widely used in the animal manure stabilization. Anaerobic digestion utilized in biogas production neutralizes pathogens and parasites present in animal manure and reduces the incidence of waterborne and other forms of zoonotic diseases among humans. Biogas technology reduces soil and water pollution and also reduces risk of environmental contamination and health hazard.

10. **Social benefit**: For proper operation, maintenance, and repair of a biogas plant, a skilled labor is required. It also provides employment benefits through requirement of bricks, cement, steel, stone chips, etc. (Purohit and Kandpal 2007). Therefore, biogas technology produces employment opportunities for unskilled, skilled, unemployed youth and entrepreneurs. Biogas production also cuts LPG costs, and villagers get extra income from bioelectricity production (Sinsuw et al. 2021). Besides this, rural people also gain profit out of biofertilizers. In addition, people using biogas experience less health issues than the conventional sources like firewood, cow dung, etc. Table 12.2 lists the benefits and drawbacks of biogas technology as follows:

12.3 Composting

Composting is defined as the process of biological decomposition of organic materials carried under controlled aerobic and thermophilic (>50 °C) conditions which aims at the following objectives: to produce stabilized organic matter with respect to nitrogen, oxygen demand, and other elements, eliminate phytotoxicity, eradicate pathogens, and weed seeds (Haug 1980; Larney and Hao 2007; Raviv 2005). Mature compost is used as soil conditioner without any adverse environmental effects. Generally, composting is performed on solid and semi-solid organic wastes such as sludge, animal manures, agricultural residues, and municipal refuse (Polprasert 2007). Aerobic composting is a preferred approach because it releases more heat energy resulting in a rapid decomposition of animal manure and forms stabilized end products, whereas anaerobic composting is a slow process and produces unpleasant gases including CH_4, CO_2, NH_3, and other trace gases.

Table 12.2 Benefits and drawbacks of biogas technology (NAS 1977)

Advantages	Disadvantages
• Methane gas is generated in enormous amount and can be kept easily at room temperature	• Likelihood of explosion
• Reduces waste materials by 30–50% and produces a stabilized sludge which can be used as soil fertilizer	• High capital cost
• Pathogens and weed seeds are either eliminated entirely or reduced significantly	• If handled improperly, liquid sludge poses a risk for water contamination
• Digestate does not attract rodents or flies	• For maximum biogas production, the digester's working conditions must be kept optimal which is therefore a cost-intensive process
• Pests and vermin have restricted access to waste	
• Offers a hygienic means of disposing animal waste	
• Promotes the conservation of local energy supplies	
• Have less environmental issues	
• Risk of health hazard is reduced to a greater extent. Likelihood of explosion	

12.3.1 Process

Composting is a dynamic process, accomplished by a succession of microorganisms. When optimum conditions are met, each group of microorganisms reaches its peak population. Because of many biochemical reactions, change in nutrition and temperature occurs; as a result, one group of microorganisms dies and population of another group dominates. According to Diaz et al. (2002); Keener et al. (2001); Reinikainen and Herranen (2001), composting can be divided into two different phases: The first phase is known as the active or heating phase. In this phase, there is a rapid increase in temperature up to 45–60 °C and is maintained for several weeks (González and Sánchez 2005). During first phase of composting, oxygen consumption is very high; therefore, process of aeration is increased. There is rapid reduction in odor and biodegradable solids. During this phase, waste stabilization and pathogen destruction are most effective. In the second phase, which is also known as maturation phase, rate of reactions as well as temperature decreases to 35 °C and consequently to ambient levels. In the maturation stage, aeration is no longer a limiting component. In this stage, secondary fermentation encourages the humification process and complex organic molecules are converted to hemic substances, minerals like Fe, Ca, N, and then to humus. The complete process of composting is shown in Fig. 12.2.

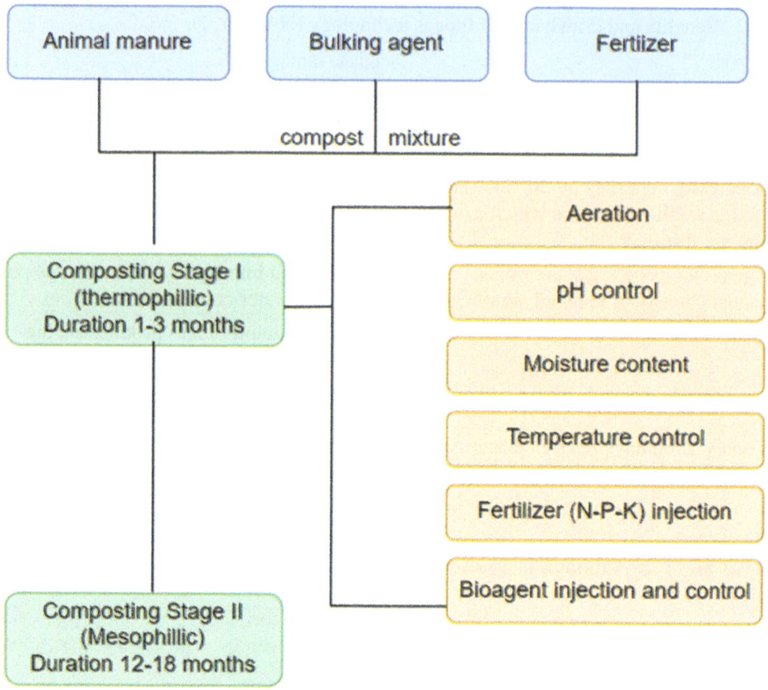

Fig. 12.2 Composting process. *Source* Aikaite-Stanaitiene et al. (2010)

Composting is a biological process that transforms complex organic waste into stabilized product by the activity of microorganisms naturally present in the organic wastes. These include nematodes, earthworms, mites, and other invertebrates as well as microorganisms including bacteria, fungus, and protozoa.

The organic wastes are inhabited by group of microorganisms such as bacteria, fungi, and actinomycetes. Mesophilic bacteria are first to appear and carry out many reactions. As the temperature increases, thermophilic bacteria replace the former and inhabit the compost heap. During final stages, temperature declines and the members of the actinomycetes become the dominant which may give white or gray appearance to the compost heap.

12.3.2 Environmental Requirements for Composting

Composting requires controlled conditions which distinguishes it from other natural biological decomposition processes like rotting and putrefaction. The process may fail if physical, chemical conditions are not met required for microbial development. The following are the main environmental parameters that must be regulated appropriately throughout the composting processes:

1. **Nutrient balance**: The C/N ratio in a composting pile should range from 20:1 (20 parts of carbon for every part of nitrogen) to 40:1 (40 parts of carbon for every part of nitrogen). The C/N ratio needs to be higher because approximately 50% of the metabolized carbon is released as carbon dioxide (Miller 1996). Adding carbon (straw, corn stalks, or woodchips) alleviates carbon content. Too much carbon (C/N ratio more than 40:1) can immobilize nitrogen, and composting process may slow down (Coyne and Thompson, 2006). C/N ratio varies greatly with the composting material. Difference in C/N ratio depends upon the bulking materials of plant origin, and in case of animal manures, it depends upon species and diet. Following Table 12.3 shows C/N ratios that various composting materials possess.
2. **Temperature**: An optimum range of 55–65 °C temperature is required for maximum efficiency of microorganisms responsible for effective composting (McKinley et al. 1985). When the temperature reaches above 70 °C, the compost piles should be aerated which can kill the composting microorganisms (Busto et al. 1997).
3. **Aeration**: Composting requires optimal aeration to absorb enough oxygen for the aerobic bacteria to grow. This is done through some non-mechanical techniques such as turning of the compost piles on regular basis, inserting perforated bamboo rods into the compost piles, or lowering compost heaps from floor to floor. The forced-air aeration method, in which air is blasted via perforated pipes and distributed into the compost heaps, is a more efficient mechanical method of turning compost. Turning homogenizes the compost and breaks up clumps. Too much aeration is wasteful and can cause a loss of heat while too little aeration would create anaerobic conditions discouraging microbial growth (Polprasert 2007).
4. **Moisture content**: An optimum moisture content of the compost mixture is essential for microbial decomposition of the organic waste. A moisture content should range from 50 to 70% and should be maintained during mesophilic and thermophilic growth. Moisture concentration of 20% or less significantly impairs biological activity, while a moisture content of 80% or above leads to the leaching

Table 12.3 C/N ratios of various composting materials (Rynk et al. 1992)

Cattle manure	19:1
Cattle carcass	10:1
Poultry carcass	4:1
Sheep manure	16:1
Dairy manure	20:1
Swine carcass	14:1
Horse manure	30:1
Turkey litter	16:1
Swine manure	12:1

of nutrients and pathogens. More water content blocks air passage creating anoxic conditions harmful for microbial growth (Polprasert 2007).

5. **Particle size of compost pile**: Composting material size should be as tiny as possible to enable effective aeration and ease the process of microbial decomposition.

12.3.3 Benefits

1. **Waste Stabilization**: Composting transforms organic waste from its complex structure into a stable inorganic form which can be safely dumped on land or utilized as fertilizer.
2. **Pathogen Inactivation**: Animal manure harbors disease-causing pathogens and parasites. When applied to fields as fertilizer, it causes soil contamination, and through surface runoff, it adds into water bodies causing biological pollution of surface water. Therefore, animal manure is a potential source of environmental pollution and human infection. Physical, chemical, and biological methods have been used for pathogen destruction. Anaerobic digestion is one of the most common biological methods used for pathogen control and stabilization (Lin et al. 2022). The efficiency of pathogens' inactivation is controlled by factors such as pathogen type, transitional products, temperature (Dominguez and Edwards 2011. At mesophilic temperatures, most of the pathogens survive, but as the temperature increases from 37 °C to 70 °C (at thermophilic temperatures), the fluidity and permeability of the cell membrane increase; therefore, diffusion of toxic chemicals into the cytoplasm increases at a rapid rate and inhibits cell growth of most pathogenic viruses, bacteria and destructs helminthic ova (Salsali et al. 2006). Mesophilic anaerobic digestion is more widely applied to practice for pathogen destruction (Gavala et al. 2003)
3. **Nutrient Availability**: Composting makes utilization of nutrients more effective. Composting converts complex form of nutrients (N, P, K) present in the organic wastes into inorganic forms such as NO_3^- and PO_4^{-3} easily available for plant uptake. The inorganic nutrients thus formed exist in the insoluble forms which less likely leach out than the complex soluble forms. It also allows long-term retention of nutrients through enhanced soil ability to retain nutrients chemically and biologically and reduces nutrient loss and eventually improves the soil quality (Polprasert 2007). The nutrient availability varies significantly with respect to the composition of the compost and can be similar or considerably higher or lower than that of the quantity of nutrients already present in the soil. Composts formed from high nutrient materials like municipal biosolids or animal wastes relatively have high nutrient content than the compost derived from cellulosic plant material which are more likely nutrient poor. For instance, municipal biosolids' compost can be used directly as a plant growth media, whereas vegetative waste or municipal refuse composts are always needed to be amended with fertilizer nutrients (Harrison 2008).

4. **Physical Effects**: Adding compost to the soil means addition of organic matter which creates the appropriate medium for vegetation growth by improving the physical properties of the soil such as aggregation, water-holding capacity, porosity, and aeration (Flavel and Murphy 2006; Harrison 2008). In addition, the soil tilth is also improved (Golueke 1982). The soil with good physical properties enhances root respiration, microorganism activity and reduces erosion beneficial for plant growth.
5. **Chemical Effects** (Harrison 2008): Other than nutrient availability, compost has a beneficial effect on soil chemical properties like buffering pH levels, increasing soil cation-exchange capacity, and reducing potentially active toxic substances. Compost produced from organic waste material is neutral or slightly alkaline in pH, with the values ranging from 5.5 to 8.0. When applied to soil, pH is either increased (in case of acidic soils) or is decreased (for alkaline soil), therefore bring the soil pH to neutral that is most ideal for plant growth. Addition of organic matter also increases cation-exchange capacity (CEC) of soil which is a measure of binding potential of a soil to hold cations that increases soil's ability to retain nutrients in an available form. Organic matter also reduces toxicity of some metal's nutrients in soils such as Fe, Cu, Al, and Mn by forming organic matter–metal complex known as chelate. Metal nutrients in chelated form are very less toxic to plants. Compost inactivates certain soil-active pesticides, forms buffer against leaching of herbicides and insecticides through the soil profile, acts as a runoff adsorbent, and therefore, reduces release of pollutants into surface water bodies.
6. **Biological Effects**: Composting reduces the prevalence of diseases in both plants and animals. The is either because of high temperatures created during microbial decomposition that destroy insect eggs (Larney et al. 2006) or the organisms present in the compost directly compete with plant pests by consuming the available nutrients. Parasite eggs causing severe health conditions in animals are also destroyed during composting due to high temperature conditions, therefore reducing the potential spread (Nielsen et al. 2007)

12.4 Vermicomposting

Due to low efficiency in nitrogen utilization (Oenema 2006), ruminants often excrete unutilized and undigested nitrogen and other nutrients in their feces and urine having adverse environmental impacts (Dijkstra et al. 2011; Hristov and Jouany 2005; Tamminga 1992). Ruminant manure is a good source of both macro and micronutrients often used as a soil fertilizer for the vegetation growth (Lazcano et al. 2008). Traditionally, livestock manure is stored for a month or longer before being applied to the soil in the majority of nations (Sommer and Hutchings 2001) or is introduced untreated directly into the farm. Untreated and excessive application of animal manure leads to environmental issues like increase in soil toxicity, water pollution, greenhouse gas emission, pathogens, and weed seeds' dispersal and could be a health hazard (Dominguez and Edwards 2011; Kulling et al. 2001). Therefore, a proper animal waste management is required so as to achieve long-term sustainability.

Vermicomposting is suggested as one of the most efficient methods for recycling animal manure and producing stabilized end products having less environmental consequences (Hao et al. 2001, 2004; Loh et al. 2005; Gutiérrez-Miceli et al. 2008). As mentioned in the previous part of this chapter, composting is one of the methods used for recycling animal waste but is highly economic activity, and during composting, noxious gas like ammonia and other greenhouse gases such as methane and nitrous oxide are released. Because of their potential to promote global warming, methane and nitrous oxide emissions are a worry for the environment (IPCC 2007). To overcome the cost of composting, vermicomposting is advised so as to reduce composting costs and provide high-quality products. In vermicomposting, organic material is bio-oxidized by mesophilic bacteria along with the joint action of earthworm to form nutrient rich humified organic product known as vermicompost (Nasiru et al. 2014). During vermicomposting, substrate is grinded into smaller particles in the earthworm gizzard. The material is further acted upon by earthworm gut microbial flora and converted into fine granular particles enriched with nutrients, hormones, and humic substances (Dominguez 2011; Hussain et al. 2018). Besides modification in physical and biochemical properties of organic matter, earthworms also facilitate uniform mixing of the material (Yadav and Garg 2011). Compared with composting, vermicomposting promotes nitrogen retention and produces more stabilized products (Frederickson et al. 2007; Lazcano et al. 2008). Vermicomposting is one of the best animal manure recycling options as it offers an environmentally and economically sound strategy to get enriched end product.

12.4.1 Vermicomposting Technique

Earthworms

There are about 4000 species of earthworms broadly divided into three groups: endogeic, anecic, and epigeic. All the earthworm species cannot be utilized for vermicomposting; only the epigeic earthworms comprising *Eisenia fetida, Eisenia andrei, Eudrilus eugeniae*, and *Perionyx excavatus* are most effective for vermicomposting. Owing to their high reproduction rates, tolerance to a wide range of environmental conditions, rapid rate of vermiconversion, and capacity to feed on a wide range of organic wastes make epigeic earthworms most effective organisms for vermicomposting (Sharma and Garg 2019). Monoculture of *Eisenia fetida* (Sharma and Garg 2019) and polycultures of *P. excavatus, E. eugeniae,* and *Dichogaster annae* (Martin and Eudoxie 2018) were suggested best for vermicomposting of animal manures.

Bulking Agent and Precomposting

Organic waste used as feedstock during vermicomposting should be biodegradable, pH should range between 5 and 8, amount of moisture is between 60 and 80%, and

C/N ratio is approximately 30 (Edwards and Bohlen 1996). Organic wastes utilized for vermicomposting should be mixed with some mixing agent known as bulking material. Utilizing mixing agent during vermicomposting improves earthworm efficiency, lowers toxic effects on earthworms, keeps waste's moisture content constant, and makes waste more palatable to earthworms (Sharma and Garg 2019). Animal excrement, kitchen waste like fruit and vegetables, agricultural remains, and trash paper are some of the most frequently employed bulking agent during vermicomposting. In addition to this, dried leaves, compost, and vermicompost are also used as bulking substrate (Alidadi et al. 2016; Varma et al. 2015). Organic waste lacking bulking substrate causes earthworm mortality due to anaerobic conditions and toxic gases' emission (Fu et al. 2015). Mixing the organic waste with bulking substrate is followed by precomposting. Precomposting eliminates anaerobic conditions and removes volatile gases harmful for earthworms.

12.4.2 Process

Vermicomposting is an aerobic process that involves joint action of microorganisms and earthworms. The process begins with the ingestion of organic waste by earthworm, that acts as a bioreactor where a variety of physical and biological reactions takes place. Enzymes such as proteases, lipases, amylases, cellulases and chitinases present in the gut of earthworms degrade cellulosic and proteinaceous materials present in organic substrate (Gupta and Garg 2009). All these metabolic activities taking place in gut of earthworm converts waste into vermicast. The vermicast is then converted into a nutrient-enriched humus-like product known as vermicompost by the action of microorganisms.

The entire process of vermicomposting is divided into two distinct phases: (1) an active phase, during which the earthworms act upon the biomass, altering its physical, chemical, and biological characteristics stimulating and enhancing biological activity through fragmentation, which increases the surface area exposed to microorganisms and promotes mineralization (Dominguez 2011; Lores et al., 2006). Most of these activities take place in the gut of earthworms; therefore, active phase is also known as the gut-associated process. (2) A maturation phase is also known as cast-associated processes (CAPs) during which earthworm's gut-associated microbes and the microbes present in the vermicast produce enzymes necessary for biochemical breakdown of organic matter and convert it into vermicompost (Dominguez 2011; Dominguez and Edwards 2011; Gómez-Brandón and Domínguez 2014). Additionally, earthworms also act as an turning and aerating agent (Ndegwa and Thompson 2001). According to Nair et al. (2006), animal manure should be pretreated before vermicomposting in order to eliminate gases and acidic compounds poisonous to earthworms. Garg et al. (2006); Lazcano et al. (2008); Loh et al. (2005); Velasco-Velasco et al. (2011) recommended vermicomposting as the best technique to stabilize ruminant manure. Lazcano et al. (2008) suggested that pre-treatment of cattle manure with composting then followed with vermicomposting is the most effective in stabilizing cattle manure.

12.4.3 Factors Governing Vermicomposting

The vermicomposting process is influenced by a variety of biotic variables which include temperature, moisture, aeration, pH, feedstock quality, and C/N ratio (Sharma and Garg 2019). Vermicomposting works best with moisture contents of 60–80%, pH values of 5–8, and C/N ratios of roughly 30. Vermicomposting is also affected by biotic parameters such earthworm species, stocking density, growth and reproductive efficiency, and feeding rate (Lim et al. 2015). As a result, these parameters must be optimized for effective conversion of organic wastes into vermicompost.

12.4.4 Advantages

1. Vermicomposting is the clean, sustainable, and zero-waste approach of ruminant manure management. This eco-friendly method is cost-effective and the best of all remediation processes.
2. Vermicomposting converts organic materials into more stabilized form. Vermicasts extruded by epigeic earthworm species are more stable than the vermicast produced by endogeic species. (Dlamini and Haynes 2004).
3. Vermicomposting transforms necessary nutrients into readily accessible and soluble forms. For instance, majority of agricultural soils have high levels of phosphorus, but not easily accessed by plants. During vermicomposting, P-solubilizing bacteria residing in the stomach of the earthworms convert insoluble P into soluble P (Sharma and Garg 2019).
4. Vermicomposting plays a crucial in K (potassium) enhancement. Earthworms' digestive enzymes transform organic potassium into exchangeable potassium, increasing the K content of vermicast (Suthar 2010).
5. Vermicompost has rich macro and micronutrients such as Ca, Mg, Cu, Fe, Mg, Mn, and Zn than compost (Yuvaraj et al. 2018). It acts as a source of nutrients, plant growth hormones, humic substances that promotes plant growth. Vermicomposting in general plays a vital role in sustainable agriculture.
6. During vermicomposting, the C/N ratio is reduced improving soil health and crop output.
7. Vermicast, a product of vermicomposting is a good soil fertilizer, improves soil structure, reduces erosion, stabilizes soil pH. In addition, vermicompost enhances moisture infiltration and improves moisture holding capacity of soil.
8. Vermicast can be exploited as an additional food source for ruminants. Since decades, efforts have been made to turn excrement into components for animal

feed (Woesttyne and Verstrate 1995). Cattle dung is being converted into feedstuff using a variety of processes, comprising dehydration, compost silage, protein synthesis, pelleting, deep stacking, chemical preservation, and chemical enhancement (Bórquez et al. 2009; Sarwar et al. 2011). The advantage of utilizing cattle manure as feedstuff is reduction in pollution and feeding cost (Martí-nez-Avalos et al. 1998). Vermicomposting reduces C:N ratio, fiber content, pH value of vermicast which is within the neutral range; therefore, vermicast can be a good feed supplement to ruminants (Bernal et al. 1993). Fermentation of cattle manure with urea (source of nitrogen) and cane molasses (source of carbohydrate) is the most common method of producing feed resource from cattle manure (Martínez-Avalos et al. 1998; Sarwar et al. 2006), improves feed quality, and increases feed intake, nutrients digestibility, and microbial protein supply (Bórquez et al. 2009; Hassan et al. 2011; Martínez-Avalos et al. 1998).

9. Vermicomposting enhances microbial profile of the soil and harbors many beneficial microbes such as biofertilizer, antibiotic-producing microorganisms which acts as boosting factor for increased decomposition, nutritive enrichment, and plant growth.
10. Vermicomposting reduces pathogenic microorganisms and antibiotic resistance genes (Hill et al. 2013; Huang et al. 2020).

12.5 Algae Production

The animal waste is also used as a nutritive medium for the algae growth. Algal cells have high protein value; therefore, algae production can be used for human and animal consumption. Pathogens present in waste water containing animal excrement is eliminated due to diurnal variations in pH caused by algal photosynthesis and toxins released by algae cells that unfavorable conditions are preventing pathogen growth. Animal waste stabilization and pathogen elimination are therefore beneficial for the waste recycling process. Figure 12.3 shows some potential uses for the algal mass cultures.

12.5.1 Algal Pond Systems

In accordance with the raw material utilized and the exploitation of algal biomass, Becker (1981) devised three systems for the cultivation and processing of algae given as follows:

1. **TYPE-I**: Selected algal species are fed with nutrients and carbon to grow artificially in freshwater systems. The algae produced in these systems is used as a source of human food.

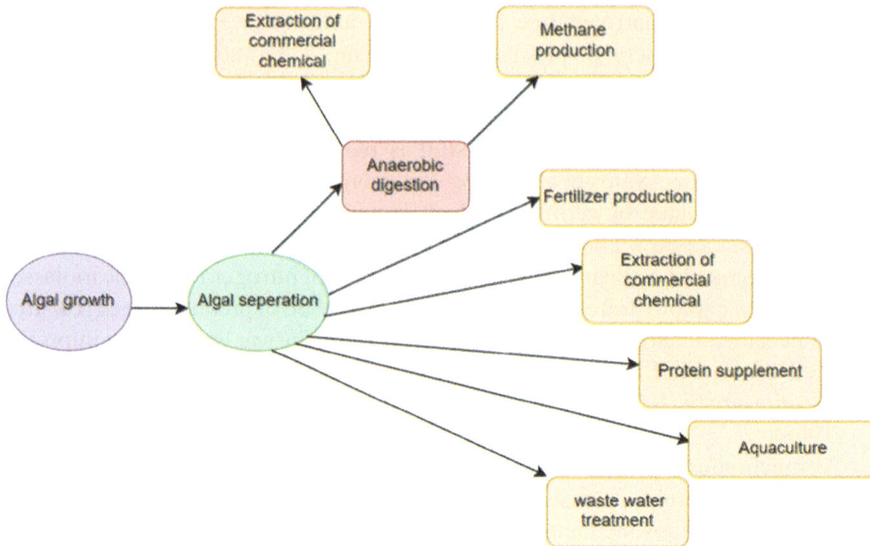

Fig. 12.3 Possible applications of algal biomass (Goldman 1979)

2. **TYPE-II**: Sewage or industrial effluent is used as a culture medium for the growth of various algae species without the addition of any supplementary minerals or carbon from external sources. These systems are crucial for the treatment of wastewater, and the biomass thus generate can be utilized as animal feed or as a substrate for energy generation.
3. **TYPE-III**: Algae cultivation is done in an enclosed fermentation unit under natural or artificial light.

High-rate algal pond (HRAP) systems

HRAP systems were developed by Oswald and co-workers at the University of California at Berkeley (Craggs 2005; Oswald 1991, 1996). HRAP system is best utilized for low-energy wastewater treatment, biofuel production from the harvested algal biomass, and energy recovery from wastewater solids (Craggs et al. 2014). HRAP systems work in the series of four processes which are as follows:

1. Wastewater solids are settled and concentrated at the bottom of the CAP (covered anaerobic ponds) and removed through anaerobic digestion
2. The supernatant from CAPs is treated in HRAP. Variables that affect operation of HRAP includes loading rate of organic materials that should be between 100 and 150 kg BOD_5 ha^{-1} d, depth (0.2–0.6 m), hydraulic retention time (3–4 days in summer and 7–9 days in winter), and horizontal mixing velocity (0.15–0.30 m s^{-1}). Under aerobic conditions and natural sunlight, algae is grown on the supernatant and later harvested as a source of animal feed, biofuel, fertilizer or used as a substrate for the chemical industry

3. Harvesting of algal biomass and subsequent conversion to biofuel.

One of the main contributors to nutrient overloading that leads to eutrophication, which affects lake and pond health, is the application of animal dung to crops. The adverse ecological impacts due to eutrophication are categorized as (1) biodiversity loss, (2) replacement of dominant species, (3) increase in amount of toxic substances, (4) increased turbidity, (5) decreased lifespan of the lake. Therefore, removing nutrients and harmful metals from wastewater to acceptable levels is a crucial requirement before water is discharged or reused. Chemical and physical techniques used in traditional wastewater treatment technologies are not cost-effective for treating agricultural wastewater. Biological methods like cultivation of algae have been found efficient in removing nitrogen, phosphorus, and toxic metals from wastewaters (Boelee et al. 2012; Sturm and Lamer 2011; Zhou et al. 2012) and have the following advantages: (1) algae can grow at a faster rate and do not compete with other plants for growth requirements, (2) have high oil contents of 20–50% on a dry weight basis (Chisti 2007), (3) have the ability to fix carbon dioxide, thus reducing greenhouse gas emissions, (4) provides a biological method for treating wastewater by using nutrients from the majority of wastewaters, and (5) algae biomass can be used as animal feed or fertilizer for crops as a source of nitrogen (Spolaore et al. 2006) (6) can be used for biofuel production.

12.6 Manure Management by Fish Cultivation

Utilization of human or animal wastes for fish production has been investigated widely. This practice of using animal wastes for fish culture is called as "coprology" or "the science of dung and filth" (Jhingran 1991). According to Edwards (1985), in many Asian countries, animal manure is used widely for fish production. There are three ways to reuse organic wastes in fish culture: fertilizing fish ponds with animal waste, sludge, or manure; raising fish in effluent-fertilized fish ponds; and directly raising fish in waste stabilization ponds (Polprasert 2007). Fish feed and fertilizers cost approximately 60% of the total expenditure. Simply by using animal manures for fertilizing fish ponds can sharply reduce these costs. This can also lead to higher fish yields as it provides all the nutrients required for the metabolic cycle of fish culture.

12.7 Conclusion

With significant increase in animal waste production, there is an urgent need to handle the wastes in an efficient manner in order to lessen the negative impacts on the environment and human health. Biogas production, composting, vermicomposting are the most ideal, clean, and sustainable methods for converting animal waste into the products that have no or less bad environmental effects. In addition to biogas production, animal manure can also be used for generating high-value products like protein synthesis, fish production. Adoption of these methods not only curtails greenhouse gas emission, environmental contamination but also provides additional revenue from biogas synthesis, biofertilizer production. Furthermore, animal waste management practices stabilize animal waste and inactivate disease-causing pathogens. Animal waste management practices are the best tools for achieving sustainable goals and climate change mitigation. There is also a need to provide incentives to developing nations for implementation of animal waste management practices in an effective manner.

References

Aikaite-Stanaitiene J, Grigiškis S, Levišauskas D et al (2010) Development of fatty waste composting technology using bacterial preparation with lipolytic activity. J Environ Eng Landsc Manag 18(4):296–305

Alidadi H, Hosseinzadeh A, Najafpoor AA et al (2016) Waste recycling by vermicomposting: maturity and quality assessment via dehydrogenase enzyme activity, lignin, water soluble carbon, nitrogen, phosphorous and other indicators. J Environ Manage 182:134e40

Barbosa DBP, Nabel M, Jablonowski ND (2014) Biogas-digestate as nutrient source for biomass production of Sida hermaphrodita, *Zea mays L.* and *Medicago sativa L.* Energy Procedia 59:120e126

Bartlett AA (1998) Reflections on sustainability, population growth and the environment revisited. Renew Resour J 15(4):6–23. http://www.hubbertpeak.com/bartlett/reflections.htm

Becker EW (1981) Algae mass cultivation—production and utilization. Process Biochem 16:10–14

Bernal MP, Lopez-Real JM, Scott KM (1993) Application of natural zeolites for the reduction of ammonia emissions during the composting of organic wastes in a laboratory composting simulator. Biores Technol 43(1):35–39. https://doi.org/10.1016/0960-8524(93)90079-q

Bischofsberger W, Dichtl N, Rosenwinkel KH et al (2005) Anaerobtechnik. Berlin Heidelberg, Springer-Verlag

Boelee NC, Temmink H, Janssen M et al (2012) Scenario analysis of nutrient removal from municipal wastewater by microalgal biofilms. Water 4:460–473

Bonten LTC, Zwart KB, Rietra RPJJ et al (2014) Bio-slurry as fertilizer: is bio-slurry from household digesters a better fertilizer than manure? A literature review. University and Research centre, Alterra Wageningen, Wageningen

Bórquez JL, González-Muñoz SS, Pinos-Rodríguez JM et al (2009) Feeding value of ensiling fresh cattle manure with molasses or bakery byproducts in lambs. Livest Sci 122(2–3):276–280. https://doi.org/10.1016/j.livsci.2008.09.009

Braun R (2007) Anaerobic digestion: a multi-faceted process for energy, environmental management and rural development. In: Ranalli P (ed) Improvement of crop plants for industrial end uses. Springer publication, p 335e416

Brown NL, Tata PBS (1985) Biomethanation, ENSIC review no. 17/18. Environmental sanitation information center, Asian Institute of Technology, Bangkok

Brusseau ML, Artiola JF (2019) Chemical contaminants. In: Environmental and pollution science. Academic Process. https://doi.org/10.1016/B978-0-12-814719-1.00012-4

Busto MD, Ortega N, Pérez MM (1997) Stabilisation of cellulases by crosslinking with glutaraldehyde and soil humates. Bioresour Technol 60(1):27–33

Capela I, Rodrigues A, Silva F et al (2008) Impact of industrial sludge and cattle manure on anaerobic digestion of the OFMSW under mesophilic conditions. Biomass Bioenergy 32(3):245–251

Caradonna JL (2014) Sustainability: a history. Oxford University Press
Chisti Y (2007) Biodiesel from microalgae. Biotechnol Adv 25:294–306
Coyne MS, Thompson JA (2006). Math for soil scientists. Thomson Delmar Learning, Clifton Park, New York, pp 176–190, 199–208
Craggs R, Park J, Heubeck S et al (2014) High-rate algal pond systems for low-energy wastewater treatment, nutrient recovery and energy production. NZ J Bot 52(1):60–73
Craggs RJ (2005) Advanced integrated wastewater ponds. In: Shilton A (ed) Pond treatment technology. IWA scientific and technical report series. London, IWA pp 282–310
Daly HE (1973) Introduction. In: Daly HE (ed) Toward a steady-state economy. San Francisco, FH Freeman
Daniels ME, Shrivastava A, Smith WA et al (2015) Cryptosporidium and giardia in humans, domestic animals, and village water sources in rural India. Am J Trop Med Hyg 93(3):596–600. https://doi.org/10.4269/ajtmh.15-0111
Diaz LF, Savage GM, Golueke CG (2002) Composting of municipal solid wastes. In: Handbook of solid waste management, pp 117–150
Dijkstra J, France J, Ellis JL et al (2011) Effects of nutritional strategies on simulated nitrogen excretion and methane emission in dairy cattle. In: Sauvant D, Milgen JV, Faverdin P et al (eds) Modelling nutrient digestion and utilisation in farm animals. Wageningen Academic Publishers, Netherlands
Dlamini TC, Haynes RJ (2004) Influence of agricultural land use on the size and composition of earthworm communities in northern KwaZulu-Natal, South Africa. Appl Soil Ecol 27(1):77–88
Dominguez J (2011) The microbiology of vermicomposting. In: Edwards CA, Arancon NQ, Sherman R (eds) Vermiculture technology earthworms, organic wastes, and environmental management. CRC Press, Boca Raton, pp 53–66
Dominguez J, Edwards CA (2011) Relationships between composting and vermicomposting. In: Edwards CA, Arancon NQ, Sherman R (eds) Vermiculture technology earthworms, organic wastes, and environmental management. CRC Press, Boca Raton, pp 11–26
Economic and Social Commission for Asia and the Pacific of the United Nations (2007) Recent developments in biogas technology for poverty reduction and sustainable development. United Nations ESCAP, Beijing
Edwards CA, Bohlen PJ (1996) Biology and ecology of earthworms. Chapman and Hall, London
Edwards P (1985) Aquaculture; a component of low-cost sanitation technology. UNDP project management report 3. World Bank, Washington, D.C
FAO (2004) Animal health and production division, Avian Influenza -Questions and Answers
Feng T, Cheng S, Min Q (2009) Productive use of bioenergy for rural household in ecological fragile area, Panam County, Tibet in China: the case of the residential biogas model. Renew Sustain Energy Rev 13:2070e2078
Flavel TC, Murphy DV (2006) Carbon and nitrogen mineralization rates after application of organic amendments to soil. J Environ Qual 35:183–193
Frederickson J, Howell G, Hobson AM (2007) Effect of precomposting and vermicomposting on compost characteristics. Eur J Soil Biol 43:S320–S326. https://doi.org/10.1016/j.ejsobi.2007.08.032
Fu X, Huang K, Chen X et al (2015) Feasibility of vermistabilization for fresh pelletized dewatered sludge with earthworms *Bimastus parvus*. Bioresour Technol 175:646e50
Garg V, Yadav Y, Sheoran A et al (2006) Livestock excreta management through vermicomposting using an epigeic earthworm *Eisenia foetida*. Environmentalist 26(4):269–276. https://doi.org/10.1007/s10669-006-8641-z
Gavala HN, Yenal U, Skiadas IV et al (2003) Mesophilic and thermophilic anaerobic digestion of primary and secondary sludge. Effect of pre-treatment at elevated temperature. Water Res 37(19):4561–4572
Gerardi MH (2003) The microbiology of anaerobic digesters. Wiley
Gerber P, Chilonda P, Franceschini G et al (2005) Geographical determinants and environmental implications of livestock production intensification in Asia. Biores Technol 96:263–276

Goldman JC (1979) Outdoor algal mass culture-I, applications. Water Res 13:1–19
Golueke CG (1982) Composting: a review of rationale, principles and public health. In: Composting: theory and practice for city, industry and farm. The JG Press, Emmaus
Gómez-Brandón M, Domínguez J (2014) Recycling of solid organic wastes through vermicomposting: microbial community changes throughout the process and use of vermicompost as a soil amendment. Crit Rev Environ Sci Technol 44(12):1289–1312
González JL, Sánchez M (2005) Treatment of poultry mortalities on poultry farms. Compost Sc Util 13(2):136–140
Gronauer A, Andrade D, Bauer C et al (2009) Prozessoptimierung-Ein Zusammenspiel von Technik und Mikrobiologie. Gülzower Fachgespräche 32:120–140
Gunnerson CG, Stuckey DC (1986) Anaerobic Digestion: principles and practices for biogas systems. UNDP project management report no. 5. The World Bank, Washington, D.C
Gupta R, Garg VK (2009) Vermiremediation and nutrient recovery of non-recyclable paper waste employing Eisenia fetida. J Hazard Mater 162:430e9
Gutiérrez-Miceli FA, Moguel-Zamudio B, Abud-Archila M et al (2008) Sheep manure vermicompost supplemented with a native diazotrophic bacteria and mycorrhizas for maize cultivation. Biores Technol 99(15):7020–7026. https://doi.org/10.1016/j.biortech.2008.01.012
Hao X, Chang C, Larney FJ et al (2001) Greenhouse gas emissions during cattle feedlot manure composting. J Environ Qual 30(2):376–386. https://doi.org/10.2134/jeq2001.302376x
Hao X, Chang C, Larney FJ (2004) Carbon, nitrogen balances and greenhouse gas emission during cattle feedlot manure composting. J Environ Qual 33(1):37–44. https://doi.org/10.2134/jeq2004.370
Harrison B (2008) Composting and formation of humic substances. In: Jørgensen SE, Fath BD (eds) Encyclopedia of ecology. Academic Press, pp 713–719
Hassan Z, Nisa M, Shahzad M, Sarwar M (2011) Replacing concentrate with wheat straw treated with urea molasses and ensiled with manure: effects on ruminal characteristics, in situ digestion kinetics and nitrogen metabolism of Nili-Ravi buffalo bulls. Asian Austral J Anim Sci 24(8):1092–1099
Haug RT (1980) Compost engineering: principles and practice. Ann Arbor Science, Michigan
Headon DR, Walsh G (1994) Biological control of pollutants. In: Cole DJA, Wiseman J, Varley MA (eds) Principles of pig science. Nottingham University Press, Leicestershire, pp 196–213
Heinberg R (2007) Peak everything: waking up to the century of declines. Gabriola Island, BC, New Society
Heinberg R, Lerch D (2010) What is sustainability. The post carbon reader, pp 11–19
Herrero M, Grace D, Njuki J et al (2013) The roles of livestock in developing countries. Anim 7:3–18
Hill GB, Baldwin SA, Vinnerås B (2013) Evaluation of Solvita compost stability and maturity tests for assessment of quality of end-products from mixed latrine style compost toilets. Waste Manag 33(7):1602–1606
Hristov AN, Jouany JP (2005) Factors affecting the efficiency of nitrogen utilization in the rumen. In: Pfeffer E, Hristov AN (eds). Nitrogen and phosphorus nutrition of cattle reducing the environmental impact of cattle operations. CABI Publishing, Cambridge
Huang H, Khanna M, Onal H et al (2013) Stacking low carbon policies on the renewable fuels standard: economic and greenhouse gas implications. Energy Policy 56:5–15
Huang K, Xia H, Zhang Y et al (2020) Elimination of antibiotic resistance genes and human pathogenic bacteria by earthworms during vermicomposting of dewatered sludge by metagenomic analysis. Bioresour Techno 297:122451
Hussain N, Das S, Goswami L et al (2018) Intensification of vermitechnology for kitchen vegetable waste and paddy straw employing earthworm consortium: assessment of maturity time, microbial community structure, and economic benefit. J Clean Prod 182:414e26
Ilyas SZ (2006) A case study to bottle the biogas in cylinders as source of power for rural industries development in Pakistan. World Appl Sci J 1(2):127–130

IPCC (2007) Climate Change 2007: impacts, adaptation and vulnerability. Summary for policy makers
Jhingran VG (1991) Fish and fisheries of India. Hindustan Publishing Corporation (India), Delhi
Kamp LM, Forn EB (2016) Ethiopia's emerging domestic biogas sector: current status, bottlenecks and drivers. Renew Sustain Energy Rev 60:475e488
Kapdi SS, Vijay VK, Rajesh SK et al (2006) Upgrading biogas for utilization as a vehicle fuel. As J Ener Env 7(04):387–393
Katuwal H, Bohara AK (2009) Biogas: a promising renewable technology and its impact on rural households in Nepal. Renew Sustain Energy Rev 13:2668e2674
Keener HM, Elwell DL, Monnin MJ (2001) Procedures and equations for sizing of structures and windrows for composting animal mortalities. Appl Eng Agri 16(6):681
Khanal SK, Nindhia TGT, Nitayavardhana S (2019) Biogas from wastes: processes and applications. In: Taherzadeh MJ, Bolton K, Wong J et al (eds) Sustainable resource recovery and zero waste approaches. Elsevier publication, pp 165–174
Kim J, Chulhwan P, Tak-Hyun K et al (2003a) Effects of various pretreatments for enhanced anaerobic digestion with waste activated sludge. J Biosci Bioeng 95(3):271–275
Kim M, Gomec CY, Ahn Y et al (2003b) Hydrolysis and acidogenesis of particulate organic material in mesophilic and thermophilic anaerobic digestion. Environ Technol 24(9):1183–1190
Kugelman IJ, McCarty PL (1965) Cation toxicity and stimulation in anaerobic waste treatment. J Water Pollut Control Fed 37(1):97e116
Kulling DR, Menzi H, Krober TF et al (2001) Emissions of ammonia, nitrous oxide and methane from different types of dairy manure during storage as affected by dietary protein content. J Agric Sci 137(02):235–250. https://doi.org/10.1017/S0021859601001186
Laramee J, Davis J (2013) Economic and environmental impacts of domestic biodigesters: evidence from Arusha, Tanzania. Energy Sustain Dev 17:296e304
Larney FJ, Hao X (2007) A review of composting as a management alternative for beef cattle feedlot manure in southern Alberta, Canada. Bioresour Technol 98:3221–3227
Larney FJ, Buckley KE, Hao X et al (2006) Fresh, stockpiled, and composted beef cattle feedlot manure: nutrient levels and mass balance estimates in Alberta and Manitoba. J Environ Qual 35:1844–1854
Lazcano C, Gómez-Brandón M, Domínguez J (2008) Comparison of the effectiveness of composting and vermicomposting for the biological stabilization of cattle manure. Chemosphere 72(7):1013–1019. https://doi.org/10.1016/j.chemosphere.2008.04.016
Lebuhn M, Liu F, Heuwinkel H et al (2008) Biogas production from mono-digestion of maize silageelong-term process stability and requirements. Water Sci Technol 58(8):1645e1651. https://doi.org/10.2166/wst.2008.495
Li Y, McCrory DF, Powell JM et al (2005) A survey of selected heavy metal concentrations in Wisconsin dairy feeds. J Dairy Sci 88(8):2911–2922
Lim SL, Wu TY, Lim PN (2015) The use of vermicompost in organic farming: overview, effects on soil and economics. J Sci Food Agric 95(6):1143e56
Lin M, Wang A, Ren L, Qiao W et al (2022) Challenges of pathogen inactivation in animal manure through anaerobic digestion: a short review. Bioengineered 13(1):1149–1161
Loh TC, Lee YC, Liang JB et al (2005) Vermicomposting of cattle and goat manures by *Eisenia foetida* and their growth and reproduction performance. Biores Technol 96(1):111–114. https://doi.org/10.1016/j.biortech.2003.03.001
Lores M, Gómez-Brandón M, Pérez-Díaz D et al (2006) Using FAME profiles for the characterization of animal wastes and vermicomposts. Soil Biol Biochem 38(9):2993–2996
Lupindu AM, Olsen JE, Ngowi HA et al (2014) Occurrence and characterization of Shiga toxin-producing Escherichia coli O157:H7 and other non-sorbitol-fermenting *E. coli* in cattle and humans in urban areas of Morogoro, Tanzania. Vector-Borne Zoonotic Dis 14(7):503–510. https://doi.org/10.1089/vbz.2013.1502
MAFF (1996) Agriculture in the United Kingdom. Ministry of agriculture, fisheries and food, United Kingdom, p 110

Manyi-Loh CE, Mamphweli SN, Meyer EL et al (2016) An overview of the control of bacterial pathogens in cattle manure. Int J Environ Res Public Health 13(9):843

Martin M, Eudoxie G (2018) Feedstock composition influences vermicomposting performance of Dichogaster annae relative to *Eudrilus eugeniae* and *Perionyx excavatus*. Environ Sci Pollut Res 1e10

Martínez-Avalos AMM, Mendoza GD, Cobos MA et al (1998) Nutritional evaluation of cattle manure silage with molasses for ruminants. Anim Feed Sci Technol 70(3):257–264. https://doi.org/10.1016/s0377-8401(97)00007-2

McInerney MJ, Bryant MP (1981) Review of methane fermentation fundamentals. In: Wise DL (ed) Fuel gas production from biomass, vol 1. CRC Press, Boca Raton, pp 19–46

McKinley VL, Vestal JR, Eralp AE (1985) Microbial activity in composting. Biocycle 26:47–50

Mengistu MG, Simane B, Eshete G et al (2016) The environmental benefits of domestic biogas technology in rural Ethiopia. Biomass Bioenergy 90:131–138

Miller RM (1996) Biological processes affecting contaminant fate and transport. In: Pepper IL, Gerba CP, Bruddeau ML (eds) Pollution science. Academic Press, San Diego, Calif, pp 77–91

Nadimpalli M, Delarocque-Astagneau E, Love DC et al (2018) Combating global antibiotic resistance: emerging one health concerns in lower-and middle-income countries. Clin Infect Dis 66:963–969

Nair J, Sekiozoic V, Anda M (2006) Effect of pre-composting on vermicomposting of kitchen waste. Biores Technol 97(16):2091–2095. https://doi.org/10.1016/j.biortech.2005.09.020

NAS (1977) Methane generation from human, animal, and agricultural wastes, U.S. National Academy of Sciences, Washington, D.C

Nasiru A, Ibrahim MH, Ismail N (2014) Nitrogen losses in ruminant manure management and use of cattle manure vermicast to improve forage quality. Int J Recycl Org Waste Agric 3(2):1–7

Ndegwa PM, Thompson SA (2001) Integrating composting and vermicomposting in the treatment and bioconversion of biosolids. Biores Technol 76(2):107–112. https://doi.org/10.1016/s0960-8524(00)00104-8

Nettmann E, Bergmann I, Pramschufer S et al (2010) Polyphasic analyses of methanogenic archaea communities in agricultural biogas plants. Appl Environ Microbiol 76(8):2540e2548

Nielsen MK, Kaplan RM, Thamsborg SM et al (2007) Climatic influences on development and survival of free-living stages of equine strongyles: implications for worm control strategies and managing anthelmintic resistance. Vet J 174:23–32

Odagiri M, Schriewer A, Daniels ME et al (2016) Human faecal and pathogen exposure pathways in rural Indian villages and the effect of increased latrine coverage. Water Res 100:232–244. https://doi.org/10.1016/j.watres.2016.05.015

Oenema O (2006) Nitrogen budgets and losses in livestock systems. Int Congr Ser 1293:262–271. https://doi.org/10.1016/j.ics.2006.02.040

Oswald WJ (1991) Introduction to advanced integrated wastewater ponding systems. Water Sci Technol 24:1–7

Oswald WJ (1996) A syllabus on advanced integrated wastewater pond systems. American society of civil engineers course notes on wastewater treatment with advanced integrated wastewater pond systems (AIWPS) and constructed wetlands, p 323

Parawira W, Murto M, Zvauya R et al (2004) Anaerobic batch digestion of solid potato waste alone and in combination with sugar beet leaves. Renew Energy 29(11):1811–1823

Pathak H, Jain N, Bhatia A (2009) Global warming mitigation potential of biogas plants in India. Environ Monit Assess 157:407e418

Pobeheim H, Munk B, Johansson J et al (2010) Influence of trace elements on methane formation from a synthetic model substrate for maize silage. Bioressour Technol 101(2):836–839

Polprasert C (2007) Organic waste recycling: technology and management. IWA publishing

Price EC, Cheremisinoff PN (1981) Biogas: production and utilization. Ann Arbor Science Publishers, Ann Arbor

Purohit P, Kandpal TC (2007) Techno-economics of biogas-based water pumping in India: an attempt to internalize CO_2 emissions mitigation and other economic benefits. Renew Sustain Energy Rev 11(6):1208–1226

Raviv M (2005) Production of high-quality composts for horticultural purposes—a mini-review. HortTechnology 15:52–57

Reinikainen O, Herranen M (2001) Different methods for measuring compost stability and maturity. Acta Hortic 549:99–104

Rodríguez-Eugenio N, McLaughlin M, Pennock D (2018) Soil pollution: a hidden reality. Food and Agriculture Organization. https://www.fao.org/3/I9183EN/i9183en.pdf

Rynk R, Van de KM, Willson GB et al (1992) On-Farm composting. Northeast Regional Agricultural Engineering Service, Ithaca

Sahm H (1981) Biologie der Methan-Bildung (Biology of Methane Formation). Chemie Ingenieur Technik 53(11):854–863

Salsali HR, Parker WJ, Sattar SA (2006) Impact of concentration, temperature, and pH on inactivation of Salmonella spp. by volatile fatty acids in anaerobic digestion. Can J Microbiol 52(4):279–286

Sarwar M, Nisa M, Hassan Z et al (2006) Influence of urea molasses treated wheat straw fermented with cattle manure on chemical composition and feeding value for growing buffalo calves. Livest Sci 105(1–3):151–161. https://doi.org/10.1016/j.livsci.2006.05.021

Sarwar M, Shahzad M, Nisa M et al (2011) Feeding value of urea molasses-treated wheat straw ensiled with fresh cattle manure for growing crossbred cattle calves. Trop Anim Health Prod 43(3):543–548. https://doi.org/10.1007/s11250-010-9745-5

Schattauer A, Abdoun E, Weiland P et al (2011) Abundance of trace elements in demonstration biogas plants. Biosys Eng 108(1):57–65

Schnurer A, Jarvis A (2010) Microbiological handbook for biogas plants. Swed Waste Manag U 2009:1–74

Schriewer A, Odagiri M, Wuertz S et al (2015) Human and animal faecal contamination of community water sources, stored drinking water and hands in rural India measured with validated microbial source tracking assays. Am J Trop Med Hyg 93(3):509–516. https://doi.org/10.4269/ajtmh.14-0824

Seyfried CF, Bode H, Austermann-Haun U et al (1990) Anaerobe Verfahren zur Behandlung von Industrieabwassern. Korrespondenz Abwasser 37(10):1247–1251

Sharma K, Garg VK (2019) Recycling of lignocellulosic waste as vermicompost using earthworm *Eisenia* fetida. Environ Sci Pollut Res 26(14):14024–14035

Singh A, Rashid M (2017) Impact of animal waste on environment, its managemental strategies and treatment protocols to reduce environmental contamination. Vet Sci Res J 8(1):12

Sinsuw AAE, Wuisang CE, Chu CY (2021) Assessment of environmental and social impacts on rural community by two-stage biogas production pilot plant from slaughterhouse wastewater. J Water Process Eng 40:101796

Sommer SG, Hutchings NJ (2001) Ammonia emission from field applied manure and its reduction—invited paper. Eur J Agron 15(1):1–15. https://doi.org/10.1016/s1161-0301(01)00112-5

Sorathiya LM, Fulsoundar AB, Tyagi KK et al (2014) Eco-friendly and modern methods of livestock waste recycling for enhancing farm profitability. Int J Recycl Org Waste Agric 3(1):1–7

Spolaore P, Joannis-Cassan C, Duran E et al (2006) Commercial applications of microalgae. J Biosci Bioeng 101:87–96

Sturm BSM, Lamer SL (2011) An energy evaluation of coupling nutrient removal from wastewater with algal biomass production. Appl Energy 88:3499–3506

Subedi M, Matthews RB, Pogson M (2014) Can biogas digesters help to reduce deforestation in Africa? Biomass Bioenergy 70:87e98

Suthar S (2010) Recycling of agro-industrial sludge through vermitechnology. Ecol Eng 36(8):1028–1036

Tamminga S (1992) Nutrition management of dairy cows as a contribution to pollution control. J Dairy Sci 75:345–357

Tamminga S, Verstegen MWA (1992) Implications of nutrition of animals on environmental pollution. In: Garnsworthy PC, Haresign W, Cole DJA (eds) Recent advances in animal nutrition. Nottingham University Press, Nottingham, pp 342–351

The Natural Step. the four systems conditions. www.thenaturalstep.Org

Varma VS, Yadav J, Das S et al (2015) Potential of waste carbide sludge addition on earthworm growth and organic matter degradation during vermicomposting of agricultural wastes. Ecol Eng 83:90e5

Velasco-Velasco J, Parkinson R, Kuri V (2011) Ammonia emissions during vermicomposting of sheep manure. Biores Technol 102(23):10959–10964. https://doi.org/10.1016/j.biortech.2011.09.047

Victor P (2008) Managing without growth: slower by design, not disaster. Edward Elgar Publishing, Cheltenham, UK

Vijay VK (2011) Biogas enrichment and bottling technology for vehicular use. Biogas Forum 1(1):12–15

Weiland P (2006) Anforderungen an Pflanzen seitens des Biogasanlagenbetreibers. In: Thuringer Bioenergietag; Schriftenreihe der TLL. pp 26–32

Woesttyne MV, Verstrate W (1995) Biotechnology in the treatment of animal manure. In: Biotechnology in animal feeds and animal feeding. Wiley-VCH Verlag GmbH, pp 311–327. https://doi.org/10.1002/9783527615353.ch16

World Commission on Environment and Development (1987) Our common future. Oxford University Press, USA. http://www.un-documents.net/wced-ocf.htm

Yadav A, Garg VK (2011) Recycling of organic wastes by employing *Eisenia fetida*. Bioresour Technol 102(3):2874–2880

Yu L, Yaoqiu K, Ningsheng H et al (2008) Popularizing household-scale biogas digesters for rural sustainable energy development and greenhouse gas mitigation. Renew Energy 33(9):2027–2035

Yu X, Liu Y, Wang Y et al (2020) Role of bioengineering and laborers in integration of farmland resources toward to improve dimension of sustainable agriculture in China. Bioengineered 11(1):559–571

Yu H-Q, Fang HHP (2002) Acidogenesis of dairy wastewater at various pH levels. Water Sci Technol 45(10):201e6

Yuvaraj A, Karmegam N, Thangaraj R (2018) Vermistabilization of paper mill sludge by an epigeic earthworm *Perionyx excavatus*: mitigation strategies for sustainable environmental management. Ecol Eng 120:187–197

Zhang LX, Wang CB, Song B (2013) Carbon emission reduction potential of a typical household biogas system in rural China. J Clean Prod 47:415e421

Zhang C, Xu Y (2020) Economic analysis of large-scale farm biogas power generation system considering environmental benefits based on LCA: a case study in China. J Clean Prod 258:120985

Zhou W, Min M, Li Y et al (2012) A hetero-photoautotrophic two-stage cultivation process to improve wastewater nutrient removal and enhance algal lipid accumulation. Bioresour Technol 110:448–455

Chapter 13
Application of Farmyard Manure in Sustainable Utilization of Animal Wastes to Reclaim Salt Degraded Lands

Muhammad Talha Bin Yousaf, Muhammad Farrakh Nawaz, Sadaf Gul, Muhammad Sajjad Haider, Irfan Ahmed, Ghulam Yasin, and Muhammad Zahid Farooq

Abstract Animal manures are the central source of providing organic matter. All resources using animal wastes are not utilized adequately. Farmyard manure, poultry manure and slurry are commonly used in agriculture as fertilizers and soil amendments. These soil amendments not only improve and sustain soil fertility but also play a vital role in increasing soil productivity. Previously, animal wastes were converted into composts, but composts produce unpleasant odor. In recent years, biochar prepared from manure has got much recognition. In this chapter, we will briefly describe the direct use of animal wastes in agriculture and conversion of farmyard manure into biochar for their sustainable use to ameliorate degraded lands and enhance plants productivity. This chapter mainly focuses on the use of farmyard manure to reclaim saline soils.

M. T. B. Yousaf · M. S. Haider
Department of Forestry, University of Sargodha, Sargodha, Pakistan

M. T. B. Yousaf · M. F. Nawaz (✉) · I. Ahmed
Department of Forestry and Range Management, University of Agriculture, Faisalabad, Pakistan
e-mail: kf_uaf@yahoo.com

S. Gul
Department of Botany, University of Karachi, Karachi, Pakistan

G. Yasin
Department of Forestry & Range Management, Bahauddin Zakariya University, Multan, Pakistan

M. Z. Farooq
Department of Animal Sciences, Jhang-Campus, University of Veterinary and Animal Sciences Lahore, Lahore, Pakistan

M. F. Nawaz
Institute of Environmental Studies, University of Karachi, Karachi, Pakistan

M. Arshad (ed.), *Climate Changes Mitigation and Sustainable Bioenergy Harvest Through Animal Waste*, https://doi.org/10.1007/978-3-031-26224-1_13

Graphical Abstract

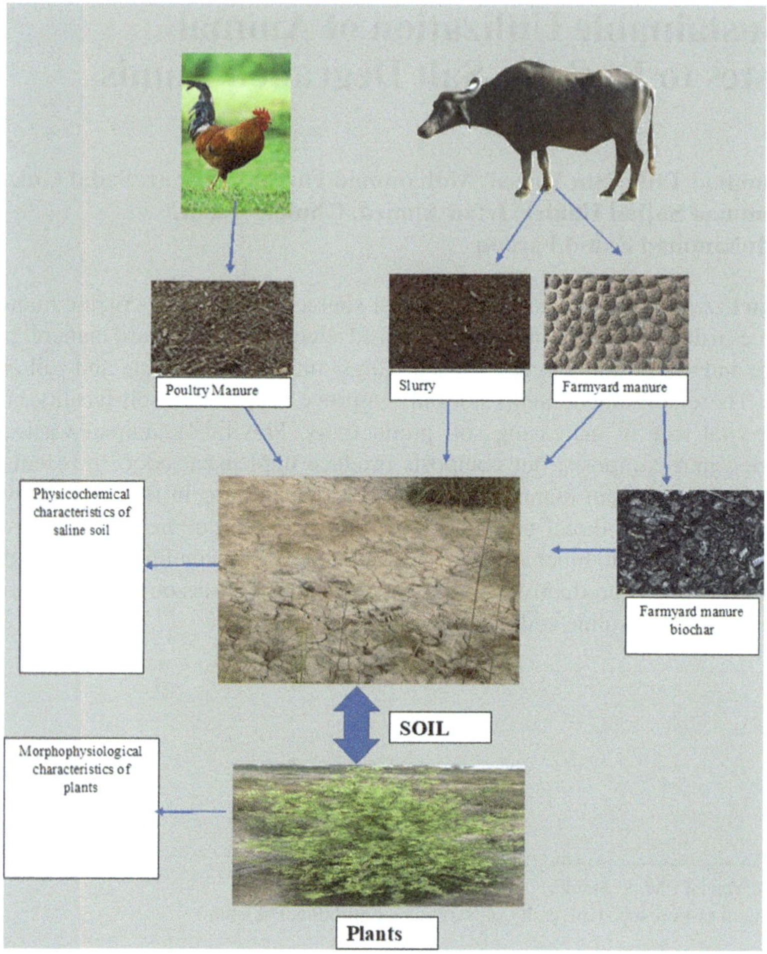

Application of farmyard manure in sustainable utilization of animal wastes to reclaim salt degraded lands

13.1 Introduction

13.1.1 Significance of Animal Wastes

Animal manures are the central source of providing organic matter. A lot of different types of animals are present worldwide but the main source of animal manures are cattle, chicken, sheep, duck and pig. In the whole world, almost 7×10^9 Mg yr^{-1} of animal manure is produced by different sources (Khan et al. 2017; Reed et al. 2017). Due to greater organic matter content, these manures are widely used for the restoration of degraded soils (Wang et al. 2014; Yousaf et al. 2021a).

13.1.2 Distribution of Animals

In Pakistan, buffalos are present in abundant quantity but in the whole world cows are the major farm animals. Almost 4.2 million of dung is produced as a waste material by world 1.4 billion cows annually (Reddy et al. 2014). A number of methods and techniques are used to solve the dumping issue of this waste material. One of them is to convert this waste into fertilizer through composting or to use it as fuel for the production of biogas (Reddy et al. 2014; Yokoyama et al. 2007). Proteins, cellulose and hemicellulose are existent in greater quantity in cow dung. Dung is also rich in other essential nutrients. So, dung can be used on sustainable basis to improve physicochemical characteristics of soils (Feng et al. 2018; Yusefi et al. 2018).

13.1.3 Different Types of Animal Wastes

Manures, mulches, cover crops and composts are some of the common organic amendments which provides nutrients to various crops by recycling of these nutrients with the help of microbial activity (Mekuria and Noble 2013; Tully and McAskill 2019). Farmyard manure, poultry manure and slurry are commonly used in agriculture as fertilizers and soil amendments. Previously, animal wastes were converted into composts, but composts produce unpleasant odor. In this chapter, we will study about the direct use of animal wastes in agriculture and conversion of animal wastes into biochar for their sustainable use.

13.2 Use of Unprocessed Animal Wastes as Soil Amendments

13.2.1 Use of Farmyard Manure as Soil Amendment

The application of farmyard manure (FYM) to ameliorate degraded soil is not a new practice. Degraded soils have been treated to improve their fertility with farmyard manure for centuries. Studies have revealed that farmyard manure as fertilizer not only ameliorates the deleterious impacts of salinity on crops but also results in greater biomass production and yield (Nawab et al. 2018). FYM has provided countless benefits like improvement in soil structure, crop yield, microbial populations and soil organic matter (Kundu et al. 2007). Agroforestry trees were grown under saline conditions. Salinity negatively affected all the growth parameters. Figure 13.1 depicts the results when trees were grown without any use of fertilizer.

When applied with different organic amendments, farmyard manure showed best results for the growth of *Eucalyptus camaldulensis*, (Yousaf et al. 2021a) *Dalbergia sissoo* and *Vachellia nilotica* also showed better results in growth with FYM (Yousaf et al. 2022). Figure 13.2 depicts the effect of farmyard manure on the growth of *Eucalyptus camaldulensis, Vachellia nilotica and Dalbergia sissoo*. Clear difference can be observed as compared to Fig. 13.1.

Farmyard manure was applied in soils where concentration of salts in the soil was 3.3 g kg^1. Organic manures enhanced the enzymatic activities and increased respiration rate of soil (Liang et al. 2003). Farmyard manure (FYM), due to the presence of essential growth nutrients and higher amount of nitrogen (N), can enhance plant growth and biomass to a significant level. Nitrogen is good in enhancing buffering capacity of amended soils. Mobility of toxic heavy metals farmyard manure amended soils was also reduced by improving the binding capacity of toxic heavy metals (Nawab et al. 2018). A significant decrease in sodium adsorption ratio (SAR) was found when saline soil was treated with farmyard manure, sulfuric acid and gypsum. This study supports that farmyard manure has potential to decrease SAR and remediate salt affected soil (Hussain et al. 2001).

Animal wastes such as manure, compost and press mud were found to be very efficient in increasing growth of plants as they pose very positive effect on various properties of saline sodic and sodic soils (Katkar et al. 2019). Rapeseed meal, pig

Fig. 13.1 Effect of salinity on trees growth

Fig. 13.2 Effect of FYM on agroforestry tree species

manure at the rate of 50 g kg^{-1} were used as organic amendments to remediate coastal saline soils and studied the wheat growth under the effect of applied manures on that saline soil. Biomass of wheat was increased by 48.4% in comparison with control. Moreover, soil pH was decreased to 7. 06 from 8.29 (Yang et al. 2018). Ovine manure, solid waste and sewage sludge were used at the rate of 24 Mg ha^{-1}. The results showed that the amount soil organic carbon increased significantly under these amendments (Albiach et al. 2001).

The combination of rice straw and pig manure was discovered to be helpful in remediating toxic effect of salts in saline soils (Liang et al. 2005). Exchangeable sodium percentage was reduced by 50% by the application of composted and non-composted manure (Tejada et al. 2006). So, we can conclude from this section that farmyard manure has the ability to ameliorate salt affected and heavy metal affected degraded lands (Table 13.1).

13.2.2 Use of Poultry Manure as Soil Amendment

Poultry manure has showed encouraging outcomes in boosting vegetative cover on saline soils (Tejada et al. 2006).

It is reported that hazardous effects of salinity were reduced in wheat and rice by using organic fertilizers and chicken manure. There was increase in uptake of nutrients and K/Na ration while decrease in Na ion accumulation (Al-Erwy et al. 2016). Poultry manure and sheep manure were found to be very beneficial in reducing

Table 13.1 Review table showing benefits of farmyard manure

FYM showed best results for the growth of *Eucalyptus camaldulensis* among different organic amendments under at 20.5 dS/m EC	Yousaf et al. (2021a)
FYM showed best results for the growth of *Vachellia nilotica* and *Dalbergia sissoo* among different organic amendments under at 20.5 dS/m EC	Yousaf et al. (2022)
FYM enhanced enzymatic activities and respiration rate	Liang et al. (2003)
Farmyard manure (FYM) increased nitrogen amount in soil	Liu et al. (2007)
FYM reduced heavy metals mobility	Nawab et al. (2018)
FYM decreased SAR of saline soil	Hussain et al. (2001)
Pig manure increased wheat growth by 48.4%	Yang et al. (2018)
Ovine manure significantly increased organic carbon	Albiach et al. (2001)
Straw and pig manure reduced salt toxicity	Liang et al. (2005)
FYM reduced ESP up to 50%	Tejada et al. (2006)

ESP and increasing CEC of soil. The addition of these amendments resulted in lowering the level of sodification in treated soils (Jalali and Ranjbar 2009).

Poultry manure enhanced the growth of agroforestry tree species (*Eucalyptus camaldulensis, Vachellia nilotica and Dalbergia sissoo*) grown under the salt stress (Yousaf et al. 2021a, 2022). Figure 13.3 shows the results of poultry manure on the growth of above mentioned tree species. You can see the clear difference between Figs. 13.2 and 13.3.

Poultry manure and cotton gin crushed compost were applied to a saline soil at the rate of 5 t ha^{-1} and 10 t ha^{-1}. After one year, the plots which were treated with higher poultry manure dose spontaneous vegetation started to appear. After five years, the results showed that the plots which were treated with organic amendments were covered with almost 80% vegetation cover while there was only 8% vegetation cover on untreated saline soils. These amendments also improved soil characteristics (Tejada et al. 2006).

13.2.3 Use of Slurry in Agriculture

Biogas slurry is a by-product of biogas plant. In biogas plant, methane rich gas is produced by using dung or other biomass. The material is digested and slurry is obtained as by-product (Myrna et al. 2019; Xi et al. 2020). The number of biogas plants have been increased in all Asian countries. It is a huge concern to dispose of that amount of slurry (Pandyaswargo et al. 2019). But by using it as an organic amendments, slurry is used to improve physicochemical characteristics of degraded soils and enhance productivity of plants. It is beneficial to grow plants in saline soils (Jabeen and Ahmad 2017).

Fig. 13.3 Effect of PM on selected tree species

Biogas slurry and vermicompost were tested on the growth of ginger plants. It was found that growth of ginger enhanced while toxic effect of Na was reduced. Fresh and dry biomass along with chlorophyll, proteins and carbohydrates was enhanced by using these amendments (Ahmad et al. 2009). Biogas slurry and vermicompost were used as organic amendments to analyze the growth of sunflower. Both organic amendments increased the growth of sunflower under saline water irrigation (EC = 8.6 dS/m). Moreover, organic manure increased K^+ contents and decreased Na^+ contents in leaves and seed coat of sunflower which helps in developing the tolerance to saline conditions in plants (Jabeen and Ahmad 2017).

A field study was performed to check the outcome of biogas slurry and fly ash on the growth and yield of wheat (*Triticum aestivum*). These amendments improved physical and biological properties of soil and also increased the wheat yield (Garg et al. 2005). Biogas slurry increased maize yield production and improved soil health under salt stress (Ahmad et al. 2014). Slurry has showed positive results in afforestation of degraded lands project under various organic amendments (Yousaf et al. 2021a, 2022). Figure 13.4 shows the beneficial effects of slurry on selected agroforestry tree species.

Fig. 13.4 Effect of slurry on agroforestry tree species

13.2.4 Soil Characteristics

Animal wastes significantly improve all soil characteristics. The effect of these animal wastes when used as fertilizers on physicochemical characteristics of soil are given in Table 13.2. Three agroforestry tree species were grown in the soil severely affected by salt toxicity having an EC of 20.5 dS/m. The results of the study revealed that EC, TSS and SAR of saline soil decreased while organic matter percentage and moisture percentage increased.

13.2.5 Limitations of Using Unprocessed Organic Amendments

The above literature shows that organic amendments have been used from centuries to ameliorate degraded soils. But there is one limitation of using these organic amendments. The application rate of these amendments should be high and these should be applied continuously at regular time intervals. In arid and semi-arid regions, temperatures are usually very high. High temperature results in increased decomposition rate of organic matter. Nutrients are released quickly and take part in higher growth but for long-term benefits organic amendments are needed to be used on regular intervals (Alessandro and Nyman 2017; Gregorich et al. 2017). Generally, manures are converted into composts, and then composts are used as organic fertilizer. During the process of composting emission of greenhouse gases takes place and the odor which is released during process of composting is very bad (Awasthi et al. 2019; Idrees et al. 2016). Conversion of manure to biochar instead of compost has gained importance. Biochar is prepared from lot of different types of feedstock material. Animal wastes have been used as feedstock material to prepare biochar (Kosheleva et al. 2019).

13.3 Use of Animal Wastes as Biochar

13.3.1 Origin and Production Process

Biochar is one of the well-recognized organic amendments. The concept of using biochar to ameliorate degraded lands is relatively new as compared to farmyard manure, poultry manure and slurry. Production of biochar and its use to ameliorate degraded soil along with enhancement in crop growth and yield is relatively a new concept but origin of biochar is very old. Its origin is centuries old. It is assumed that

Table 13.2 Effect of different types of organic amendments and selected agroforestry trees species on post-harvest soil characteristics

	Control			FYM		
	Sp.1	Sp.2	Sp.3	Sp.1	Sp.2	Sp.3
pH	7.73a ± 0.04	7.77a ± 0.04	7.77a ± 0.04	7.93a ± 0.04	7.87a ± 0.04	7.88a ± 0.04
EC (dS/m)	8.29b ± 0.18	8.71b ± 0.18	12.41a ± 0.18	3.77e ± 0.18	3.94e ± 0.18	8.42b ± 0.18
TSS ($mmol_c$/L)	82.90b ± 1.81	87.10b ± 1.81	124.10a ± 1.81	32.73e ± 1.81	32.70e ± 1.81	84.20b ± 1.81
CO_3^{2-} ($mmol_c$/L)	(−)	(−)	(−)	(−)	(−)	(−)
HCO_3^- ($mmol_c$/L)	3.20 cd ± 0.24	4.44ab ± 0.24	4.51a ± 0.24	2.82d ± 0.24	3.23 cd ± 0.24	3.54bcd ± 0.24
Cl^- ($mmol_c$/L)	40.00c ± 1.63	51.00b ± 1.63	60.00a ± 1.63	16.00i ± 1.63	19.33hi ± 1.63	27.33efg ± 1.63
$Ca^{2+} + Mg^{2+}$ ($mmol_c$/L)	3.73b ± 0.75	3.73b ± 0.75	3.83b ± 0.75	6.67a ± 0.75	7.53a ± 0.75	7.57a ± 0.75
Na^+ ($mmol_c$/L)	48.00b ± 1.27	51.00b ± 1.27	62.33a ± 1.27	17.00 h ± 1.27	18.33gh ± 1.27	35.00d ± 1.27
SAR	35.23b ± 1.38	37.41b ± 1.38	45.50a ± 1.38	9.37f ± 1.38	9.50f ± 1.38	18.08 cd ± 1.38
K^+	6.60b ± 0.48	6.53b ± 0.48	6.03b ± 0.48	10.87a ± 0.48	11.33a ± 0.48	11.40a ± 0.48
OM (%)	0.51d ± 0.01	0.51d ± 0.01	0.50d ± 0.01	0.70a ± 0.01	0.69ab ± 0.01	0.66abc ± 0.01
TOC (%)	0.30d ± 0.01	0.29d ± 0.01	0.29d ± 0.01	0.41a ± 0.01	0.40ab ± 0.01	0.38abc ± 0.01

(continued)

Table 13.2 (continued)

	Control			FYM		
	Sp.1	Sp.2	Sp.3	Sp.1	Sp.2	Sp.3
Saturation (%)	34.19c ± 0.44	0.028d ± 0.44	34.53c ± 0.44	39.40 a ± 0.44	0.041ab ± 0.44	38.83ab ± 0.44
	PM			SL		
	Sp.1	Sp.2	Sp.3	Sp.1	Sp.2	Sp.3
pH	7.94a ± 0.04	7.87a ± 0.04	7.89a ± 0.04	7.82a ± 0.04	7.85a ± 0.04	7.94a ± 0.04
EC (dS/m)	6.51c ± 0.18	6.83c ± 0.18	8.75b ± 0.18	3.88de ± 0.18	4.13d ± 0.18	8.15b ± 0.18
TSS ($mmol_c/L$)	65.07c ± 1.81	68.30c ± 1.81	87.50b ± 1.81	37.50de ± 1.81	41.33d ± 1.81	81.47b ± 1.81
CO_3^{2-} ($mmol_c/L$)	(–)	(–)	(–)	(–)	(–)	(–)
HCO_3^- ($mmol_c/L$)	3.03 cd ± 0.24	3.61abcd ± 0.24	3.73abc ± 0.24	2.82d ± 0.24	3.23 cd ± 0.24	3.83abc ± 0.24
Cl^- ($mmol_c/L$)	23.00fgh ± 1.63	31.33de ± 1.63	34.67 cd ± 1.63	19.67hi ± 1.63	22.67gh ± 1.63	28.67ef ± 1.63
$Ca^{2+} + Mg^{2+}$ ($mmol_c/L$)	7.37a ± 0.75	7.17a0.75	6.67a ± 0.75	6.07ab ± 0.75	6.40ab ± 0.75	6.67a ± 0.75
Na^+ ($mmol_c/L$)	22.67 fg ± 1.27	28.00ef ± 1.27	40.33c ± 1.27	19.33gh ± 1.27	25.00e ± 1.27	37.33 cd ± 1.27
SAR	11.82ef ± 1.38	14.99de ± 1.38	20.63c ± 1.38	11.09ef ± 1.38	13.98def ± 1.38	22.10c ± 1.38
K^+	11.10a ± 0.48	11.30a ± 0.48	11.80a ± 0.48	11.20a ± 0.48	11.83a ± 0.48	11.50a ± 0.48
OM (%)	0.66abc ± 0.01	0.64bc ± 0.01	0.63c ± 0.01	0.67abc ± 0.01	0.62c ± 0.01	0.63c ± 0.01
TOC (%)	0.39abc ± 0.01	0.37bc ± 0.01	0.37c ± 0.01	0.39abc ± 0.01	0.36c ± 0.01	0.37c ± 0.01
Saturation (%)	37.27b ± 0.44	0.038bc ± 0.44	37.65b ± 0.44	38.40ab ± 0.44	0.037c ± 0.44	38.49ab ± 0.44

* Mean value ($n = 3$) ± Standard error different alphabetical letters showed significant ($P \leq 0.05$) difference among treatments and vice versa

Sp. 1 = *Eucalyptus camaldulensis*, Sp. 2 = *Vachellia nilotica*, Sp. 3 = *Dalbergia sissoo* it was developed first in the Amazon region by Amerindian population (Lehmann et al. 2011; McBeath et al. 2014).

Materials or plant biomass containing greater amount carbon are used to produce biochar. Agricultural residues are generally used as source material for biochar production. Biochar is produced under limited supply of oxygen (Luo et al. 2017). If it is produced when oxygen is not present at all, the process is called as pyrolysis but if oxygen is present is present to some extent while preparing biochar, the procedure is called as gasification (Lehmann et al. 2011). Pyrolysis temperature for the production of biochar deviates from 200 to 1000 °C. Biochar can be produced through slow pyrolysis or fast pyrolysis but preferred method is slow pyrolysis (Lian and Xing 2017).

13.3.2 Significance of Biochar

Biochar has the ability to convert carbon which is being captured by living plants into solid form of char. In this way, biochar acts as a very important tool to capture and sequester carbon. Carbon sequestration is very helpful to fight against global climate change. Production of biochar has helped in converting waste material into very useful organic fertilizers. Plants yield and growth has increased to a significant extent by using various kinds of biochar. It has been found to be very helpful in ameliorating soils affected by different abiotic stresses. These abovementioned and numerous other benefits make biochar as one of the most important discoveries in history of mankind (Awasthi et al. 2019; Wan et al. 2018; Wang et al. 2014; Yousaf et al. 2021b; Yuan et al. 2019).

In recent years, biochar has got much attention across the globe. Apart from these benefits biochar has capacity to immobilize the pollutants of organic and inorganic origin (Shahid et al. 2018; Shakya and Agarwal 2020). Applying biochar under salt and drought stress has the capability to improve growth and increases biomass as well as yield in plants (Kim et al. 2016; Xiao et al. 2020). In addition to that, biochar improves physicochemical characteristics of soil significantly (Getahun et al. 2020; Lim et al. 2016).

13.3.3 Feedstock Material

A lot of different varieties of feedstocks can be used for biochar production. Physical and chemical properties of biochar differ according to feedstocks used, pyrolysis temperature and method of pyrolysis (Weber and Quicker 2018). Although properties of biochar are dependent on number of different factors, generally biochar shows very promising results in ameliorating degraded soils (Dai et al. 2017; Xiao et al. 2018). Feedstock is also used for the preparation of biochar. Biochar characteristics are determined by the feedstock material. Manure is available in large amount. Biochar prepared from manures is found to be very rich in nutrients. If lignocellulose is the main component of feedstock such as wood to prepare biochar, then biochar produced will be very high in carbon contents. These different characteristics in various situations permit biochars to be selectively picked as particular elemental fertilizers (Ojeda et al. 2015; Zhang et al. 2020).

If soil is deficient in certain nutrients and feedstock is not available for that nutrient in abundant quantity, biochar can also be engineered to fulfill that requirement (Luo et al. 2017). Farmyard manure, biochar and mineral nitrogen were used as amendments to treat problematic soil. The interaction effect of these three amendments increased available phosphorus, potassium and nitrogen after six months and affected physicochemical characteristics of soil (Sarfraz et al. 2017).

13.3.4 Properties of Biochar

One of the very significant properties of biochar is that it releases nutrients in very slow fashion. It is very beneficial function. There can be more than one reasons for this slow release. Unique structure of biochar is the primary cause of that slow release. Pore size distribution in biochar is generally very high, and other networks in biochar structure create hindrance for easy release of nutrients by wrapping them (nutrients) physically. This slow release of nutrients is due to chemical bonding of carbon molecules and presence of various functional groups in biochar in a great amount (Purakayastha et al. 2019; Xiao et al. 2018).

13.3.5 Biochar and Salinity

In saline soils, biochar can reduce the toxicity by lowering the concentration of sodium ions. Initially, biochar acts as barrier and reduces the sodium (Na^{+1}), uptakes and increases potassium (K^{+}) uptake. This results in less availability of sodium ion to plants, thus reducing its toxic effects (Ali et al. 2017; Yousaf et al. 2022). The leaching process of salts increased to many folds under biochar. Higher leaching of salts results in less concentration of salts in root zone, ultimately enhancing the plant growth and yield (Yu et al. 2019).

Biochar helps in ameliorating salt degraded lands by improving soil quality and production crop production. Biochar is helpful in improving stomatal conductance and other physiological characteristics of leaves (Akhtar et al. 2015; Dai et al. 2017; Yu et al. 2019).

Biochars can reduce pH values of saline soils. Higher pH value is not good for soil health (Liao et al. 2019; Yousaf et al. 2021b). CEC of biochars is generally high. In amended soils, biochar enhances the cations (potassium, calcium and magnesium) uptake chances to manyfold and improve the soil health and plant growth (Lashari et al. 2015; Wu et al. 2019). Sodium in the soil can be replaced with several important ions like potassium, magnesium and calcium under the biochar application (Rostamian et al. 2015; Yousaf et al. 2022; Zheng et al. 2018). It provides habitat for soil microorganisms and also improves the physical properties of soils, such as porosity, surface area and water retention capacity (Yu et al. 2017; Zheng et al. 2018).

Molecules of other organic amendments are easily degradable as compared to biochar. Soil aggregate stabilization for a very long term can be obtained by using biochar (Bhadha et al. 2017; Mia et al. 2017). It can increase microporosity and field available water. It also reduces the bulk density of saline sodic bauxite wastes (Reed et al. 2017). Moreover, it is beneficial in minimizing the effects of salinity by means of sorption. In addition to that, when biochar is applied in reforestation programs of degraded soils, the species diversity enhanced to a great extent (Drake et al. 2015). Biochar reduced activities of antioxidant enzyme activities in bean seedlings which in turns reduced oxidative stress and enhanced growth. Stomatal conductance was also increased (Farhangi-Abriz and Torabian 2017).

13.3.6 Farmyard Manure Biochar

The conversion of dung into biochar has gained significant importance in recent times. Biochar was prepared from cow dung and its effect was checked on the growth of lettuce plants. Growth, yield and phosphorus concentration was significantly increased by using engineered biochar. Capacity of engineered biochar to adsorb phosphorus reached up to 345 mg/g (Chen et al. 2018). Farmyard manure and poultry manure were used to produce biochar and their effectiveness was tested to adsorb toxic heavy metal from water. Both of these biochars showed positive results in adsorbing cadmium ions (Idrees et al. 2016).

In another study, cow dung and its biochar were used at a rate of 5 t/ha to check their effect on leafy vegetable kalmi (*Ipomoea aquatica*). Results showed that the CEC of biochar treated soil was more (Niazi et al. 2018). Biochar was prepared by using cow dung as feedstock material which showed the ability to adsorb perchlorate from

Fig. 13.5 Effect of FYMB on growth of agroforestry tree species

water and remove it from water (Wan et al. 2018). *Albizia lebbeck* commonly known as siris was treated with two levels of biochar under cadmium stress. It is reported that biochar was helpful in combating cadmium stress and enhancing growth of *A. lebbeck* (Yousaf et al. 2019).

Farmyard manure biochar was applied at different rates 3, 6 and 9% to study its impact on the growth of *Eucalyptus camaldulensis*, *Dalbergia sissoo* and *Vachellia nilotica*. All levels of biochar showed better results for growth with 6% biochar giving the best results for growth and photosynthetic parameters (Yousaf et al. 2022). In another various types of biochar prepared from animal wastes, agricultural wastes and tree wastes were used to reclaim degraded lands and check their effect on the growth and photosynthetic parameters of various agroforestry tree species under saline stress. Although all types of biochar exhibited improved results as compared to unamended soil, farmyard manure biochar showed the best results. Figure 13.5 shows that FYMB showed best results for all three species among biochars prepared from different feedstocks (Fig. 13.6).

13.4 Conclusion

In this chapter, from the above discussion, we can conclude that animal wastes are not waste material at all. They can be converted into very useful products like biochar. These wastes are very good source of organic matter. Organic agriculture is directly linked and heavily dependent on these manures. This material can be used to enhance productivity and reclaim degraded lands especially salt affected soils. These materials can also be used in various afforestation projects.

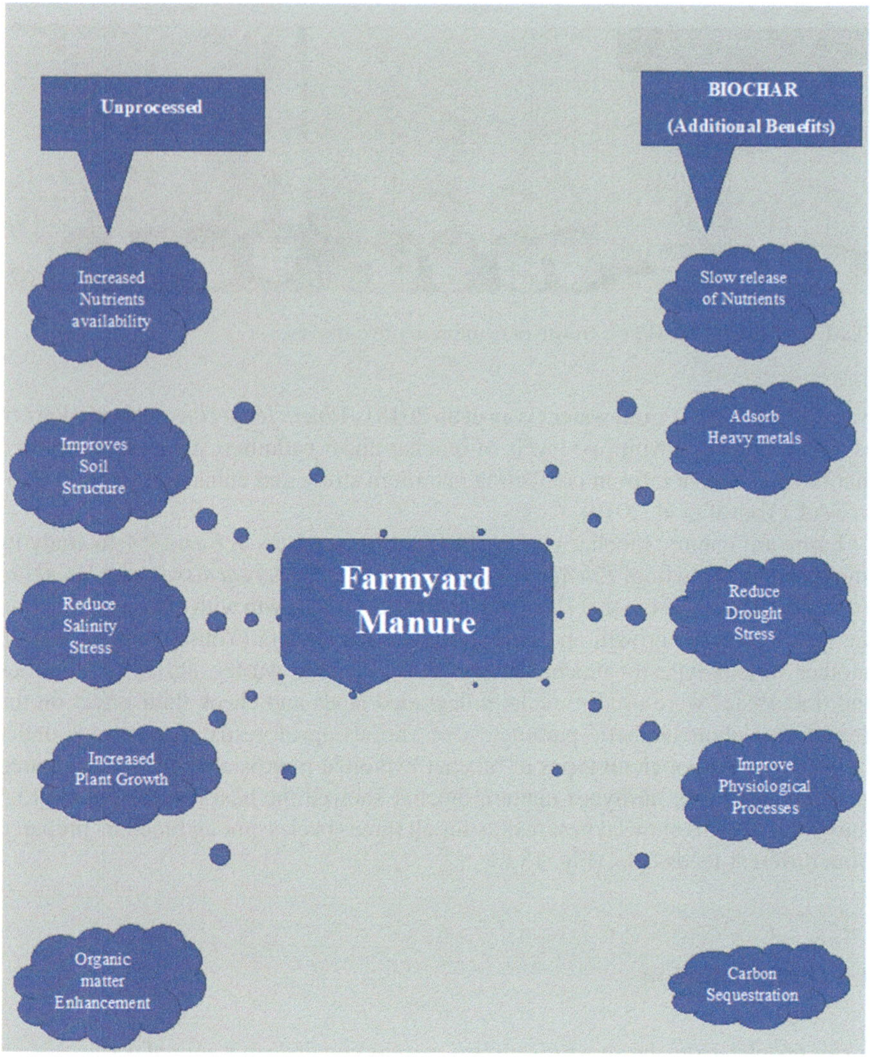

Fig. 13.6 Benefits of farmyard manure

References

Ahmad M, Ahmad Zahir Z, Jamil M, Nazli F, Latif M, Fakhar-U-Zaman AM (2014) Integrated use of plant growth promoting rhizobacteria, biogas slurry and chemical nitrogen for sustainable production of maize under salt-affected conditions. Pakistan J Bot 46:375–382

Ahmad R, Azeem M, Ahmed N (2009) Productivity of ginger (Zingiber officinale) by amendment of vermicompost and biogas slurry in saline soils. Pakistan J Bot 41:3107–3116

Akhtar SS, Andersen MN, Liu F (2015) Biochar mitigates salinity stress in potato. J Agron Crop Sci 201:368–378. https://doi.org/10.1111/jac.12132

Al-Erwy AS, Bafeel SO, Al-Toukhy A (2016) Effect of chemical, organic and bio fertilizers on germination, growth and yield of wheat (Triticum aestivum. L) plants irrigated with sea water. Agric Biol J North Am 7:121–3. https://doi.org/10.5251/abjna.2016.7.3.121.133

Albiach R, Canet R, Pomares F, Ingelmo F (2001) Organic matter components and aggregate stability after the application of different amendments to a horticultural soil. Bioresour Technol 76:125–129. https://doi.org/10.1016/S0960-8524(00)00090-0

Alessandro O, Nyman P (2017) Aridity indices predict organic matter decomposition and comminution processes at landscape scale. Ecol Indic 78:531–540. https://doi.org/10.1016/j.ecolind.2017.03.049

Ali S, Rizwan M, Qayyum MF, Ok YS, Ibrahim M, Riaz M et al (2017) Biochar soil amendment on alleviation of drought and salt stress in plants: a critical review. Environ Sci Pollut Res 24:12700–12712. https://doi.org/10.1007/s11356-017-8904-x

Awasthi MK, Sarsaiya S, Wainaina S, Rajendran K, Kumar S, Quan W et al (2019) A critical review of organic manure biorefinery models toward sustainable circular bioeconomy: technological challenges, advancements, innovations, and future perspectives. Renew Sustain Energy Rev 111:115–131. https://doi.org/10.1016/j.rser.2019.05.017

Bhadha JH, Capasso JM, Khatiwada R, Swanson S (2017) Raising soil organic matter content to improve water holding capacity. Uf/Ifas 1–5

Chen Q, Qin J, Sun P, Cheng Z, Shen G (2018) Cow dung-derived engineered biochar for reclaiming phosphate from aqueous solution and its validation as slow-release fertilizer in soil-crop system. J Clean Prod 172:2009–2018. https://doi.org/10.1016/j.jclepro.2017.11.224

Dai Z, Zhang X, Tang C, Muhammad N, Wu J, Brookes PC et al (2017) Potential role of biochars in decreasing soil acidification—a critical review. Sci Total Environ 581–582:601–611. https://doi.org/10.1016/j.scitotenv.2016.12.169

Drake JA, Carrucan A, Jackson WR, Cavagnaro TR, Patti AF (2015) Biochar application during reforestation alters species present and soil chemistry. Sci Total Environ 514:359–365. https://doi.org/10.1016/j.scitotenv.2015.02.012

Farhangi-Abriz S, Torabian S (2017) Antioxidant enzyme and osmotic adjustment changes in bean seedlings as affected by biochar under salt stress. Ecotoxicol Environ Saf 137:64–70. https://doi.org/10.1016/j.ecoenv.2016.11.029

Feng H, Ge Z, Chen W, Wang J, Shen D, Jia Y et al (2018) Carbonized cow dung as a high performance and low cost anode material for bioelectrochemical systems. Front Microbiol 9. https://doi.org/10.3389/fmicb.2018.02760

Garg RN, Pathak H, Das DK, Tomar RK (2005) Use of flyash and biogas slurry for improving wheat yield and physical properties of soil. Environ Monit Assess 107:1–9. https://doi.org/10.1007/s10661-005-2021-x

Getahun A, Muleta D, Assefa F, Kiros S, Hungria M (2020) Biochar and other organic amendments improve the physicochemical properties of soil in highly degraded habitat. Eur J Eng Res Sci 5:331–8. https://doi.org/10.24018/ejers.2020.5.3.1735

Gregorich EG, Janzen H, Ellert BH, Helgason BL, Qian B, Zebarth BJ et al (2017) Litter decay controlled by temperature, not soil properties, affecting future soil carbon. Glob Chang Biol 23:1725–1734. https://doi.org/10.1111/gcb.13502

Hussain N, Hassan G, Arshadullah M, Mujeeb F (2001) Evaluation of amendments for the improvement of physical properties of sodic soil. Int J Agric Biol 3:319–322

Idrees M, Batool S, Hussain Q, Ullah H, Al-Wabel MI, Ahmad M et al (2016) High-efficiency remediation of cadmium (Cd^{2+}) from aqueous solution using poultry manure—and farmyard manure–derived biochars. Sep Sci Technol 51:2307–2317. https://doi.org/10.1080/01496395.2016.1205093

Jabeen N, Ahmad R (2017) Growth response and nitrogen metabolism of sunflower (Helianthus annuus L.) to vermicompost and biogas slurry under salinity stress. J Plant Nutr 40:104–14. https://doi.org/10.1080/01904167.2016.1201495

Jalali M, Ranjbar F (2009) Effects of sodic water on soil sodicity and nutrient leaching in poultry and sheep manure amended soils. Geoderma 153:194–204. https://doi.org/10.1016/j.geoderma.2009.08.004

Katkar RN, Kharche VK, Konde NM (2019) Characterization and problems of saline/sodic vertisols and their management options. In: Research developments in saline agriculture. Springer Singapore, Singapore, pp 357–68. https://doi.org/10.1007/978-981-13-5832-6_11

Khan MA, Khan S, Khan A, Alam M (2017) Soil contamination with cadmium, consequences and remediation using organic amendments. Sci Total Environ 601–602:1591–1605. https://doi.org/10.1016/j.scitotenv.2017.06.030

Kim H-S, Kim K-R, Yang JE, Ok YS, Owens G, Nehls T et al (2016) Effect of biochar on reclaimed tidal land soil properties and maize (Zea mays L.) response. Chemosphere 142:153–9. https://doi.org/10.1016/j.chemosphere.2015.06.041

Kosheleva RI, Mitropoulos AC, Kyzas GZ (2019) Synthesis of activated carbon from food waste. Environ Chem Lett 17:429–438. https://doi.org/10.1007/s10311-018-0817-5

Kundu S, Bhattacharyya R, Prakash V, Ghosh B, Gupta H (2007) Carbon sequestration and relationship between carbon addition and storage under rainfed soybean–wheat rotation in a sandy loam soil of the Indian Himalayas. Soil Tillage Res 92:87–95. https://doi.org/10.1016/j.still.2006.01.009

Lashari MS, Ye Y, Ji H, Li L, Kibue GW, Lu H et al (2015) Biochar-manure compost in conjunction with pyroligneous solution alleviated salt stress and improved leaf bioactivity of maize in a saline soil from central China: a 2-year field experiment. J Sci Food Agric 95:1321–1327. https://doi.org/10.1002/jsfa.6825

Lehmann J, Rillig MC, Thies J, Masiello CA, Hockaday WC, Crowley D (2011) Biochar effects on soil biota—a review. Soil Biol Biochem 43:1812–1836. https://doi.org/10.1016/j.soilbio.2011.04.022

Lian F, Xing B (2017) Black carbon (Biochar) in water/soil environments: molecular structure, sorption, stability, and potential risk. Environ Sci Technol 51:13517–13532. https://doi.org/10.1021/acs.est.7b02528

Liang Y, Si J, Nikolic M, Peng Y, Chen W, Jiang Y (2005) Organic manure stimulates biological activity and barley growth in soil subject to secondary salinization. Soil Biol Biochem 37:1185–1195. https://doi.org/10.1016/j.soilbio.2004.11.017

Liang Y, Yang Y, Yang C, Shen Q, Zhou J, Yang L (2003) Soil enzymatic activity and growth of rice and barley as influenced by organic manure in an anthropogenic soil. Geoderma 115:149–160. https://doi.org/10.1016/S0016-7061(03)00084-3

Liao F, Liu Y, Li Q, Xue J, Li Y, Huang D et al (2019) Effect of biochar on growth, photosynthetic characteristics and nutrient distribution in sugarcane. Sugar Tech 21:289–95. https://doi.org/10.1007/s12355-018-0663-6

Lim TJ, Spokas KA, Feyereisen G, Novak JM (2016) Predicting the impact of biochar additions on soil hydraulic properties. Chemosphere 142:136–144. https://doi.org/10.1016/j.chemosphere.2015.06.069

Liu B, Tu C, Hu S, Gumpertz M, Ristaino JB (2007) Effect of organic, sustainable, and conventional management strategies in grower fields on soil physical, chemical, and biological factors and the incidence of Southern blight. Appl Soil Ecol 37:202–214. https://doi.org/10.1016/j.apsoil.2007.06.007

Luo X, Liu G, Xia Y, Chen L, Jiang Z, Zheng H et al (2017) Use of biochar-compost to improve properties and productivity of the degraded coastal soil in the yellow river delta, China. J Soils Sediments 17:780–789. https://doi.org/10.1007/s11368-016-1361-1

McBeath AV, Smernik RJ, Krull ES, Lehmann J (2014) The influence of feedstock and production temperature on biochar carbon chemistry: a solid-state 13C NMR study. Biomass Bioenerg 60:121–129. https://doi.org/10.1016/j.biombioe.2013.11.002

Mekuria W, Noble A (2013) The role of biochar in ameliorating disturbed soils and sequestering soil carbon in tropical agricultural production systems. Appl Environ Soil Sci 2013:1–10. https://doi.org/10.1155/2013/354965

Mia S, Singh B, Dijkstra FA (2017) Aged biochar affects gross nitrogen mineralization and recovery: a 15 N study in two contrasting soils. GCB Bioenergy 9:1196–1206. https://doi.org/10.1111/gcbb.12430

Myrna O, Odening M, Ritter M (2019) The influence of wind energy and biogas on farmland prices. Land 8:19. https://doi.org/10.3390/land8010019

Nawab J, Ghani J, Khan S, Xiaoping W (2018) Minimizing the risk to human health due to the ingestion of arsenic and toxic metals in vegetables by the application of biochar, farmyard manure and peat moss. J Environ Manage 214:172–183. https://doi.org/10.1016/j.jenvman.2018.02.093

Niazi NK, Bibi I, Shahid M, Ok YS, Shaheen SM, Rinklebe J et al (2018) Arsenic removal by Japanese oak wood biochar in aqueous solutions and well water: investigating arsenic fate using integrated spectroscopic and microscopic techniques. Sci Total Environ 621:1642–1651. https://doi.org/10.1016/j.scitotenv.2017.10.063

Ojeda G, Mattana S, Àvila A, Alcañiz JM, Volkmann M, Bachmann J (2015) Are soil–water functions affected by biochar application? Geoderma 249–250:1–11. https://doi.org/10.1016/j.geoderma.2015.02.014

Pandyaswargo AH, Jagath Dickella Gamaralalage P, Liu C, Knaus M, Onoda H, Mahichi F, et al (2019) Challenges and an implementation framework for sustainable municipal organic waste management using biogas technology in emerging Asian countries. Sustainability 11:6331. https://doi.org/10.3390/su11226331

Purakayastha TJ, Bera T, Bhaduri D, Sarkar B, Mandal S, Wade P et al (2019) A review on biochar modulated soil condition improvements and nutrient dynamics concerning crop yields: pathways to climate change mitigation and global food security. Chemosphere 227:345–365. https://doi.org/10.1016/j.chemosphere.2019.03.170

Reddy TRK, Rao TS, Suvarna R P (2014) Studies on thermal characteristics of cow dung powder filled glass–polyester hybrid composites. Compos Part B Eng 56:670–2. https://doi.org/10.1016/j.compositesb.2013.08.059

Reed EY, Chadwick DR, Hill PW, Jones DL (2017) Critical comparison of the impact of biochar and wood ash on soil organic matter cycling and grassland productivity. Soil Biol Biochem 110:134–142. https://doi.org/10.1016/j.soilbio.2017.03.012

Rostamian R, Heidarpour M, Mousavi SF, Afyuni M (2015) Characterization and sodium sorption capacity of biochar and activated carbon prepared from rice husk. J Agric Sci Technol 17:1057–1069

Sarfraz R, Shakoor A, Abdullah M, Arooj A, Hussain A, Xing S (2017) Impact of integrated application of biochar and nitrogen fertilizers on maize growth and nitrogen recovery in alkaline calcareous soil. Soil Sci Plant Nutr 63:488–498. https://doi.org/10.1080/00380768.2017.1376225

Shahid SA, Zaman M, Heng L (2018) Soil salinity: historical perspectives and a world overview of the problem. In: Guideline for salinity assessment, mitigation and adaptation using nuclear and related techniques. Springer International Publishing, Cham, pp 43–53. https://doi.org/10.1007/978-3-319-96190-3_2

Shakya A, Agarwal T (2020) Potential of biochar for the remediation of heavy metal contaminated soil. In: Biochar applications in agriculture and environment management. Springer International Publishing, Cham, pp 77–98. https://doi.org/10.1007/978-3-030-40997-5_4

Tejada M, Garcia C, Gonzalez JL, Hernandez MT (2006) Use of organic amendment as a strategy for saline soil remediation: influence on the physical, chemical and biological properties of soil. Soil Biol Biochem 38:1413–1421. https://doi.org/10.1016/j.soilbio.2005.10.017

Tully KL, McAskill C (2019) Promoting soil health in organically managed systems: a review. Org Agric. https://doi.org/10.1007/s13165-019-00275-1

Wan D, Wu L, Liu Y, Zhao H, Fu J, Xiao S (2018) Adsorption of low concentration perchlorate from aqueous solution onto modified cow dung biochar: effective utilization of cow dung, an agricultural waste. Sci Total Environ 636:1396–1407. https://doi.org/10.1016/j.scitotenv.2018.04.431

Wang L, Sun X, Li S, Zhang T, Zhang W, Zhai P (2014) Application of organic amendments to a coastal saline soil in North China: effects on soil physical and chemical properties and tree growth. PLoS ONE 9:e89185. https://doi.org/10.1371/journal.pone.0089185

Weber K, Quicker P (2018) Properties of biochar. Fuel 217:240–261. https://doi.org/10.1016/j.fuel.2017.12.054

Wu L, Wei C, Zhang S, Wang Y, Kuzyakov Y, Ding X (2019) MgO-modified biochar increases phosphate retention and rice yields in saline-alkaline soil. J Clean Prod 235:901–909. https://doi.org/10.1016/j.jclepro.2019.07.043

Xi B, Dang Q, Wei Y, Li X, Zheng Y, Zhao X (2020) Biogas slurry as an activator for the remediation of petroleum contaminated soils through composting mediated by humic acid. Sci Total Environ 730:139117. https://doi.org/10.1016/j.scitotenv.2020.139117

Xiao L, Yuan G, Feng L, Bi D, Wei J, Shen G et al (2020) Coupled effects of biochar use and farming practice on physical properties of a salt-affected soil with wheat–maize rotation. J Soils Sediments. https://doi.org/10.1007/s11368-020-02616-0

Xiao X, Chen B, Chen Z, Zhu L, Schnoor JL (2018) Insight into multiple and multilevel structures of biochars and their potential environmental applications: a critical review. Environ Sci Technol 52:5027–5047. https://doi.org/10.1021/acs.est.7b06487

Yang Y-J, Tong Y-G, Yu G-Y, Zhang S-B, Huang W (2018) Photosynthetic characteristics explain the high growth rate for eucalyptus camaldulensis: implications for breeding strategy. Ind Crops Prod 124:186–191. https://doi.org/10.1016/j.indcrop.2018.07.071

Yokoyama H, Waki M, Ogino A, Ohmori H, Tanaka Y (2007) Hydrogen fermentation properties of undiluted cow dung. J Biosci Bioeng 104:82–85. https://doi.org/10.1263/jbb.104.82

Yousaf MTB, Nawaz MF, Khawaja HF, Gul S, Ali S, Ahmad I et al (2019) Ecophysiological response of early stage Albizia lebbeck to cadmium toxicity and biochar addition. Arab J Geosci 12:134. https://doi.org/10.1007/s12517-019-4296-1

Yousaf MTB, Nawaz MF, Yasin G, Ahmad I, Gul S, Ijaz M et al (2022) Effect of organic amendments in soil on physiological and biochemical attributes of vachellia nilotica and dalbergia sissoo under saline stress. Plants 11. https://doi.org/10.3390/plants11020228

Yousaf MTB, Nawaz MF, Yasin G, Cheng H, Ahmed I, Gul S et al (2022) Determining the appropriate level of farmyard manure biochar application in saline soils for three selected farm tree species. PLoS ONE 17:1–17. https://doi.org/10.1371/journal.pone.0265005

Yousaf MTB, Nawaz MF, Zia M, Rasul F, Tanvir MA (2021a) Ecophysiological response of early stage eucalyptus camaldulensis to biochar and other organic amendments under salt stress 58:999–1006. https://doi.org/10.21162/PAKJAS/21.1012

Yousaf MT Bin, Nawaz MF, Zia ur Rehman M, Gul S, Yasin G, Rizwan M et al (2021b) Effect of three different types of biochars on eco-physiological response of important agroforestry tree species under salt stress. Int J Phytoremediation 0:1–11. https://doi.org/10.1080/15226514.2021.1901849

Yu F, Zhou H, Zhu Z, Sun J, He R, Bao J et al (2017) Three-dimensional nanoporous iron nitride film as an efficient electrocatalyst for water oxidation. ACS Catal 7:2052–2057. https://doi.org/10.1021/acscatal.6b03132

Yu H, Zou W, Chen J, Chen H, Yu Z, Huang J et al (2019) Biochar amendment improves crop production in problem soils: a review. J Environ Manage 232:8–21. https://doi.org/10.1016/j.jenvman.2018.10.117

Yuan P, Wang J, Pan Y, Shen B, Wu C (2019) Review of biochar for the management of contaminated soil: preparation, application and prospect. Sci Total Environ 659:473–490. https://doi.org/10.1016/j.scitotenv.2018.12.400

Yusefi M, Khalid M, Yasin FM, Abdullah LC, Ketabchi MR, Walvekar R (2018) Performance of cow dung reinforced biodegradable poly (Lactic acid) biocomposites for structural applications. J Polym Environ 26:474–486. https://doi.org/10.1007/s10924-017-0963-z

Zhang Y, Ding J, Wang H, Su L, Zhao C (2020) Biochar addition alleviate the negative effects of drought and salinity stress on soybean productivity and water use efficiency. BMC Plant Biol 20:288. https://doi.org/10.1186/s12870-020-02493-2
Zheng H, Wang X, Chen L, Wang Z, Xia Y, Zhang Y et al (2018) Enhanced growth of halophyte plants in biochar-amended coastal soil: roles of nutrient availability and rhizosphere microbial modulation. Plant Cell Environ 41:517–532. https://doi.org/10.1111/pce.12944

Chapter 14
Applications of Low-Capital-Cost Technologies for Bioconversion of Slaughter Wastes

Sahar Fazal, Rabbiah Manzoor Malik, Sher Zaman Safi, Ghulam Mustafa, Muhammad Anjum Zia, and Muhammad Arshad

Abstract Waste from slaughterhouses is a possible source of pathogens that can infect both people and animals, including bacteria, viruses, prion diseases, and parasites. Therefore, it is crucial to find a quick, affordable, and secure disposal solution to lower the risk of disease after animal slaughter. Composting, Anaerobic Digestion (AD), Alkaline Hydrolysis (AH), rendering, incineration, and burning are some of the different ways that such wastes might be disposed of. Composting is a method of disposal that enables the recycling of nutrients from slaughterhouse waste into the soil. However, due to their high fat and protein content, slaughterhouse wastes are a great substrate for AD processes, which produce methane while both disposing of trash and recovering nutrients (by amending the soil with sludge). There are questions about the ability of AD and composting procedures to render pathogens inactive. In contrast, practically all known microbes can be rendered inactive by AH. Blood from the slaughter of livestock carries a heavy burden of organic pollutants and poses dangers. It increases the organic pollutant load on wastewater treatment plants by 35–50% if it is discharged untreated to sewer systems. Blood can be digested anaerobically as a single substrate, although being difficult, if the culture-reactor system is managed based on a thorough characterization and comprehension

S. Fazal
Capital University of Science and Technology, Islamabad, Pakistan

R. M. Malik
Department of Biochemistry, Wah Medical College, National University of Medical Sciences, Rawalpindi, Pakistan

S. Z. Safi
Faculty of Medicine. Bioscience and Nursing, MAHSA University, Selangor, Malaysia

G. Mustafa
Department of Biochemistry, Government College University Faisalabad, Faisalabad 38000, Pakistan

M. A. Zia
Department of Biochemistry, University of Agriculture Faisalabad, Faisalabad, Pakistan

M. Arshad (✉)
Department of Basic Sciences, University of Veterinary and Animal Sciences, Jhang-Campus, Lahore, Pakistan
e-mail: Muhammad.arshad@uvas.edu.pk

M. Arshad (ed.), *Climate Changes Mitigation and Sustainable Bioenergy Harvest Through Animal Waste*, https://doi.org/10.1007/978-3-031-26224-1_14

of the microbial population and its metabolic activities. If the mixtures are properly created, co-digestion of blood and other feedstock continues successfully. In general, the specific methane yield from digesting blood alone ranges between zero to 0.45 $m^3\ kg^{-1}$ protein, but the yield varies between 0.1 and 0.7 $m^3\ kg^{-1}$ volatile solids when digesting blood along with other substrates. To properly regulate the process, more study is needed on the microbiology and kinetics, the function of adsorbents, reactor structure, and culture adaptation during anaerobic digestion of blood.

Graphical Abstract

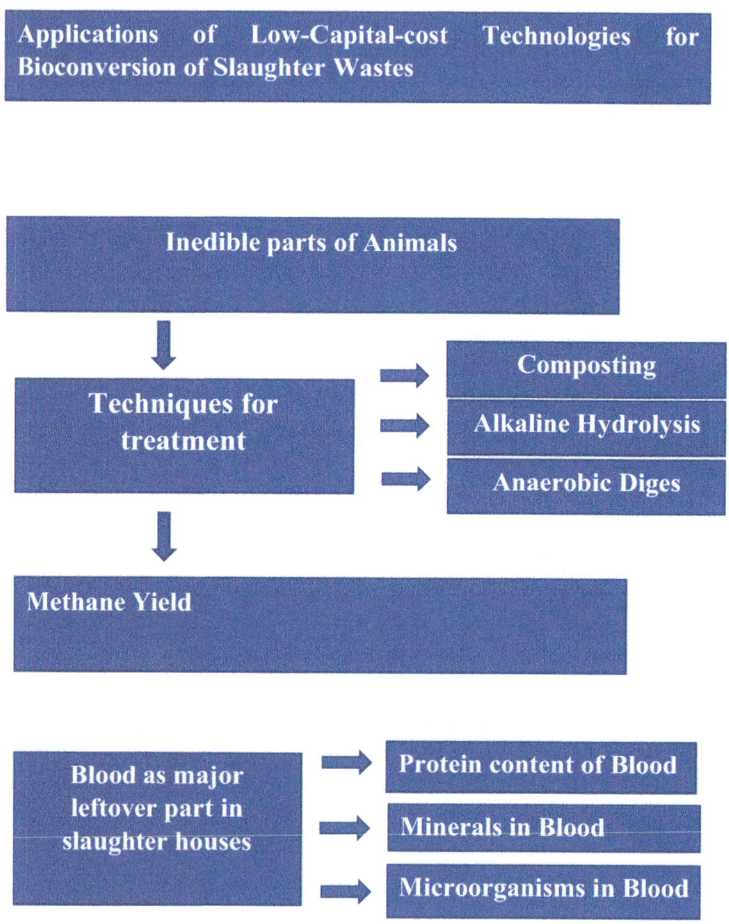

Keywords Slaughtering waste · Low cost technologies · Anaerobic digestion · Pyrolysis

14.1 Introduction

Organs, tendons, ligaments, integuments, bones, blood vessels, and feathers are examples of inedible animal tissues. These tissues can account for almost 45% or more from the remains of slaughtered animal. The remaining portion, on the other hand, is not only suitable for disposal but also suitable for purchase by pet food companies in large quantities. According to Salminen and Rintala (2002a), these can be utilized either by themselves or as an addition to animal feed. Despite this, there is a significant amount of waste generated by slaughterhouses worldwide, and the logistics of disposing of this waste present a significant logistical task for poultry and meat processing units. Animal waste is improperly and unsafely disposed off due to the high magnitude of waste, rising wastage treatment expenditures and legal restrictions. According to Arvanitoyannis and Ladas (2008), such practices may also result in serious environmental issues. Slaughterhouse wastes, on the other hand, have the potential to be used as replacement for some source of energy and can support cut down on the amount of fuels built on petroleum needed to meet Earth's current energy requirements (Srivastava and Prasad 2000).

There will be a number of inquiries about the safety of produced products as these alternative energy sources emerge. Because slaughterhouse wastes could get contaminated with a significant number of microorganisms, like viruses, bacteria, fungi, yeasts, prions, and related microbial poisons (Urlings et al. 1992), the quality of the product formed is of the utmost concern when using them to produce biogas. If these wastes are not handled and treated properly, they could be harmful to the health of animals and humans.

In animal-related facilities like farms, slaughterhouses, and animal production operations, the major concern has always been the management of animal remains. According to Gwyther et al. (2011), the most common methods of disposing of on-farm dead have been burning to a little extent and majorly burial throughout history. However, the Animal By-Product Regulation (EC) of the European Union does not1774/2002 (Anon 2002) prohibits these practices within the Europe and restricts dumping options to rendering, incineration, licensed waste collectors, disposal at maggot farms, and high temperature/pressure AH. According to Anon, the supposed hazard of pathogens contaminating the food chain of animals as a result of incomplete pathogen destruction during burial and burning was the impetus behind the prohibition on their disposal. The slaughtering of animals carries the same risks. In comparison with burning, properly run incineration facilities are less likely to cause pollution. Additionally, viruses and bacteria, including those that produce spores, should not remain alive after the process of incineration (NABC 2004). However, if incineration processes are not carried out at a temperature that is high enough, transmissible spongiform encephalopathies (TSEs) like BSE (bovine spongiform encephalopathy) may survive (NABC 2004).

Another technique for disposing of products related to animal waste is rendering. It includes transforming animal waste to three products: water, melted fat/tallow, and carcass meal (proteinaceous solids, Fig. 14.1). Mechanical processes like mixing, grinding, separating, pressing, and decanting, thermal processes like drying, evaporation, and cooking, along with the chemical processes like solvent extraction are used to accomplish this (NABC 2004). The material is simultaneously dried out and protein and fat are separated during the rendering process. Among raw materials, tallow is one of the important which are used in the rolling industry of steel, because it provides the lubrication required for compression of sheets of steel. Additionally, the obtained fat could be utilized as a raw material of low cost for the manufacture of animal feed, grease, biodiesel, candles, and soap. Feed for animals can be made with the protein meal. As a result, the products bring in a lot more money for the slaughterhouse. However, feeding cattle with bone meal and meat and is currently against the law in the developed nations due to BSE issues. As a result, rendering plants do not play as much of a role in today's animal waste disposal as they did in the past.

The execution of the Landfill Directive for better environmentally complex waste disposal techniques has resulted in new risks. According to Noble and Roberts (2004), there is indication that some pests and pathogens are able to survive from composting or some other processes for the waste treatment, sometimes because these procedures are ineffective or fail. In addition to composting two other techniques have been introduced which are Alkaline Hydrolysis (AH) and Anaerobic Digestion (AD). The European Union's (EU) Animal By-Product Regulation (EC) number 1774/2002 defines three categories for animal byproducts (ABP). While ABP of category 1 cannot be treated, ABP of categories 2 and 3 can be processed by AD in permitted biogas plants or composting plants. Allowances for AH are not mentioned in the guidelines. The majority of waste from slaughterhouses falls into categories 2 and 3.

According to Dees and Ghiorse (2001), composting is an aerobic procedure in which successive assemblies of microorganisms reduce organic materials. Various inorganic and organic wastes, sludge, sewage, pig, cattle, or poultry manure, municipal waste, and garden waste are examples of substrates. Numerous authors have reported that composts have beneficial effects on arable soil (Ibekwe et al. 2001; Bailey and Lazarovits 2003).

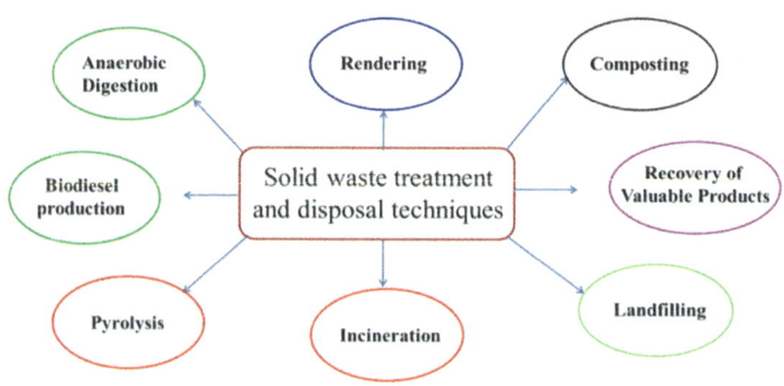

Fig. 14.1 Different processes being operated for disposal of animal slaughter wastes

A high microbial diversity is necessary for an effective and satisfying composting process (Beffa et al. 1996). During the four phases that follow, distinct microbial communities dominate (Blanc et al. 1999; Alfreider et al. 2002; Ryckeboer et al. 2003) involving aerobic microorganisms that are mesophilic, thermotolerant, and thermophilic (Beffa et al. 1996). Different organisms go through exponential growth stages and stationary growth phases depending on the environmental conditions of the compost—pH, temperature, moisture, aeration, substrates availability—that are constantly changing. Utilizing tools of molecular biology like the COMPOCHIP microarray (Franke-Whittle et al. 2009) we are rapidly gaining knowledge of the variety of microbial communities found in compost (Danon et al. 2008).

The survival and spread of pathogens that are harmful to humans, animals, and plants are one of the issues that have been raised about the process of composting as well as the application of composts in horticulture and agriculture. Therefore, any composting procedure should be capable of removing any potential health hazards from the product (Strauch et al. 2002). The dispersion and proliferation of possibly allergenic and/or pathogenic thermophilic/thermotolerant bacteria and fungi can result from a composting process that is not properly managed (Böhnel et al. 2002).

There are windrow systems and in-vessel systems for composting. According to Cekmecelioglu et al., windrow composting possesses several disadvantages over composting, including requiring more space, providing poor control, and offering a low process efficiency. On the other hand, window composting's higher temperatures have been shown to reduce pathogens and bacteria more effectively (Cekmecelioglu et al. 2005). According to Noble and Roberts (2004), pathogen survival is a concern in cool regions of in-vessel composting systems where there is little or no turning. As a result, windrow composting for slaughterhouse waste is the best strategy.

Waste from slaughterhouses can be disposed of in an environmentally friendly and cost-effective manner by composting. The majority of pathogens can be killed or greatly reduced by the temperatures reached during decomposition, reducing the likelihood of disease transmission. Despite the fact that the heat generated during composting reduces the number of microorganisms and pathogens, it does not fully sterilize the material produced, leaving some room for pathogens to survive and regrow. The temperatures attained and the length of time for maintaining the temperature determine the amount of pathogenic bacteria that remain at the conclusion of the process of composting. The lowest temperature of 65 °C at least for six successive days or two to three-day periods above 65 °C is required by various legal requirements (Kompostverordnung 2002). Pathogenic bacteria, viruses, helminth ova, and protozoa can typically be reduced to an acceptable level by maintaining an average of temperature at 55–60 °C for one to two days. However, spore-bearing bacteria like Bacillus and Clostridium endospores cannot be destroyed by these conditions (Urlings et al. 1992). The composting process typically results in pH values above 8 as well (Reuter et al. 2011)), which aids in the inactivation of pathogens as well. Antibiotics produced by various compost-associated microorganisms have also been shown to inactivate pathogens (Hoitink and Boehm 1999).

Slaughterhouse waste can potentially be disposed of in an acceptable manner through composting. However, in order to eliminate some pathogens from the final product, additional treatments would be required. Post-treatment of AD wastes can also be done by composting. This post-treatment improves the compost in rapports of nutrients and ensures a more efficient degradation of organic material. The anaerobic digestate's composting would also help lower the number of pathogens in these products.

14.2 Use of Technologies in Disposal of Waste Materials

For the elimination of pathogenic wastes and carcass material from animal, AH is a relatively new method (Urlings et al. 1992; Kaye et al. 1998). Proteins, nucleic acids, carbohydrates, lipids, and other biological components are hydrolyzed with a sterile aqueous solution that contains amino acids, tiny peptides, soaps, and sugars, using sodium or potassium hydroxide as a catalyst. Normally, pressure and temperature are used in the process to hasten the hydrolysis. Waste carcasses must be heated up to 100 °C and compressed for almost 3 h at 103 kPa to inactivate microbiological pathogens. They should be heated up to 150 °C and pressured at 486 kPa for 6–8 h in order to eliminate prion-containing material (http://ssl-edss.tamu.edu/disposal/handbook/04Alkaline.pdf).

Despite the fact that there are few studies examining the inactivation of microorganisms using AH, those that do show the way.

14.3 Anaerobic Digestion (AD) and Alkaline Hydrolysis (AH)

Anaerobic digestion is a biotic process in which organic wastes decompose without oxygen to make sludge that can be used in agriculture and biogas that can be used to make energy (Lastella et al. 2002; Insam and Wett 2008) (Fig. 14.2). AD technology has been commercially applied to the treatment of municipal, industrial, and farm, wastes in recent years. Additionally, AD is a different way to get rid of animal feces and waste from slaughterhouses. It has the double advantage of making energy and getting rid of waste, which saves carbon credits (Insam et al. 2010). However, there are questions regarding whether the process of AD can effectively kill pathogens. Prior to AD, a pasteurization step has been frequently used for assisting in the inactivation of pathogens. As an additional measure, it is common as well, to use a secondary heat treatment process, such as pasteurization or composting, and a minimum storage period for the digestate at the end of the process (Sahlström 2003). To guarantee a hygienically tolerable final product, Swedish law, for instance, mandates that biogas plants which utilize animal waste must pasteurize the received substrate for 60 min at 70 °C (Sahlström et al. 2008).

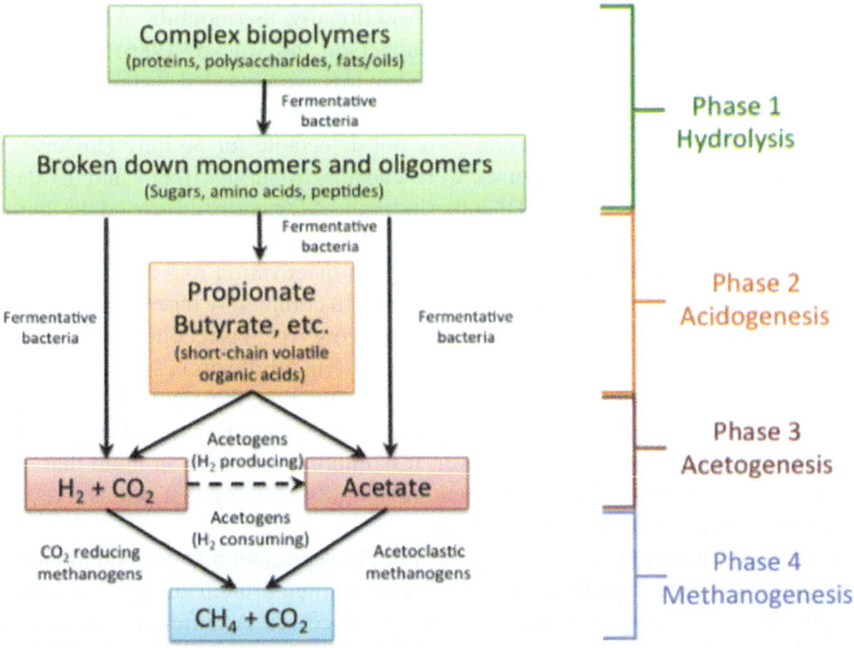

Fig. 14.2 Schematic sketch of anaerobic digestion

The anaerobic process necessitates specific environmental conditions and close and effective syntrophism between the involved bacteria and archaea (Schink 1997). Hydrolysis, acetogenesis, acidogenesis, and methanogenesis are the four major phases of digestion, and complex polymers are broken down step by step to produce CO_2 and CH_4.

Both mesophilic conditions (at 35 °C) and thermophilic conditions (at 55 °C) can be used to conduct AD, which has different effects on pathogen destruction and duration. A mesophilic process typically lasts between 15 and 30 days, whereas a thermophilic procedure typically lasts between 12 and 14 days (Vandeviviere et al. 2002). The finished product would probably offer a bigger pathogen risk if sprayed directly to the field, however, as gas generation is reduced (Urlings et al. 1992). Compared to thermophilic processes, mesophilic AD processes are thought to be more resilient and less susceptible to changes in process parameters. Compared to other bacteria, methane-producing microorganisms are more sensitive to temperature variations, and it has been reported that changes in temperature as low as 2 degrees Celsius can have negative effects on mesophilic processes of AD (Gunnerson and Stuckey 1986).

On the other hand, thermophilic AD is better at inactivating pathogens and produces more methane at a faster rate. The thermophilic digestion process has advantages, but it also needs much expensive technology with higher levels of monitoring and operation (Vandeviviere et al. 2002). Additionally, thermophilic AD processes can be affected by even 0.5 °C changes (Gunnerson and Stuckey 1986).

In a study, a bovine parvovirus and enterovirus were seeded in liquified cattle manure. The researchers found that both groups of viruses were quickly inactivated after a thermophilic AD process and were not detectable for 30 min. However, in mesophilic conditions, these viruses were capable to survive for 8 and 13 days, respectively. Another study found that thermophilic AD rapidly destroyed the animal viruses like coxsackievirus B5 and rotavirus. According to these findings, viruses undergo inactivation during AD process, but the inactivation rate is influenced by the temperature, virus, and duration of digestion. To completely inactivate pathogens like spore-formers which have capability of surviving AD, it is recommended to have an extra heat treatment after the conclusion of a thermophilic process of AD. Before spreading sludges on land, regulations 1774/2002 and 208/2006 mandate that animal waste undergo a 70 °C/60 min pasteurization step (Monteith et al. 1986; Spillman et al. 1987; Bagge et al. 2005). However, it is important to keep in mind that even if there were an additional heat treatment, it is highly unlikely that prions like BSE would be inactivated (Urlings et al. 1992).

Due to the formation of ammonia after protein degradation and accumulation of long-chain fatty acid from breakdown of lipids, the high protein and lipid content of slaughterhouses waste products can inhibit AD processes. Additionally, the adsorption of lipids into the biomass can result in the formation of floating foam and aggregates, both of which can impede stratification (Cuetos et al. 2010). However, by allowing the microbial groups to gradually adapt to a medium rich in ammonia, AD can be carried out successfully on wastes from slaughterhouse with higher production of methane if properly managed (Edström et al. 2003).

Rendering was used to treat solid waste from slaughterhouses in the past, providing the slaughterhouses an extra revenue stream. Though the economic value of these products has decreased significantly due to the risk of TSEs, in many instances, they must be treated as waste themselves (Palatsi et al. 2011). As a result, the cost of properly disposing of waste from slaughterhouses has significantly increased in recent years. This is primarily because pathogens in such wastes pose health risks.

One option for getting rid of waste from slaughterhouses is composting. The valuable byproduct production and decreased environmental pollution, and the elimination of most pathogens are among the process's major advantages (Urlings et al. 1992). However, strict control is required for the successful alteration of slaughterhouse wastes into high-quality compost. The finished product should not be a threat to animal or human health when carried out under strict supervision (Gale 2004). However, some pathogens, like spore-forming bacteria and prions, cannot be eradicated by composting.

The alkaline hydrolysis of slaughterhouse wastes is a moderately new procedure (Fig. 14.3). In order to hydrolyze the biological materials and convert them into an aqueous solution that is sterile and containing amino acids, peptides, soaps, and sugars, it makes use of a potent base, heat, and temperature. Although it can be discharged into a sanitary sewer, this waste, which is extremely abundant in nutrients and alkaline, may also cause issues (Urlings et al. 1992). This has been observed to

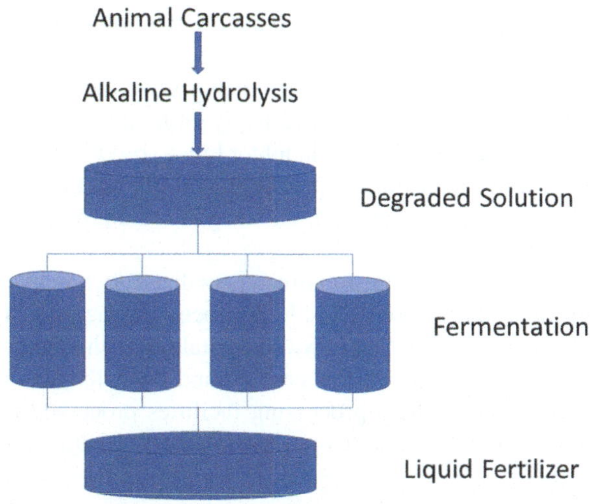

Fig. 14.3 Schematic diagram for alkaline hydrolysis (AH) (Kang et al. 2019)

be very efficient at getting rid of a lot of prions and pathogens from animal waste and carcasses. The loss from the cycle is anyway extremely wealthy in supplements and would in this manner provide high biogas age potential.

Promotion is today among the most encouraging techniques for the removal of slaughterhouse squander. In addition to producing degraded material, which could be utilized as useful fertilizer, this process also produces biogas and heat, both of which could be used to generate energy. Additionally, due to having higher nitrogen and protein content, wastes from slaughterhouse make excellent AD substrates. The effectiveness of AD in eradicating various pathogens is the target area of various studies.

14.4 Blood

The process of slaughtering livestock animals produces a lot of blood (Tables 14.1 and 14.2) as a result of the high chemical oxygen demand (COD) and high nitrogen content. The COD population equivalent (PE) of blood from a carcass of cattle is 50 g per L, and the total nitrogen content of blood is approximately 30 g L1 (Marcos et al. 2017). Untreated blood has been found to increase the wastewater load on organic pollution by 35–50% when discharged directly into the sewer system (Wang 2015; USEPA 1973). As a result, the high organic pollution load of blood poses environmental risks if it is disposed of or discharged without proper treatment (Bah et al. 2013; Hoyo et al. 2008). Discharging untreated blood into streams of water reduces dissolved oxygen levels, increases the amount of nutrients in the water, results in septic conditions, and causes spread of viral and microbial waterborne diseases (Saidu and Musa 2012; Nwachukwu et al. 2011). Therefore, the most effective strategies for managing, controlling, and reducing the pollution brought on by wastewater from slaughterhouses are blood collection and treatment (Tritt and Schuchardt 1992). In 2003, 2014, and 2016, the Food and Agriculture Organization (FAO) of the United Nations estimated that respectively, more than 49.2, 70.3, and 74.2 million animals were killed. The Global Slaughter Index (FAOSTAT 2017) indicates that this number is steadily increasing. Depending on the animal's type, size, slaughtering method, and blood collection duration, all slaughtered animals bleed to varying degrees during the process.

Blood processing requires specialized technical infrastructure. In the EU, for instance, there are only 11 blood-processing facilities (Commission 2005). Denmark, Belgium, France, Spain, Sweden, Italy, and the Netherlands, each have one blood-processing plant, while Germany and the UK each have two plants (Commission 2005). These blood-processing facilities processed approximately 300,109 metric tons of blood in total (Commission 2005). Treatment of liquid blood was somewhat

Table 14.1 Quantities of blood generated from animal slaughtering

Type of animals slaughtered	%age of Live Weight (%)	Kg per head	Volume in liter
	2.5	10–22	10–20
	2.5	4–6	2.4
	0.6	4–6	1.5
	3.2	0.04	0.5
Mixed animals	2.4–8.0		

Table 14.2 Quantity of blood produced from animal slaughtering

Animal	Live weight (Kg)	Blood (Kg)
Beef cattle	454	14.5
Beef cattle	453	14.7
Beef cattle	450	16
Cattle	270	19
Cow	250	12.7
Heavy livestock	250	21
Light livestock	40	1.2
Pig	28	3
Sheep and Goat	22	0.72
Sheep	20	0.48
Poultry	1.5	0.045
Chicken	2.28	0.024

challenging in Spain, for instance, due to the absence of such facilities near slaughterhouses or within a practical distance of them. Following heat dewatering, coagulated blood is typically shipped to be burned (Cuetos et al. 2009). Blood, contrarily, is an organic substance primarily composed of protein. It could be used as a feedstock for the production of biogas after some proper management, pretreatment, and control of the fermenting processes. High potential yield of methane (500 L per Kg) of blood is contributed by its highly degradable content of protein (94–96%) along with its fat/lipid content (Hejnfelt and Angelidaki 2009; Salminen and Rintala 2002b). Afazeli and others (2014) have estimated that approximately 31% of Iran's potential annual yield of biogas from waste of slaughterhouse comes from blood. Wang (Arvanitoyannis and Ladas 2008) found that Anaerobic Digestion (AD) of 1:3 (v/v) of poultry blood of and wastewater obtained from poultry processing and slaughtering about 200,000 birds/day will produce 146.2 m^3 CH_4 per d. In a bioreactor with a relatively small volume (530 m^3), the recovery of net energy will be 1.5 GJ per d, while the recovery of phosphorus (P), nitrogen (N), and potassium (K) would be 3.0, 251.5, and 3.7 kg per day, respectively. An anaerobic digester can burn biogas produced from blood, which is organic biomass.

In 2010, approximately 959 sheep and goats, 304 cattle, and 1374 pigs were slaughtered and processed in world for the production of meat (Wang 2015; FAOSTAT 2017) Bah and co. Based on the assumption that the volume in liters of blood that is recovered by per head slaughtered is about 15 for cattle while 2–3 for pigs, it has been estimated that the volume in liters of blood obtained from these animals was approximately 4.56×10^9 (Fallows and Verner Wheelock 1982). According to the European Community report (Commission 2005), 10–20 L for cattle and about 2–4 L (for pigs), respectively, of blood are taken from each of these animals that is killed (Commission 2005). Approximately, 200,000 broilers are slaughtered by about 162 processing plants every day in the United States (Wang 2015; Cuetos et al. 2009). When the bird is slaughtered, only 75% of its live weight is turned into meat, with the remaining 25% being organic, unpalatable byproducts like feather, blood, and intestine residues (Cuetos et al. 2009; Hejnfelt and Angelidaki 2009; Yoon et al. 2014). This is true for nearly every kind of animal that is killed. The typical bleeding time following livestock slaughter is 6 min for cattle, 4–5 min for sheep, 6 min for pigs, and 3–4 min for calves (Kiepper et al. 2008; Salminen and Rintala 2002b). As a result, the United States could only collect 0.5–0.9 109 kg of blood yearly (Wang 2015). 2518 and 2462 tons of blood waste were produced by slaughtering 237,300 goats and 181,770 cattle, respectively (Aniebo et al. 2009). At the Tamale abattoir in Ghana, 125 livestock animals, including 55 cattle, 20 goats, and 50 sheep, were slaughtered each day, producing 0.7 metric tons of blood per day (Fearon et al. 2014). Organic byproducts of the slaughtering process are inedible. On the basis of the type of the animal, blood makes up between 2.4 and 8.0% of the live weight of animal (Jayathilakan et al. 2012; Banks and Wang 2006) and between 6 and 7% of the carcass's lean meat content (Banks and Wang 2006). Cows make up 2.5% of their live weight, while sheep and goats make up 0.6%. Afazeli and others (2014) reported that slaughtered poultry (1.5 kg), heavyweight (250 kg), and lightweight (40 kg) animals had blood that made up 7.0%, 3.0%, and 3.0% of their total weight, respectively. There are almost 10–22 kg of the blood taken from each of the head of slaughtered cattle and approximately 40 g taken from each broiler (2% of the live weight) Nevertheless, Jayathilakan et al. (2012) stated that approximately 3.2–3.7% of the slaughtered poultry's live weight of is made up of the blood.

The species of an animal along with its live weight determine the blood mass. In Iran, heavy livestock account for 17%, poultry account for 14%, and light livestock account for 5% of the blood produced at slaughterhouses (Afazeli et al. 2014). In most cases, slaughterhouses only recover a small amount of the blood that is produced. Pigs, cattle, and lambs each yield an average of 3.0–4.0%, 3.0–4.0%, and 3.5–4.0% of blood, respectively (Jayathilakan et al. 2012). Following the slaughter of an animal, the typical volume (in liters) of blood collected is 2–4 for pigs, 10–20 for cattle, and 1.5 for sheep, or approximately 10% of the chicken's body weight (Commission 2005; Salminen and Rintala 2002b; Banks and Wang 2006). For instance, a typical commercial steer weighing 450 kg produces approximately 16 kg of blood (Banks and Wang 2006). Singleton (Jedrzekewska et al. 2006) measured about 13.3 L per slaughtered head of the cattle. After the animal has been killed, about half of the blood that is bleeding is taken from the carcass (Salminen and Rintala 2002b). About 30–50% of blood of broiler drains and can be collected, while 70–50% of the blood is left back in the carcass and is washed out partially when the cutting is in process (Kiepper et al. 2008; Owens et al. 2000). Because vacuum pressure shortens the time required for the collection of blood while simultaneously recovering more amount of blood from carcass, using it to collect the blood could result in an increase in the volumes of blood collected.

Biochemical oxygen demand (BOD) ranges from 140 to 200 g per L for protein-rich blood (Jedrzekewska et al. 2006; Johns 1995) and as much high as 400 g per L for carbon dioxide (Banks and Wang 2006). Bovine blood in raw form, for instance, has a COD of up to 375 g per L. As a result, slaughterhouse wastewater contains a substantial amount of the organic pollutants. The BOD and COD of slaughterhouse wastewater are determined by the efficacy of collection, separation, and recovery facilities (Jayathilakan et al. 2012). According to Wu and Mitta, 62% of Ontario slaughterhouses collect blood separately, while 13% of them compost it on site, of which 87% is sent to execution plants. The blood's ration of BOD5/COD decreased from 0.53 to 0.46 for beef, 0.67 to 0.47 for sheep, 0.42 to 0.31 for pork, 0.46–0.21 for poultry, and 0.38 to 0.33 for mixed after the wastewater from slaughterhouse was separated from the blood. Consequently, it reduced the wastewater's odor (Wu and Mittal 2012).

Most of the time, blood is utilized as a raw material for making fertilizer, animal feed, blood meal, blood serum, or fibrin. Blood can also be utilized as a feedstock for recovering the energy for AD, lost to protein content of blood by converting that into heat, electricity, and biogas. Blood is full of nutrients and carbon.

14.5 Waste Blood's Composition

Cells, water, proteins, enzymes, and other inorganic and organic components make up the red fluid of blood. Blood gets coagulated in one to two minutes once it leaves the blood vessels (See et al. 2009; Oladele and Samuel 2014). The blood components which are heavier cause the blood to stratify over time, during storage. As a result, mixing is inevitable during AD treatment procedures. Red blood cells and white blood cells, along with platelets, make up 30–40% of the wet mass of blood, and they are suspended in a liquid called plasma (60–63%). The majority of the protein that makes up the cellular matter is hemoglobin. Plasma, a yellow liquid, contains over 100 smaller proteins and 6–8% proteins like albumin, globulins, and protein fibrinogen (Bah et al. 2013; Jayathilakan et al. 2012). Blood typically has a density of about 1.0 kg per L. Bovine blood, however, can have a density of about 1.5 kg per L (Spillman et al. 1987). The percentage of cellular material in blood of pigs is 43.5%, while cattle have 2.5%, and sheep has 23%. However, as far as the plasma content (percent of blood by weight) is concerned, cattle contain 67.5%, pigs contain 56.5%, and sheep has 72%. The majority of blood proteins are composed of hemoglobin and albumin. In the blood of sheep, cattle, and pigs, hemoglobin accounts for 14.2%, 10.3%, and 9.3% of the proteins, respectively, while albumin accounts for an average of 3.8% (Bah et al. 2013; Gorbatov 1990).

14.6 Protein Content of the Blood

The composition of the blood varies slightly from animal to animal. Volatile solids (VS) make up 91–96.6% of total solids (TS) of blood, which range from 10.8 to 23%. Blood typically contains a lot of protein (up to 965 g per kg dry organic matter). As a result, its COD is quite high. The sterility and abundance of amino acids found in healthy animal blood are unparalleled. Blood from sheep, poultry, cattle, swine, and mixed animals contains 17–19% proteins, 18–20% dry solids, 78–79% moisture, and 0.8–1.25 ash, kg m^3 of blood dry solids contain 13–15% protein. For comparison, dairy cow slurry that has not been diluted contains 4 kg N m^3 (Smith et al. 2007). There are about 3.5 g of NH3-N L1 in blood; this is approximately 270 times higher than domestic wastewater, which ranges from 4 to 13 mg L1 (Moskowitz 2012; USEPA 2002; Tchobanoglous et al. 1991; Sedlak 1991). As a result, the population equivalent of nitrogen pollution caused by blood is extremely high. Blood-containing wastewater from slaughterhouses has total nitrogen levels that are 2.3–4.4 times higher than typical wastewater of domestic origin (USEPA 2002; Tchobanoglous et al. 1991; Sedlak 1991). The abundance of blood in the wash water is primarily to blame for the elevated levels of proteins and nitrogen in slaughterhouse wastewater, which contribute to the elevated pollution load. Meat rendering and processing, for instance, consumes water at rates of 6–30 for poultry, 1.5–10 for pigs, and 2.5–40 for cattle (Omole and Longe 2008). This blood-contaminated water must be treated before being disposed of after processing.

14.7 Blood Mineral Content

Potassium, Phosphorus, calcium, iron, sodium, magnesium, cobalt, sulfur, copper, cadmium, nickel, chromium, zinc, and lead are among the minerals found in livestock blood, according to research published in the literature (Table 14.3). They make up about 4.8% of animal blood in total (by dry weight) (Hansen and West 1992; Zhang and Banks 2012; Alvarez et al. 2006; Banks and Zhang 2012; Okanović et al. 2009). Blood is a very nutrient-rich medium for mixed culture microorganisms growing in anaerobic conditions, that mediate the chemical reactions of bioreactors of AD because of its protein content and mineral content.

The biodegradable substance known as theoretical methane potential of blood protein has the chemical formula of $C_cH_hO_oN_nS_s$. It possesses a methane potential that is higher than that of carbohydrates and fats. The half-reactions demonstrate that content of protein yields the highest specific molar amount of methane by employing the characteristic chemical formulas for fats $C_8H_{16}O$, carbohydrates CH_2O, and proteins $C_{16}H_{24}O_5N_4$. Protein yields 8.25 mol CH_4 per mol of stoichiometric specific methane, while fat yields 5.75 mol CH_4 per mol and carbohydrates yield 0.5 mol CH_4 per mol (Rittmann and McCarty 2001).

The biogas production of blood is 18% DM ranges from 0.3 to 0.6 m_3/kg TS (Avcioglu and Turker 2012; Deublein and Steinhauser 2008). Steffen and others described a biogas yield for animal blood of 0.65 m_3/kg with a VS/TS ratio of 0.95 and a TS/VS concentration of 9.7%. In a biogas plant, blood plasma is fed as a co-substrate with a ratio of mass of 10–15% produced biogas yields of 0.40–0.60 m_3/kg VS (Steffen et al. 1998). The blood effect was blamed for the less methane production by use of ruminal waste, according to Banks and Wang (Banks and Wang 2006). High concentrations of ammonia were accumulated after the breakdown of the nitrogen-rich compounds of the blood. A mixture containing biodegradable waste

Table 14.3 The amount of micronutrient contents in blood of slaughtered animals

Element	Broiler	Sheep	Cow	Swine	Pig	Mixed
P	118	164				183
K	92.7	731	668	2380	118	798
Ca	44.9		130	90	1221	55
Mg	5.4				224	27
Na	148.3		2763	1650	94	818
Fe	48				368	164
Co	0.01				0.1	< 0.2
S	Not reported				4000	300
Cd	Not reported	< 0.20				0.05
Cr	Not reported	< 0.40				0.3
Cu	0.1	1.32			14.6	0.7
Ni	1	< 1.0			1.6	
Pb	Not reported	< 2.0				
Zn	1.7	3.2			13	13

from municipal, and a co-substrate was jointly digested by Banks and Zhang (2012). With a methane production of 0.357 m^3 CH_4 /kg VS at STP, the highest content was produced by sheep blood in biogas (58.7%) in comparison with the gut content of pig that comprised of biodiesel byproduct, floatation fat, and poultry litter.

14.8 Methods for Degrading Proteins

The structure of proteins is composed of chains of subunits amino acids that are linked by peptide bonds (–CO–NH–). There are four atoms or active molecules are attached to the central carbon atom of each amino acid: a hydrogen atom (H), a carboxylic group (–COOH), an amino group (–NH_2), and a fourth group (R) that is unique to each amino acid. There are 20 different amino acids that are utilized in the synthesis of proteins, each of which belongs to a specific group (R) (Ivanov 2015). Extracellular enzymes called proteases hydrolyze proteins into free amino acids and polypeptides. Nitrogenous compounds like NH_3, short- or branched-chain volatile fatty acids (VFAs), sulfides, phosphates, CO_2, and H_2 are all products of amino acid decomposition (Gray 2004). Ammonia and VFAs are formed by the deamination reaction, and the VFAs are further oxidized to form CO_2. Additionally, some reactions can result in the production of amines, which have a pungent odor (Ivanov 2015; Hogg 2005). As a result, blood waste can have an unpleasant odor. Amino acid metabolism can be carried out in one of two ways: Stickland reactions are required for the breakdown of paired amino acids, while H_2-utilizing microorganisms are required for the breakdown of single amino acids, as is the case with carbohydrate fermentation.

First amino acid is oxidized and used as an energy source in the Stickland reaction, and a second amino acid accepts electrons. Compared to fermenting uncoupled amino acids, the Stickland reaction is quicker and simpler (Willey et al. 2013).

14.9 Microbiology

During AD, stress factors act as pickers that encourage the domination of particular bacterial species among the microbes and impose inhibition on other microorganisms (Vrieze et al. 2015; Muller et al. 2016). During AD of protein-rich compounds, ammonia is a stressor. Frank and co (2016) reported that oxidation reaction of the syntrophic acetate, i.e., the formation of H_2 and CO_2 from acetate and their generation of methane by hydrogenotrophic methanogenesis are promoted by high levels of ammonia caused due to the breakdown of protein content. A biogas plant's anaerobic mixed culture that was digesting maize silage and protein-rich pig slurry contained Clostridia-like proteolytic bacteria (Kovacs et al. 2015). From anaerobic digesters, the following species of protein-specific, chemo-organotrophic, and proteolytic anaerobic bacteria are isolated: the species Proteiniborus ethanoligenes nov.,sp.nov., which is a member of Clostridia (Niu et al. 2008); c. thiosulfatireducens (Hernandez-Eugenio et al. 2002; Thabet et al. 2004) Clostridium tunisiense; acetatigenes Proteiniphilum (Chen and Dong 2004). A group of bacteria that only eat protein is known as Proteiniborus. They convert protein into CO_2, acetic acid, H_2, propionic acid, and ethanol in trace amounts (Xu et al. 2018). In recent times, researchers have observed the attributes of the microbial culture anaerobically, during blood AD of pig using metagenomic analyses on the basis of next-generation sequencing, and they found Alkaliphilus oremlandii, and Dethiosulfovibrio peptidovorans, and Alkaliphilus metalliredigens increased in abundance by 2.0, 11.4, and 8.7 times, respectively, over a 12-week period, whereas Candidatus Cloacamonas acidaminovorans and Anaerobaculum hydrogeniformans vanished from the culture. Frank et al. (2016) used metagenomic sequencing and small subunit rRNA genes reported that an uncultured phylotype with metabolic dominance and acetate oxidation capabilities comprised of the 16S rRNA genes by 5%, in a biogas digester being used commercially, that was ammonia-tolerant. The homoacetogenic-characterizing carbon monoxide dehydrogenase/acetyl coenzyme A (Acs) synthase (Acs) operon was encoded by UnFirm 1. They suggested that protein-digesting biogas reactors' long-term stability and success depend on UnFirm 1. Extracellular enzymes are secreted by a number of bacteria, including Proteus, Vibrio, Clostridium, Bacillus, Peptococcus, and Bacteroides, to degrade proteins to amino acids (Gonzalez-Fernandez et al. 2015). Lee and others (2016) used real-time quantitative polymerase chain reaction (QPCR) and pyrosequencing to measure the microbial populations in protein-based thermophilic anaerobic digesters by using the 16S rRNA genes of archaea and bacteria. They discovered that the digester was dominated by the genera Defluviitoga and Keratinibaculum (10.4% and 8.1%, respectively). Digesters that were fed stillage (Röske et al. 2014) and food waste (Guo et al.

2013) showed a similar dominance. A known bacterium that consumes proteins is Keratinibaculum paraultunense (Huang et al. 2013). Fisher et al. examined the immediate effects of protein-rich feed-induced ammonia shock on the expression and structure of microbiome of a biogas reactor (Fischer et al. 2019) by sequencing the amplicons of the 16S rRNA gene. The digester's first and second most dominant classes were Clostridia and Bacteroidia, respectively. After 10 days of incubation, the abundance of Clostridia was reduced from 63 to 40%, while the abundance of Bacteroidia increased from 15 to 23% in the ammonia-treated reactors. They demonstrated that MBA08 and the Caldicoprobacteraceae family of clostridial taxa were the most prevalent in the digester. Next-generation sequencing microbial community analysis of blood as a single substrate revealed that the phyla Proteobacteria, Firmicutes, Bacteroidetes, Actinobacteria, and Synergistetes accounted for 99.6% of the sequences. Bacillales, Clostridiales, Lactobacillales, Natranaerobiales, MBA08, Thermoanaerobacterales and two unclassified orders (FB_2) and three (SC_3) were identified among the Firmicutes (Plácido and Zhang 2018). In addition, it has been demonstrated that protein can be broken down in anaerobic digesters by other Firmicutes genera like Aminomonas, Aminobacterium, Peptoniphilus, Gelria, Clostridium, Thermanaerovibrio, Sporanaerobacter, and Proteiniborus. However, anaerobic digesters can be used to metabolize amino acids by species of the Cloacimonetes, Bacteroidetes (Fermentimonas and Proteiniphilum), and Fusobacteria (Westerholm and Schnürer 2019; Hahnke et al. 2016; Stolze et al. 2018). The genus Sporanaerobacter sp. belongs to the Clostridiales order produces VFAs from blood proteins. Hydrolytic bacteria have a wide range of phylogenetic relationships, but the most well-known species are found in the phyla Firmicutes and Bacteroidetes (Venkiteshwaran et al. 2015). In recent times, Xu et al. (2013) reported that Bacteroidetes, Firmicutes, Chloroflexi, and Proteobacteria dominated a substrate rich in protein fed into an anaerobic digester.

Anaerobic digestion of waste having blood has been stigmatized by previous studies due to its long hydraulic retention time (HRT), low organic loading rate (OLR), and low methane and biogas yields. The digester's volume is increased by the last two operational parameters, which results in higher operating and capital costs. The AD of blood has been studied with a variety of reactors. Batch, semi-continuous, and continuous reactors are examples of these.2–3600 L (semi-continuous), 2–47 L (batch), and 1–4000 L (continuous) were the reactor volumes. Normal and sub-atmospheric pressures were used in the operation of batch reactors. An upflow anaerobic filter reactor (UAFR), upflow anaerobic sludge blanket reactor (UASBR), a completely stirred tank reactor (CSTR), and two-stage reactors were all used for the digestion of blood. Biochar and bamboo cylinders, for example, were used as immobilizing materials in some studies. Only a single study provided data from a biogas plant of capacity 7600 m^3 using blood and manure (83% and 17%, respectively) as fuel (Guo et al. 2013). In a semi-continuous 3.6 m^3 working volume continuously mixed digester operated at 35 °C and with an HRT of 18 days, Ozturk (Vrieze et al. 2015) investigated AD using a single substrate of blood. A strong, obnoxious odor was produced, but no methane was observed. Therefore, feeding another organic waste as a co-digestate was suggested, and it was concluded that AD of blood alone is difficult. In a similar manner, Cuetos et al. (2010) reported that 0.047 m^3 biogas per kg VS was produced when batch-digesting poultry blood. This yield is very low in comparison with the theoretical methane potential. The experiment also had a high pH of 8.8 and a lot of ammonia (4500 mg/L). The culture was repressed by the average amount of free ammonia, which was 1813 mg/L. Despite this, Kovács et al. in a reactor at 37 °C for 56 days, digesting precipitated blood protein yielded a specific methane product of 0.447 m^3/kg dry matter of protein. The theoretical yield is fairly close to this yield. Provided that the microbial colony is controlled, well-characterized, and understood, these scientists accentuated that continuous biogas production from the blood protein is possible. Dilution was used by Bauer to weaken blood. Within ten days of incubation, yield of a particular biogas of 0.733 m^3 per kg VS and a specific yield of methane of 0.117 m^3 per kg VS were obtained by consuming blood water mixture from a slaughterhouse in a batch reactor at 35 °C and fed with 0.6 g VS/L (Bauer 2011). Dilution with water increased the yield, despite the fact that the methane yield remained low in comparison with the hypothetical yield of the methane from blood. Cuetos et al., on the other hand, (Cuetos et al. 2010) used co-digestion for getting around the inhibition of ammonia that happened when blood was digested as a single substrate. In a group assay at 34 °C, poultry maize and blood residues (70% and 30% by mass of VS, respectively) were mixed for diluting the nitrogen, resulting in 0.188 m^3 CH_4 per kg VS. However, due to the accumulation of VFAs and non-digested materials, a semi-continuous reactor only produced 0.065 m^3 CH_4 per kg VS. Ozturk (2013) co-digested cattle blood, combining the ideas of co-digestion and dilution: urea from cattle: water with a mass ratio of 30:20:50 in a semi-continuous continuously mixed digester with a working volume of 3.6 m^3 and an HRT of 18 days at 35 °C. In a 30-day study, they reported a COD removal rate of 34% and a production rate of biogas of 0.64 L min^{-1} m^3. By co-digesting condensated plant waste (98% by weight) and blood (2% by weight) in an upflow anaerobic sludge blanket (UASB) reactor at 35 degrees Celsius and an HRT of 15.6 days, Hansen and West (1992) produced 0.14 m^3 biogas m^3 digester. Marcos et al. co-digested a mixture of wastewater and water (1:49 w/w)0.93] at 38 °C in a discontinuous digester, they were able to remove 56.9% COD at an OLR of 0.17 kg COD m^3 per day. Cuetos and others used a semi-continuous anaerobic digester at 32 °C that digested poultry blood with the small organic fraction of municipal solid waste (OFMSW) for avoiding inhibition caused by ammonia accretion. They got specific yields of methane of 0.33 m^3 per kg VSS fed, a production rate of methane of 0.5 m^3 per d, and 60% CH_4 in biogas at an OLR of 1.5 VSS fed m^3 per d. The process was quickly destabilized when the OLR was raised to 2.0 kg VSS fed m^3 per d and the specific yield of methane was reduced to 0.20 m^3 per kg VSS fed. The yield of methane for mixed slaughterhouse waste containing blood increased to 0.33 m^3 per kg VS in a continuous reactor from 0.117 m^3 per kg VS in a batch (Bauer 2011).

14.10 Problems Associated with Anaerobic Digestion of Livestock Blood

The EU has approximately 8000 biogas plants. Despite the fact that such kind of waste is continuously produced in very large quantities approximately around 106 metric tons annually from slaughtering of poultry and cattle throughout the world, none of them digests it as a mono-substrate. This suggests that issues exist with the AD of this kind of waste. The animals' blood killed in an abattoir is included in the category of animal byproducts (ABPs) and is considered safe for being used by humans. So, before being put into a biogas process, blood should be pasteurized and cleaned properly (Kirchmayr et al. 2003). Both the net energy yield and the cost of AD are impacted by this. Overloading and inhibition are two operational issues associated with anaerobic blood digestion. Proteins, which are nitrogen-rich compounds, are abundant in slaughterhouse blood. Carbon to nitrogen ratio is low in it as compared to some other raw biomasses. It has between 81.25 and 197.3 g proteins per kilogram. Nitrogenous compounds are released when protein breaks down, inhibiting the microorganisms responsible for AD's bioreactions. Free (unionized) NH_3-N (FAN) and ionized ammonium N (NH_4), which are collectively referred to as total ammonia nitrogen (TAN), are released as a result of the breakdown of proteins during the process of hydrolysis as well as the acidogenic and acetogenic fermentation pathways. The levels of TAN and FAN are linked to AD inhibition (Prochazka et al. 2012). During blood AD, high total ammonia (NH_3 and NH_4+) levels are observed. Higher concentrations of the NH_3 cause inhibition of the microbes that mediate the AD bioreactors, despite the fact that optimal concentrations of ammonia (NH_3) provide the AD medium with sufficient buffer capacity. Cuetos and others reported that concentrations of ammonium and free ammonia during the AD of remaining poultry blood in a batch reactor at 37 °C and a 1:1 inoculum-to-substrate ratio were 1813 mg per L and 4500 mg per L, respectively. Due to the severe inhibition imposed by this level of free ammonia, VFAs (acetic acid = 2180 mg per L, C3–C5 about 350 mg per L each) accumulated. According to what has been recorded and reported in the researches (Yenigün and Demirel 2013), the broad range of inhibitory levels of NH_3, approximately a 1500–7000 mg TAN per L, is caused by varieties of inoculant substrates, pH, temperature, operation mode, adaptation, and reactor configuration. Methane-producing compounds are primarily inhibited by FAN at higher nitrogen concentrations (Fernandes et al. 2012) when FAN is present in resolution, as opposed to NH_4 (McCarty 1964). NH_3 dissociation and solution concentrations are affected by temperature. At thermophilic, psychrophilic, or mesophilic temperatures, anaerobic digestion can occur. By raising the temperature to 55 °C, the FAN concentration was increased by six times (Kayhanian 1999). As a result, it is suggested that livestock blood at a low temperature regulates NH_3 inhibition during AD. At high VFA and TAN concentrations, anaerobic type of digestion of poultry blood and poultry wastewater (PBPW) was inhibited (USEPA 1973) which measured 15.25 g per L of VFA and 3.1 g per L of TAN, respectively. Methanogens were inhibited by this level of concentration of the TAN, which increased the buffering capability and sustained a pH of 6.6 to 7.7. The concentrations of TAN and VFAs were decreased when the substrate was diluted, but methane yield was not increased (USEPA 1973). According to Nielsen and Angelidaki, feedstock made up of manure approximately 83% by mass and blood up to 17% by mass was fed to a real biogas plant with a capacity of 7600 m^3 and a thermophilic operating temperature of 53 degrees Celsius for 17 to 18 days. They came to the conclusion that the blood caused an imbalance in the performance of the plant and decreased biogas manufacture by 32% with increasing levels of ammonia and VFAs. VFAs accumulated over approximately 2.5 of the HRT, or 45 days, while ammonia concentrations rose immediately. The reactor resumed and recovered its original yield of methane with nearly one HRT after blood feeding was stopped. Mesophilic and thermophilic reactors were used in the laboratory to further simulate this actual occurrence. After feeding blood, the thermophilic reactor saw a significant increase in VFA concentrations (equal to 0.16 g N per L per day), while in contrast, a mesophilic reactor, which also produced more methane due to the fact that blood has a higher methane potential than the cattle manure it replaces, saw a slight increase only in concentrations of VFA. Six days after the blood feeding was started, thermophilic reactor showed a lower methane yield. Methane production was not resumed till the experiment was over, despite the fact of stopping blood feeding. Two weeks after the blood was fed into the mesophilic reactor, there were inhibition signs and a reduction in methanogenesis. However, the inhibition concentration of 0.7–1.0 g N per L suggested by Hansen et al. and Angelidaki and Ahring (1993) was lower than its free NH_3 concentration (Hansen et al. 1998). The operation temperature had a significant effect on the stability of the co-digestion of manure and blood, and Angelidaki and Nielsen discovered that both the free NH_3 and few of the components of the blood could destabilize the reactor. They came to the conclusion that when blood weighs 17% by weight, combined digestion of manure and blood is impossible at mesophilic or thermophilic temperatures. In order to avoid inhibition, they suggested feeding blood at a lower percentage.

Sulfide production when sulfur-containing proteins are broken down is another problem. During the AD of blood, sulfides are corrosive, toxic, and produce an unpleasant and strong odor (Ozturk 2013). The primary cause of the smell is the hydrogen sulfide (H_2S), which smells like rotten eggs. Additionally, the occurrence of sulfur enables sulfate-reducing bacteria (SRB) to switch electron equivalent entities from pathways that produce methane toward those that produce H_2S. Methanogens are well-known to be strong competitors for the SRBs. For every gram of sulfide produced, approximately 0.7 L methane is lost (Rittmann and McCarty 2001). Methanogenic cultures typically expand slowly (with the doubling duration at 35 °C is around four days) (Aniebo et al. 2009). In addition, only 8% of the electron equivalent of the substrate—protein—is used to synthesize new microbial biomass, while 28% of the electron equivalent of carbohydrates is used (Rittmann and McCarty 2001; Mosey and Fernandes 1989; Mosey 1983). Because of this, the AD of proteins results in the formation of fewer new cells of microorganisms. A culture of methanogenic bacteria develops slowly. It will be susceptible to disturbances and have a difficult time recovering when it gets exposed to the toxicity of sulfides and ammonia. Additionally, the SBRs' faster growth rate than methanogens enhances the difficulty. In addition to sulfate reduction, acetate is produced by some SRBs through proton-reducing and fermentative reactions (Raskin et al. 1996; Widdel and Hansen 1992). As a result, they are more adaptable and tolerant of reactor changes than the meticulous methanogens. Desulfomicrobium, Desulfobulbus, and Desulfobotlus are examples of SRB microorganisms that produce acetate (Saady 2011). In addition, when sulfate is absent, SRB groups like Desulfosarcina, Desulfococcus, and Desulfobotulus are highly viable for H_2 (Raskin et al. 1996). As a result, they divert hydrogen away from methane formation and deprive hydrogenotrophic methanogens of it.

14.11 Strategies for Anaerobic Digestion of Blood

There has been a number of attempts to reduce and manage the inhibition caused by blood protein degradation byproducts. Optimizing the C by acclimating the higher levels of NH_3 by microorganisms levels. Strategies to control NH3 inhibition include maintaining a low N ratio and controlling pH (Kayhanian 1999). Cuetos and others looked into how the addition of activated carbon affected the mass ratios of blood to carbon of 4.5, 3.0, and 1.5. They discovered that the specific yield of methane (l per kg VS) increased from 46.5 to 317.4 for batch reactor and 216 for semi-continuous reactor, at a minimum ratio of 1.5. However, a study found that the high cost of activated carbon would make such an application unprofitable on an industrial scale. Temperature increases activated carbon's ability to adsorb proteins (Silva et al. 2017) in a thermophilic operation, this may offer a potential solution for blood digestion. Because it increases the surface area, adding more activated carbon increases protein adsorption, resulting in less protein degradation and less ammonia available to inhibit microorganisms. Cuetos et al. (2010) found that the low adsorptive capacity and uneven pattern of the experimental affinity between activated acetic acid and carbon (Niu et al. 2008) eliminated the possibility that the observed improvement in blood AD was due to VFAs adhering to the activated carbon. They, on the other hand, recommended that promoting metabolism of microbes could be the factor. To lessen the inhibitory effects by TAN on methanogens, dilutions of the substrate rich in nitrogen have been used. Wang (USEPA 1973) digested a mixture of poultry blood and poultry wastewater (PBPW) with a biochar-filled anaerobic filter (BFAF). The substrate was diluted with the biochar. It was determined that, at moderate OLRs, diluting the substrates of nitrogen slightly increased removal of COD. However, at high OLR and short HRT values, this improvement was lost (USEPA 1973). Because it may have shielded some methanogens from the inhibition that was imposed by TAN, the porous structure of biochar may have been the reason for this improvement. Utilizing granulated microbial culture may also provide such protection, which is relevant to the speculated mechanism. It has been established that culture granulation can be used to control inhibition caused due to long-chain fatty acids. In recent times, Xu et al. (2020) discovered that biochar derived from sewage sludge had little effect on the AD of substrates rich in protein substrates in comparison with carbohydrate-

and lipid-rich substrates. In protein-rich substrate AD, it was discovered that there is no promotion of the growth of methanogens by use of biochar. Jedrzejewska and others examined the effect of reduced pressure or vacuum on production of blood methane in a digester at laboratory-scale operating at 32 °C. This digester operated at sub-atmospheric level of pressure and an average pH of 7.92 was maintained regardless of the applied OLR, in contrast to the digester which was controlled, and was maintained at atmospheric pressure. Within 30 days of the experiment, the pH of the control digester dropped to 5.53 as organics increased systematically. The percentage of the CH_4, NH_3, and H_2S in biogas increased by level of 5–15%, 3–4%, and 1%, respectively, due to the decreased headspace pressure (Jedrzekewska et al. 2006). A successful study was done by controlling the inhibition imposed by VFAs and NH_3 accumulated by the digestion of rumen and cattle blood using two-stage anaerobic digesters. The first stage had a shorter retention of time for liquids but had a longer retention time for the solids. It also reduced solids by 87%, whereas a single-stage digester only reduced solids by 50% (the same retention time for liquids and solids). The second-stage digester yielded methane specifically of about 0.12–0.25 m^3 CH^4 per kg. COD was removed at HRTs of 10 and 2 days and OLRs of 0.58–7.0 kg from COD m^3 per day. Given that the methane yield was reported per COD deleted, the upper perimeter was low. Optimizing the C by acclimating the microorganisms to high NH_3 levels.

14.12 Strategies to Control NH3 Inhibition

These include maintaining a low N ratio, controlling pH, and (Kayhanian 1999). Alkalinity rises in proteins like blood that are affected by AD. When blood protein reacts with CO_2 produced by anaerobic bioreactions, it releases free NH3 , which results in the formation of ammonium bicarbonate. The alkalinity is raised by ammonium bicarbonate (Khanal 2008; Padilla-Gasca et al. 2011).

Blanc et al. (1999) found that when blood is digested with a substrate high in carbohydrates, it increases the buffering capacity. Protein and nitrogen are abundant in blood, therefore, in order to adjust the carbon, it should be degraded alongside some substrates with a high content of carbon ratio of nitrogen within the limits that are best for AD. Blood and feedstocks like paunch, manure, and wheat straw could be digested together. Anaerobic microbial cultures receive powerful and nutrient-rich components from blood.

Only a number of studies have looked into the microbiology of AD of blood to figure out how different trophic groups of microorganisms work, how their dynamics change, and eventually what is the cause. As a result, studies from engineering reactor of blood AD must be combined with advanced molecular biology techniques. Further research and quantification of the kinetics of the AD bioreactions of blood like acidogenesis, hydrolysis, methanogenesis, and acetogenesis are required. In particular, it is necessary to investigate the inhibition and enzymes kinetics that mediate such bioreactions. Although previously found studies demonstrated that it might be an approach to accomplish the inhibition executed by the nitrogen containing compounds released after degradation of blood proteins, there are no researches for blood AD at low temperatures. Adsorbents' ability to boost yield of methane and lessen inhibition in the processing of blood AD must be demonstrated and deciphered through additional research. It is necessary to quantify the blood AD kinetics underneath the influence of an adsorbent.

14.13 Conclusion

Rendering was once the most popular method for treating solid slaughterhouse wastes, and it gave the facilities a second source of income. The commercial worth of these products has been severely diminished due to the TSE danger, and in many instances, they must now be disposed of as waste on their own. Thus, the price of safely disposing of slaughterhouse waste has significantly increased in recent years. This is primarily because such wastes include microorganisms that pose health concerns.

One option for disposing slaughterhouse wastes is composting. The procedure has a number of advantages, including less environmental side effects, the creation of a useful byproduct, and the elimination of the majority of infections. However, careful control is necessary for the fruitful conversion of these wastes into high-quality compost. When carried out under strict monitoring, the finished product should not be harmful to either human or animal health. However, some diseases, such as spore-forming bacteria and prions, cannot be eliminated by composting. The alkaline hydrolysis of slaughterhouse wastes procedure is a recent development. It hydrolyzes the content of biological materials and converts it to a sterile water-based aqueous solution that is composed of amino acids, peptides, soaps, and sugars by using heat, a strong base, and temperature. Although this effluent can be discharged into a sanitary sewer, due to having high alkalinity and high content of nutrients, it may also cause issues. The removal of numerous infections and prions from corpses and animal waste has been found to be exceedingly effective. However, because the process's waste is so nutrient-rich, it has a significant potential for producing biogas. One of the most promising current techniques for removing waste from slaughterhouses is AD. In addition to a digestate that can be utilized as valuable fertilizer, this process also generates heat and biogas, both of which can be used to create energy. Additionally, waste from slaughterhouses is abundant in nitrogen and proteins. Anaerobic digestion of blood using it as a single substrate is challenging but is achievable with cautious system control, based on accurate identification and correct comprehension of the colonies of microbes and its metabolizing processes. A specific yield of methane of 0.45 m^3 per kg blood protein content has been attained as a result. When the mixture is properly constructed, co-digestion process of blood along with other feedstock continues successfully. According to current data, the specific output of methane from co-digestion of blood with other substrates varies between 0.7 and 0.1 m^3 per kg VS. For improved control and management, more study is needed in the areas of microbiology, kinetics, adsorbents' function, and how it affects the dynamics of growth pattern of the microbes, reactor operation and designs, and adaptation of culture during anaerobic digestion of blood.

References

Afazeli H, Jafari A, Rafiee S, Nosrati M (2014) An investigation of biogas production potential from livestock and slaughterhouse wastes. Renew Sustain Energy Rev 34:380–386. [CrossRef]
Alfreider A, Peters S, Tebbe CC, Rangger A, Insam H (2002) Microbial community dynamics during composting of organic matter as determined by 16S ribosomal DNA analysis. Compost Sci Util 10:303–312
Alvarez R, Riera VH, Lidén G (2006) Batch co-digestion of manure, solid slaughterhouse waste, and fruit and vegetable waste. Boliv. J. Chem 23:8
Angelidaki I, Ahring BK (1993) Thermophilic anaerobic digestion of livestock waste: the effect of ammonia. Appl Microbiol Biotechnol 38:560–564. [CrossRef]
Aniebo AO, Wekhe SN, Okoli IC (2009) Abattoir blood waste generation in rivers state and its environmental implications in the Niger Delta. Toxicol Environ Chem 91:619–625. [CrossRef]
Anon (2002) The animal by-products regulations (EC) no. 1774/2002. European Commission, Brussels
Arvanitoyannis IS, Ladas D (2008) Meat waste treatment methods and potential uses. Int J Food Sci Tech 43:543–559
Avcioglu AO, Turker U (2012) Status and potential of biogas energy from animal wastes in Turkey. Renew Sustain Energy Rev 16:1557–1561. [CrossRef]
Bagge E, Sahlström L, Albihn A (2005) The effect of hygienic treatment on the microbial flora of biowaste at biogas plants. Water Res 39:4879–4886
Bah CSF, Bekhit AEA, Carne A, McConnell MA (2013) Slaughterhouse blood: an emerging source of bioactive compounds. Compr Rev Food Sci 12:314–331. [CrossRef]
Bailey KL, Lazarovits LG (2003) Suppressing soil-borne diseases with residue management and organic amendments. Soil till Res 72:169–180
Banks CJ, Wang Z (2006) Treatment of meat wastes. In: Waste treatment in food processing industry, Wang L, Hung YT, Lo HH, Yapijakis K (eds). Taylor and Francis Group, New York, NY, USA, pp. 67–100. ISBN 042919109. [CrossRef]
Banks C, Zhang Y (2012) Optimising inputs and outputs from anaerobic digestion processes. DEFRA, London, UK
Bauer A (2011) Investigation into the biochemical methane potential of abattoir wastewater. Bachelor's Thesis, University of Southern Queensland, Darling Heights, Australia. Available online: https://core.ac.uk/download/pdf/11049339.pdf (Accessed on 16 Jun 2021)
Beffa T, Blanc M, Lyon PF, Vogt G, Marchiani M, Fischer JL, Aragno M (1996) Isolation of Thermus strains from hot composts (60 to 80 degrees C). Appl Environ Microbiol 62:1723–1727
Blanc M, Marilley L, Beffa T, Aragno M (1999) Thermophilic bacterial communities in hot composts as revealed by most probable number counts and molecular (16S rDNA) methods. FEMS Microbiol Ecol 28:141–149
Böhnel H, Briese B-H, Gessler F (2002) Methods for health risk assessment by Clostridium botulinum in biocompost. In: Insam H, Riddech N, Klammer S (eds) Microbiology of composting. Springer, Berlin, Germany, pp 517–526
Cekmecelioglu D, Demirci A, Graves RE, Davitt NH (2005) Applicability of optimized in-vessel food waste composting for windrow system. Biosyst Eng 91:479–486
Chen SY, Dong XZ (2004) Acetanaerobacterium elongatum gen. nov., sp. nov., from paper mill waste water. Int J Syst Evolut Microbiol 54:2257–2262. [CrossRef]

Cuetos MJ, Gómez X, Otero M, Morán A (2010) Anaerobic digestion and co-digestion of slaughterhouse waste (SHW): influence of heat and pressure pre-treatment in biogas yield. Waste Manag 30:1780–1789

Cuetos MJ, Moran A, Otero M, Gomez X (2009) Anaerobic co-digestion of poultry blood with OFMSW: FTIR and TG-DTG study of process stabilization. Environ Technol 30:571–582. [CrossRef]

Danon M, Franke-Whittle IH, Insam H, Chen Y, Hadar Y (2008) Molecular analysis of bacterial community succession during prolonged compost curing. FEMS Microbiol Ecol 65:133–144

Dees PM, Ghiorse WC (2001) Microbial diversity in hot synthetic compost as revealed by PCR-amplified rRNA sequences from cultivated isolates and extracted DNA. FEMS Microbiol Ecol 35:207–216

Del Hoyo P, Rendueles M, Diaz M (2008) Effect of processing on functional properties of animal blood plasma. Meat Sci 78:522–528. [CrossRef] [PubMed]

Deublein D, Steinhauser A (2008) Biogas from Waste and renewable resources: an introduction. Wiley-VCH, Weinheim, Germany

De Vrieze J, Saunders AM, He Y, Fang J, Nielsen PH, Verstraete W, Boon N (2015) Ammonia and temperature determine potential clustering in the anaerobic digestion microbiome. Water Res 75:312–323. [CrossRef]

Edström M, Nordberg A, Thyselius L (2003) Anaerobic treatment of animal byproducts from slaughterhouses at laboratory and pilot scale. Appl Biochem Biotechnol 109:127–138

European Commission (2005) Reference document on best available techniques in the slaughterhouses and animal by-products industries. European Commission, Brussels, Belgium. Available online: https://eippcb.jrc.ec.europa.eu/sites/default/files/2020-01/sa_bref_0505.pdf. Accessed on 18 May 2018

Fallows SJ, Verner Wheelock J (1982) By-products from the U.K. food system 2. The meat industry. Conserv Recycl 5:173–182. [CrossRef]

FAOSTAT (2017) Food and agriculture organization of the United Nations; FAO: Rome, Italy. Available online: http://www.fao.org/faostat/en/#home (Accessed on 20 May 2018)

Fearon J, Mensah B, Boateng V (2014) Abattoir operations, waste generation and management in the Tamale Metropolis: case study of the Tamale slaughterhouse. J Public Health Epidemiol 6:14–19. [CrossRef]

Fernandes TV, Keesman KJ, Zeeman G, van Lier JB (2012) Effect of ammonia on the anaerobic hydrolysis of cellulose and tributyrin. Biomass Bioenergy 47:316–323. [CrossRef]

Fischer MA, Ulbricht A, Neulinger SC, Refai S, Waßmann K, Künzel S, Schmitz RA (2019) Immediate effects of ammonia shock on transcription and composition of a biogas reactor. Microbiome 10:2064. [CrossRef]

Frank JA, Arntzen MØ, Sun L, Hagen LH, McHardy AC, Horn SJ, Eijsink VGH, Schnürer A, Pope PB, Summers ZM (2016) Novel syntrophic populations dominate an ammonia-tolerant methanogenic microbiome. Msystems 1:e00092–16. [CrossRef] [PubMed]

Franke-Whittle IH, Knapp BA, Fuchs J, Kaufmann R, Insam H (2009) Application of COMPOCHIP microarray to investigate the bacterial communities of different composts. Microb Ecol 57:510–521

Gale P (2004) Risks to farm animals from pathogens in composted catering waste containing meat. Vet Rec 155:77–82

Gonzalez-Fernandez C, Sialve B, Molinuevo-Salces B (2015) Anaerobic digestion of microalgal biomass: Challenges, opportunities and research needs. Bioresour Technol 198:896–906. [CrossRef] [PubMed]

Gorbatov VM (1990) Collection and utilization of blood and blood proteins for edible purposes in the USSR. Adv Meat Res 5:167–195

Gray NF (2004) Biology of wastewater treatment. Imperial College Press, London, UK. ISBN 1860943284

Gunnerson, CG, Stuckey DC. (1986). Anaerobic digestion: principles and practice of biogas systems (World Bank Technical Paper No. 49). Washington, DC, World Bank

Guo X, Wang C, Sun F, Zhu W, Wu W (2013) A comparison of microbial characteristics between the thermophilic and mesophilic anaerobic digesters exposed to elevated food waste loadings. Bioresour Technol 152:420–428. [CrossRef]

Gwyther CL, Williams AP, Golyshin PN, Edwards-Jones G, Jones DL (2011) The environmental and biosecurity characteristics of livestock carcass disposal methods: a review. Waste Manag 31:767–778

Hahnke S, Langer T, Koeck DE, Klocke M (2016) Description of Proteiniphilum saccharofermentans sp. nov., Petrimonas mucosa sp. nov. and Fermentimonas caenicola gen. nov., sp. nov., isolated from mesophilic laboratory-scale biogas reactors, and emended description of the genus Proteiniphilum. Int J Syst Evolut Microbiol 66:1466–1475. [CrossRef]

Hansen CL, West GT (1992) Anaerobic-digestion of rendering waste in an upflow anaerobic sludge blanket digester. Bioresour Technol 41:181–185. [CrossRef]

Hansen KH, Angelidaki I, Ahring BK (1998) Anaerobic digestion of swine manure: Inhibition by ammonia. Water Res 32:5–12. [CrossRef]

Hejnfelt A, Angelidaki I (2009) Anaerobic digestion of slaughterhouse by-products. Biomass Bioenergy 33:1046–1054. [CrossRef]

Hernandez-Eugenio G, Fardeau ML, Cayol JLA, Patel BKC, Thomas P, Macarie H, Garcia JL, Ollivier B (2002) Clostridium thiosulfatireducens sp nov., a proteolytic, thiosulfate- and sulfur-reducing bacterium isolated from an upflow anaerobic sludge blanket (UASB) reactor. Int J Syst Evolut Microbiol 52:1461–1468. [CrossRef]

Hogg S (2005) Essential microbiology, 2nd ed. Wiley, New York, NY, USA. ISBN 0471497533

Hoitink H, Boehm M (1999) Biocontrol within the context of soil microbial communities: a substrate-dependent phenomenon. Annu Rev Phytopathol 37:427–446

Huang Y, Sun Y, Ma S, Chen L, Zhang H, Deng Y (2013) Isolation and characterization of Keratinibaculum paraultunense gen. nov., sp. nov., a novel thermophilic, anaerobic bacterium with keratinolytic activity. FEMS Microbiol Lett 345:56–63. [CrossRef] [PubMed]

Ibekwe AM, Papiernik SK, Gan J, Yates SR, Crowley DE, Yang CH (2001) Microcosm enrichment of 1,3-dichloropropene-degrading soil microbial communities in a compost-amended soil. J Appl Microbiol 91:668–676

Insam H, Franke-Whittle IH, Goberna M (2010) Microbes at work. From wastes to resources. Springer, Heidelberg

Insam H, Wett B (2008) Control of GHG emission at the microbial community level. Waste Manag 28:699–706

Ivanov V (2015) Environmental microbiology for engineers. CRC Press, New York, NY, USA. ISBN 9780429172861

Jayathilakan K, Sultana K, Radhakrishna K, Bawa AS (2012) Utilization of byproducts and waste materials from meat, poultry and fish processing industries: a review. J Food Sci Technol 49:278–293. [CrossRef]

Jedrzekewska M, Kozak K, Krzemieniewski M (2006) Effect of reduced pressure on anaerobic blood biodegradation. Pol J Environ Stud 15:87–93

Johns MR (1995) Developments in wastewater treatment in the meat processing industry: a review. Bioresour Technol 54:203–216. [CrossRef]

Kang S, Jeong C, Seo D, Kim SY, Cho J (2019) Liquid fertilizer production by alkaline hydrolysis of carcasses and the evaluation of developed fertilizer in hot pepper cultivation. Process Safety Environ Protect 122:307–312. ISSN:0957-5820. https://doi.org/10.1016/j.psep.2018.12.017

Kaye G, Weber P, Evans A, Venezia R (1998) Efficacy of alkaline hydrolysis as an alternative method for treatment and disposal of infectious animal waste. Contemp Top Lab Anim Sci 37:43–46

Kayhanian M (1999) Ammonia inhibition in high-solids biogasification: an overview and practical solutions. Environ Technol 20:355–365. [CrossRef]

Khanal SK (2008) Environmental factors. In: Anaerobic biotechnology for bioenergy production, Khanal SK (ed). Wiley, Des Moines, IA, USA

Kiepper BH, Merka WC, Fletcher DL (2008) Proximate composition of poultry processing wastewater particulate matter from broiler slaughter plants. Poult Sci 87:1633–1636. [CrossRef]

Kirchmayr R, Scherzer R, Baggesen DL, Braun R, Wellinger A (2003) Animal by-products and anaerobic digestion. BIOEXELL—Biogas center of excellence, Vienna, Austria. Available online: https://provisioncoalition.com/Assets/ProvisionCoalition/Documents/FoodWasteManagementSolutions/IEA_ABP-Brochure_en_2.pdf (Accessed on 13 Jun 2018)

Kompostverordnung (2002) Österreichisches Bundesgesetzblatt II, Nr. 292/2001. (Austrian Compost ordinance)

Kovacs E, Wirth R, Maroti G, Bagi Z, Nagy K, Minarovits J, Rakhely G, Kovacs KL (2015) Augmented biogas production from protein-rich substrates and associated metagenomic changes. Bioresour Technol 178:254–261. [CrossRef] [PubMed]

Lastella G, Testa C, Cornacchia G, Notornicola M, Voltasio F, Sharma VK (2002) Anaerobic digestion of semi-solid organics waste: biogas production and its purification. Energ Convers Manage 43:63–75

Lee J, Han G, Shin SG, Koo T, Cho K, Kim W, Hwang S (2016) Seasonal monitoring of bacteria and archaea in a fullscale thermophilic anaerobic digester treating food waste-recycling wastewater: Correlations between microbial community characteristics and process variables. Chem Eng J 300:291–299. [CrossRef]

Marcos AC, Al-Kassir A, Cuadros F, Yusaf T (2017) Treatment of slaughterhouse waste water mixed with serum from lacteal industry of extremadura in Spain to produce clean energy. Energies 10:765. [CrossRef]

McCarty PL (1964) Anaerobic waste treatment fundamentals III. Public Works 50:91. Available online: https://citeseerx.ist.psu.edu/viewdoc/download?doi=10.1.1.456.3796&rep=rep1&type=pdf (Accessed on 13 Jun 2018)

Monteith HD, Shannon EE, Derbyshire JB (1986) The inactivation of a bovine enterovirus and a bovine parvovirus in cattle manure by anaerobic digestion, heat treatment, gamma irradiation, ensilage and composting. J Hyg (lond) 97:175–184

Mosey FE (1983) Mathematical modelling of the anaerobic digestion process: Regulatory mechanisms for the formation of short-chain volatile acids from glucose. Water Sci Technol 15:209–232. [CrossRef]

Mosey FE, Fernandes XA (1989) Patterns of hydrogen in biogas from the anaerobic-digestion of milk-sugars. Water Sci Technol 21:187–196. [CrossRef]

Moskowitz S (2012) Anaerobic digestion of slaughterhouse waste. University of California, Davis, CA, USA

Muller B, Sun L, Westerholm M, Schnurer A (2016) Bacterial community composition and fhs profiles of low- and high-ammonia biogas digesters reveal novel syntrophic acetate-oxidising bacteria. Biotechnol Biofuels 9:1–18. [CrossRef] [PubMed]

NABC (2004) Carcass disposal: a comprehensive review. Report written for the USDA animal and plant health inspection service. National Agricultural Biosecurity Centre, Kansas State University, USA

Niu L, Song L, Dong X (2008) Proteiniborus ethanoligenes gen. nov., sp. nov., an anaerobic protein-utilizing bacterium. Int J Syst Evol Microbiol 58:12–16. [CrossRef]

Noble R, Roberts SJ (2004) Eradication of plant pathogens and nematodes during composting: a review. Plant Pathol 53:548–568

Nwachukwu MI, Akinde SB, Udujih OS, Nwachukwu IO (2011) Effect of abattoir wastes on the population of proteolytic and lipolytic bacteria in a recipient water body (Otamiri River). Glob. Res. J. Sci 1:40–42

Okanović Đ, Ristic M, Kormanjos S, Filipovic S, Zivkovic B (2009) Chemical characteristics of poultry slaughterhouse by products. Biotechnol Anim Husb 25:143–152. [CrossRef]

Oladele SB, Samuel J (2014) Whole blood coagulation time, haematocrit, haemoglobin and total protein of turkeys reared in Zaria, Nigeria. Sokoto J Vet Sci 12:13–17. [CrossRef]

Omole DO, Longe EO (2008) An assessment of the impact of abattoir effluents on River Illo, Ota, Nigeria. J Environ Sci Technol 1:56–64. [CrossRef]

Owens, C.M.; Alvarado, C.; Sams, A.R. (2000). Poultry Meat Processing, 1st ed.; CRC Press, Boca Raton, FL, USA. ISBN 0849301203

Ozturk B (2013) Evaluation of biogas production yields of different waste materials. Earth Sci Res 2:165. [CrossRef]

Padilla-Gasca E, López-López A, Gallardo-Valdez J (2011) Evaluation of stability factors in the anaerobic treatment of slaughterhouse wastewater. J Bioremediat Biodegrad 2:114. [CrossRef]

Palatsi J, Viñas M, Guivernau M, Fernandez B, Flotats X (2011) Anaerobic digestion of slaughterhouse waste: main process limitations and microbial community interactions. Bioresour Technol 102:2219–2227

Plácido J, Zhang Y (2018) Production of volatile fatty acids from slaughterhouse blood by mixed-culture fermentation. Biomass Convers Biorefinery 8:621–634. [CrossRef]

Prochazka J, Dolejs P, Maca J, Dohanyos M (2012) Stability and inhibition of anaerobic processes caused by insufficiency or excess of ammonia nitrogen. Appl Microbiol Biotechnol 93:439–447. [CrossRef] [PubMed]

Raskin, L.; Rittmann, B.E.; Stahl, D.A. (1996). Competition and coexistence of sulfate-reducing and methanogenic populations in anaerobic biofilms. Appl Environ Microbiol 62:3847–3857. [CrossRef] [PubMed]

Reuter T, Alexander TW, McAllister TA (2011) Viability of Bacillus licheniformis and Bacillus thuringiensis spores as a model for predicting the fate of bacillus anthracis spores during composting of dead livestock. Appl Environ Microbiol 77:1588–1592

Rittmann BE, McCarty PL (2001) Environmental biotechnology: principles and applications. McGraw-Hill, Boston, MA, USA. ISBN 0071181849

Röske I, Sabra W, Nacke H, Daniel R, Zeng AP, Antranikian G, Sahm K (2014) Microbial community composition and dynamics in high-temperature biogas reactors using industrial bioethanol waste as substrate. Appl Microbiol Biotechnol 98:9095–9106. [CrossRef]

Ryckeboer J, Mergaert J, Coosemans J, Deprins K, Swings J (2003) Microbiological aspects of biowaste during composting in a monitored compost bin. J Appl Microbiol 94:127–137

Saady NMC (2011) Effects of long-chain fatty acids on culture dynamics in hydrogen fermentation. Ph.D. Thesis, University of Windsor, Windsor, ON, Canada. Available online: https://scholar.uwindsor.ca/cgi/viewcontent.cgi?article=5773&context=etd (Accessed on 13 Jun 2018)

Sahlström L (2003) A review of survival of pathogenic bacteria in organic waste used in biogas plants. Bioresour Technol 87:161–166

Sahlström L, Bagge E, Emmoth E, Holmqvist A, Danielsson-Tham ML, Albihn A (2008) A laboratory study of survival of selected microorganisms after heat treatment of biowaste used in biogas plants. Bioresour Technol 99:7859–7865

Saidu M, Musa JJ (2012) Impact of abattoir effluent on river Landzu, Bida Nigeria. J Chem Biol Phys Sci 2:5

Salminen E, Rintala J (2002b) Anaerobic digestion of organic solid poultry slaughterhouse waste—a review. Bioresour. Technol 83:13–26. [CrossRef]

Salminen E, Rintala J (2002a) Anaerobic digestion of organic solid poultry slaughterhouse waste–a review. Bioresour Technol 83:13–26

Schink B (1997) Energetics of syntrophic cooperation in methanogenic degradation. Microbiol Mol Biol Rev 61:262–280

Sedlak RI (1991) Phosphorus and nitrogen removal from municipal wastewater: principles and practice, 2nd ed. Taylor and Francis, Abingdon, UK. ISBN 9780873716833

See AM, Swindells KL, Sharman MJ, Haack KL, Goodman D, Delaporta A, Robertson I, Foster SF (2009) Activated coagulation times in normal cats and dogs using MAX-ACT tubes. Aust Vet J 87:292–295. [CrossRef]

Silva KCG, Amaral TN, Junqueira LA, de Oliveira Leite N, de Resende JV (2017) Adsorption of protein on activated carbon used in the filtration of mucilage derived from Pereskia aculeata Miller. S Afr J Chem Eng 23:42–49. [CrossRef]

Smith K, Grylls J, Metcalfe P, Jeffrey B, Sinclair A (2007) Nutrient value of digestate from farm-based biogas plants in Scotland, SAC Commercial Ltd, Edinburgh, Scotland. Available online: http://docshare01.docshare.tips/files/25098/250982493.pdf (Accessed on 18 Jun 2018)

Spillman SK, Taub F, Schwyzer M, Wyler R (1987) Inactivation of animal viruses during sewage sludge treatment. Appl Environ Microbiol 5:2077–2081
Srivastava A, Prasad R (2000) Triglycerides-based diesel fuels. Renew Sust Energ Rev 4:111–133
Steffen R, Szolar O, Braun R (1998) Feedstocks for anaerobic digestion. Institute for Agrobiotechnology, Vienna, Austria
Stolze Y, Bremges A, Maus I, Pühler A, Sczyrba A, Schlüter A (2018) Targeted in situ metatranscriptomics for selected taxa from mesophilic and thermophilic biogas plants. Microb Biotechnol 11:667–679. [CrossRef] [PubMed]
Strauch D (2002) Occurrence of microorganisms pathogenic for man and animals in source separated biowaste and compost importance, control, limits and epidemiology. In: De Bertoldi M, Sequi P, Lemmes B, Papi T (eds) The sciences of composting. Blackie Academic and Professional, Glasgow, UK, pp 224–232
Tchobanoglous G, Burton FL, Eddy M (1991) Wastewater engineering: treatment, disposal, and reuse. McGraw-Hill, New York, NY, USA
Thabet OB, Fardeau ML, Joulian C, Thomas P, Hamdi M, Garcia JL, Ollivier B (2004) Clostridium tunisiense sp. nov., a new proteolytic, sulfur-reducing bacterium isolated from an olive mill wastewater contaminated by phosphogypse. Anaerobe 10:185–190. [CrossRef]
Tritt WP, Schuchardt F (1992) Materials flow and possibilities of treating liquid and solid-wastes from slaughterhouses in Germany—a review. Bioresour Technol 41:235–245. [CrossRef]
Urlings HA, van Logtestijn JG, Bijker PG (1992) Slaughter by-products: problems, preliminary research and possible solutions. Vet Q 14:34–38
USEPA (1973) In-process pollution abatement—upgrading poultry processing facilities to reduce pollution; USEPA: Washington, DC, USA, EPA 625-3-73-001
USEPA (2002) Development document for the proposed effluent limitations guidelines and standards for the meat and poultry products industry point source category. EPA Number: 821-B-01-007. USEPA, Washington, DC, USA. Available online: https://nepis.epa.gov/Exe/ZyPDF.cgi/20002F0Q.PDF?Dockey=20002F0Q.PDF (Accessed on 18 Jun 2018)
Vandeviviere P, De Baere L, Verstraete W (2002) Types of anaerobic digester for solid wastes. In: Mata-Alvarez J (ed) Biomethanization of the organic fraction of municipal solid wastes. IWA Publishing, London, pp 111–137
Venkiteshwaran K, Bocher B, Maki J, Zitomer D (2015) Relating anaerobic digestion microbial community and process function: supplementary issue: Water microbiology. Microbiol Insights 8(Suppl. 2):37–44. [CrossRef]
Wang S (2015) Anaerobic digestion of poultry processing wastes for bioenergy and nutrients recovery; University of Georgia: Athens. GA, USA
Westerholm M, Schnürer A (2019) Microbial responses to different operating practices for biogas production systems. In: Anaerobic Digestion, Banu JR (ed). IntechOpen, London, UK, pp 1–36. ISBN 9781838818500. Available online: https://www.intechopen.com/chapters/65614 (Accessed on 11 May 2018)
Widdel F, Hansen TL (1992) The dissimilatory sulfate- and sulfur-reducing bacteria. In: The prokaryotes, 2nd ed, Balows A, Trüper HG, Dworkin M, Harder W, Schleifer KH (eds), vol 1. Springer, New York, NY, USA, pp 583–624
Willey J, Sherwood L, Woolverton C (2013) Prescott's microbiology. McGraw-Hill Education, New York, NY, USA. ISBN 9780073402406
Wu P, Mittal GS (2012) Characterization of provincially inspected slaughterhouse wastewater in Ontario, Canada. Can Biosyst Eng 54:6.9–6.18. Available online: https://library.csbe-scgab.ca/docs/journal/54/C12062.pdf (Accessed on 8 Sept 2021). [CrossRef]
Xu R, Zhang K, Liu P, Khan A, Xiong J, Tian FK, Li XK (2018) A critical review on the interaction of substrate nutrient balance and microbial community structure and function in anaerobic co-digestion. Bioresour Technol 247:1119–1127. [CrossRef]
Xu Q, Liao Y, Cho E, Ko JH (2020) Effects of biochar addition on the anaerobic digestion of carbohydrate-rich, protein-rich, and lipid-rich substrates. J Air Waste Manag Assoc 70:455–467. [CrossRef]

Yenigün O, Demirel B (2013) Ammonia inhibition in anaerobic digestion: a review. Process Biochem 48:901–911. [CrossRef]
Yoon YM, Kim SH, Oh SY, Kim CH (2014) Potential of anaerobic digestion for material recovery and energy production in waste biomass from a poultry slaughterhouse. Waste Manag 34:204–209. [CrossRef]
Zhang Y, Banks CJ (2012) Co-digestion of the mechanically recovered organic fraction of municipal solid waste with slaughterhouse wastes. Biochem Eng J 68:129–137. [CrossRef]

Chapter 15
Role of Microbes in Sustainable Utilization of Animal Wastes

Iram Liaqat, Faiza Bashir, Urooj Zafar, Uzma Hanif, Saiqa Andleeb, Sadiah Saleem, and Muhammad Arshad

Abstract Various human, industrial, and agricultural operations generate a wide variety of biomass wastes, ample with organic and inorganic resources along with pathogenic microorganisms. Environmental conservation is of the utmost significance. Researchers have been looking for naturally occurring technologies to improve the regulation of agricultural and animal wastes. The biggest threat to mankind is the persistent release of hazardous wastes and toxins as a byproduct of faulty industrialization. Urban and industrial wastes, toxins and animal wastes have been improperly and unscientifically managed, putting the ecology and ecosystem in jeopardy of viability. In order to establish a safe and habitable ecology for future generations, it is now necessary to repair and clean up the contaminated environment. It is well known that waste creation and economic growth are significantly associated, both in developed Western nations and in emerging nations, such as Pakistan. More efficient approaches are required for the treatment of potentially toxic wastes. Microorganisms have a potential future in this area. The distinctive characteristics of microorganisms can be efficiently employed to revive the environment. Microorganisms are used as "miracle cures" for biodegradation and the repair of polluted environments. Likewise, the application of genetically modified organisms (GMOs) in highly contaminated environments makes the microorganisms advantageous for human well-being. This chapter provides information on various types of wastes and elaborates how microorganisms can be employed productively for waste management and more eco-sustainable environments.

I. Liaqat (✉) · F. Bashir
Microbiology Lab, Department of Zoology, Government College University, Lahore, Pakistan
e-mail: dr.iramliaqat@gcu.edu.pk

U. Zafar · S. Saleem
Department of Microbiology, University of Karachi, Karachi, Pakistan

U. Hanif
Department of Botany, Government College University, Lahore, Pakistan

S. Andleeb
Department of Zoology, University of Azad Jammu and Kashmir, Muzaffarabad, Pakistan

M. Arshad
Jhang-Campus, University of Veterinary and Animal Sciences, Lahore, Pakistan

M. Arshad (ed.), *Climate Changes Mitigation and Sustainable Bioenergy Harvest Through Animal Waste*, https://doi.org/10.1007/978-3-031-26224-1_15

Graphic Abstract

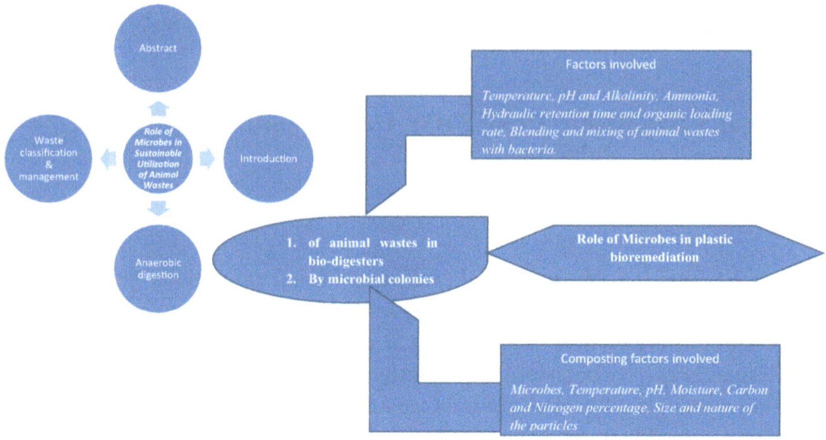

Keywords Microorganisms · Sustainable utilization · Animal wastes · Industrialization · Contaminated environments

15.1 Introduction

The global environment is under a significant strain due to rapid population growth, unplanned urbanization, industrialization, and the enormous population demand on natural resources. Multiple erroneous activities in contemporary civilization yield massive amounts of trash that causes pollution. Additionally, large-scale industries have increased the substantial quantities of radioactive and chemical wastes into the environment, causing irreversible harm to the whole biosphere (Raj et al. 2018). The primary cause of a loss of resources and energy is waste creation, which also has negative effects on the environment and costs society money to collect, treat, and manage. In India and Pakistan, the production of hazardous waste is strongly correlated with urbanization and varies greatly between cities. It is believed that as industrialization accelerates, waste production would follow suit unless scientific management practices are implemented (Jhariya et al. 2018) (Fig. 15.1).

Due to the absence of adequate trash collection and elimination facilities in poor countries, the waste management (WM) scenario in developed and developing countries is very different. WM is now given top emphasis as a result of growing concern over environmental deterioration and longevity (Brewer 2001). Regarding the management of animal wastes after 1973, the number of allegations per farm doubled, with the exception of those against cattle farms. Overall, 2590 complaints have been documented in past. The majority of them (61.6%) dealt with odor concerns, which were preceded by water pollution (40.6%), insects (7.8%),

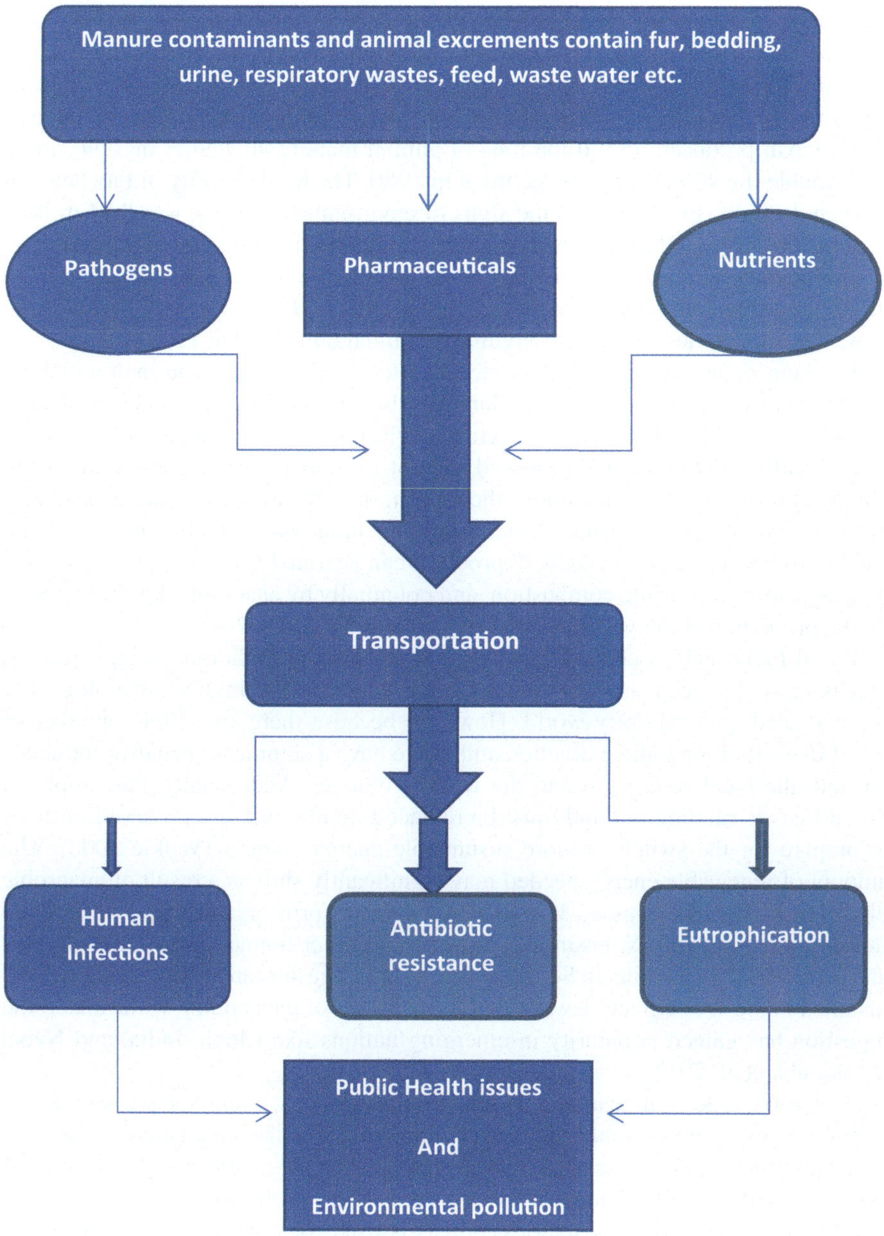

Fig. 15.1 Environmental and public health inference of animal manure

and others (5.9%). Pork production (34.4%), dairy (32.4%), poultry (18.7%), beef (12.2%), and other sources (2.3%) were the main culprits. The majority of these issues were the outcome of poor waste management. However, it is challenging for most farms to pay the necessary expenditures for manure cleanup. According to calculations, Japan produced 95 million tons of animal manure altogether in 1996, more than double the 43 million tons recorded in 1960. The total quantity of biochemical oxygen demand (BOD), one of the signs of environmental stress, was 0.84 million tons in 1960; by 1996, it had nearly tripled to 2.5 million tons. Animal excrement is significantly increasing the environmental load. Although the anticipated volume of human excrement recorded was 44 million tons, its BOD was only 0.44 million tons, significantly lesser than the volume of animal dung (Nakai 1995).

Unplanned, unsystematic, and illogical methods of dumping waste on the outskirts of cities and villages lead to spilling landfills that are not only impossible to restore to a suitable condition but also have grave environmental consequences in the form of soil and groundwater pollution and a grant to global warming. Materials from plants, animals, humans, including their trash, are all included in the category of biomass. Food manufacturing, agriculture, and industrial effluents are additional sources of biomass wastes. These byproducts can be transformed into energy or fuel via gasification, co-firing, combustion, and potentially by anaerobic digestion, based on the properties of the wastes (Sam-Anyaoma and Anjorin 2018).

Fossil fuels such as coal, oil, and natural gas are the traditional energy sources that have so far been used to construct and maintain the highly technologically sophisticated contemporary world. However, because there are a finite number of fossil fuels, their ongoing extraction and usage have a significant negative influence on both the local ecosystem and the world's climate. Additionally, the supply of oil and gas is running out and must be replaced, reinforced, or conserved in order to prepare for the switch to more sustainable energy sources (Wilkie 2008). The amount of renewable energy needed may significantly shift as a result of anaerobic digestion of biomass wastes. It works best to transform organic byproducts from farming, raising animals, businesses, towns, and other human operations into fuel and fertilizers. Meanwhile, in South Africa, biogas digesters are mostly designed and installed in the Western and Kwa-Zulu Natal regions of the country, while anaerobic digestion has gained popularity in emerging nations like China, India, and Nepal (Mukumba et al. 2012).

Anaerobic digestion generally lessens biomass wastes, counteracts a wide range of environmental undesirables, improves sanitation, aids the management of air and water pollution, and lowers emission of greenhouse gases. Additionally, it offers a superior fertilizer full of nutrients as well as energy in the form of biogas. From developing to industrialized nations, biogas is used for a wide range of purposes. Biogas has been used as electricity, fuel, and heat at farms (Liu et al. 2009).

Animal manures have been recognized as appropriate sources of biogas generation in Africa due to the significant roles played by rumen bacteria in anaerobic digestion, while in Denmark and Germany they are co-digested with crop residues (Kröber et al. 2009). Co-digestion is the term employed to describe the parallel anaerobic digestion of many organic wastes in a single digester. Due to beneficial synergisms formed in the digestive medium, bacterial diversity in various wastes, and the co-substrate provision of deficient nutrients, this concept increases methane output (Li et al. 2011). As a result of primary variations in the digestive physiology of the various species, the components and types of diet, the stage of growth of the animal, and consequently the management system of waste accumulation and storage, the wastes generated from various animals also differ in chemical makeup and natural substances (Anunputtikul 2004).

Significant chunks of the agricultural sector in both emerging and established nations are focused to raise poultry and cattle, thus generating tremendous quantities of animal excrement which raises public, ecological, and social issues. Currently, several digesters are designed and operated on farms for the effective handling of the wastes. It is crucial to clean up a contaminated environment in a sustainable manner; in this light, the relevance of microorganisms in wastewater cleaning and the biodegradation of pollutants have grown recently. Numerous biotechnological methods reliant on microorganisms, including bioremediation, biodegradation, biocomposting, and biotransformation, have been employed to adequately accumulate and destroy a wide variety of pollutants. *Cladophora sp.* (green algae) is a robust and efficient potential WM agent and has a high bioaccumulation capacity for hazardous metals (Maghraby and Hassan 2018).

Finland and polar arctic regions have been extensively using archaea and bacteria in the bioreactors to treat WW. As part of a process known as nano-bioremediation, nanoparticles are now successfully used to increase the activity of microbes. Since because of the radioactive resistance and potential to withstand radiation naturally, the extremophilic bacteria (*Deinococcus radiodurans*) is commonly used in radioactive waste extraction methods (Brim et al. 2000; Varma et al. 2017).

The primary goal of this study is to suggest and encourage the use of the most practical and environmentally responsible way for treating contaminated animal waste using various microbiological agents in order to achieve environmental sustainability. In this context, biomass materials have been regarded as a means of increasing energy generation, reducing the world's rising reliance on fossil fuels, and also mitigating the environmental and health risks associated with the usage of fossil fuels in both developing and wealthy nations (Uzodinma et al. 2008).

15.2 Waste

Human activities are mostly responsible for the creation of waste material. The created trash has been worsened by the unplanned and rapid development and alteration of livelihoods around the world. The global emission of dangerous contaminants from many occupations causes the overall biosphere to progressively deteriorate. A significant amount of biomedical and agricultural waste is created as a consequence of the sudden expansion of healthcare institutions and the automation of agricultural practices, which has a detrimental impact on environment. Wastes may be divided into three major categories: solid, liquid, and gaseous waste.

15.2.1 Classification of Wastes

Waste might be in the form of solid, liquid, gas, or heat and produced from four distinct sites, such as industrial, municipal, biomedical, and electronic sources. Waste may be characterized using a variety of factors, including the kind of material, how easily it degrades, how it will affect the environment, and the source. Each category could include several kinds (Fig. 15.2).

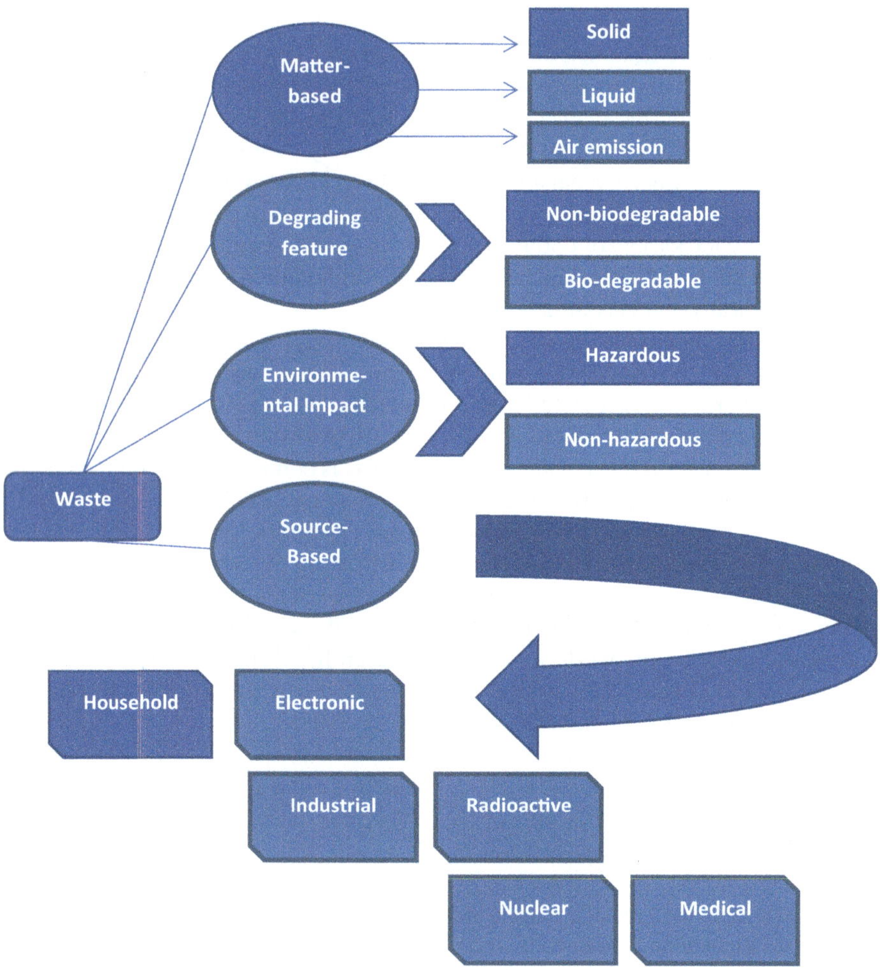

Fig. 15.2 Classification of wastes: an inference

15.2.2 Waste Management

Storage, collection, disposal, and management of waste materials are the main components of WM. The primary goal of WM is to lessen the negative impact that wastes have on the environment and human health. Rapid population growth, industrialization, and the enormous population demand on the NR have all contributed to this problem becoming more and more of a concern for the world's environment (Fig. 15.3).

15.3 Microbes in Waste Management

The approach to use contemporary scientific methods and procedures that employ a wide range of microorganisms under controlled conditions without disrupting the ecosystem is known as microbial biotechnology in WM. Composting, biodegradation, bioremediation, and biotransformation are the most widespread and effective WM techniques. Numerous bacteria, including *Bacillus* sp., *Corynebacterium* sp., *Staphylococcus* sp., *Streptococcus* sp., *Scenedesmus platydiscus, S. quadricauda, S. capricornutum,* and *Chlorella vulgaris,* have been successfully employed for WM (Liaqat et al. 2022) (Fig. 15.4).

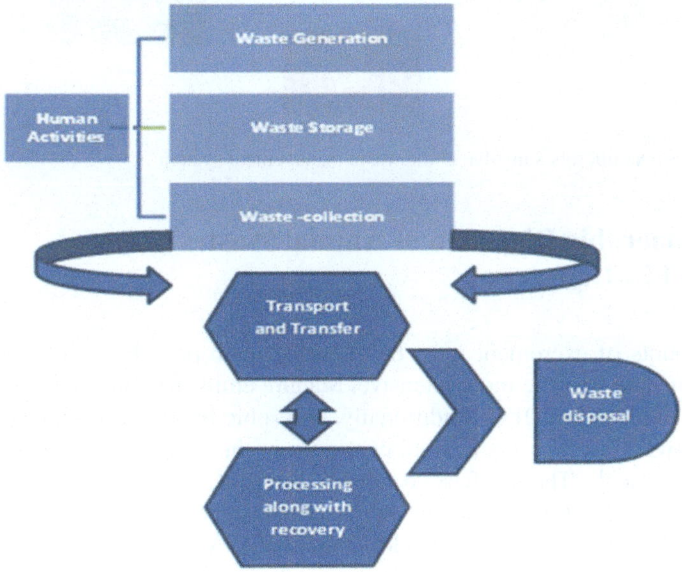

Fig. 15.3 Waste management (WM) system: A schematic representation

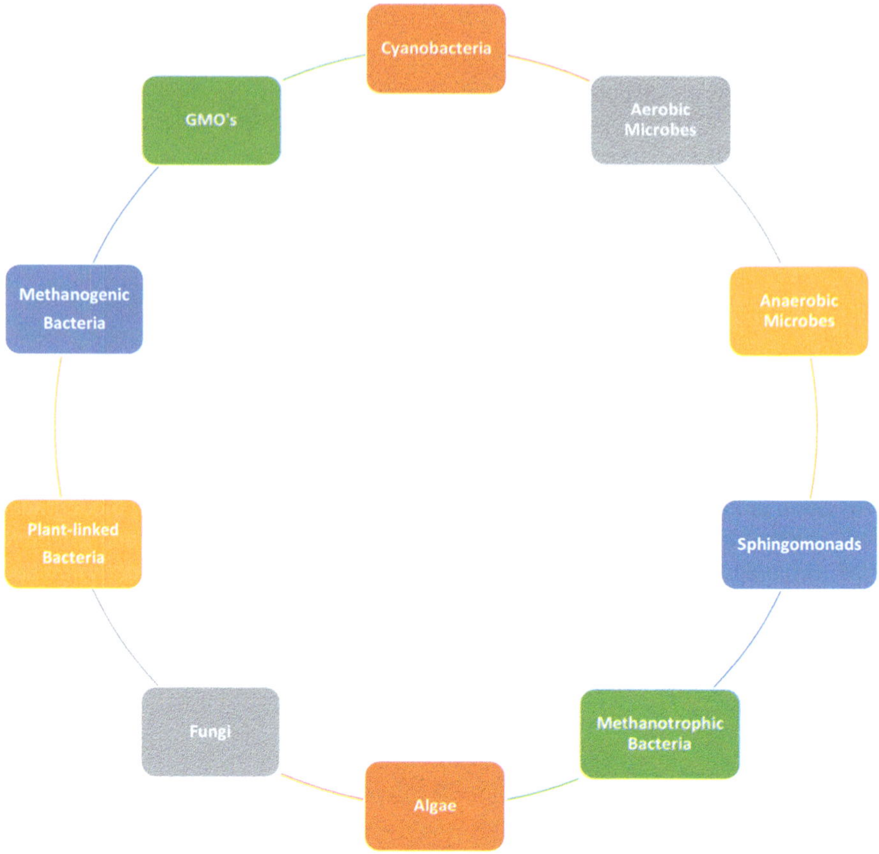

Fig. 15.4 Various microbes involve in the waste management system

15.4 Anaerobic Digestion of Animal Wastes in Bio-Digesters

Large amounts of excrement must be managed appropriately as livestock operations expand and become more intensive. Manure emits methane, a greenhouse gas (St-Pierre and Wright 2013). Additionally, anaerobic breakdown often starts in the animal's lower digestive system and continues in the dung heaps, producing foul-smelling chemicals. These offensive substances result from inadequate decomposition of organic materials in manure by anaerobic microorganisms in uncontrolled environmental circumstances (Husfeldt et al. 2012). An option to the correct treatment of these wastes is a farm-based anaerobic digester. Around the world, there are countless on-farm digester facilities, including Blue Spruce Farm, Green Mountain Diary, Chaput Family Farm, Cantabria Diary Plant, Buttermilk Hall Farm, Bulcote

Farm, Minnesota Mid-sized Diary Farm, etc. (Husfeldt et al. 2012; Rico et al. 2011) (Fig. 15.5).

There are a few on-farm anaerobic digesters in South Africa. However, the manure collected, varies in accordance with the nature of animal food, on-farm activities, and the type of digester utilized (Manyi-Loh et al. 2013). In light of the variations in animal management approaches among the farms, the crucial process of solid–liquid isolation of the manure mixture may be carried out either prior to or following anaerobic process (Tucker 2008). Additionally, for anaerobic digestion, the obtained manure may be combined with milk house waste. Manure is often collected together

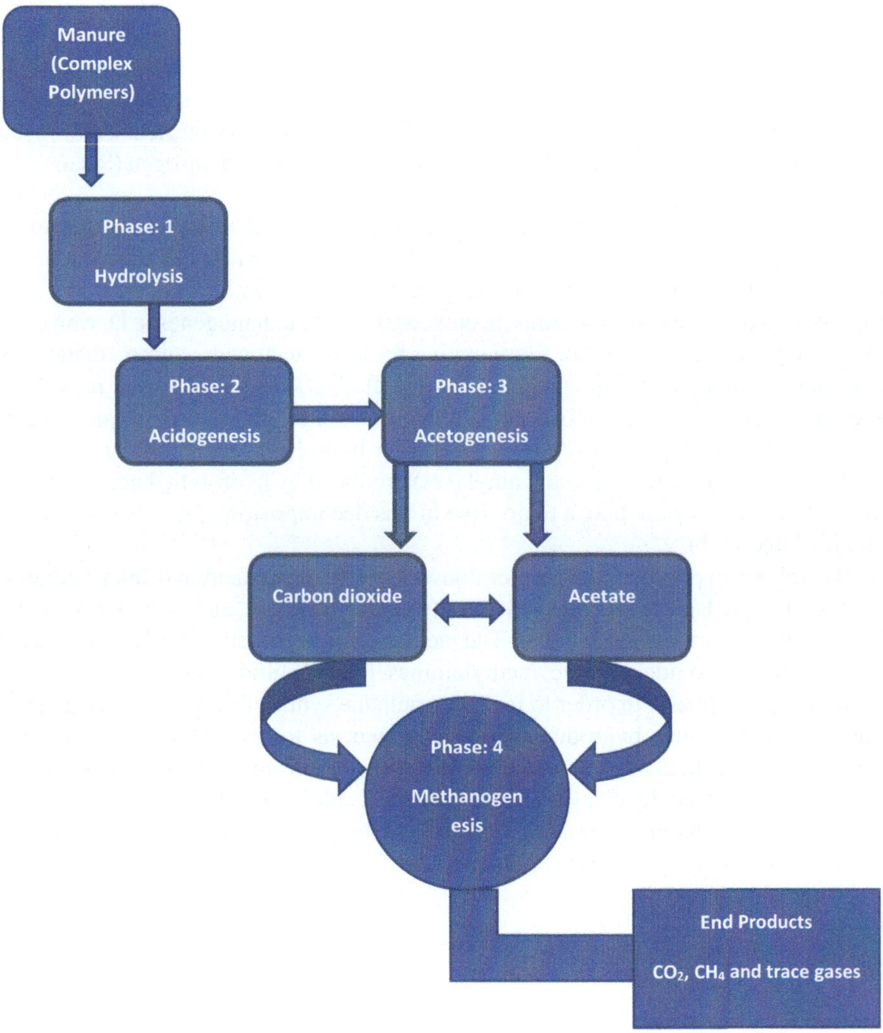

Fig. 15.5 Anaerobic digestion: A schematic layout

Table 15.1 Different components of biogas emitted by the anaerobic digestion process; average composition (De Graaf and Fendler 2010)

Component	Sign	%
Oxygen	O_2	< 2
Hydrogen	H_2	1–2
Carbon dioxide	CO_2	23–45
Methane	CH_4	50–75
Hydrogen sulfide	H_2S	< 1
Water vapor	H_2O	< 1
Ammonia	NH_3	2–7

with or without milk house trash and turned into slurry by adding water. Pumping the slurry to the separator allows the mixture to be divided into liquid and solid parts by screening. The digester then receives the filtered liquid proportion, while the solid fraction might be de-watered and distributed to places in need of nutrients, employed as bedding, or compost to provide an extra supply of carbon and nitrogen (Sakar et al. 2009).

Intricate assemblages of bacteria in the digester catabolize larger molecules in animal manure during the course of anaerobic digestion, subsequently producing methane and carbon dioxide. Essentially, this process may be broken down into four parts: hydrolysis, acidogenesis, acetogenesis, and methanogenesis. During each of the step, certain hydrolytic, fermentative bacteria, acetogens, and methanogens contribute a vital part for the process (Lozano et al. 2009). Complicated polymers, such as carbohydrates, proteins, and lipids, are broken down into simple sugars, amino acids, and long-chain fatty acids during hydrolysis. Extracellular enzymes (cellulases, lipases, proteases, and amylases) produced by hydrolytic bacteria linked to a polymeric substrate play a major role in this decomposition process (Song et al. 2005) (Table 15.1).

Hydrolysis byproducts are further converted into acetic acid and intermediates, such as ethanol, lactic acid, short-chain fatty acids, hydrogen, and carbon dioxide by fermentative or acidogenic bacteria. The methanogenesis process can directly use the acetate, carbon dioxide, acetone, methylamines, methyl sulfide, and methanol generated during this phase. In order to increase methane synthesis, syntrophic acetogens transform the various byproducts from acidogenesis to acetate, formate, or CO_2 and H_2. Finally, methanogens manufacture methane during methanogenesis in two different ways: either by cleaving acetic acid molecules to form methane and carbon dioxide, or by reducing carbon dioxide with hydrogen, depending on whether they are acetotrophic or hydrogenotrophic methanogens (Franke-Whittle et al. 2009).

15.5 Anaerobic Digestion-Related Microbial Colonies

The particular bacteria and their metabolic processes during anaerobic digestion are influenced by the chemical makeup of the feedstock, ambient variables, and digester operation parameters. There are four types of concerned bacteria, and these groups are biologically closely linked. The early steps of digestion result in decreased intermediates which are used by acetogens and methanogens (Franke-Whittle et al. 2009). But the interplay between acetogens and methanogens is quite intricate. Since these bacteria are anaerobes, oxygen poses a hazard by disrupting cellular metabolism and triggering the oxidation of cellular components that commonly arise in compact forms. Contrarily, new research has shown that certain methanogens can adapt to oxygen because their genomes contain genes that produce enzymes that protect them against oxygen toxicity, such as catalase and superoxide dismutase (Brioukhanov et al., 2006). Methanogens, such as *Methanobacterium thermoautotrophicum, Methanobrevibacter arboriphilus*, and *Methanosarcina barkeri*, have been found to be very tolerant to oxygen and desiccation, according to a number of reports (Kiener and Leisinger 1983; Fetzer et al. 1993).

Due to the production of thick exterior cell layers made of extracellular polysaccharide (EPS), which were combined with the buildup of cyclic 2,3-diphosphoglycerate, *M. barkeri* exhibits the natural capacity to endure prolonged periods of exposure to air and deadly temperatures. In a digester system, the microbial population may generally be divided into three types: acidogens, syntrophic acetogens, and methanogens (McInerney et al. 2009).

15.5.1 Acidogens

According to documentation, the bacterial species that are active during the polymer hydrolysis stage are likewise active during the acidogenic phase. As a result, fermentative bacteria can also be referenced as acidogenic and hydrolytic bacteria. They can be either stringent anaerobes or facultative anaerobes, which means that they can exist in both aerobic and anaerobic environments. One of the microorganisms in charge of the initial stage in the biotransformation of carbohydrates to CH_4 is the family *Enterobacteriaceae*, sometimes known as enteric bacteria. This category of bacteria lives in the intestines of humans and other animals (Carbone et al. 2002).

15.5.2 Syntrophic Acetogens

Alcohols, short-chain fatty acids (C3–C6), certain amino acids, and aromatic compounds are syntrophically metabolized by syntrophic acetogens, such as *Syntrophobacter wolinii, Syntrophomonas wolfei,* and *Smithella sp.,* to produce methanogenesis precursors. The thermodynamics of converting the aforementioned substrates to produce methanogenesis precursors is unfavorable, but they are made favorable by the inclusion of a syntrophic mate such as hydrogenotrophs (McInerney et al. 2008; Hori et al. 2011). However, the buildup of eruptive fatty acids forces the pH to drop, the acidification to worsen, the destruction of methanogen functionality, and finally the collapse of the digester. *Homoacetogens* also make acetate from the conversion of carbon dioxide via the acetyl Co-A reductase reaction, whereas syntrophic acetogens transform intermediate metabolites to acetate and other methanogenesis fuels (Siriwongrungson et al. 2007).

In principle, acetotrophic, hydrogenotrophic, and methylotrophic channels all contribute to the production of methane through methanogenesis. Particular methanogens belonging to the order *Methanosarcinales* are in command of the acetate breakdown to generate methane via the acetoclastic path. On the other hand, a subset of acetate-oxidizing bacteria exist in syntrophic interactions with hydrogenotrophic methanogens, where they work together to oxidize acetate to produce methane. Both mesophilic and thermophilic bacteria are included in this class of bacteria, which is also known as syntrophic acetogens. Syntrophic acetate-oxidizing bacteria, which intricate in the reversed reductive acetogenesis, are being identified, using integrating flow measurement with transcriptional profiling of the formyltetrahydrofolate synthetase (FTHFS) gene, an ecological bioindicator essential for reductive acetogenesis (Hori et al. 2011).

15.5.3 Methanogens

Methanogens have been discovered in a variety of reactive environments, such as freshwater and marine ecosystems, sewage digesters, the gastrointestinal tracts of herbivores and animals, and insects that feed on humus and wood. They are members of the archaea domain and play a crucial role in the anaerobic process of digestion since it is at this stage that the lucrative methane is created (Zhu et al. 2004; Manyi-Loh et al. 2013). The methanogenic communities are particularly susceptible to pH, fatty acid levels, free NH_3, and NH_4^+ in the digesting medium during an inconsistent anaerobic digestion operation in a malfunctioning anaerobic digester (Westerholm et al. 2012). Additionally, there are six major orders of methanogens, *Methanococcales, Methanomicrobiales, Methanosarcinales, Methanocellales,* and *Methanopyrales.* Acetate, which has long been recognized as the primary source of more than 70% of the methane, generated in the majority of designed anaerobic digesters, is employed by the members of the order *Methanosarcinales* (Batstone 2006).

The two families *Methanosarcinaceae* and *Methanosaetaceae*, which make up the *Methanosarcinales*, are also characterized as being acetoclastic. The morphology, biokinetics, and growth conditions of these two groups of acetoclastic methanogens vary, depending on the acetate content. In summation, the linkages among the numerous anaerobic microbe groups are very complex, and the harmony of these interconnections is crucial to the ability of biological process to function well (Amani et al. 2010).

15.6 Factors Affecting Anaerobic Digestion of Animal Manure

Performance parameters such as hydraulic retention time (HRT), temperature, pH, organic loading rate (OLR), free ammonia concentration, medium characteristics, biodegradability, and bio-digester layout are primary elements, impacting the effectiveness of anaerobic digester. Moreover, the anaerobic digestion of animal dung is substantially influenced by temperature, biodegradability, OLR, and HRT. Despite this, it is important to remember other factors as well (Giesy et al. 2005; Cioabla et al. 2012).

15.6.1 Temperature

Anaerobic bacteria can be subdivided into psychrophiles (20 °C), mesophiles (25–37 °C), and thermophiles (55–65 °C) according to their preferred temperature ranges. Some methanogenic organisms are categorized as hyperthermophilic methanogens because they enjoy extremely hot temperatures (90–100 °C). The most significant environmental element impacting the proliferation of microorganisms might be regarded as temperature. *Methanocaldococcus jannaschii* and *Methanococcus vulcanius* are two instances. All bacteria can only reproduce and expand within a specific temperature range. The enzymatic and chemical interactions speed up as the temperature rises within a particular band, and thus, accelerates growth (Ver Eecke et al. 2012; Saleh and Mahmood 2004).

Conversely, enzymes have a protein-like nature and get permanently destroyed beyond their appropriate temperature, important chemical processes which take place in many biosynthetic processes, some of which are catalyzed by enzymes, thus, cannot take place. As they enable organisms to drive desired energy-dependent events by combining them with spontaneous reactions which release energy, enzymes are essential for metabolism. Microbes will, therefore, stop growing, as well. To sudden temperature fluctuations, many microbial species react in different ways. In addition, temperature influences the metabolic rates of the microorganisms and other processing variables including OLR and ammonia content (El-Mashad et al. 2004).

Biomass residues may typically be digested anaerobically at both mesophilic (25–37 °C) and thermophilic (55–65 °C) temperatures. At thermophilic temperature ranges, though, the proportion of free ammonia (NH_3) to total ammonium ion is greater. In order to minimize ammonia-mediated suppression of methanogenesis, animal wastes that include nitrogen and ammonia molecules are digested at mesophilic temperatures (25–37 °C) (Garcia and Angenent 2009). Furthermore, thermophilic processing utilizes a significant amount of energy, which might lower the net energy gained from the entire digesting process. Despite the previously mentioned negatives, thermophilic fermentation probably destroys pathogens and weed seeds significantly while also boosts metabolism and CH_4 output (Campos Pozuelo et al. 1999).

15.6.2 pH and Alkalinity

It is more pertinent to speak about alkalinity and pH in relation to anaerobic digestion because the former may be employed to modulate pH, buffering the system acidity resulting from the acidogenesis phase. As an outcome, the buffering capability of an anaerobic digester is represented by the quantity of alkalinity in the system (Gerardi 2003). The OLR (which relies on reactor system) and the substrate buffering ability determine the pH range of anaerobic digestion, which typically takes place in the vicinity of neutral pH value. Cow, swine, and poultry dung are examples of livestock wastes with good buffering potential because, when microbes break them down, they create alkalinity (Molinuevo-Salces et al. 2010). The anaerobic digestion of these wastes is frequently kept at elevated pH levels of 7.6, though. Increased volatile fatty acid buildup owing to increased acidity of the digesting media can be caused by an increase in OLR with a matching reduction in HRT (Veeken et al. 2000).

15.6.3 Ammonia

Biological breakdown of organic nitrogen induces high percentage of total ammonium ions and free ammonia. The amount of ammonia generated during digestion is influenced by the nitrogen content of the medium, reactor loads, the C/N proportion, buffering ability, and temperature (Benabdallah et al. 2009). In water solutions, inorganic ammonia and nitrogen may be found in two main states which rely on pH, ammonium ions (NH_4^+), and unionized ammonia or free ammonia (NH_3). The operational pH and heat have a significant impact on ammonia toxicity. Since a greater portion of total ammonia nitrogen will be in the form of free ammonia, which is known to be hazardous, a rise in pH will make the ammonia poisoning of the system more worse (Chen et al. 2008).

A decrease in pH value, on the contrary end, will assist to balance out the levels of free ammonia and bringing it closer to the ideal pH range desired for the growth of microbes. But at the other perspective, process, destabilization brought on by oxidative stress frequently causes a rise in volatile fatty acid levels and a commensurate reduction in methane output. Elevated ammonia content causes deficient biogas integrity, lower COD competence, impaired biogas output, and stinking, in addition to stifled operation. As free ammonia inhibits methanogen growth, it has often been linked to poor performance characteristics and a higher likelihood of process failure. As a result of greater ammonia levels in a fermenter, the biomethanation process switches from acetoclastic methanogenesis (carried out by methanogens that use acetate) to syntrophic acetate oxidation, which is carried out by syntrophic acetogens working with hydrogenotrophs (Westerholm et al. 2012).

Additionally, the operating temperature (mesophilic and thermophilic temperatures) of the digester unit might have an impact on chemical equilibriums, notably of free ammonia concentration at a fixed total ammonium concentration. Despite the fact that temperature plays a crucial role in the kinetics and thermodynamics of microbial activities in methanogenesis, treating biomass components rich in ammonium, urea, and proteins at thermophilic temperatures (56–65 °C) might be difficult because of a larger supply of free ammonia (Angelidaki and Ahring 1993). Since free ammonia inhibits methane production, the ratio of total ammonium to free ammonia gets substantially higher at warmer temperatures. Nevertheless, the ammonia lethality of digester system got reduced by a rise in temperature within the mesophilic ranges. Therefore, mesophilic range of temperatures gives greater process stability to anaerobic digestion of animal manure and a better performance than thermophilic temperatures (Campos Pozuelo et al. 1999).

It has been shown that combining animal manure with carbon-rich co-substrates would aid the avoidance of inhibition, imposed by both volatile fatty acids and ammonia. However, due to inconsistent findings from various investigations carried out in various environments with distinct substrates and buffers, along with the intricate nature of the process of anaerobic digestion and acclimation times, the inhibitory ammonia threshold concentration is not standardized (Chen et al. 2008).

15.6.4 Hydraulic Retention Time (HRT) and Organic Loading Rate (OLR)

The HRT refers to the average time the substrate spends in the anaerobic digester, while the OLR measures the quantity of organic matter added to the digester per reactor volume and unit time. HRT and OLR have an inversely proportional relationship and convey valuable information on the design and operation of the reactor (Bolzonella et al. 2005).

The holding duration of animal manure in the reactor has a significant impact on its biological breakdown. Heat, the kind of reactor utilized for the processing, and the solid concentration of the excrement, all affect retention duration. More specifically, fixed film reactors often have a short residence period of a few hours to a few days, whereas CSTR and plug flow reactors for animal manure treatment generally take retention duration of 20–30 days (Karim et al. 2005). The holding period for covered lagoons must be 60 days. The integrity of waste is likewise impacted by HRT in terms of nutrient concentration, methane output, and microbial load. It has been looked at how HRT affects anaerobic batch process reactors and found that when HRT increases, methane output and effluent clarity both improve (Umaña et al. 2008).

In contrast, OLR is influenced by HRT and temperature. System failure is triggered by a rapid rise in OLR and is attributable to declines in pH, methane generation rate, and COD removal effectiveness. More specifically, a larger OLR above the optimum potential increases the rate at which acidogenic and hydrolytic bacteria produce intermediates such as fatty acids. Owing to the sluggish pace at which these fatty acids are consumed by methanogens, they would eventually build, causing the pH to decrease and reducing methanogenic operations. The elevated OLR affects the microbial communities in the digester circuit. With the genus *Clostridium* predominating at reduced OLR, the classes and phyla *Gammaproteobacteria, Deferribacteres, Actinobacteria,* and *Bacteroidetes* prevail at high OLR (Rincón et al. 2008).

15.6.5 Heavy Metals and Features of Substrate

The contents of excrement strongly influence the rate of biological pathways which occurs in the digester unit and the generation of biogas. The nutrition, waste handling, and storage strategies used in farming will all have an impact on the makeup of the manure. There is a demand for easily accessible energy sources including carbon for the creation of fresh biomaterials, inorganic materials like sulfur, nitrogen, phosphorus, magnesium, potassium, and calcium as well as organic nutrients, for the effective functioning and continuous reproductive success of microbes involved in the anaerobic digestion. Therefore, before the digestion process begins, the chemical and physical features of the feedstock, including all the moisture contents, total solid subject matter, volatile solids substances, phosphorus, nitrogen, and carbon values, must be assessed (Ganorkar et al. 2014).

Volatile manure matter is a highly important factor since it is made up of two sections: the biodegradable half, which comprises carbohydrates, lipids, and proteins, while the resistive or lignocellulosic portion, which cannot be decomposed anaerobically. The phrase "biodegradability of manure" is determined by the production of biogas or methane and the proportion of solids (total or volatile solids) which have been eliminated. For maximum growth and functioning, microorganisms need a trace quantity of certain metals, such as iron, nickel, copper, zinc, cobalt, and molybdenum (Zhang et al. 2007). Coenzymes and cofactors contain these essential trace nutrients, which are additionally considered as the stimulatory micronutrients. Methane output, substrate utilization, and unit consistency all raise as a result of these metals as they have stimulatory impacts on the functionality of the biogas operation. The researchers have determined that variances in OLR, pH, HRT, substrate properties, and the intricate biochemical and biological mechanisms involved to regulate trace metal accessibility are to blame for the large variations in the amount of these metals which have stimulatory affects for biogas production (Zhang et al. 2007).

Likewise, various kinds of trace metals have unique enhancing effects. It is confirmed that adding a well-measured trace metal solution (made up of Ni, Co, and Mo) promotes a rise in methane yields. Yet, the removal of Ni from the solution resulted in a larger reduction in methane production and process stability. Additional findings indicate that methane productivity grew by 10% at 0.4–2 μm values of Co but did not substantially change with the addition of molybdenum (Mo). It is indeed worthwhile to know that animal dung has been shown to have high levels of micro and macronutrients. However, anaerobic digestion of single feedstock like maize silage has shown that digestion disruption (induced by the lack of trace elements) can occur.

Certain single feedstock such as maize silage, potato, or even food wastes cannot offer both the micronutrients and macronutrients necessary for the development of anaerobic bacteria that are essential for the anaerobic digestion. As a rule, essential nutrients must be supplied before the digestive process can start. Even better, they can be co-decomposed with animal manure such that the animal excrement offers strong buffering capacity and necessary nutrients while the energy crop boosts the fuel production (Facchin et al. 2013). Alterations in the composition of the microbial population can result from a lack of certain metals. Furthermore, a proportion of these heavy metals that is too high would be hazardous to the system and impede the biological mechanisms via disrupting the structure and function of relevant enzymes. They reportedly have the potential to replace naturally existing metals in the prosthetic group of enzymes or by interacting with the Sulfhydryl groups on enzymes (Chen et al. 2008).

15.6.6 Blending and Mixing of Animal Wastes with Bacteria

The degree of interaction between the flowing animal waste and a viable bacterial population and consequence of blending in the reactor are crucial for a better anaerobic digestion of animal manure. The advantages of mixing the entire content of the fermenter during the anaerobic treatment have been noted by several authors, and they include: It inhibits the emergence of filth inside the digester, guarantees homogenous distribution of microorganisms and substrate throughout the mixture and intensifies contact between them, hinders stratification within the digester so that balanced dispersion of heat is conceivable, and lastly, it aids in the discharge of gas from the concoction (Rojas et al. 2010). Nonetheless, as anaerobic digestion advances, stirring may cause a decrease in the substrate particle size. When contemplating the possibility of using various modalities (mechanical mixers and recirculation pumps), the intensity and duration of mixing are what remain uncertain. Total and volatile solids play a critical role in the classification of manure because, beyond a certain point, the manure ceases to be slurry, which complicates mixing and pumping activities. It has been noted as a conclusion that combining anaerobic digestion of dairy manure with low volatile solids added to extended HRT reduces the necessity for doing so (Rico et al. 2011).

15.7 Composting

A diversified population of microbes aids the composting, which is an aerobic breakdown activity. The metabolic reactions of microbiota aids to decompose various types of wastes. Organic waste has been transformed and consolidated by composting into a form that may be beneficial for a variety of agricultural techniques. It is a waste management strategy that is both inexpensive and sustainable. Humus and plant nutrients are the primary end products of composting, whereas carbon dioxide, water, and heat are the contaminants. This process involves a variety of microorganisms, including bacteria, *actinomycetes*, yeasts, and fungus. The three stages of composting are the mesophilic phase, the thermophilic phase, and the cooling and maturation phase. The kinds of composting organic matter (OM) and the efficacy of the technique, which is determined by the level of aeration and agitation, are the two variables that control the longevity of the composting stages (García-Gómez et al. 2005; Abbasi et al. 2000).

15.7.1 Factors Influencing the Rate of Composting

15.7.1.1 Microbes

Various biochemical molecules can be effectively oxidized or digested by a number of microorganisms into more obvious and stable byproducts. A pile of biodegradable solid waste may be colonized by specific microorganisms, including mesophilic bacteria, *actinomycetes*, fungus, and protozoa (Gajalakshmi and Abbasi 2008). These microorganisms can flourish between 10 and 45 °C and efficiently break down biodegradable materials. The active stage of composting is known as the thermophilic phase, and it can be present for several weeks. The majority of the waste is decomposed in the thermophilic stage (Meena et al. 2018).

15.7.1.2 Temperature

According to reports, optimum composting occurs between the ranges of 52–60 °C and proceeds at temperatures as high as 60–70 °C, when the majority of microorganisms are less active. Below 20 °C, the composting process might halt or become much more stable. Additionally, it has been shown that temperatures exceeding 60 °C can lower microbial activity since they go beyond the optimal thermophilic boundary for microbes (Gajalakshmi and Abbasi 2008).

15.7.1.3 pH

The pH value has a significant impact on the composting. Varying pH ranges are preferred by various composting bacteria. For the growth of bacteria, a pH range from 6.0 to 7.5 is optimum, whereas fungi prefer a pH range between 5.5 and 8.0. If the pH value somehow exceeds 7.5, nitrogen is lost (Gajalakshmi and Abbasi 2008). A large variety of bacteria presumably thrive best at a pH somewhere around 6.5 and 7.5. Bacterial activity is reported substantially hindered or even abolished below pH 5.0.

15.7.1.4 Moisture Content

The recommended and optimal moisture level for composting is often between 60 and 70%. However, the ideal moisture content is between 50 and 60% at the finishing stage. Higher than 75% and lower than 30% moisture ratio greatly lower the microbial activity. The moisture content is efficiently managed by striking a balance between microbial activity and the amount of oxygen accessible. Anaerobic conditions caused by too much moisture produce unwanted compounds and a foul smell (de-Bertoldi et al. 1983).

15.7.1.5 Carbon and Nitrogen (%)

Both C and N are vital for microbes. The primary form of energy is carbon, and N is crucial for microbial development. A substrate ability to humify quickly and completely is largely dependent on the C to N ratio, which is generally from 25 to 35% (Gajalakshmi and Abbasi 2008).

15.7.1.6 Size and Nature of the Particles

During composting, nature and particle size are crucial factors. Both oxygen saturation in the heap and microbiological infiltration to the substrates are impacted by particle size. Smaller particles need more surface space for microbial assault, whereas bigger particles reduce the surface area available for microbial invasion, slowing down or even stopping the composting equipment (Zia et al. 2003).

15.8 Role of Microbes in Biodegradation of Plastic

Plastics are regarded as a significant waste material. Plastic trash recycling is a significant issue nowadays. Plastic is a polymer, and depending on the nature of the causative agents, polymers can degrade in a variety of ways, including thermally, photo-oxidatively, mechanochemically, catalytically, and biologically. The biodegradation method, among these, is the most promising because of its efficiency and ecological beneficial approach. The inherent capacity of microorganisms to start the process of breaking down via enzymatic activity is known as biodegradation (Albertsson et al. 1987) (Table 15.2).

Microbes play a big part in how natural and manmade polymers disintegrate and deteriorate. Plastics degrade gradually, and a variety of environmental conditions, including temperature and pH, are necessary for this cycle. The principal organisms that break down plastic are bacteria and fungus. One of most significant of the successive enzymatic processes that occur during the biodegradation of plastic is hydrolysis. Typically, several variables, such as the availability of microbial enzymes and adequate abiotic conditions, affect how biodegradable polymeric compounds behave (Gu and Gu 2005; Schink et al. 1992). The impurities are used by microbes for growth, feeding, and development. This is the primary driver underlying the microbial transformation of many organic pollutants. Carbon is obtained by microbes from organic compounds. C is crucial for bacteria because it serves as the foundation for new cells (Chapelle 1993).

15.9 Bioremediation

A natural procedure known as bioremediation employs microorganisms to clean trash or pollution from soil and water (Fig. 15.6). This technique enables eco-friendly bacteria to treat solid waste, making it biodegradable and beneficial to the environment (Kensa 2011). There are two types (Fig. 15.6).

15.9.1 In-Situ *Bioremediation*

In this, wastes are removed from the soil or water without extraction or transportation. Bacteria conduct biological treatment on the waste interface. It is an alternate way of treating groundwater and soil. Non-toxic microorganisms are used in this method. There are three categories of this sort of bioremediation.

15.9.1.1 Biosparging

It is a procedure for treating waste at locations where petroleum products like diesel, gasoline, and lubricating oils are present. This technique involves pumping compressed air below ground water to raise the oxygen content. To prevent the release of volatile particles into the environment, which causes air pollution, the air pressure needs to be properly managed.

Table 15.2 Numerous bacterial, fungi, and algae strains that degrade plastics

Type of Plastic	Bacteria	Fungi	Algae	References
Polyethylene bags	*Pseudomonas aeruginosa, Pseudomonas putida, Bacillus subtilis*	*Phanerochaete chrysosporium, Aspergillus niger*	Algae is not involved in the biodegradation of polyethylene bags	Nwachukwu et al. (2010), Aswale and Ade (2009)
Low density Polyethylene	*Rhodococcus ruber C208, Brevibacillus borstelensis* 707, *Rhodococcus ruber* C208, *Staphylococcus epidermidis, Bacillus cereus* C1	*Aspergillus niger, Penicillium sp., Chaetomium globosum, Pullularia pullulans, Fusarium sp. AF4, Aspergillus oryzae*	Not Involved	(Gilan et al. 2004) Shah et al. (2009) Chatterjee et al. (2010) Sivan et al. (2006)
High density Polyethylene	*Bacillus sp., Micrococcus sp., Vibrio sp., Arthrobacter sp., Pseudomonas sp.*	*Aspergillus terreus* MF12, *Trametes sp.*	Not Involved	Balasubramanian et al., (2014) Fontanella et al. (2013) Iiyoshi et al. (1998)
Polyurethane	*Corynebacterium sp., Pseudomonas sp., Arthrobacter globiformis, Bacillus sp.*	*Chaetomium globosum, Aspergillus terreus, Curvularia senegalensis, Fusarium solani*	*Acinetobacter calcoaceticus, A. gerneri*	Howard et al. (2012) Crabbe et al. (1994)
Degradable Plastic	*Pseudomonas sp., Micrococcus luteus, Bacillus subtilis, Streptococcus lactis, Proteus vulgaris*	*Phanerochaete chrysosporium, Penicillium sp., Aspergillus sp.*	*Streptomyces sp.*	El-Shafei et al. (1998) Seneviratne et al. (2006) Priyanka and Archana (2011)
Degradable Polyethylene bags and Polyethylene carry bags	*Serratia marcescens, Bacillus cereus, Pseudomonas aeruginosa, Streptococcus aureus, Micrococcus lylae, Pseudomonas sp., Micrococcus luteus, Bacillus subtilis, Streptococcus lactis, Proteus vulgaris*	*Phanerochaete chrysosporium, Aspergillus niger, A. glaucus, Pleurotus ostreatus*	Not Involved	Priyanka and Archana (2011) Aswale and Ade (2009)

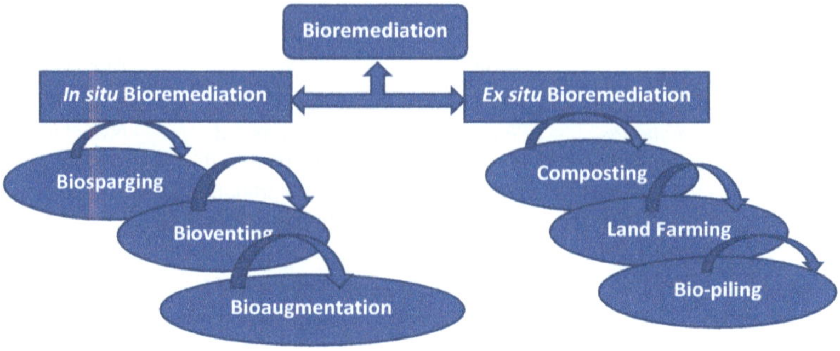

Fig. 15.6 Flow diagram of bioremediation

15.9.1.2 Bioventing

It is an aerobic method for the degradation of waste materials. When oil resources are mined for petroleum and gasoline, various solid wastes are produced that can be treated via bioventing. To speed up the cleanup procedure, oxygen, nutrients like phosphorus and nitrogen are delivered to the polluted spot during the procedure.

15.9.1.3 Bio-augmentation

In this instance, cultivated microorganisms are introduced to the contaminated area with the intention of causing the pollutants in a particular environment to degrade. As a result of this process, pollutants in the groundwater and soil are converted to non-toxic compounds by microbes.

15.9.2 Ex-Situ *Bioremediation*

It defines the procedure of removing polluted soil or water. The many forms of ex situ bioremediation are listed here.

15.9.2.1 Composting

Composting is an aerobic process that involves mixing polluted soil with safe organic fillers. Microorganism community is greatly increased with the use of organic additives.

15.9.2.2 Land Farming

In this bioremediation technique, polluted soil is combined with green manure before being tilled into the ground. Enhancing native bio-degradative bacteria is the major goal in order to allow for the aerobic breakdown of pollutants.

15.9.2.3 Bio-Piling

Bio-piling is a composite method that combines composting with on-site gardening. This method creates an environment that is ideal for both aerobic and anaerobic microbial growth. With the aid of biodegradation, bio-piles are used to reduce the quantities of petroleum elements (Fig. 15.6).

15.10 Conclusions

The protection and sustainable development of the environment are acknowledged as having the greatest degrees of significance and calling for immediate assistance on a worldwide scale. The main areas which require a focus of concentration are waste management, conservation of NR and biodiversity, and treatment of contaminants and pollutants if sustainability is to be ensured. To protect the environment from degradation nowadays, it is not only necessary to remove toxins and pollutants, but also necessary to recycle and control hazardous chemicals by converting various wastes into a wealth of usable items in an aesthetically pleasing and environmentally beneficial way. As human struggles to find a durable way to clean up contaminated surroundings and garbage, awareness in the employment of microbes has grown and gained importance in the recent decades. The potential of microbes for specific applications has drawn more interest and curiosity with the development of biotechnology. The nature of microorganisms is unusual and even unforeseen. Numerous environmental issues may be effectively solved by using microorganisms. The ethical and scientific reliance of microorganisms results in a stunning progression of knowledge and cutting-edge equipment that offers a practical solution to protect our world as well as contemporary methods of biological WM and environmental sensing. Finally, it can be said that the application of microorganisms and microbiological techniques has created new opportunities for sustainable prosperity, notably in the fields of the environment and other significant health issues. Animal manure digestion by anaerobic bacteria is viewed as a potent alternative for properly recycling animal wastes or turning them into useful products and fuels. The potential advantages of the biodegradation method that occurs in a confinement include lessening the biological oxygen demand (BOD) and chemical oxygen demand (COD) of wastes, wrecking pathogenic microorganisms to mitigate the microbial load to a level that humans could handle safely with marginal health risks, and destroying volatile fatty acids and many odorous compounds present in the feedstock and lowering emissions. Fundamentally, it promotes the idea of turning waste into riches by producing biogas and high-quality, nutrient-rich fertilizer from animal dung. Likewise, the other methods for the controlled management of animal wastes discussed in the chapter have a promising future to overcome the public and environmental health concerns.

References

Abbasi S, Ramasamy E, Gajalakshm S et al (2000) A waste management project involving engineers and scientists of a university, a voluntary (nongovernmental) organization, and lay people—a case study. In: Proceedings of international conference on transdisciplinarity, Swiss Federal Institute of Technology, Zurich, pp 1–3.

Albertsson A-C, Andersson SO, Karlsson S (1987) The mechanism of biodegradation of polyethylene. Polym Degrad Stab 18:73–87

Amani T, Nosrati M, Sreekrishnan T (2010) Anaerobic digestion from the viewpoint of microbiological, chemical, and operational aspects—a review. Environ Rev 18:255–278

Angelidaki I, Ahring B (1993) Thermophilic anaerobic digestion of livestock waste: the effect of ammonia. Appl Microbiol Biotechnol 38:560–564

Anunputtikul W (2004). Laboratory scale experiments for biogas production from cassava tubers

Aswale P, Ade A (2009) Effect of pH on biodegradation of polythene by *Serretia marscence*. The Ecotech 1:152–153

Balasubramanian V, Natarajan K, Rajeshkannan V et al (2014) Enhancement of in vitro high-density polyethylene (HDPE) degradation by physical, chemical, and biological treatments. Environ Sci Pollut Res 21:12549–12562

Batstone DJ (2006) Mathematical modelling of anaerobic reactors treating domestic wastewater: Rational criteria for model use. Rev Environ Sci Biotechnol 5:57–71

Benabdallah El Hadj T, Astals S, Gali A et al (2009) Ammonia influence in anaerobic digestion of OFMSW. Water Sci Technol 59:1153–1158

Bolzonella D, Pavan P, Battistoni P et al (2005) Mesophilic anaerobic digestion of waste activated sludge: influence of the solid retention time in the wastewater treatment process. Process Biochem 40:1453–1460

Brewer LJ (2001) Maturity and stability evaluation of composted yard debris.

Brim H, Mcfarlan SC, Fredrickson JK et al (2000) Engineering *Deinococcus radiodurans* for metal remediation in radioactive mixed waste environments. Nat Biotechnol 18:85–90

Brioukhanov AL, Netrusov AI, Eggen RI (2006) The catalase and superoxide dismutase genes are transcriptionally up-regulated upon oxidative stress in the strictly anaerobic archaeon *Methanosarcina barkeri*. Microbiology 152:1671–1677

Campos Pozuelo E, Palatsi Civit J, Flotats Ripoll X (1999) Codigestion of pig slurry and organic wastes from food industry. In: Proceedings of the II International symposium on anaerobic digestion of solid waste, pp 192–195

Carbone S, Da Silva F, Tavares C et al (2002) Bacterial population of a two-phase anaerobic digestion process treating effluent of cassava starch factory. Environ Technol 23:591–597

Chapelle F (1993) Ground-water geochemistry and microbiology. Wiley, New York

Chatterjee S, Roy B, Roy D et al (2010) Enzyme-mediated biodegradation of heat treated commercial polyethylene by Staphylococcal species. Polym Degrad Stab 95:195–200

Chen Y, Cheng JJ, Creamer KS (2008) Inhibition of anaerobic digestion process: a review. Bioresour Technol 99:4044–4064

Cioabla AE, Ionel I, Dumitrel G-A, Popescu F (2012) Comparative study on factors affecting anaerobic digestion of agricultural vegetal residues. Biotechnol Biofuels 5:1–9

Crabbe JR, Campbell JR, Thompson L et al (1994) Biodegradation of a colloidal ester-based polyurethane by soil fungi. Int Biodeterior Biodegradation 33:103–113

De Bertoldi MD, Vallini GE, Pera A (1983) The biology of composting: a review. Waste Manag Res 1:157–176

De Graaf D, Fendler R (2010) Biogas production in Germany. SPIN Background Paper, 24

El-Mashad HM, Zeeman G, Van Loon WK et al (2004) Effect of temperature and temperature fluctuation on thermophilic anaerobic digestion of cattle manure. Bioresour Technol 95:191–201

El-Shafei HA, Abd El-Nasser NH, Kansoh AL et al (1998) Biodegradation of disposable polyethylene by fungi and Streptomyces species. Polym Degrad Stab 62:361–365

Facchin V, Cavinato C, Fatone F et al (2013) Effect of trace element supplementation on the mesophilic anaerobic digestion of foodwaste in batch trials: the influence of inoculum origin. Biochem Eng J 70:71–77

Fetzer S, Bak F, Conrad R (1993) Sensitivity of methanogenic bacteria from paddy soil to oxygen and desiccation. FEMS Microbiol Ecol 12:107–115

Fontanella S, Bonhomme S, Brusson J-M et al (2013) Comparison of biodegradability of various polypropylene films containing pro-oxidant additives based on Mn, Mn/Fe or Co. Polym Degrad Stab 98:875–884

Franke-Whittle IH, Goberna M, Pfister V et al (2009) Design and development of the ANAEROCHIP microarray for investigation of methanogenic communities. J Microbiol Methods 79:279–288

Gajalakshmi S, Abbasi S (2008) Solid waste management by composting: state of the art. Crit Rev Environ Sci Technol 38:311–400

Ganorkar R, Rode P, Bhambhulkar A et al (2014) Development of water reclamation package for wastewater from a typical railway station. Int J Innov Technol Res 2:841–846

Garcia ML, Angenent LT (2009) Interaction between temperature and ammonia in mesophilic digesters for animal waste treatment. Water Res 43:2373–2382

García-Gómez A, Bernal M, Roig A (2005) Organic matter fractions involved in degradation and humification processes during composting. Compost Sci Util 13:127–135

Gerardi MH (2003) The microbiology of anaerobic digesters. Wiley

Giesy R, Wilkie AC, De Vries A (2005) Economic feasibility of anaerobic digestion to produce electricity on Florida dairy farms. UF (University of Florida), IFAS Extension, AN, p 159

Gilan I, Hadar Y, Sivan A (2004) Colonization, biofilm formation and biodegradation of polyethylene by a strain of Rhodococcus ruber. Appl Microbiol Biotechnol 65:97–104

Gu J-G, Gu J-D (2005) Methods currently used in testing microbiological degradation and deterioration of a wide range of polymeric materials with various degree of degradability: a review. J Polym Environ 13:65–74

Hori T, Sasaki D, Haruta S et al (2011) Detection of active, potentially acetate-oxidizing syntrophs in an anaerobic digester by flux measurement and formyltetrahydrofolate synthetase (FTHFS) expression profiling. Microbiology 157:1980–1989

Howard GT, Norton WN, Burks T (2012) Growth of Acinetobacter gerneri P7 on polyurethane and the purification and characterization of a polyurethanase enzyme. Biodegradation 23:561–573

Husfeldt A, Endres M, Salfer J et al (2012) Management and characteristics of recycled manure solids used for bedding in Midwest freestall dairy herds. J Dairy Sci 95:2195–2203

Iiyoshi Y, Tsutsumi Y, Nishida T (1998) Polyethylene degradation by lignin-degrading fungi and manganese peroxidase. J Wood Sci 44:222–229

Jhariya M, Yadav D, Banerjee A (2018) Plant mediated transformation and habitat restoration: phytoremediation an eco-friendly approach. Metallic contamination and its toxicity. Daya Publishing House, A Division of Astral International Pvt. Ltd, New Delhi, pp 231–247

Karim K, Hoffmann R, Klasson KT et al (2005) Anaerobic digestion of animal waste: effect of mode of mixing. Water Res 39:3597–3606

Kensa VM (2011) Bioremediation-an overview. I Cont Pollut 27:161–168

Kiener A, Leisinger T (1983) Oxygen sensitivity of methanogenic bacteria. Syst Appl Microbiol 4:305–312

Kröber M, Bekel T, Diaz NN et al (2009) Phylogenetic characterization of a biogas plant microbial community integrating clone library 16S-rDNA sequences and metagenome sequence data obtained by 454-pyrosequencing. J Biotechnol 142:38–49
Li J, Jha AK, He J et al (2011) Assessment of the effects of dry anaerobic co-digestion of cow dung with waste water sludge on biogas yield and biodegradability. Int J Phy. Sci 6:3679–3688
Liaqat I, Ali S, Butt A, Durrani AI et al (2022) Purification and characterization of keratinase from *Bacillus licheniformis* dcs1 for poultry waste processing. J Oleo Sci 71(5):693–700
Liu F, Wang S, Zhang J et al (2009) The structure of the bacterial and archaeal community in a biogas digester as revealed by denaturing gradient gel electrophoresis and 16S rDNA sequencing analysis. J Appl Microbiol 106:952–966
Lozano CJS, Mendoza MV, De Arango MC et al (2009) Microbiological characterization and specific methanogenic activity of anaerobe sludges used in urban solid waste treatment. J Waste Manag 29:704–711
Maghraby D, Hassan J (2018) Heavy metals Bioaccumulation by the green alga *Cladophora herpestica* in Lake Mariut, Alexandria, Egypt. J Pollut 1
Manyi-Loh CE, Mamphweli SN, Meyer EL et al (2013) Microbial anaerobic digestion (bio-digesters) as an approach to the decontamination of animal wastes in pollution control and the generation of renewable energy. Int J Environ Res Public Health 10:4390–4417
Mcinerney MJ, Struchtemeyer CG, Sieber J et al (2008) Physiology, ecology, phylogeny, and genomics of microorganisms capable of syntrophic metabolism. Ann N Y Acad Sci 1125:58–72
Mcinerney MJ, Sieber JR, Gunsalus RP (2009) Syntrophy in anaerobic global carbon cycles. Curr Opin Biotechnol 20:623–632
Meena H, Meena RS, Lal R et al (2018) Response of sowing dates and bio regulators on yield of clusterbean under current climate in alley cropping system in eastern UP, India. Legume Res: An Int J 41
Molinuevo-Salces B, García-González MC, González-Fernández C et al (2010) Anaerobic co-digestion of livestock wastes with vegetable processing wastes: a statistical analysis. Bioresour Technol 101:9479–9485
Mukumba P, Makaka G, Mamphweli S et al (2012) An insight into the status of biogas digesters technologies in South Africa with reference to the Eastern Cape Province. Fort Hare Pap 19:5–29
Nakai Y (1995) Animal production environment and manure treatment. New Handbook of Anim Sci 455–486
Nwachukwu S, Obidi O, Odocha C (2010) Occurrence and recalcitrance of polyethylene bag waste in Nigerian soils. Afr J Biotechnol 9:6096–6104
Priyanka N, Archana T (2011) Biodegradability of polythene and plastic by the help of microorganism: a way for brighter future. J Environ Anal Toxicol 1:1000111
Raj A, Jhariya M, Bargali S (2018) Climate smart agriculture and carbon sequestration. Climate change and agroforestry: adaptation mitigation and livelihood security. New India Publishing Agency (NIPA), New Delhi, pp 1–19
Rico C, Rico JL, Muñoz N et al (2011) Effect of mixing on biogas production during mesophilic anaerobic digestion of screened dairy manure in a pilot plant. Eng Life Sci 11:476–481
Rincón B, Borja R, González J et al (2008) Influence of organic loading rate and hydraulic retention time on the performance, stability and microbial communities of one-stage anaerobic digestion of two-phase olive mill solid residue. Biochem Eng J 40:253–261
Rojas C, Fang S, Uhlenhut F et al (2010) Stirring and biomass starter influences the anaerobic digestion of different substrates for biogas production. Eng Life Sci 10:339–347
Sakar S, Yetilmezsoy K, Kocak E (2009) Anaerobic digestion technology in poultry and livestock waste treatment—a literature review. Waste Manag Res 27:3–18
Saleh MM, Mahmood UF (2004) Anaerobic digestion technology for industrial wastewater treatment. Proceedings of the eighth international water technology conference, IWTC, Alexandria, Egypt, Citeseer, pp 26–28
Sam-Anyaoma C, Anjorin S (2018) An investigation into the energy potential of abattoir waste and palm oil mill effluent. Eur J Eng Sci Tech

Schink B, Brune A, Schnell S (1992) Anaerobic degradation of aromatic compounds. Microbial Degradation Natl Prod 219–242.
Seneviratne G, Tennakoon N, Weerasekara M et al (2006) Polyethylene biodegradation by a developed Penicillium-Bacillus biofilm. Curr Sci 90:20–21
Shah AA, Hasan F, Hameed A et al (2009) Isolation of Fusarium sp. AF4 from sewage sludge, with the ability to adhere the surface of polyethylene. Afr J Microbiol Res 3:658–663
Siriwongrungson V, Zeng RJ, Angelidaki I (2007) Homoacetogenesis as the alternative pathway for H2 sink during thermophilic anaerobic degradation of butyrate under suppressed methanogenesis. Water Res 41:4204–4210
Sivan A, Szanto M, Pavlov V (2006) Biofilm development of the polyethylene-degrading bacterium Rhodococcus ruber. Appl Microbiol Biotechnol 72:346–352
Song H, Clarke WP, Blackall LL (2005) Concurrent microscopic observations and activity measurements of cellulose hydrolyzing and methanogenic populations during the batch anaerobic digestion of crystalline cellulose. Biotechnol Bioengineer 91:369–378
St-Pierre B, Wright A-DG (2013) Metagenomic analysis of methanogen populations in three full-scale mesophilic anaerobic manure digesters operated on dairy farms in Vermont, USA. Bioresource Technol 138:277–284
Tucker MF (2008) Cow power-farm digesters for small dairies in Vermont. Biocycle 49:44–48
Umaña O, Nikolaeva S, Sánchez E et al (2008) Treatment of screened dairy manure by upflow anaerobic fixed bed reactors packed with waste tyre rubber and a combination of waste tyre rubber and zeolite: effect of the hydraulic retention time. Bioresource Technol 99:7412–7417
Uzodinma E, Ofoefule A, Eze J et al (2008) Effect of some organic wastes on the biogas yield from carbonated soft drink sludge
Varma D, Meena RS, Kumar S et al (2017) Response of mungbean to NPK and lime under the conditions of Vindhyan Region of Uttar Pradesh. Legume Res-an Int J 40:542–545
Veeken A, Kalyuzhnyi S, Scharff H et al (2000) Effect of pH and VFA on hydrolysis of organic solid waste. J Environ Eng 126:1076–1081
Ver Eecke HC, Butterfield DA, Huber JA (2012) Hydrogen-limited growth of hyperthermophilic methanogens at deep-sea hydrothermal vents. Proc Natl Acad Sci USA 109:13674–13679
Westerholm M, Levén L, Schnürer A (2012) Bioaugmentation of syntrophic acetate-oxidizing culture in biogas reactors exposed to increasing levels of ammonia. Appl Microbiol Biotechnol 78:7619–7625
Wilkie AC (2008) Biomethane from biomass, biowaste, and biofuels. Bioenergy. Wiley Online Library, pp 195–205
Zhang R, El-Mashad HM, Hartman K et al (2007) Characterization of food waste as feedstock for anaerobic digestion. Bioresource Technol 98:929–935
Zhu W, Reich CI, Olsen GJ et al (2004) Shotgun proteomics of Methanococcus jannaschii and insights into methanogenesis. J Proteome Res 3:538–548
Zia M, Khalil S, Aslam M et al (2003) Preparation of compost and its use for crop-production. Sci Technol Dev 22:32–44

Chapter 16
Biofertiliser from Animal Wastes

Raees Khan, Hammad Ahmad Jan, Baseerat Shaheen, Sheikh Zain Ul Abidin, Asad Ullah, Abdul Basit, Zeeshan Ahmad, and Shujaul Mulk Khan

Abstract The increase in demand for animal products is a major and growing source of pollution and diseases globally. Due to constrained disposal places and firmer regulations on burning, animal waste is a worldwide problem. Handling animal waste has certain significant dangers that are connected to the condition of the land, water, and air. As a regenerative and sustainable supply of plant nutrients, biofertilisers based on animal waste are an advantageous tool in the agricultural sector. In order to promote and encourage their usage as well as develop the supply chain, it is necessary for different stakeholders and governments to strengthen animal waste-based biofertilisers. Animal waste-based biofertilisers will mitigate hazards from global population food needs and production and will slow down/stop the widespread chemicalization in agroecosystems.

R. Khan (✉) · Z. Ahmad · S. M. Khan (✉)
Department of Plant Sciences, Faculty of Biological Sciences, Quaid-I-Azam University, Islamabad 45320, Pakistan
e-mail: raeeskhan@bs.qau.edu.pk

S. M. Khan
e-mail: smkhan@qau.edu.pk

Z. Ahmad
e-mail: zahmad@bs.qau.edu.pk

H. A. Jan
Department of Botany, University of Buner, Swari, Khyber Pakhtunkhwa, Pakistan

B. Shaheen
Higher Education Department, Khyber Pakhtunkhwa, Pakistan

S. Z. U. Abidin
Institute of Biological Sciences (IBS), Gomal University Dera Ismail Khan, Khyber Pakhtunkhwa, Pakistan

A. Ullah
Centre of Plant Biodiversity, University of Peshawar, Khyber Pakhtunkhwa, Pakistan

A. Basit
Government Post Graduate College Nowshera, Khyber Pakhtunkhwa, Pakistan

R. Khan · S. M. Khan
Member Pakistan Academy of Sciences, Islamabad, Pakistan

M. Arshad (ed.), *Climate Changes Mitigation and Sustainable Bioenergy Harvest Through Animal Waste*, https://doi.org/10.1007/978-3-031-26224-1_16

Graphical Abstract

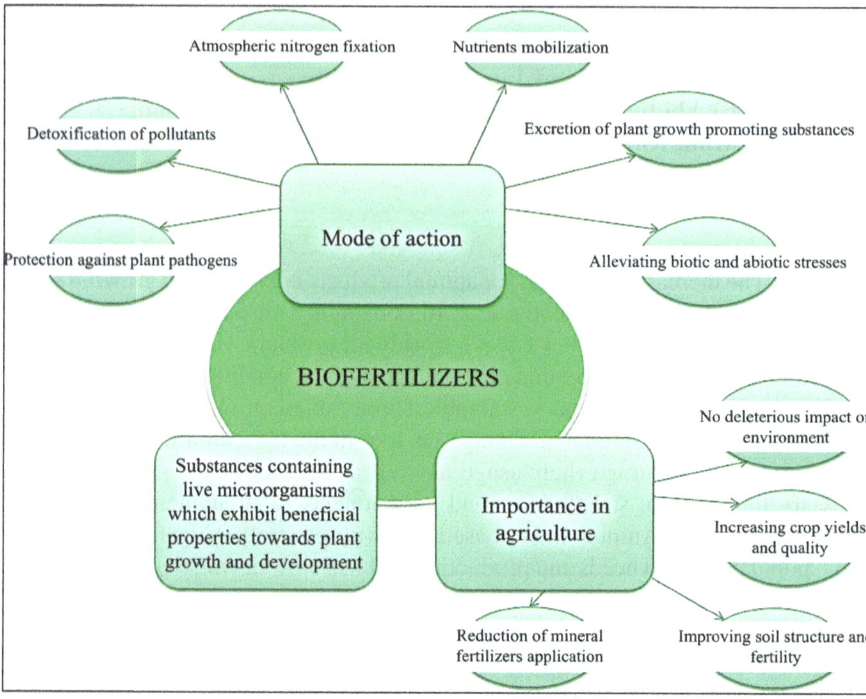

Keywords Agroecosystems · Environment · Industrialisation · Livestock · Chemical fertilisers · UN sustainable development goals

16.1 Introduction

The farming of animals in agriculture has a substantial role in both social and economic well-being of a nation by helping to supply the nation with food, create jobs, increase household income, generate revenue, pay taxes, preserve assets, use animals for traction, diversify agriculture, improve soil fertility, and provide transportation (Malomo et al. 2018). The term "animal wastes" refer to the solid, semisolid, and liquid by-products (faeces, urine, bedding materials including straw, sawdust, and rice hulls) produced by animals. Typically, the animals are raised to provide food for human use, such as meat, milk, and eggs (Sims and Maguire 2018). The faeces of cows, pigs, and chickens are the most common types of animal waste. Given their potential to pollute both surface and groundwater, animal wastes are a source of growing concern. (Gerba and Pepper 2009). Pathogens in animal faeces may contaminate food or water, or they can enter the body directly via inhalation, skin sores, and other paths that are open to pathogen entrance, which contributes to the spread of many zoonotic illnesses (Cavin and Butler 2016). Various approaches have been investigated in recent years for the reduction of organic waste materials as well as recycling and reusing these resources for the manufacturing of goods with additional value. Converting organic waste into biofertilisers is one of the newest strategies (Du et al. 2020).

Organic fertilisers known as "biofertilisers" also have microorganisms that have qualities that are advantageous for plant growth and development. These fertilisers also promote nutrient absorption, soil fertility, and crop yields (Mącik et al. 2020). Due to their sustainability and environmental friendliness, biofertilisers are seen to be an effective substitute for synthetic chemical fertilisers (SCF), which are expensive and harmful to the food and soil health (Jaffri et al. 2021). For intensive agriculture to provide large yields, several synthetic fertilisers are used. Synthetic fertilisers are expensive and dangerous for the nutrition and health of the soil (Singh et al. 2021). The eutrophication of water bodies, the greenhouse effect, and the heavy metals accumulation such as arsenic, cadmium, and plumbum are all strongly related to the excessive use of synthetic fertilisers (Mącik et al. 2020). Additionally, intensive chemical fertiliser usage reduces soil fertility and biodiversity (Singh et al. 2021). One of the finest alternatives to chemical fertilisers is biofertilisers derived from animal manure. Biofertilisers made from animal waste assist in restoring normal soil fertility and enhance the structure and functionality of the soil (Singh et al. 2021). Use of biofertiliser is primarily intended to support plant development without having a negative impact on the environment and to increase agricultural yields (Mishra et al. 2013). The yearly production of manure by domesticated cattle, pigs, and poultry, excluding animals on pasture, is around 120 million tonnes. (Loyon 2018). By 2023, the biofertiliser market is anticipated to grow at a CAGR of over 14%. The size of the worldwide biofertiliser market was USD 1106.4 million in 2016, and by the end of 2024, it is expected to have increased by USD 3124.5 million at a pace of 14.2% (Joshi and Gauraha 2022).

According to projections, the world's population will reach 9.7 billion by 2050, and the globe currently faces a climate emergency as a result of rapid urbanisation, industrialisation, and agricultural production using synthetic chemicals (Joshi and Gauraha 2022). Animal waste-based biofertilisers are amongst alternatives to the established farming system and help to decrease dependency on artificial plant protection inputs in crop production. This also helps to keep the environment clean, reduces population, and helps to protect the environment.

16.2 Biofertiliser from Animal Wastes

The quickest and best approach to make use of organic wastes is as substitute fertilisers and soil additives. Potentially safe to use as a broad-acre biofertiliser is the acquired low-cost biodegradable end product of organic waste. The development of an affordable technology for the treatment of industrial animal wastes is made possible by several methods for using animal wastes (Fig. 16.1). The European rendering market gathered and processed 15 million tonnes of animal waste in 2002 (Hall and Sullivan 2001). Utilisation of animal remains by different methods which is a special field of waste management. Some of these are following.

Fig. 16.1 Animal waste produced in a local dairy farm in Peshawar, Pakistan

16.2.1 Vermicompost

Vermicompost is a term used to describe how earthworms break down organic waste into a uniform mixture that resembles humus. The faecal by-products of bacteria and earthworms are combined to form vermicompost, a complex material. As voracious eaters, earthworms change the makeup of organic materials and progressively transform it into more nutritious elements. The nutrient content of vermicompost is greater than the traditional compost, hence converting it into valuable fertilisers. Earthworm increases the surface area of materials, making them more favourable for microbial activity. These have the ability to consume different types of excreta from livestock and cattle dung (Pandit et al. 2012).

16.2.2 Digestate Biofertiliser

Anaerobic digestion is a low-cost method of producing digestive biofertiliser. The technique makes use of a variety of raw materials, including commercial, agricultural, and household trash. The generation of food waste has significantly increased as a result of the growing world population (Curry and Pillay 2012). According to Johansen et al. (2013), digestate biofertiliser increases the diversity of soil microbes. The biofertiliser quality of digestate produced by the digestion of chicken droppings and cow dung was evaluated by Alfa et al. in 2014. Garfi et al. (2011) investigated the characteristics of digestate made from guinea pig dung. The soil's fertility may be increased by applying digestate biofertiliser, which is produced by the anaerobic digestion of human excreta. The digestate may be used as an effective biofertiliser for crop growth and yield due to the presence of organisms that fix nitrogen and solubilise phosphate (Owamah et al. 2014).

16.2.3 Poultry Waste-Based Biofertiliser

Animal waste is defined as bones or animal parts that are not primarily intended for human consumption and are regarded as high-risk items. Composting and anaerobic digestion are ecologically friendly methods for disposing of this garbage and treating it to get rid of any potential microbes. It reduces pollutants, stabilises sludge, and generates biogas, making it a potential option for poultry slaughterhouses processing organic waste. The hydrolysis of organic matter, which occurs as part of digestion, is the rate-limiting stage since the organics in chicken slaughterhouse wastes degrade slowly (Park et al. 2017). Poultry slaughterhouse wastes are combined with agricultural wastes, sewage sludge, wood dust, and activated compost over a 90-day composting cycle. The creation of mature and stable compost was made possible by the study of microbiological and physicochemical factors throughout the composting process (Asses et al. 2019).

16.3 Manure Management

A critical component of the agricultural waste management system is the treatment of animal manure (Malomo et al. 2018). Manure has long been valued as a soil amendment for growing crops. Manure management, taken in its broadest meaning, refers to the effective use of animal waste in accordance with each farm's capabilities and objectives in order to improve the soil's quality, the nutrition of the crops, and farm profitability. Manure management is described as a process of making decisions that aims to maximise agricultural productivity whilst minimising nutrient loss from manure, both now and in the future (Karmakar et al. 2007). Due to the growth of the livestock business, the increase in the number of livestock animals, and the adoption of environmental legislation and standards, appropriate manure management systems (MMS) are becoming more and more crucial. Management planning now has to take into account more decision factors as a result of growing environmental and sustainability concerns (Li et al. 1994). Environmental laws that aim to avoid pollution of the air, water, and land have a greater impact on the choice of manure management and treatment alternatives. Examples include how housing management, manure storage and treatment, and land application methods may be impacted by controlled decreases in ammonia emissions (Westerman and Bicudo 2005).

Enacted in 1997 and adopted in 2005, the Kyoto Protocol to the United Nations Framework Convention on Climate Change (UNFCCC) is a global agreement (Boehm 2005). For the first commitment period (from 2008 to 2012), parties of this protocol agreed on a legally enforceable greenhouse gas (GHG) emission reduction target. These objectives were created to lower GHG emissions worldwide by around 5% from 1990 levels. A potential avenue for reducing GHG emissions is manure management in livestock husbandry. At every step of managing manure, including collection, handling/storage, treatment, and application to the ground, greenhouse gases are produced. Utilizing the right decision support system (DSS) might help livestock owners discover MMS or other elements that contribute to the decrease of GHG emissions (Karmakar et al. 2007).

16.4 Importance of Livestock Waste Management

Animal waste is most often a source of concern since it may produce a lot of CO_2 and ammonia, which can cause acid rain and the greenhouse effect. It may contaminate water supplies and aid in the spread of infectious illnesses. The discharge of smells and the contamination of water sources might lead to societal unrest if sufficient planning is not made. The following factors show the necessity, benefits, and significance of proper management of animal waste (Font-Palma 2019):

1. In soils deficient in organic matter, livestock dung aids in maintaining soil fertility. The physical state of the soil is improved by adding manure, which also improves soil structure and improves the soil's ability to store water.
2. Additionally, animal manure contributes to the improvement of the soil's microclimate for fauna and flora.
3. Manure is also used as fuel.
4. Manure waste and other organic wastes from animal ranches have the potential to be a significant source of energy.
5. Utilizing animal manure might help with resource management, crop and livestock production, and post-harvest loss reduction.
6. As a sustainable energy resource, bio-energy sources are becoming more and more popular which might help address issues like the growing cost of fuel and the demand for energy by serving as an alternative to pricey fossil fuels.
7. Sustainable agriculture is supported by biogas produced from animal waste and by-products, which is renewable and ecologically benign. Additionally, the digesters provide superior-quality organic waste.
8. Lower the risk of infection for both the animal and human populations.
9. Until it is decomposed and changed into a soluble form, the nitrogen in manure remains bound up in its organic state (ammonium nitrate). Ammonium nitrate is added to soil to increase fertility.
10. Decreases prohibited waste discharge which might endanger the quality of the water and soil.

16.4.1 Traditional Method of Livestock Waste Management

Dung cake: In less developed nations, cow dung is physically collected and spread out on the proper racks to sundry before being used as fuel for heating and cooking (Font-Palma 2019).

Dumping into heaps or pits: It is the most common and traditional waste management method, in which all garbage is dumped into a pit on a farm or field (Font-Palma 2019).

Composting: The first stage in the composting of organic waste is the thermophilic stage (45–65 °C), in which microorganisms, mostly bacteria, fungus, and actinomycetes create heat, carbon dioxide, and water. By combining or aerating the varied organic material, a uniform and stable humus-like product is produced. Composting is the process of aerobically breaking down biodegradable organic waste. The biodegradation process is rather quick, requiring just 4–6 weeks to stabilise the material. Compostable material may be utilised as organic fertiliser since it is odourless, fine-textured, and low in moisture. It is furnished with a fork, a brush, and a tiny cart with four wheels. The brush is rotated by a gear motor with a belt drive and pulley system (Font-Palma 2019).

16.4.2 Advance Methods of Livestock Waste Management

16.4.2.1 Biogas Production

Animal waste, household garbage, and agricultural waste may all be used to produce biogas, a clean, environmentally friendly fuel that is readily accessible in rural areas. Under anaerobic circumstances, bacteria convert organic materials to gases to produce biogas. Methane makes up 55–65% of biogas, along with carbon dioxide (35–45%), hydrogen sulphide (0.5–1.0%), and water vapour in very small amounts. Biogas has an average calorific value of 20 MJ/m^3 (4713 kcal/m^3) (Font-Palma 2019).

16.4.2.2 Vermicomposting

The earthworm consumes the organic material and excretes "vermicompost," which is a small, pelleted substance. Important plant nutrients like N, P, K, and Ca that are contained in organic waste are released during vermicomposting and changed into forms that are more soluble and useful to plants. Additionally, vermicompost has physiologically active ingredients like plant growth regulators. In addition, the worms themselves serve as a source of protein for animal feed. Composting period is shortened to 60–75 days, whilst N, P, and K content are increased by three to four times (Font-Palma 2019).

16.4.2.3 Pyrolysis

The process of pyrolysis involves heating condensed organic molecules in a reactor, often without oxygen, and causing chemical degradation. Straw, twigs, sawdust, and other agricultural and forestry waste are some of the main raw materials utilised in pyrolysis. These raw materials are converted into a variety of products under high pressure and temperature. Manure may be pyrolyzed by heating it to between 480 and 830°F whilst maintaining a low oxygen level in the air. In a closed system at temperatures between 400 to 1472°F, waste is chemically broken down by a thermochemical process. Gases, oil, and ash are the by-products. H_2, H_2S, CH_3, CO, and ethylene are amongst the gases. When compared to other animal manures, dairy faeces produced the most gas per unit of dry solids, followed by chicken, beef, and swine faeces (Brugger and Windisch 2015).

16.4.2.4 Soldier Fly Breeding

The larvae or "grubs" of black soldier flies (BSF) are well adapted to handle animal manure. The BSF larvae may survive for many weeks, whereas the adults only have a short lifespan, and during that period, they can absorb enormous amounts of food waste or manure. This process yields two beneficial by-products: the castings or waste, which may be used to enhance the soil, and the larvae, which are a fantastic source of food for a variety of creatures, including fish, birds, reptiles, and amphibians (Font-Palma 2019).

16.4.2.5 Litter Management

The excreta, bedding, leftover feed, and feathers all make up poultry litter. Wood shavings, sawdust, straw, peanut hulls, and other fibrous materials may be used as bedding. The majority of the chicken litter comes from the broiler industry. The litter may have accumulated across numerous bird harvests or may have come from a single crop of broilers. 20–25% of the moisture in the litter is typical. Mostly, beef cows and stocker cattle are fed poultry litter. On a dry matter basis, broiler litter includes 25–50% crude protein and 55–60% TDN and is a good source of important minerals. The nutritional content is thus on par with or better than that of high-quality legume hay. Instead of being an issue with waste, poultry litter can and should be a source of nutrition and energy (Schlegel et al. 2015).

16.4.2.6 Ammonia Recycling

Using a gas permeable membrane to recycle ammonia from animal wastewater is a conventional technique (Maglinao et al. 2015). This membrane is impermeable to water and only permits the passage of gases. Gaseous ammonia is captured and concentrated in a stripping solution with the use of a microporous hydrophobic membrane. Organic acid and mineral acid of 1 normalcy make up the stripping solution. Polypropylene and polyurethane are membranes that are used for filtration. Manure pH affects ammonia recovery, increasing it by 1.2% per hour at pH 8.3 and 13% per hour at pH 10, and the average removal rate is 45–153 mg of ammonia litre/day (Parihar et al. 2019).

16.4.2.7 Enzymatic Fermentation into Ethanol

Due to its relatively high (up to 50%) fibre content, manure comes within this group. Fibre makes up the majority of manure's resource component, therefore converting it to biochemical using a sugar platform offers a method for this higher degree of manure use. This procedure comprises the hydrolysis of the cellulose and hemicellulose found in fibre into simple sugars, which may then be processed chemically or biologically to produce ethanol for use as a fuel or other compounds (Chen et al. 2005). With total sugar conversions nearing 79%, diluted sulphuric acid pre-treatment and enzyme hydrolysis are a preferable method. The feasibility of turning feedlot cow dung into ethanol (70% efficiency) was confirmed by fermentation tests using the resulting C6 hydrolysates. This procedure might provide higher yields with advancements (such as the fermentation of C5 sugars), thus enhancing its viability as a feedstock for biofuels (Vancov et al. 2015).

16.5 Value-Added Products from Animal Waste

External components including beddings, urine, wash water, spilt feed, and water are also included in manure made from animal waste. Animal dung was essential for improving soil fertility before organic fertilisers were developed (Malomo et al. 2018). Through technology, several items from animal excrement are produced (and bio-waste). Processing of animal manure often includes digestion, which frequently occurs concurrently with the digestion of other bio-wastes. The majority of methods are also useful for digestates and the separation products they produce (Ehlert and Schoumans, 2015). According to Malomo et al. (2018), major animal waste component includes nutrients (manure, fertiliser, biomass conversion; animal feed, soil alterations, compost; etc.), organic matter (soil amendments/structuring), solids (bedding), energy (biogas, bio-oil, and syngas), and fibre (peat substitute, paper, and building materials).

A range of value-added products is produced via processing methods. Ammonium sulphate solutions in water (lightly acidic), ash (PK fertiliser, liming material), biochar, compost (organic fertiliser or organic soil amendment), and digestate are the primary products, and their potential for recycle and reuse (organic fertiliser or organic soil amendment) and mineral concentrations of potassium and nitrogen organo-mineral fertilisers, which are NPK fertilisers embedded in organic matter and have relatively high nutrient contents, are organic fertilisers with relatively low nutrient contents (Table 16.1). Precipitated salts include magnesium ammonium phosphate (Mg-struvite), potassium ammonium phosphate (K-struvite), magnesium phosphates, and calcium phosphates (Ehlert and Schoumans, 2015).

16.6 Application of Biofertilisers in Agriculture Practices

The world's population is growing, and it is predicted that by 2050, there will be 9.7 billion people on the planet (Ehrlich and Harte 2015). Industrialisation, urbanisation, and agricultural productivity are all intimately related to the growing world population (Gizaki et al. 2015; Mahanty et al. 2017; Santos et al. 2012). The nutritional needs of mankind cannot be satisfied by traditional agriculture. For plant nutrition and disease management, traditional farming practices employ a tonne of synthetic

Table 16.1 Categorisation of different types of livestock manure (Wen et al. 2007)

Parameters	Dairy	Beef	Feedlot
Solid content (% of fresh manure)			
Total solids (dry matter)	13.39	12.56	26.61
Total volatile solids	11.21	9.97	22.78
Elements (% of dry matter)			
Carbon	45.37	43.81	43.56
Nitrogen	3.03	1.94	2.72
Phosphorus	0.48	0.42	0.81
Potassium	2.86	1.44	0.92
Calcium	1.2	1.06	0.69
Magnesium	0.55	0.3	0.34
Sodium	0.47	0.25	0.12
Copper	0.003	0.0002	0.0018
Zinc	0.032	0.0042	0.0087
Iron	0.03	0.059	0.055
Sulphur	0.31	0.25	0.21
Aluminium	0.014	0.017	0.021
Cobalt	0.0009	0.0002	0.0002
Chromium	0.0002	0.0002	0.0001
Manganese	0.051	0.06	0.012
Molybdenum	0.0003	0.0002	0.0008
Nickel	0.001	0.001	–
Vanadium	0.0005	0.0006	–

fertilisers and pesticides, which has increased crop growth, productivity, and quality as well as farmers' revenue. On the other hand, by polluting water, air, and soil reservoirs, the rising use of artificial assistance has seriously harmed the natural environment (Rahman and Zhang 2018). These agrochemicals have accumulated belowground as a result of their improper application and inability to biodegrade. These build up underground, changing the soil's properties negatively in terms of structure, fertility, and water-holding capacity resulting in the greenhouse effect and eutrophication of water basins. An alternative to traditional agriculture, organic farming promotes crop development whilst preserving the soil's high quality and biodiversity. Biofertilisers contribute to the preservation of a soil environment rich in micronutrients as well as macronutrients via nitrogen fixation, phosphate and potassium solubilisation and mineralisation, the release of plant growth-regulating compounds, and the synthesis of antibiotics. Farmers have utilised legumes to improve soil fertility since prehistoric times. Whilst Nobbe and Hiltner developed "Nitragin," a Rhizobia laboratory culture, in 1895, it was followed by the discovery of Azotobacter and blue-green algae, which marked the beginning of the industry's usage of biofertilisers (Ramesh 2008).

16.7 Animal Waste-Based Biofertilisers in Aquaculture

What occurred in the past, current trends, and potential future developments are three key issues that are discussed in relation to the development of ecologically sustainable aquaculture. Traditional aquaculture mostly benefits the environment since it feeds farmed aquatic animals with trash and by-products from the farm and the neighbourhood, such as leftover food from open water bodies, animal or human excrement, or agricultural leftovers (Zajdband 2011).

Traditional and contemporary aquaculture production methods are compared, with a focus on the kind of nutrient inputs, in natural ecosystems and man-made agroecosystems. Natural ecosystems and man-made agroecosystems are studied in relation to terrestrial and aquatic, coastal/offshore, land- and waterscapes, and aquaculture. The failure of traditional integrated aquaculture, along with increased environmental sustainability concerns, has resulted in a substantial change. The two-way impacts of aquaculture on the environment are explored, and environmental issues are shown via case studies of genuine traditional and modern inland and coastal aquaculture techniques in temperate and tropical settings (Zajdband 2011).

Due of its widespread availability on farms, animal dung is the most widely used organic fertiliser. In reality, one significant element that accounts for the sharp rise in Chinese inland aquaculture productivity over the last several decades is the increasing availability of organic fertilisers as a consequence of increased chicken and pig production (Weimin 2010). The only method for the majority of small farmers to increase pond production is to employ animal dung, even though wealthy farmers often use artificial fertilisers (Ahmed et al. 2010). But combining mineral and organic fertilisers is another strategy (Zajdband 2011). Feed that has been spilled, bedding material, and farm animal excrement are all considered to be manure. Organic fertilisers also include green manures with high nitrogen concentration and fresh or composted agricultural by-products such as pressmud and a by-product of sugarcane (Keshavanath et al. 2006). Fish ponds have been fertilised using a wide range of animal manures. Its composition, which may change over time, determines the effectiveness of the manure as fertiliser. For effective use, the manure must be applied to the pond in a certain order (Zajdband 2011).

16.7.1 Pond Water for Irrigation

Aquaculture effluents can also be utilised to water land-based crops. In reality, rather than being used for fish farming, pond building on farms may be primarily for irrigation reasons (Fernando and Halwart 2000; Nhan et al. 2007). Because it includes both dissolved and suspended inorganic and organic components from fish culture (such as feeds and fertilisers, as well as other external nutrients, including materials produced via soil erosion, run-off, and leaching), pond-outlet water is often nutrient-rich. However, pesticide or antibiotic chemical residues may be present in the aquaculture wastes. Commonly, the composition of pond effluent changes with the season, being richer in the summer when fertilisation and feeding rates are greater (Zajdband 2011).

16.8 Factors Affecting the Quantity and Quality of Animal Manure

Animal dung varies in terms of both volume and nutritional content and depends on a wide range of factors that can be categorised into four groups: (1) animal-specific factors; (2) factors relating to animal feed and feeding; (3) factors relating to housing, bedding materials, and waste collection; and (4) factors relating to the transportation, processing, and storage of waste. Manure is considered to be of good quality if it contains enough amounts of the nutrients, such as nitrogen and phosphorus that are often to be responsible for limiting pond primary production (Zajdband 2011).

16.9 Environment and Economic Significance of Animal-Based Biofertilisers

According to Malomo et al. (2018), some of the major associated benefits of comprehensive manure management through animal-derived biofertilisers include as follows:

- Prevents the environmental impacts on air, water, soil, wildlife, and the marine ecosystem.
- Reduces the risks associated with animal waste and protects human health by preventing and spread of diseases.
- Increases productivity, lowers medical expenses, improves environmental quality, and maintains ecosystem services. It also contributes to economic stability by reducing costs via environmental and human health benefits.
- Contributes to economic stability by creating appealing and amusing human settlements, and employment, including low, medium, and high-skilled jobs.

16.10 Animal-Based Biofertilisers and Sustainable Development Goals

Due to population expansion and intense pressure to raise agricultural output to meet the needs of the expanding population, the world's need for food has grown. For the last several decades, chemical fertilisers have been the easy fix, but their excessive and careless use has resulted in food contamination, weed resistance, new illnesses, severe environmental effects, and a major negative impact on human health (Sansinenea 2021).

Animal waste has increased turbidness through the movement of soil particles into streams, rivers, and lakes (Gerba and Pepper 2009). The possibility to generate and utilise animal waste-based biofertilisers as an alternative to chemical fertilisers in agriculture has been made possible by rising demands for sustainable agricultural objectives, declining reliance on agrochemicals, and producing more nutritious and organic foods (Joshi and Gauraha 2022). Sustainable handling of animal waste may benefit both farmers and the broader population in a number of ways (Malomo et al. 2018). The use of animal wastes as biofertilisers will help and contribute significantly towards achieving UN sustainable development goals (SDGs). The waste of animals poses significant problems and risks to the public's health, but with the right management, it may become a useful resource.

Certain significant barriers to biofertiliser use include farmer unawareness, performance benefits over chemical counterparts, and supply chain, which will lead to stable and sustained growth for biofertilisers in the future (Joshi and Gauraha 2022). The application of biofertilisers is helpful, especially in increasing sustainable agricultural practices along with enhancing the yield, protecting the plants from various biotic and abiotic stresses, and improving the content of pharmaceutically vital secondary metabolites (Tripathi and Singh 2021). Strict policies and legal frameworks are needed for sustainable animal waste management through the production of biofertilisers (Malomo et al. 2018). Such practices will contribute to reducing the detrimental impacts of animal waste and increase organic agricultural practice and global strive for zero emissions.

16.11 Conclusions

The significance of sustainable animal waste management through the production of biofertilisers cannot be over-emphasised. Although some local and international practices of biofertiliser production from animal waste are gaining popularity, still more comprehensive policies and practices are needed for better management of animal waste to produce biofertilisers. Pressure from global policymakers, international environmental organisations, scientist, climate activists, and movements is needed to increase and encourage appropriate activities and actions to promote animal waste-based biofertilisers.

References

Ahmed N, Allison EH, Muir JF (2010) Rice fields to prawn farms: a blue revolution in Southwest Bangladesh? Aquacult Int 18(4):555–574

Alfa MI, Adie DB, Igboro SB, Oranusi US, Dahunsi SO, Akali DM (2014) Assessment of biofertilizer quality and health implications of anaerobic effluent of cow dung and chicken droppings. Renewable Energy 63:681–686

Asses N, Farhat W, Hamdi M, Bouallagui H (2019) Large scale composting of poultry slaughterhouse processing waste: microbial removal and agricultural biofertilizer application. Process Saf Environ Prot 124:128–136

Boehm M (2005) Post Kyoto protocol—what next. In: Special seminar in Department of Soil Science, University of Saskatchewan, November, vol 14

Brugger D, Windisch WM (2015) Environmental responsibilities of livestock feeding using trace mineral supplements. Animal Nutrition 1(3):113–118

Cavin RM, Butler DR (2016) Animal hazards: their nature and distribution. Biol Environ Hazards Risks Disasters 153–180

Chen S, Wen Z, Liao W, Liu C, Kincaid RL, Harrison JH, Stevens DJ et al (2005) Studies into using manure in a biorefinery concept. Appl Biochem Biotechnol 124(1):999–1015

Curry N, Pillay P (2012) Biogas prediction and design of a food waste to energy system for the urban environment. Renewable Energy 41:200–209

Du C, Munir S, Abad R, Lu D (2020) Valorization of organic waste into biofertilizer and its field application. Waste Biorefinery 179–198

Ehlert PAI, Schoumans OF (2015) Products, by-products and recovered secondary materials from processed animal manure (No. 2668). Alterra, Wageningen-UR

Ehrlich PR, Harte J (2015) To feed the world in 2050 will require a global revolution. Proc Natl Acad Sci 112(48):14743–14744

Fernando CH, Halwart M (2000) Possibilities for the integration of fish farming into irrigation systems. Fish Manage Ecol 7(1–2):45–54

Font-Palma C (2019) Methods for the treatment of cattle manure—a review. C 5(2):27

Garfí M, Gelman P, Comas J, Carrasco W, Ferrer I (2011) Agricultural reuse of the digestate from low-cost tubular digesters in rural Andean communities. Waste Manage 31(12):2584–2589

Gerba CP, Pepper IL (2009) Wastewater treatment and biosolids reuse. In: Environmental microbiology. Academic Press, pp 503–530

Gizaki LJ, Alege AA, Iwuchukwu JC (2015) Farmer's perception of sustainable alternatives to the use of chemical fertilizers to enhance crop yield in Bauchi state Nigeria. Int. J. Sci. Res. Sci. Technol. 1:242–250

Hall B, Sullivan P (2001) Alternative soil amendments: horticulture technical notes. ATTRA-465 National Sustainable Agriculture Information Service, p 12

Jaffri SB, Ahmad KS, Jabeen A (2021) Biofertilizers' functionality in organic agriculture entrenching sustainability and ecological protection. In: Biofertilizers. Woodhead Publishing, pp 211–219

Johansen A, Carter MS, Jensen ES, Hauggard-Nielsen H, Ambus P (2013) Effects of digestate from anaerobically digested cattle slurry and plant materials on soil microbial community and emission of CO_2 and N_2O. Appl Soil Ecol 63:36–44

Joshi SK, Gauraha AK (2022) Global biofertilizer market: emerging trends and opportunities. Trends Appl Microbiol Sustain Econ 689–697

Karmakar S, Laguë C, Agnew J, Landry H (2007) Integrated decision support system (DSS) for manure management: a review and perspective. Comput Electron Agric 57(2):190–201

Kashavanath P, Gangadhara SB (2006) Evaluation of sugarcane by. product press mud as a manure in carp culture. Biores Technol 97:628–634

Li E, Moncrieff IJ, Ariyaratnam A (1994) Decision support software for on-farm application: views from a panel of Australian professionals. In: Computers in Agriculture. Proceedings of the Fifth International Conference Orlando. Florida, February, pp 6–9

Loyon L (2018) Overview of animal manure management for beef, pig, and poultry farms in France. Frontiers Sustain Food Syst 2:36
Mącik M, Gryta A, Frąc M (2020) Biofertilizers in agriculture: an overview on concepts, strategies, and effects on soil microorganisms. Adv Agron 162:31–87
Maglinao AL Jr, Capareda SC, Nam H (2015) Fluidized bed gasification of high tonnage sorghum, cotton gin trash and beef cattle manure: evaluation of synthesis gas production. Energy Convers Manage 105:578–587
Mahanty T, Bhattacharjee S, Goswami M, Bhattacharyya P, Das B, Ghosh A, Tribedi P (2017) Biofertilizers: a potential approach for sustainable agriculture development. Environ Sci Pollut Res 24(4):3315–3335
Malomo GA, Madugu AS, Bolu SA (2018) Sustainable animal manure management strategies and practices. Agricult Waste Residues 119
Mishra D, Rajvir S, Mishra U, Kumar SS (2013) Role of bio-fertilizer in organic agriculture: a review. Res J Recent Sci 39–41
Nhan DK, Phong LT, Verdegem MJ, Duong LT, Bosma RH, Little DC (2007) Integrated freshwater aquaculture, crop and livestock production in the Mekong delta, Vietnam: determinants and the role of the pond. Agric Syst 94(2):445–458
Owamah HI, Dahunsi SO, Oranusi US, Alfa MI (2014) Fertilizer and sanitary quality of digestate biofertilizer from the co-digestion of food waste and human excreta. Waste Manage 34(4):747–752
Pandit NP, Ahmad N, Maheshwari SK (2012) Vermicomposting biotechnology an eco-loving approach for recycling of solid organic wastes into valuable biofertilizers. J Biofertil Biopestic 3:1–8
Parihar SS, Saini KPS, Lakhani GP, Jain A, Roy B, Ghosh S, Aharwal B (2019) Livestock waste management: a review. J. Entomol. Zool. Stud 7:384–393
Park S, Yoon YM, Han SK, Kim D, Kim H (2017) Effect of hydrothermal pre-treatment (HTP) on poultry slaughterhouse waste (PSW) sludge for the enhancement of the solubilization, physical properties, and biogas production through anaerobic digestion. Waste Manage 64:327–332
Rahman KA, Zhang D (2018) Effects of fertilizer broadcasting on the excessive use of inorganic fertilizers and environmental sustainability. Sustainability 10(3):759
Ramesh P (2008) Organic farming research in M.P. Organic farming in rainfed agriculture: central institute for dry land agriculture, Hyderabad, pp 13–17
Sansinenea E (2021) Application of biofertilizers: current worldwide status. In: Biofertilizers. Woodhead Publishing, pp 183–190
Santos VB, Araújo AS, Leite LF, Nunes LA, Melo WJ (2012) Soil microbial biomass and organic matter fractions during transition from conventional to organic farming systems. Geoderma 170:227–231
Schlegel AJ, Assefa Y, Bond HD, Wetter SM, Stone LR (2015) Corn response to long-term applications of cattle manure, swine effluent, and inorganic nitrogen fertilizer. Agron J 107(5):1701–1710
Sims JT, Maguire RO (2018) Manure management. In: Hillel D (ed) Encyclopedia of soils in the environment. Elsevier, pp 402–410
Singh D, Thapa S, Geat N, Mehriya ML, Rajawat MVS (2021) Biofertilizers: mechanisms and application. In: Biofertilizers. Woodhead Publishing, pp 151–166
Tripathi P, Singh A (2021) Biofertilizers: "An ace in the hole" in medicinal and aromatic plants cultivation. In: Biofertilizers. Woodhead Publishing, pp 253–263
Vancov T, Schneider RCS, Palmer J, McIntosh S, Stuetz R (2015) Potential use of feedlot cattle manure for bioethanol production. Biores Technol 183:120–128
Weimin M (2010) Recent developments in rice-fish culture in China: a holistic approach for livelihood improvement in rural areas. Success Stories Asian Aquacult 15–40
Wen Z, Liao W, Liu C, Chen S (2007) Value-added products from animal manure. Bioprocess Value-Added Prod Renew Resour 629–651

Westerman PW, Bicudo JR (2005) Management considerations for organic waste use in agriculture. Biores Technol 96(2):215–221

Zajdband AD (2011) Integrated agri-aquaculture systems. In: Genetics, biofuels and local farming systems. Springer, Dordrecht, pp 87–127

Chapter 17
Valorization of Animal Waste for the Production of Sustainable Bioenergy

Mehnaz Hashim, Ali Akbar, Sher Zaman Safi, Muhammad Arshad, and Zareen Gul

Abstract The alarming rise in our dependency on fossil fuels like coal, oil, and natural gas has put these resources in a state of depletion. Due to the rapid industrialization and population growth, there is also an enormous growth in energy demand. Access to clean and green energy is crucial for the sustainable growth of human society in order to meet this steadily increasing demand. A possible alternative source for producing clean energy is biomass. One of the highly promising renewable energy options that lead to a more sustainable energy system is the use of animal waste (blood, bone, sludge, manure, lipids, and meat effluent). These wastes are rich in carbohydrates, proteins, and lipids. Animal wastes can contribute significantly to an appropriate waste management system since it does not endanger food security and minimize environmental impacts. The generation of bioenergy from animal waste is linked to reducing the generation of solid waste and its disposal on the land, specifically minimizing all risks and negative effects in all sectors during waste disposal. A sustainable bioenergy industry that improves fuel security covers the difficult issue of climatic change as well. In order to meet domestic demand and to export, intensification of animal waste conversion strategies to bioenergy (biogas, biodiesel, bio-alcohol, and bioelectricity) production is established. The energy-rich bioenergy is produced by the anaerobic digestion and fermentation processes, which also reduce pollution and global warming. In this chapter, we have explored the applications of animal waste as potential source to be used for sustainable energy production.

M. Hashim · A. Akbar (✉)
Department of Microbiology, University of Balochistan, Quetta 87300, Pakistan
e-mail: ali.akbar@um.uob.edu.pk

S. Z. Safi
Faculty of Medicine. Bioscience and Nursing, MAHSA University, Jenjarom, Selangor, Malaysia

M. Arshad
Department of Basic Sciences, Jhang-Campus, University of Veterinary and Animal Sciences Lahore, Lahore, Pakistan

Z. Gul
Department of Botany, University of Balochistan, Quetta, Pakistan

M. Arshad (ed.), *Climate Changes Mitigation and Sustainable Bioenergy Harvest Through Animal Waste*, https://doi.org/10.1007/978-3-031-26224-1_17

Graphical Abstract

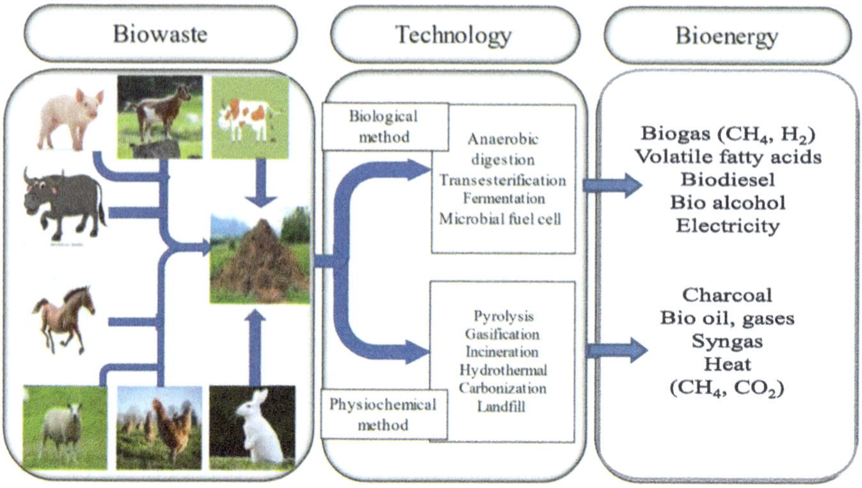

Overall plan for producing bioenergy from various biowastes

Keywords Food waste · byproducts · Sustainable energy · Biogase

17.1 Introduction

Manure and other byproducts of meat processing are produced in massive amounts worldwide, especially on dairy and poultry farms in third-world nations (Zhu and Hiltunen 2016). The quantity of waste collected from farms typically varies depending on several factors including its type, size, age, diet, feeding, and reproductive practices (Khalil et al. 2019). For instance, estimates place the daily yields of animal manure from cattle (10–20 kg), sheep (2 kg), and chickens (0.1 kg), respectively (Kaygusuz 2002; Omrani 1996). Additionally, the above parameters can also affect the quantity of blood and rumen material. The average weight of large and small ruminants and poultry is 250 kg, 40 kg, and 1.5 kg, respectively (Abdeshahian et al. 2016; Afazeli et al. 2014). Blood and rumen usually make up about 8.4% and 12% of large ruminants (cattle and buffalo) body weight while small ruminants (goat and sheep) account for 3% and 25%, respectively. Numerous applications, including fuels, internal combustion energy generators, microturbines, fuel cells, etc., can directly utilize it. The animal waste from the meat processing industry is comprised of animal fat waste, bone, feathers, hair, meat, and skin, and these include a lot of organic stuff that generate odors and, if not treated, can serve as a medium for the growth of several pathogenic microorganisms. Its management is a significant issue for livestock producers. Animal farms frequently store it in a lagoon for future use as manure (Mathias 2014). Methane, one of the most dangerous pollutants to air than CO_2, is produced as a result of the natural decomposition process. Animal waste storage tank runoff has the potential to contaminate groundwater by penetrating the water table. Over the past decade, there has been an increase in the conversion of animal waste into biofuel, although this practice is still limited by the expense and maintenance of digesters (Gebrezgabher et al. 2010). The waste digestion capability, the amount of protein and fibrous matter, all affect the quality of animal feces. Garbage that contains between 20 and 25% solids is treated as such and is simple to stack and pick up with a loader, whereas semi-solid waste is processed as a liquid. Animal excreta makes an excellent starting point for the production of biogas since it already includes the majority of the necessary bacteria in a waste-to-energy technical process.

17.2 Effects of Animal Wastes

Wastes have a serious threat to the environment and pose a potential risk to human life. It leads to disease outbreaks, reduced life expectancy, and dangerous environments, which have an impact on both humans and animals. While some wastes might decay, those that do not produce methane gas, which greatly add to global warming and unpleasant odor. Wastes cause environmental pollution including land, water, and air pollution. Dust, smoke, and stink are the key elements of air pollution. Burning solid waste releases CO_2 and N_2O, which contribute to ozone layer depletion and the greenhouse effect (Bhat et al. 2018). Additionally, CH_4 and H_2S are released into the environment. These compounds are poisonous to living things. Polluted water is an additional undesirable impact of waste on the ecosystem. According to reports, 1400 individuals each day lose as a result of water-related illnesses and disorders (Khan et al. 2019). Water sources, such as springs, rivers, groundwater, streams, and oceans, are adversely affected by wastes by reducing pH levels and posing health risks to aquatic life and individuals who use this water. Among these contaminants, some are extremely hydrophilic and less water-soluble (Varjani et al. 2017). The use of receiving water that has been polluted by waste from one region could be permitted at another. Ineffective waste management may result in soil pollution. Wastes that are discarded carelessly breed disease-carrying vectors and are unsightly. Iron-derived metals, radioactive wastes, and other substances are dangerous to organisms and plants of soil habitat, which lowers crop yield (Mani and Kumar 2014). Mosquitoes thrive in stagnant bodies of water, clogged drains, tires where rainwater is collected, empty food containers, polythene bags, and other materials. The risks faced by the waste management force also include tissue injury, respiratory tract infections, wounds from glass, contaminated razor blades, and needles, in addition to parasitic diseases brought on by interaction with the waste on the skin (Alam and Ahmade 2013). Although workers wear safety gear including gloves and nasal masks, modern automated methods should be promoted to reduce the risk of waste treatment-related injuries to refuse workers. Langdon et al. (2019) conducted research on the risk evaluation of organic pollutants in composted solid municipal waste in Australia. Based on the potential effects on health, they can classify the risk levels in their study into low, moderate, and high priority. As a result, steps are taken to ensure that each of the waste's harmful components is disposed of safely and effectively. The study of the electronic waste impact on public health by Ayilara et al. (2020) revealed that the discharge of numerous toxic metals into the air causes air pollution and affects human health and the environment badly.

17.3 Protein from Animal Source

Furthermore, the meat industry consistently produces large amounts of byproducts that are unavoidable, such as wastewater that is frequently highly contaminated with chemicals and organic debris, as well as blood, skeletons, epidermis, hair, feathers, viscera, and hoofs. According to reports of Abdilova et al. (2021), Jayawardena et al. (2022), the worldwide production of meat byproducts is estimated to be at 52.6 million tons per year or 20% of total meat production. Although some metabolites are eatable and rich in nutrients including polypeptides, vitamins, essential amino acids, and minerals. Their consumption as human food for nourishment is waning and is still confined to a small number of limited regions (Baysal and Ülkü 2022). The improper disposal of meat byproducts potentially pollutes the surroundings and endanger human health by spreading encephalopathies. Its disposal is both difficult and expensive (Jayawardena et al. 2022). Due to these factors, considerable work has gone into creating procedures to effectively revalue some of these byproducts, which are highly proteinaceous, such as cartilage, blood, epidermis, and skeletons. A wide variety of products for human nutrition, bird and pet food, animal feed, and composts, as well as for the chemical, biomedical, and pharmacological uses, were developed as a result of these processes (Arvanitoyannis and Ladas 2008; Lynch et al. 2018; Shirsath and Henchion 2021; Toldrá et al. 2016, 2021).

17.3.1 Source of Collagen

The most prevalent protein in mammals and the principal component of numerous meat byproducts, such as skin, cartilage, and bones, is collagen. Collagen, which is made up of bundles of fibrillar structures, gives various bodily parts of living beings special mechanical strength and stability. Collagen-rich byproducts are instead valued for the isolation of bioactive amino acids due to collagen's poor nutritional summary and deficiency of important amino acids, which are predominantly made up of glycine, proline, and hydroxyproline (Shoulders and Raines 2009). Several purification techniques have been developed to efficiently separate this protein from many byproducts so that it is used in food applications, cosmetics, and biomedicine to benefit from its technologically functional properties. This protein can also be used in partially hydrolyzed forms, such as in gelatin (Cao et al. 2021; Gómez-Guillén et al. 2011; Jayawardena et al. 2022; Lynch et al. 2018; Toldrá et al. 2016, 2021).

Although the amount of blood produced globally is not known, this byproduct, estimated to be produced at 2.5 billion liters annually, contains enough protein to satisfy the needs of nearly 17 million people. The potential of this waste is not being fully realized because only 4% of these proteins are valued for human use. Different industrial uses of blood for value-added products include coloring agents in textiles, a spray adjuvant, and fertilizer. In addition, it is also utilized for animal and human nutritional and pharmaceutical applications. Proteins account for 35% and 8% of the quantities of red blood cells, which make up 40% of blood, and plasma, which makes up 60% of blood. Up to 60% of the protein in plasma is made up of albumin, which is the most common protein in the body. When purified, albumin is a valuable component for use in medical applications because it stabilizes vaccines, tests for antibiotic sensitivity, and Rh factor evaluations (Toldrà et al. 2019).

17.3.2 Source of Keratin

Hair, nails, hooves, and feathers are among the byproducts which are often exceptionally rich in keratin but are hardly used as food ingredients due to their poor nutritional value and limited digestibility. Keratin's filamentous structure, which is similar to collagen and is characterized by strong power and chemical strength, confers extraordinary resistance to most physical and biological mediators (Bragulla and Homberger 2009). Annually, 40 million tons of keratin-rich trash are generated, posing waste management issues due to a shortage of suitable dumping techniques (Tesfaye et al. 2017). Conventional keratin waste incineration secrets hazardous gases which are ironic in sulfur because of the high level of cysteine, whereas a high amount of disulfide cross-linking and condensed structures causes delayed biodegradation rates, increasing these slow-degrading byproducts and so posing unmanageable issue for the atmosphere and human health worldwide (Korniłłowicz-Kowalska and Bohacz 2011; Peng et al. 2020).

Due to the little digestibility of keratinous constituents, hydrolysis procedures are employed for the creation of soluble peptides or amino acids for human and animal diets, or as useable fertilizer. Pure keratin extracts possess numerous uses in biotechnology, cosmetics, dermatology, and medicine as biocompatible materials such as sponges, fibers, and microcapsules after treatment. On the other hand, the mentioned approaches to recover the protein are expensive and unproductive till now, and hence, long-term solutions to valorize these byproducts are immediately required. Various organs and glands are counted edible in various parts of the world, although their use for human ingesting is still limited (Chojnacka et al. 2011; Holkar et al. 2018; Reddy et al. 2021). They do, however, supply valuable pharmacological substances in a few circumstances, like melatonin from the pineal gland, heparin from liver, insulin from the pancreas, and estrogen and progesterone hormones from the ovaries (Nollet and Toldrá 2011).

To repurpose leftover meat parts, interpreting is used for removing fat from solid animal waste at high temperatures, resulting in edible fats like lard and tallow (Jayathilakan et al. 2012). Collagen and gelatin are isolated and extracted from the residual insoluble meal using diluted citric acid, hydrothermal treatments, and enzymes that involve high pressure and temperature; other tissue proteins, like myofibrillar proteins, are extracted using a base, salt, acid, and biological catalyst (enzymes) (Dhillon 2016; Jayathilakan et al. 2012). Process effluents not only contain significant amounts of solid waste, but they also contain sufficient amounts of blood and muscle proteins, which are possibly recovered using isoelectric precipitation after NaOH solubilization (Bethi et al. 2020; Lynch et al. 2018; Mokrejs et al. 2009).

Keratin extraction is problematic because of its resistance to disintegration via conventional solvents. Although tough extraction conditions involving strong acids and base, sodium sulfide, and reducing mediators help disturb the condensed structure of keratin and assisting dissolving, their use is limited due to the hazardous nature of such substances (Bethi et al. 2020; Vineis et al. 2019). Aqueous extraction employing catalysts, microwave-assisted extraction, reducing agents, ionic liquids, steam flash explosion, and enzymes are some other less harmful extraction techniques that have been researched (Ji et al. 2014; Plowman et al. 2014; Pourjavaheri et al. 2019; Shavandi et al. 2017; Sinkiewicz et al. 2017). By the process of high-temperature hydrolysis with base and acid or microorganism treatment, the marketable extraction of keratin results in keratin hydrolysates, afterward that are employed in hair conditioning and care products and leather bronzing (Karthikeyan et al. 2007; Shah et al. 2019).

17.3.3 Source of Gelatin

Mammalian gelatin represents more than 90% of the market share for all types of gelatin (Usman et al. 2022). Animal collagen is partially hydrolyzed to produce it. White connective tissues found in bones, muscles, and skin are primarily made up of collagen. It is a protein that is present in tissues and organs and makes up about 30% of all proteins. The primary components of collagen are found in significant amounts and include proline, glycine, and hydroxyproline amino acids. Water does not mix with gelatin, but it does with collagen. Three structural modifications are made during the conversion: (1) the breaking up of large peptide chains into smaller ones, (2) the breaking up of bonds between chains, and (3) the configuration of the peptide chain. To produce gelatin, unwanted materials like lipids and minerals must be removed because they could prevent the extraction of gelatin (Boran and Regenstein 2010).

It is primarily connective tissue protein and is widely found in mammals, birds, and fishes (Eysturskarð et al. 2009). Although pigskin is the main source of gelatin, other sources also help to provide the necessary conditions for gelatin synthesis (Boran and Regenstein 2010; Sai-Ut et al. 2012). Bovine skin (29.4%) and pork, as well as cattle bones (23.1%), all contribute to the formation of gelatin, which contains around 46% porcine (23.1%) (Gómez-Guillén et al. 2011). The alternatives to porcine gelatin have received great demand (Morrison et al. 1999). Furthermore, compared to bovine and porcine gelatin, fish gelatin also has a small market share (Choi and Regenstein 2000). However, a sizable number of scientific research are already accessible that suggest gelatin obtained from aquatic sources has superior film-forming characteristics to that of mammals (Avena-Bustillos et al. 2011). Fish gelatin was also noted to have strong sensory attribute-releasing qualities with a low melting point (Aewsiri et al. 2009). Gelatin from Halal sources thus became a crucial problem. Contrarily, gelatin generated from sources such as poultry, animals (which are regarded as Halal and slaughtered following Islamic law), and particularly marine sources, are approved as Halal and might be a viable replacement for swine gelatin (Bhat and Karim 2009).

According to the literature, 15% of edible gelatin comes from bovine hides and 80% comes from pigskins. Bovine and porcine materials, such as pig bones, pigskins, bovine bones, and hides, are the main raw materials used to manufacture gelatin on a large scale. These constituents are employed in the production of gelatin-based products on a large scale, including gelatin sheets, granules, and powder (Peters 2006). Gelatin's foaming, emulsifying, setting, and water-holding capacities are some of its further useful properties. Having it is unique gelling, stabilizing, healing, ointment, capsule, and coating properties, gelatin is also the most widely utilized biodegradable ingredient in the commercial food processing, pharmaceutical, and photography industries (Rakhmanova et al. 2018). Tons of gelatin are reportedly utilized yearly in bread goods, ice cream, meat, candy, and desserts (Djagny et al. 2001; Poppe 1992). Additionally, during cold storage, the gelatin prevents lactose sugar from recrystallizing (Jamilah and Harvinder 2002). A survey claims that the pharmaceutical business uses about 6% of all gelatin produced. It is used in the pharmaceutical business to create hard and soft capsule shells, tablets, granulation, and syrups since it acts as a natural coating material (Hidaka and Liu 2003). Gelatin is a crucial ingredient in energy drinks and plays a significant role in the synthesis of these beverages for athletes in the sports field (Phillips & Williams, 2011). After applying gelatin to a glass plate containing the sensitizing agent, it was first employed in photography in 1871. It is frequently used in the formulations of shampoo, lipstick, conditioner, cream, and fingernail products cosmetics industry.

17.4 Bioenergy Generation from Animal Wastes

17.4.1 Biogas Production via Animal Waste

One of the best choices for addressing the growing worldwide energy consumption is the conversion of waste into energy by different technologies, such as the generation of biogas. Biogas is among the alternative renewable energy resource that are utilized to replace other nonrenewable fuels in countryside regions, notably in developing nations in Asia and Africa, to conserve energy. It is a clean, effective, and sustainable source of energy. In the year 2006, biogas production was approximately 62 billion kWh in Europe, whereas Germany produced about 4300 plants producing 1600 MW of power (Rohstoffe eV 2009). The two leading Asian nations utilizing biogas technology are India and China. In the majority of emerging Asian and African nations, biogas production is mostly used for home applications (Sorathiya et al. 2014).

The production of biogas is also desired; subsequently, the solid waste product (digestate) is able to be utilized as a natural fertilizer or a substrate for growing plants in greenhouses. Anaerobic digestion (AD), often known as the production of biogas, is this process in which organic materials are broken down by microorganisms in an anoxic environment (Adekunle and Okolie 2015). This procedure often results in the creation of biogas mixes that primarily contain CH_4 (60%) and CO_2 (35–40%), with the presence of other gases like H_2, NH_3, and H_2S (dependent on the composition and source of used organic matter) (Zupančič and Grilc 2012). Approximately, 0.75–0.8 $m^3$of biogas can provide 1 unit (1 kWA) of power and a 10 kVA generator, which, if run for 8 h, may produce 80–90 units. It has been reports that the Biogas-based Power Generation Program (BPGP) is the primary strategy in India for generating biogas power, with 73 projects established with a total capacity of 461 kW (Buragohain 2012). Germany possess 7320 electricity-generating biogas plants with a combined capacity of 2997 MW, accounting for 13% of all renewable energy sources (Markard et al. 2016). Indonesia can produce 684.83 Mw of power from biogas like other Asian nations (Widodo and Hendriadi 2005). Although China has started to establish biogas-based power plants, just 3% of them were producing electricity as of 2012. They have tested biogas power units with a day-to-day capacity of 18,000–60,000 kWh based on chicken and calf waste (Fig. 17.1) (Chen et al. 2012). Many nations have already begun using biogas to generate electricity.

The production of algal biomass is achieved by thermo-chemical transformation procedures from livestock waste (Cantrell et al. 2008) and anaerobic digestion (Vijay

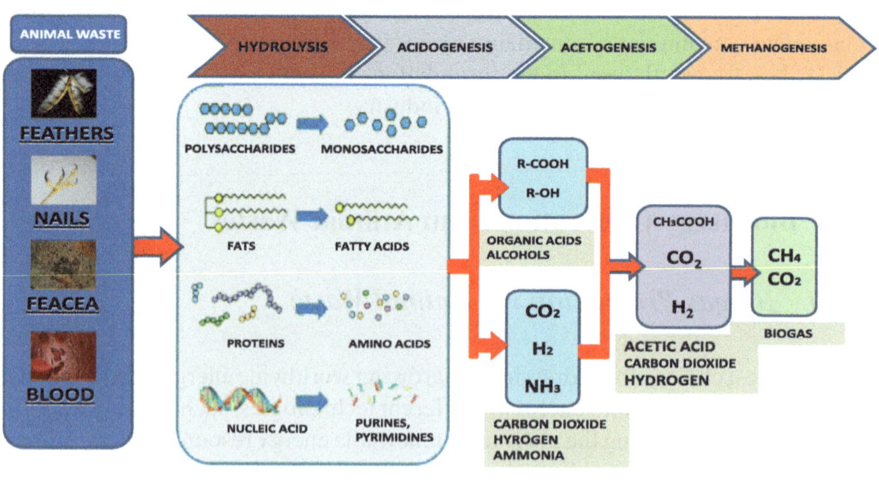

Fig. 17.1 Biogas production process

Table 17.1 Biogas generation from different animal wastes

Substrate	Biogas l/Kg	Methane l/Kg
Anaerobic digestion		
Cattle slaughterhouse waste	–	641
Animal waste	0.45	–
Chicken manure	1.08	–
Cattle manure	12.1	184
Swine slurry	–	394
Poultry slaughterhouse waste	–	595
Chicken droppings	4.6	438
Anaerobic co-digestion		
Cattle excreta	1.1	179
Cattle manure	2.4	620
Sheep dung	–	199

2011), both of which require carbon dioxide as a primary constituent of the produced gases. In addition to producing algal biomass and intracellular oil, algae possibly use carbon dioxide 10 times more effectively than land plants (Miao and Wu 2006). The growth of algae offers several advantages, including quick biomass production rates of about 50 metric tons per acre per annum (Demirbaş 2001), a high accumulation of fatty acids and hydrocarbons, and the potential to contribute to waste treatment. Bio-oil is one of the many products with added value that are made from these algal products. Therefore, it is a very potential non-crop-based raw material for the manufacturing of biofuel. In a pond nourished with biogas slurry that included 200 gm^3 of total nitrogen and 2.5 gm^3 of total dissolved phosphorus, 6.83 gm^3 of algal biomass productivity per day was noted (Table 17.1; Fig. 17.1) (Chen et al. 2012).

17.4.2 Biodiesel from Animal Fat Waste (AFW)

Three long-chain fatty acids and glycerol combine to form the triacylglycerol (TAG) that constitutes AFW. Due to their high viscosity, these fats cannot be used as motor fuel directly. Pretreatment is essential because AFW contains free fatty acids from the leather industry, meat processing industry, and slaughterhouses that affect the transesterification process (Adewale et al. 2015). Similar to the technique used to make biodiesel from fat from other feedstocks, transesterification is used to process the derived fat further. Numerous research organizations have provided evidence of the successful production of biodiesel from a variety of animal fats, including beef tallow, pig lard, chicken, fish, and sheep (Alptekin et al. 2014; Mata et al. 2014). Kwon et al. (2012) recorded a unique technique for producing biodiesel from animal fat utilizing charcoal, and CO_2 at ambient pressure achieves a biodiesel conversion efficiency of 98.5% in one minute at 350–500 °C. Free fatty acids (FFAs) affect biodiesel production and transesterification, therefore, Shi et al. (2013) designed an integrated catalytic process to convert chicken fat into biodiesel. Using a membrane catalytic reactor, biodiesel is created from the free fatty acids in chicken fat via esterification with methanol, and the resulting oil is then put through transesterification with sodium methoxide. Animal cells require additional pretreatment because of their high levels of protein and phosphoacylglycerols, therefore, using animal flesh as feedstock has certain drawbacks as well.

17.4.3 Bioelectricity from Animal Waste

Microbial fuel cells (MFCs) are anaerobic bioreactors where microorganisms consume organic waste carbon to transform chemical energy into electrical energy (Chen et al. 2018; Santoro et al. 2017). Additionally, MFCs are utilized to generate power from biowaste as a raw material. An MFC's cathode and anode (electrodes) are separated from one another by a membrane or salt bridge. When microbes decompose organic matter for growth and reproduction, a range of intermediates is produced that undergo several oxidation and reduction reactions and produce electrons and protons (Pant et al. 2016; Roy et al. 2016). Under anaerobic conditions, microbes might provide a redox mediator with electrons, and the redox mediator might subsequently oxidize the electrons to transmit them to electrodes (Ucar et al. 2017). Following its passage through an external circuit, an electron is transmitted to an electrode and diffuses through the solution to the cathodic chamber, where it combines with oxygen to synthesize water (Angelaalincy et al. 2018). The anodic chamber's potential drops as the substrate is being oxidized, and this causes a potential difference between electrodes, which produces current (Rahimnejad et al. 2015) (Fig. 17.2).

The two primary MFC designs for electricity production are single-chamber and double-chamber. A single-chamber MFC contains the waste as well as the electrodes

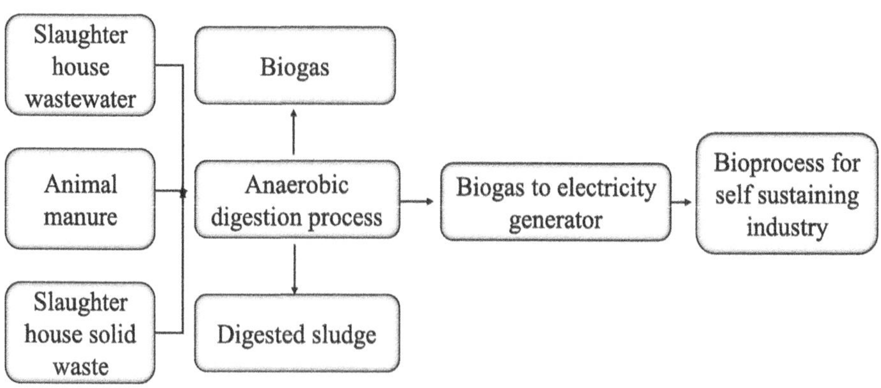

Fig. 17.2 BioElectricity generation from animal waste

Fig. 17.3 Sawdust layer on compost

in one chamber. Independent anodic and cathodic chambers make up a double-chamber MFC. Water is present in the cathodic chamber, while wastewater is present in the anodic chamber. A salt bridge connects the chambers, and a multimeter is used to gage the current. The substrate and microorganisms utilized in the MFC affect the current generated (Luo et al. 2017; Oyiwona et al. 2018; Pandey et al. 2016). MFCs heavily rely on the electron transfer mechanism, which comes in two primary flavors: direct electron transfer (DET) and mediated electron transfer (MET) (Islam, 2016; Sayed et al. 2015). While MET is carried out using an agent that can transport electrons, DET is carried out using nanowires or transmembrane-associated proteins (Fig. 17.3) (TerAvest et al. 2014).

There have been reports of many bacterial species producing electricity in an MFC, including *Aeromonas, E. coli, S. cerevisiae, C. albicans, Clostridium, K. pneumonia, and Shewanella* (Bhatia et al. 2018). The effectiveness of MFCs has been observed to be increased by some electron mediators, including potassium ferricyanide ($K_3Fe(CN)_6$) and natural redox (Sund et al. 2007). Exogenous mediators are used to enhance an MFC's performance; however, they are poisonous, may prevent microbial development, and raise costs. To address this issue, Islam et al. (2018) explored a method by using palm oil mill effluent as a raw material to co-culture both *L. starkeyi* and *K. pneumonia* bacteria in an anodic chamber. In contrast to a pure culture, an electron shuttle mediator was developed by *K. pneumonia* (2, 6 di-tert-butyl benzoquinone) that improved *L. starkeyi* performance by six times. Due to its limited ability to produce current, an MFC remains an impractical method for producing bioelectricity. Additionally, because mesophilic bacteria catalyze the majority of processes, this method requires a lower temperature.

17.5 Compost

Compost, a carbon-rich fertilizer used to enrich soils, is made from organic materials such as livestock manures and other organic components. A properly composted manure has a humus-like odor. Pathogens and weed seeds are eliminated from it by the heating process during the compost formation. Composting is a useful method for properly recycling deceased animals, but it can also leave behind feathers, teeth, and bone fragments that are removed mechanically if necessary. By composting dead birds according to a certain technique, the poultry sector in the United States has discovered a solution to dispose of that waste (Morrow and Ferket 1993). Due to the poultry sector's success in this region, the US pig business is becoming increasingly interested in starting to compost its mortalities (Kashmanian and Rynk 1995). Pig carcasses are composted over a concrete floor in containers made of treated wood, concrete, or hay bales (McCaskey et al. 1996). The corpses are laid on top of a layer of sawdust that is one foot thick, which is then one foot thick on all sides. Layers are added until the bin is full. When dealing with huge sows, the carcasses can either be composted whole or dismembered (Morrow et al. 1995).

The compost is churned manually or automatically after three months, and then, the process is finished after another three months (Imbeah 1998). The compost is not turned constantly when static-pile composting is utilized. Within 24 h of death, animal corpses are placed in a compost pile and covered with a significant amount of solid manure or dirt. Composting technique is carried out to prevent unpleasant smells, flies, and other animals (Morrow and Ferket 2001). On the other hand, this technique should not be done with deceased animals which are previously infected with neurological disorders such as cow-made disease, anthrax, and other ailments that are isolated in restrain (Belay et al. 2002). In India, pile composting is the most used method. The NADEP method is another composting technique that has been shown to provide more significant nutrients than the traditional method (Yadav 2012). In Japan, composting is primarily used for livestock waste. Composting was used to treat about 50% and 25%, respectively, of the solid waste produced by pigs and cattle. Additionally, in Japan, liquid composting was applied to 2.1% and 6.9% of the waste from pigs and cattle, respectively (Haga 1999). In Japan, there are primarily five different types of composting systems in use: piles, boxes, rotary kilns, closed vertical types, and open elongated types with turning devices. The open elongated and pile type are primarily used for large ruminants. In affluent nations, since a lot of water is used to collect pig feces, making composting is a challenging procedure. The manure's composting qualities will be improved via solids separation. It has been effective to compost separated pig manure containing 79% moisture (Lo et al. 1993). Composting high-moisture manure was proposed to be done with low-moisture bulking materials such as straw, sawdust, peat, peanut shells, and rice hulls (Georgacakis et al. 1996).

17.6 Conclusion and Future Perspectives

According to estimates, the millions of tons of animal waste produced worldwide have the potential to harm the environment by emitting CO_2. As a result, the conversion of animal waste can reduce greenhouse gas emissions, which in turn helps to prevent global warming and climatic change, and control environmental pollution. The use of animal waste does come with some challenges that call for additional research to develop trustworthy, widely used, and efficient conversion procedures. These challenges are brought on by high-moisture content, complicated combinations of changeable compositions, and poor waste management. It was discovered that anaerobic digestion was the most developed technology for processing manure into renewable energy. However, the overall growth of this technology in rural parts of developing countries was hampered by problems such as high capital costs, poor working conditions, manure transportation, and digestate management (Khoshnevisan et al. 2021). Literature indicated that to address the aforementioned issues and manage liquid and solid livestock wastes following the objectives of the circular bioeconomy and sustainable development principles, biorefinery is the best option. The implementation of biological treatments is fairly easy and requires little upfront investment. Their drawbacks include a lengthy (15–30 day) processing period and a tendency for bacterial inhibition when exposed to pollutants found in food waste. Additionally, these treatment procedures could need a pretreatment. Biological treatments are very easy to adopt and need little upfront investment. They do, however, have drawbacks, such as a lengthy (15–30 day) processing period and a propensity to restrict bacterial growth when exposed to toxins found in waste. Additionally, pretreatments are necessary for various procedures. The bacteria used in anaerobic digestion are extremely sensitive to pH, temperature, and salts. Hazardous chemicals formed during the reactions, mandating policies concerning control measures, and process condition optimization (Pham et al. 2015). Without any additional cleaning, biogas is often utilized to generate heat and, ultimately, electricity. Biogas must be upgraded if it is to be utilized for other purposes, such as fuel for cars and fuel cells. The conversion methods that can transform biogas into high-value compounds are focused on by many researchers. These chemicals could be produced by converting methane (from biogas) into syngas, which could then be utilized as a raw material to produce other chemicals with a high value-added. Future study is required to ensure the successful application of animal manure management from a technological, ecological, agricultural, commercial, and interpersonal safety point of view.

References

Abdeshahian P, Lim JS, Ho WS, Hashim H, Lee CT (2016) Potential of biogas production from farm animal waste in Malaysia. Renew Sustain Energy Rev 60:714–723

Abdilova G, Rebezov M, Nesterenko A, Safronov S, Knysh I, Ivanova I, Mikolaychik I, Morozova L (2021) Characteristics of meat by-products: nutritional and biological value. Int J Modern Agric 10(2):3895–3904

Adekunle KF, Okolie JA (2015) A review of biochemical process of anaerobic digestion. Adv Biosci Biotechnol 6(03):205

Adewale P, Dumont M-J, Ngadi M (2015) Recent trends of biodiesel production from animal fat wastes and associated production techniques. Renew Sustain Energy Rev 45:574–588

Aewsiri T, Benjakul S, Visessanguan W, Eun J-B, Wierenga PA, Gruppen H (2009) Antioxidative activity and emulsifying properties of cuttlefish skin gelatin modified by oxidised phenolic compounds. Food Chem 117(1):160–168

Afazeli H, Jafari A, Rafiee S, Nosrati M (2014) An investigation of biogas production potential from livestock and slaughterhouse wastes. Renew Sustain Energy Rev 34:380–386

Alam P, Ahmade K (2013) Impact of solid waste on health and the environment. Int J Sustain Dev Green Econ (IJSDGE) 2(1):165–168

Alptekin E, Canakci M, Sanli H (2014) Biodiesel production from vegetable oil and waste animal fats in a pilot plant. Waste Manage 34(11):2146–2154

Angelaalincy MJ, Navanietha Krishnaraj R, Shakambari G, Ashokkumar B, Kathiresan S, Varalakshmi P (2018) Biofilm engineering approaches for improving the performance of microbial fuel cells and bioelectrochemical systems. Front Energy Res 6:63

Arvanitoyannis IS, Ladas D (2008) Meat waste treatment methods and potential uses. Int J Food Sci Technol 43(3):543–559

Avena-Bustillos R, Chiou B-S, Olsen C, Bechtel P, Olson D, McHugh T (2011) Gelation, oxygen permeability, and mechanical properties of mammalian and fish gelatin films. J Food Sci 76(7):E519–E524

Ayilara MS, Olanrewaju OS, Babalola OO, Odeyemi O (2020) Waste management through composting: challenges and potentials. Sustainability 12(11):4456

Baysal SS, Ülkü MA (2022) Food loss and waste: a sustainable supply chain perspective. In: Disruptive technologies and eco-innovation for sustainable development, pp 90–108

Belay A, Claassens A, Wehner F (2002) Effect of direct nitrogen and potassium and residual phosphorus fertilizers on soil chemical properties, microbial components and maize yield under long-term crop rotation. Biol Fertil Soils 35(6):420–427

Bethi C, Narayan B, Martin A, Kudre TG (2020) Recovery, physicochemical and functional characteristics of proteins from different meat processing wastewater streams. Environ Sci Pollut Res 27(20):25119–25131

Bhat R, Karim A (2009) Ultraviolet irradiation improves gel strength of fish gelatin. Food Chem 113(4):1160–1164

Bhat RA, Dar SA, Dar DA, Dar GH (2018) Municipal solid waste generation and current scenario of its management in India. Int J Adv Res Sci Eng 7(02):419–431

Bhatia SK, Joo H-S, Yang Y-H (2018) Biowaste-to-bioenergy using biological methods–a mini-review. Energy Convers Manage 177:640–660

Boran G, Regenstein JM (2010) Fish gelatin. Adv Food Nutr Res 60:119–143

Bragulla HH, Homberger DG (2009) Structure and functions of keratin proteins in simple, stratified, keratinized and cornified epithelia. J Anat 214(4):516–559

Buragohain T (2012) Impact of solar energy in rural development in India. Int J Environ Sci Dev 3(4):334

Cantrell KB, Ducey T, Ro KS, Hunt PG (2008) Livestock waste-to-bioenergy generation opportunities. Biores Technol 99(17):7941–7953

Cao C, Xiao Z, Ge C, Wu Y (2021) Animal by-products collagen and derived peptide, as important components of innovative sustainable food systems—A comprehensive review. Crit Rev Food Sci Nutrit, pp 1–25

Chen R, Li R, Deitz L, Liu Y, Stevenson RJ, Liao W (2012) Freshwater algal cultivation with animal waste for nutrient removal and biomass production. Biomass Bioenerg 39:128–138

Chen S, Patil SA, Schröder U (2018) A high-performance rotating graphite fiber brush air-cathode for microbial fuel cells. Appl Energy 211:1089–1094

Choi SS, Regenstein J (2000) Physicochemical and sensory characteristics of fish gelatin. J Food Sci 65(2):194–199

Chojnacka K, Górecka H, Michalak I, Górecki H (2011) A review: valorization of keratinous materials. Waste Biomass Valorization 2(3):317–321

Demirbaş A (2001) Biomass resource facilities and biomass conversion processing for fuels and chemicals. Energy Convers Manage 42(11):1357–1378

Dhillon GS (2016) Protein byproducts: transformation from environmental burden into value-added products. Academic Press

Djagny KB, Wang Z, Xu S (2001) Gelatin: a valuable protein for food and pharmaceutical industries. Crit Rev Food Sci Nutr 41(6):481–492

Eysturskarð J, Haug IJ, Elharfaoui N, Djabourov M, Draget KI (2009) Structural and mechanical properties of fish gelatin as a function of extraction conditions. Food Hydrocolloids 23(7):1702–1711

Gebrezgabher SA, Meuwissen MP, Prins BA, Lansink AGO (2010) Economic analysis of anaerobic digestion—A case of Green power biogas plant in The Netherlands. NJAS Wageningen J Life Sci 57(2):109–115

Georgacakis D, Tsavdaris A, Bakouli J, Symeonidis S (1996) Composting solid swine manure and lignite mixtures with selected plant residues. Biores Technol 56(2–3):195–200

Gómez-Guillén MC, Giménez B, López-Caballero MA, Montero MP (2011) Functional and bioactive properties of collagen and gelatin from alternative sources: a review. Food Hydrocolloids 25(8):1813–1827

Haga K (1999) Development of composting technology in animal waste treatment-review. Asian Australas J Anim Sci 12(4):604–606

Hidaka S, Liu S (2003) Effects of gelatins on calcium phosphate precipitation: a possible application for distinguishing bovine bone gelatin from porcine skin gelatin. J Food Compos Anal 16(4):477–483

Holkar CR, Jain SS, Jadhav AJ, Pinjari DV (2018) Valorization of keratin based waste. Process Saf Environ Prot 115:85–98

Imbeah M (1998) Composting piggery waste: a review. Biores Technol 63(3):197–203

Islam MA, Ethiraj B, Cheng CK, Yousuf A, Thiruvenkadam S, Prasad R, Rahman Khan MM (2018) Enhanced current generation using mutualistic interaction of yeast-bacterial coculture in dual chamber microbial fuel cell. Ind Eng Chem Res 57(3):813–821

Islam K (2016) Municipal solid waste to energy generation in Bangladesh: possible scenarios to generate renewable electricity in Dhaka and Chittagong city. J Renew Energy

Jamilah B, Harvinder K (2002) Properties of gelatins from skins of fish—black tilapia (Oreochromis mossambicus) and red tilapia (Oreochromis nilotica). Food Chem 77(1):81–84

Jayathilakan K, Sultana K, Radhakrishna K, Bawa A (2012) Utilization of byproducts and waste materials from meat, poultry and fish processing industries: a review. J Food Sci Technol 49(3):278–293

Jayawardena R, Morton JD, Brennan CS, Bhat ZF, Bekhit AE-DA (2022) Meat co-products. In: Alternative proteins. CRC Press, pp 329–359

Ji Y, Chen J, Lv J, Li Z, Xing L, Ding S (2014) Extraction of keratin with ionic liquids from poultry feather. Sep Purif Technol 132:577–583

Karthikeyan R, Balaji S, Sehgal P (2007) Industrial applications of keratins–a review

Kashmanian RM, Rynk RF (1995) Agricultural composting in the United States. Compost Sci Utilization 3(3):84–88

Kaygusuz K (2002) Renewable and sustainable energy use in Turkey: a review. Renew Sustain Energy Rev 6(4):339–366
Khalil M, Berawi MA, Heryanto R, Rizalie A (2019) Waste to energy technology: the potential of sustainable biogas production from animal waste in Indonesia. Renew Sustain Energy Rev 105:323–331
KhanSA, Suleman M, Asad M (2019) Assessment of pollution load in marble waste water in Khairabad, District Nowshera, Khyber Pukhtunkhwa, Pakistan. Int J Econ Environ Geol, pp 35–39
Khoshnevisan B, Duan N, Tsapekos P, Awasthi MK, Liu Z, Mohammadi A, Angelidaki I, Tsang DC, Zhang Z, Pan J, Ma L (2021) A critical review on livestock manure biorefinery technologies: sustainability, challenges, and future perspectives. Renew Sustain Energy Rev 135:110033
Korniłłowicz-Kowalska T, Bohacz J (2011) Biodegradation of keratin waste: theory and practical aspects. Waste Manage 31(8):1689–1701
Kwon EE, Seo J, Yi H (2012) Transforming animal fats into biodiesel using charcoal and CO_2. Green Chem 14(6):1799–1804
Langdon KA, Chandra A, Bowles K, Symons A, Pablo F, Osborne K (2019) A preliminary ecological and human health risk assessment for organic contaminants in composted municipal solid waste generated in New South Wales, Australia. Waste Manage 100:199–207
Lo K, Lau A, Liao P (1993) Composting of separated solid swine wastes. J Agric Eng Res 54(4):307–317
Luo H, Xu G, Lu Y, Liu G, Zhang R, Li X, Zheng X, Yu M (2017) Electricity generation in a microbial fuel cell using yogurt wastewater under alkaline conditions. RSC Adv 7(52):32826–32832
Lynch SA, Mullen AM, O'Neill E, Drummond L, Álvarez C (2018) Opportunities and perspectives for utilisation of co-products in the meat industry. Meat Sci 144:62–73
Mani D, Kumar C (2014) Biotechnological advances in bioremediation of heavy metals contaminated ecosystems: an overview with special reference to phytoremediation. Int J Environ Sci Technol 11(3):843–872
Markard J, Wirth S, Truffer B (2016) Institutional dynamics and technology legitimacy–A framework and a case study on biogas technology. Res Policy 45(1):330–344
Mata TM, Mendes AM, Caetano NS, Martins AA (2014) Properties and sustainability of biodiesel from animal fats and fish oil. Chem Eng Trans 38:175–180
Mathias JFCM (2014) Manure as a resource: livestock waste management from anaerobic digestion, opportunities and challenges for Brazil. Int Food Agribusiness Manage Rev 17(1030-2016-83037):87–110
McCaskey T, Little J, Krotz R, Lino S, Hannah T (1996) On-farm composting feasible for disposal of swine carcasses. Highlights of agricultural research (USA)
Miao X, Wu Q (2006) Biodiesel production from heterotrophic microalgal oil. Biores Technol 97(6):841–846
Mokrejs P, Langmaier F, Mladek M, Janacova D, Kolomaznik K, Vasek V (2009) Extraction of collagen and gelatine from meat industry by-products for food and non food uses. Waste Manage Res 27(1):31–37
Morrison N, Clark R, Chen Y, Talashek T, Sworn G (1999) Gelatin alternatives for the food industry. In: Physical chemistry and industrial application of gellan gum. Springer, pp 127–131
Morrow WM, Ferket PR (1993) The disposal of dead pigs: a review. Swine Health Prod Official J Am Assoc Swine Practitioners (USA)
Morrow W, Ferket P (2001) Alternative methods for the disposal of swine carcasses. North Carolina State University Animal Science Facts
Morrow WM, O'Quinn P, Barker J, Erickson G, Post K, McCaw M (1995) Composting as a suitable technique for managing swine mortalities. Swine Health Prod Official J Am Assoc Swine Practitioners (USA)
Nollet LM, Toldrá F (2011) Handbook of analysis of edible animal by-products. CRC Press
Omrani G (1996) Basics biogas production from urban and rural waste. University of Tehran Publication, Iran

Oyiwona GE, Ogbonna JC, Anyanwu CU, Okabe S (2018) Electricity generation potential of poultry droppings wastewater in microbial fuel cell using rice husk charcoal electrodes. Bioresources Bioprocess 5(1):1–6

Pandey P, Shinde VN, Deopurkar RL, Kale SP, Patil SA, Pant D (2016) Recent advances in the use of different substrates in microbial fuel cells toward wastewater treatment and simultaneous energy recovery. Appl Energy 168:706–723

Pant D, Van Bogaert G, Alvarez-Gallego Y, Diels L, Vanbroekhoven K (2016) Evaluation of bioelectrogenic potential of four industrial effluents as substrate for low cost microbial fuel cells operation. Environ Eng Manage J (EEMJ) 15(8)

Peng Z, Mao X, Zhang J, Du G, Chen J (2020) Biotransformation of keratin waste to amino acids and active peptides based on cell-free catalysis. Biotechnol Biofuels 13(1):1–12

Peters D (2006) Raw materials. In: White biotechnology, pp 1–30

Pham TP, Kaushik R, Parshetti GK, Mahmood R, Balasubramanian R (2015) Food waste-to-energy conversion technologies: current status and future directions. Waste Manag 38:399–408. https://doi.org/10.1016/j.wasman.2014.12.004. Epub 2014 Dec 30. PMID: 25555663

Phillips GO, Williams PA (2011) Handbook of food proteins. Elsevier

Plowman JE, Clerens S, Lee E, Harland DP, Dyer JM, Deb-Choudhury S (2014) Ionic liquid-assisted extraction of wool keratin proteins as an aid to MS identification. Anal Methods 6(18):7305–7311

Poppe J (1992) Gelatin. In: Thickening and gelling agents for food. Springer, pp 98–123

Pourjavaheri F, Pour SO, Jones OA, Smooker PM, Brkljača R, Sherkat F, Blanch EW, Gupta A, Shanks RA (2019) Extraction of keratin from waste chicken feathers using sodium sulfide and l-cysteine. Process Biochem 82:205–214

Rahimnejad M, Adhami A, Darvari S, Zirepour A, Oh S-E (2015) Microbial fuel cell as new technology for bioelectricity generation: a review. Alex Eng J 54(3):745–756

Rakhmanova A, Khan Z, Sharif R, Lv X (2018) Meeting the requirements of halal gelatin: a mini review. MOJ Food Proc Technol 6:477–482

Reddy CC, Khilji IA, Gupta A, Bhuyar P, Mahmood S, AL-Japairai KAS, Chua GK (2021) Valorization of keratin waste biomass and its potential applications. J Water Process Eng 40:101707

Rohstoffe eV FFN (2009). Biogas-Messprogramm II–61 Biogasanlagen im Vergleich (1st edn). Fachagentur Nachwachsende Rohstoffe eV (Ed) Gülzo

Roy S, Schievano A, Pant D (2016) Electro-stimulated microbial factory for value added product synthesis. Biores Technol 213:129–139

Sai-Ut S, Jongjareonrak A, Rawdkuen S (2012) Re-extraction, recovery, and characteristics of skin gelatin from farmed giant catfish. Food Bioprocess Technol 5(4):1197–1205

Santoro C, Arbizzani C, Erable B, Ieropoulos I (2017) Microbial fuel cells: from fundamentals to applications. A review. J Power Sources 356:225–244

Sayed ET, Barakat NA, Abdelkareem MA, Fouad H, Nakagawa N (2015) Yeast extract as an effective and safe mediator for the baker's-yeast-based microbial fuel cell. Ind Eng Chem Res 54(12):3116–3122

Shah A, Tyagi S, Bharagava RN, Belhaj D, Kumar A, Saxena G, Saratale GD, Mulla SI (2019) Keratin production and its applications: current and future perspective. In: Keratin as a protein biopolymer, pp 19–34

Shavandi A, Silva TH, Bekhit AA, Bekhit AE-DA (2017) Keratin: dissolution, extraction and biomedical application. Biomater Sci 5(9):1699–1735

Shi W, Li J, He B, Yan F, Cui Z, Wu K, Lin L, Qian X, Cheng Y (2013) Biodiesel production from waste chicken fat with low free fatty acids by an integrated catalytic process of composite membrane and sodium methoxide. Bioresour Technol 139:316–322

Shirsath AP, Henchion MM (2021) Bovine and ovine meat co-products valorisation opportunities: a systematic literature review. Trends Food Sci Technol 118:57–70

Shoulders MD, Raines RT (2009) Collagen structure and stability. Annu Rev Biochem 78:929

Sinkiewicz I, Śliwińska A, Staroszczyk H, Kołodziejska I (2017) Alternative methods of preparation of soluble keratin from chicken feathers. Waste Biomass Valorization 8(4):1043–1048

Sorathiya L, Fulsoundar A, Tyagi K, Patel M, Singh R (2014) Eco-friendly and modern methods of livestock waste recycling for enhancing farm profitability. Int J Recycl Organic Waste Agric 3(1):1–7
Sund CJ, McMasters S, Crittenden SR, Harrell LE, Sumner JJ (2007) Effect of electron mediators on current generation and fermentation in a microbial fuel cell. Appl Microbiol Biotechnol 76(3):561–568
TerAvest MA, Rosenbaum MA, Kotloski NJ, Gralnick JA, Angenent LT (2014) Oxygen allows Shewanella oneidensis MR-1 to overcome mediator washout in a continuously fed bioelectrochemical system. Biotechnol Bioeng 111(4):692–699
Tesfaye T, Sithole B, Ramjugernath D (2017) Valorisation of chicken feathers: a review on recycling and recovery route—current status and future prospects. Clean Technol Environ Policy 19(10):2363–2378
Toldrá F, Mora L, Reig M (2016) New insights into meat by-product utilization. Meat Sci 120:54–59
Toldrá F, Reig M, Mora L (2021) Management of meat by-and co-products for an improved meat processing sustainability. Meat Sci 181:108608
Toldrà M, Lynch SA, Couture R, Álvarez C (2019) Blood proteins as functional ingredients. In: Sustainable meat production and processing. Elsevier, pp 85–101
Ucar D, Zhang Y, Angelidaki I (2017) An overview of electron acceptors in microbial fuel cells. Front Microbiol 8:643
Usman M, Sahar A, Inam-Ur-Raheem M, Rahman UU, Sameen A, Aadil RM (2022) Gelatin extraction from fish waste and potential applications in food sector. Int J Food Sci Technol 57(1):154–163
Varjani SJ, Gnansounou E, Pandey A (2017) Comprehensive review on toxicity of persistent organic pollutants from petroleum refinery waste and their degradation by microorganisms. Chemosphere 188:280–291
Vijay V (2011) Biogas enrichment and bottling technology for vehicular use. Paper presented at the Biogas forum
Vineis C, Varesano A, Varchi G, Aluigi A (2019). Extraction and characterization of keratin from different biomasses. In: Keratin as a protein biopolymer. Springer, pp 35–76
Widodo TW, Hendriadi A (2005) Development of biogas processing for small scale cattle farm in Indonesia. Paper presented at the conference proceeding: international seminar on biogas technology for poverty reduction and sustainable development. Beijing
Yadav R (2012) Innovative application of scientific fact for nutrient recovery from waste water streams for sustainable agriculture and protection of environment: a review. Hydrol Curr Res 3(5)
Zhu L-D, Hiltunen E (2016) Application of livestock waste compost to cultivate microalgae for bioproducts production: a feasible framework. Renew Sustain Energy Rev 54:1285–1290
Zupančič GD, Grilc V (2012) Anaerobic treatment and biogas production from organic waste. Manage Organic Waste 2

Chapter 18
Green Biochemicals from Manure

Sadia Javed, Nazima Anwaar, Abid Hussain, Shumaila Kiran, Muhammad Arfan Zaman, Amreen Aftab, and Saboor Gul

Abstract It was examined that various types of organic materials and wastes with distinct levels of maturity reacted chemically, chemically-physically, and biochemically. The manures are used to increase soil productivity by modifying the physical, chemical, and biological properties of the soil. They are organic in nature, of plant or animal origin, and consist of a substantial proportion of organic matter and a moderate amount of plant nutrients. Compost is an organic substance that is utilized in agriculture as organic fertilizer. By incorporating organic matter and nutrients, like nitrogen, that are held in the soil by bacteria, it helps to increase the soil's fertility. The soil food web is a cycle of life that includes fungi, bacteria, and higher species that consume the fungi and bacteria. The store of organic matter in the soil is further enhanced by the various organic manures that are frequently introduced.

S. Javed (✉) · N. Anwaar · A. Aftab · S. Gul
Department of Biochemistry, Government College University, Faisalabad, Pakistan
e-mail: sadiajaved@gcuf.edu.pk

A. Hussain
Department of Civil and Environmental Engineering, Carleton University, Ottawa, Canada

S. Kiran
Department of Applied Chemistry, Government College University, Faisalabad, Pakistan

M. Arfan Zaman
Department of Pathobiology, University of Veterinary and Animal Sciences Lahore, Jhang-Campus, Lahore, Pakistan

M. Arshad (ed.), *Climate Changes Mitigation and Sustainable Bioenergy Harvest Through Animal Waste*, https://doi.org/10.1007/978-3-031-26224-1_18

Graphical Abstract

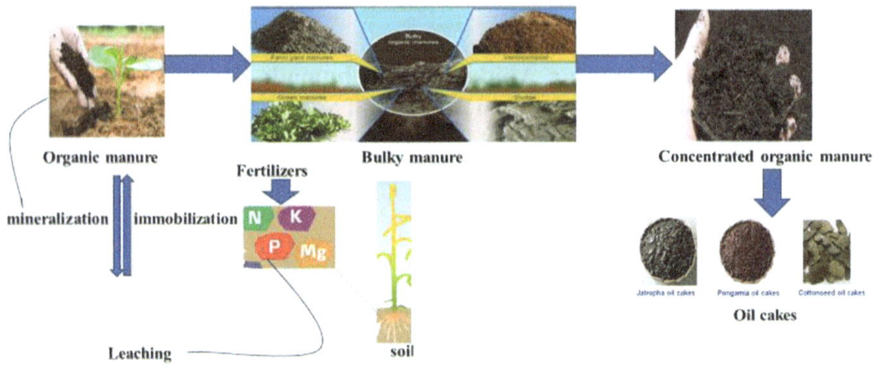

Keywords Separation of solids manure · Biological waste · Bio waste · Organic substance

18.1 Introduction

The crops production can be increased by manure which can be used to increase the fertility of soil. The manure quality can be affected by contents like species, protein, diet, digestibility as well as environment, housing, animal age, and stage of production. There are number of ways to classify the manure. Solids percentage (the amount of solids collection, storage, and utilization can be handled by these classified contents (solids may be fixed, suspended, volatile, dissolved). The land production rate and fertility are important which can be increased by contents like potassium, phosphorus, and nitrogen. Organic and inorganic components can be found in manure. For reduction of sickness and parasites, the waste material of human cannot be mixed with animal manure.

The world food creation and biodiversity are the pivotal supplies of soil. The dirt quality and well-being are impacted by the populace development and horticulture region constraint (Thangarajan et al. 2013). The natural mixtures of soil supplement in arable soil is fastly corrupting because of soil disintegration (Pu et al. 2019). Compost first and foremost acquired from creatures or plant extras, it is valuable for soil added substance that can assist with correcting soil degeneration and furthermore efficiency of harvest and ecological effect of agribusiness in the Anthropocene (Liu et al. 2020; Mandal et al. 2020). As well as expanding plushness and soundness of soil (Karami et al. 2012) and diminishing huge robustness (Edmeades 2003), excrement application increment the biochemical properties of soil (Jiang et al. 2018). Supplements, like nitrogen (N), phosphorous (P), rise when compost is applied (Ros et al. 2006; Zhang et al. 2015). For instance, 46 years of field use of cow waste improve the all-out nitrogen focuses and natural compound of soil by roughly 34% and 31%, individually (Giacometti et al. 2014). The entire P sum improved from 123 to 458 mg kg $^{-1}$ following 15 years of hoard fertilizer requesting at a pace of 41 Mg ha $^{-1}$. Tejada (2009) found that applying poultry fertilizer or treated the soil steers excrement (Zaller and Köpke 2004) expanded microbial C by in excess of 200% and 27%, separately. Despite the great advantages of fertilizer on soil microbial catalogs, the biochemical properties of the dirt change in various ways after compost application, particularly regarding compound movement. The catalysts glucosidase, urease, phosphatase, dehydrogenase, and sulfatase showed moderate impacts of green excrement (Trifolium pratense and Brassica napus, L.) or treated the soil dairy cattle fertilizer in a six-month field explore, while chitinase had a negative response to both green and treated the soil pig compost (Liu et al. 2017). As per (Giacometti et al. 2014), not at all like the positive and negative responses, there was no phosphatase movement response to a 46-year use of green/dairy cattle excrement. The potential that unmistakable C bases might be drained by preparation, which thusly influences the capability of C cycle proteins, is raised by the fluctuating impacts of fertilizer treatment on the movement of C cycle proteins specifically. These discoveries check out all alone, yet they cannot be united to frame a solitary speculation that can be utilized to make sense of an assortment of soil types and the board systems. The event of numerous unexplained reactions in certain frameworks, however, not in others proposes a critical information hole and the requirement for multi-meta-examination (Gurevitch et al. 2018). There have been not many endeavors to give a more intensive, unthinking comprehension in view of soil credits, regardless of various examinations on the effect of compost application on soil biochemical boundaries (Luo et al. 2018; Kallenbach and Grandy 2011). In the wake of looking at the effects of fertilizer application on soil microbial C and N, tracked down a significant relationship between yearly C and N consumption from fertilizer and microbial C and N in croplands. A sizable data set was utilized in a new report to all the more completely assess the impacts of compost and mineral manures on SOC, TN, and organic elements. The chance of mistake in the calculation of normal extracellular chemical exercises might raise doubt about the case made by that compost treatment supports soil compound exercises. Rather than utilizing individual proteins, the all-out movement of the significant C-or supplement procuring compounds is utilized as a substitute for the procurement of a specific substrate or supplement since it gives a more precise portrayal of generally speaking C-or supplement securing (Ringer et al. 2014). This strategy, in the meantime, darkens the various responses of explicit compounds that are associated with a similar food cycle. For example, (Saha et al. 2008) found that in some natural surroundings in the northern Himalayas, consistent treatment of treated the soil dairy cattle fertilizer at paces of 3700 and 5500 kgha^{-1} supported soluble phosphatase movement while diminishing corrosive phosphatase action. A new meta examination likewise neglected to recognize applying compost alone and blending excrement in with mineral manures (Maillard and Angers 2014). The effect of fertilizer on soil natural matter and biochemical properties was exacerbated when the two methodologies were utilized. Future exploration ought to look at the impacts of excrement synthetic arrangement, like C and N fixation, C/N proportion, yet has not yet been finished (Luo et al. 2018). These components (like sorts, creation techniques, and medicines) are fundamental for keeping up with soil ripeness, despite the fact that their capability in deciding soil qualities has not yet been entirely portrayed (Ali et al. 2019). To all the more likely comprehend the impacts of compost application on soil science, further inside and out examination from both agronomic and ecological points of view was required. This makes it more straightforward for us to comprehend what horticultural administration procedures can mean for food utilization. Impacts of applied excrement on SOC, TN, microbial C and N as well as the exercises of seven proteins (like C-cycling: 1, 4-glucosidase, dehydrogenase, urease, N-acetyl-D-glucosaminidase, corrosive and All N-cycling catalysts add to digestion) have been researched in examinations that have been distributed in various worldwide diaries. Impacts of meteorological elements, including mean yearly temperature (MAT), mean yearly precipitation (Guide), beginning soil pH, type (e.g., Alfisols, Entisols), the executives (alone or in mix with mineral composts, field or research facility, term), and excrement attributes, including type (e.g., cow, pig, and so on), were completely examined.

Extraordinary consideration was paid to the associations between the essential synthetic parts of excrement, the sum, and stoichiometry of C, N, P, and K, and their effect on soil properties. Our fundamental goals were to:

- completely assess and measure the impacts of excrement on soil biochemical properties,
- affirm the precision of involving mean extracellular protein exercises as marks of explicit substrate or supplement obtaining, and
- indicate the overall commitments of the above informative factors to fertilizer impacts.

We laid out the accompanying speculations:

- Compost properties altogether affect soil properties;
- The impact of excrement on the biochemical properties of the dirt is decreased by the expansion of mineral manures.
- Furthermore, proteins associated with the securing of a similar component might answer distinctively to fertilizer application because of their various capabilities in complex C and supplement cycles.

18.2 Historical Background

Before, livestock excrement filled in as the primary wellspring of plant supplements. Its significance for keeping up with and expanding soil efficiency has been perceived since antiquated times (Nowak and Sigmund 1998). Treating crops with supplements from dairy cattle compost has been rehearsed for centuries, as itemized in the Hebrew Scriptures Book of scriptures and other old texts. Before the approach of business compost creation, admittance to excrement was viewed as basic to the drawn-out reasonability of a farming framework. Long-haul cultivating framework preliminaries show that this is valid in any situation that does not utilize business manures.

Crop yields were just 20% higher following 40 years of compost-free administration than when excrement was added every year. Business manure creation changed supplement the executives in the USA and different nations when the new century rolled over. This strategy affected American farming and took into consideration the grouping of animals creation that we see today. In 1996, the absolute worth of all animals and poultry on Minnesota ranches was more than $2.1 billion. Domesticated animals agribusiness in Minnesota, which is generally packed in the south, south focal, and southwest districts of the state, is overwhelmed by pork, dairy, and poultry creation. The excrement of these creatures presents an exceptional mix of issues and open doors. Excrement can be utilized to make preparing blend, garden fertilizer, and other non-farming applications.

18.3 Manure

Compost is an incredible wellspring of supplements for crops and can likewise assist with soil creation. Creature species, diet, absorbability, protein and fiber content, as well as creature age, lodging, climate, and phase of creation all influence compost quality. Fertilizer can be arranged in numerous ways. Solids content (percent solids per unit fluid) and size and piece of compost solids are significant factors for excrement assortment, capacity, dealing with, and usage (solids and unstable solids, suspended solids, and broke down solids). How much nitrogen, phosphorus, and potassium in the dirt are basic since it influences soil application rates and treatment draws near. Natural and inorganic parts can be tracked down in compost. Human waste ought not to be blended in with creature fertilizer to lessen sickness and parasites.

18.3.1 Value of Manure

The essential objective for supplement contribution to agroecosystems is to accomplish high harvest yields, whether or not the yield has low worth and low peripheral yield. In current trimming frameworks, the dietary requirements of harvests are met by blending inorganic and natural supplement sources. Manures and different wellsprings of inorganic supplements are powerful wellsprings of agrarian composts that are typically promptly accessible. Natural supplement sources like creature fertilizer, biosolids, and different farming and modern buildups can likewise be utilized to enhance crop supplements. Fundamental parts in compost have been found to work on the effect of business ammonium nitrate on corn yield (Durieux 1995). Excrement significantly affects wheat and corn yields practically identical to business manures

(Weidemann 1943). The supplement content of excrement decides its worth. Supplement worth can shift emphatically. Supplement fixations are impacted by wash water weakening and stable spillover, as well as misfortunes. Water content is presumably the main variable. For instance, the worth of N, P, and K in poultry fertilizer was exclusively about $5 to $80 per ton, as per study information. Different advantages of fertilizer are connected with less substantial financial advantages. Because of the humus and natural components in the compost, it can work on the physical and organic properties of the dirt. Working on the physical and natural properties of the dirt outcomes in a better climate for plant development and less toxin spillage. For a really long time, researchers have known the advantages of applying excrement to soil, from reestablishing disintegrated destinations (Latham 1940) to further developing soil actual properties and ripeness after ages of preparation (Sandor and Eash 1991). At the point when compost is added to a sandy soil, both the water holding limit and the construction increment (Hornick 1988). Excrement application has been recommended by certain specialists to help corn drill obstruction by keeping up with higher nitrogen levels in the dirt (Allee and Davis 1996).

18.3.2 Manure Composition

Supplement content, particularly nitrogen, phosphorus, and potassium, is basic while working out soil application rates and deciding treatment strategies. The primary supplements in excrement are nitrogen, phosphorus, and potassium. There are two sorts of supplements: solvent and insoluble. Supplements that are dissolvable are all the more promptly accessible to crops. Insoluble supplements may not be accessible to crops for as long as a year or more. An estimated circulation of key supplements in excrement and pee is displayed in Fig. 18.1 Insoluble and a few solvent supplements are tracked down in the solids (dung) in the discharged compost, while dissolvable supplements are viewed as in the fluid (pee). Commonly, 80% of the phosphorus in fertilizer capacity is insoluble and tracked down in the settled solids. The fluid contains up to 80% potassium, which is incredibly solvent. Since nitrogen is similarly circulated among solids and fluids, it is roughly 50% solvent and 50% insoluble. Both natural and inorganic parts can be tracked down in fertilizer. Crops require critical measures of optional components (sulfur, calcium, and magnesium). Micronutrients, for example, zinc, boron, iron, and copper, are likewise expected in follow sums.

18.3.3 Storage and Handling

Manure can be kept for a day, a week, or a year. Increased storage gives for greater flexibility in application time to match crop demand. This decreases environmental damage in general. Nutrient and pollutant losses occur in liquid or gaseous form during storage and processing, as well as after land supplication. In general, as storage

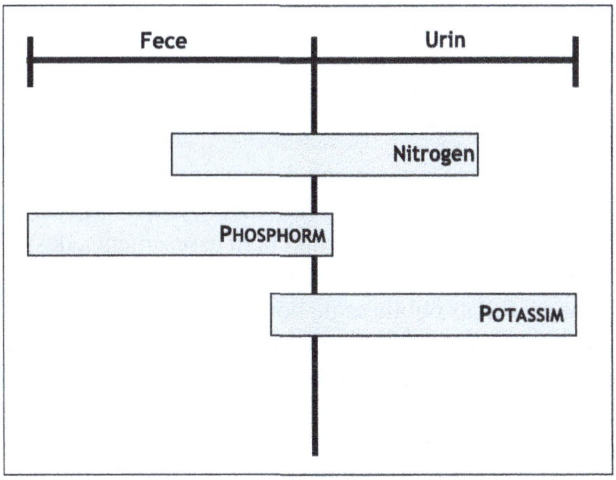

Fig. 18.1 Distribution of nutrients between feces and urine. Based on NRCS *Agricultural Waste Management Field Handbook,* Part 651

capacity expands, allowing for greater application scheduling flexibility, costs rise. The storage and processing of manure can be done in a variety of ways.

18.3.4 Soil Analysis and Nutrient Recommendation

Plant growth requires fourteen important aspects, all of which are found in the soil. Water soluble and insoluble forms of these elements can be found in soil. Plants can quickly absorb the soluble fraction of nutrients, but only a small portion of the insoluble form of different components may be absorbed. A research focus has been on establishing how much water insoluble components can potentially become available for plant growth. Soil scientists have devised chemical reagents that remove a fraction of the insoluble elements proportional to the quantity that becomes available for plant absorption throughout the course of a growing season. These indicators, as well as data from the field and the lab, are used to produce fertilizer recommendations for crop productivity.

18.3.5 Additional Nutrition Sources

Data on a variety of nutrient inputs and outputs on a regional scale are needed to address the issue of nutrient management for animal-based agriculture, including:

- sewage sludge
- atmospheric deposition
- purchased feeds
- animal manures

- commercial fertilizers
- legumes
- plant breakdown
- sewage sludge.

For instance, sewage muck is a minor wellspring of supplements on a public scale, yet can be basic for a solitary ranch. Likewise, yard excrement can be viewed as a contribution to both mass and field assessments, despite the fact that the compost fills in as a transitory store of supplements that enter the homestead from various sources. State as well as public scale nourishment spending plans have been created by various scientists. Supplements can enter an agroecosystem from better places. The intricacy of the cycles included makes evaluating the commitment of numerous assets a troublesome undertaking for farming specialists. These are normally numerous complex and frequently fundamentally unrelated cycles. Microbial change is utilized to use the supplements present in different natural manures. Plant squanders and fertilizer convert microscopic organisms and growths in the dirt into valuable supplements for crops. These cycles include the transformation of nitrogen-containing atoms to ammonium, trailed by change to nitrites and nitrates. Natural compost sources consume a large chunk of the day to decay, with something like a fourth of the nitrogen accessible in the main year. About portion of the complete nitrogen applied will be accessible in the primary year and the excess half will be accessible the next year, as a wide speculation. Real paces of mineralization are hard to decide on the grounds that a natural cycle is delicate to changes in temperature and dampness in the dirt framework. Appraisals of nitrogen and phosphate mineralization rates, as well as the accessibility of weighty metals that might be available in excrement, are fundamental for precisely anticipating application paces of natural supplement sources (Berti and Jacobs 1996). The subject of nitrogen mineralization is examined in more prominent profundity in the accompanying areas of this post. Compost deals information can be utilized to compute manure supplements. Composts bought in Minnesota are thought to be utilized in Minnesota. The substance organization and capacity of creature fertilizer have been talked about somewhere else in this record. No misfortunes of excrement supplements were determined, as no information is accessible on the extent of various fertilizers put away in various capacity frameworks. We picked fertilizer as a source since we trusted that smelling salts volatilization from excrement reenters the framework (plants and soil) inside the state limit. Future examinations could affirm this suspicion. Environmental nitrogen obsession by vegetables is one more key wellspring of nitrogen for trimming frameworks in Minnesota. On account of its environmental and monetary significance for crop production in Minnesota, it is utilized at large scale.

18.4 Handling Characteristics of Manure

The sum, structure, and consistency of excrement essentially affect the plan of animals fertilizer gear. The nature of compost taking care of fluctuates relying upon the sum and sort of solids present in manure will be helpful for soil fertility. The distinction between dealing with characterizations is not fixed, yet rather differs by arrangement. By and large, excrement can be arranged by how it should be made due. Fertilizer the executives properties change as consistency increments from fluid to strong. At one limit of the range is tidal pond liquid with an exceptionally low solids focus (under 1%) that can be taken care of with standard diffusive siphons. Huge firearms or focus turn water system frameworks with little spouts can be utilized to inundate the tidal pond fluid. Strong excrement, then again, requires the utilization of front loaders as well as forks. Strong excrement typically contains over 20% solids. In the center are harder-to-handle slurries that contain somewhere in the range of 5 and 20% solids. The key impacting factor is the dampness content of the excrement, while the size of the solids and the presence of litter can altogether influence the hardware expected for dealing with, treatment and transport. Solids will quite often settle, yet extremely thick excrement (more noteworthy than 10% solids) forestalls settling and can give a more uniform fertilizer than settled, more slender compost. Sand is one more troublesome strong that is incidentally utilized as dairy bedding. Due to its high thickness and abrasiveness, sand requires exceptional settling and taking care of cycles. Solids fixations are connected with supplement levels. The higher the solids fixation, the higher the supplement focus overall. Gauges are accessible for most kinds of excrement, yet delegate tests should be analyzed to really comprehend what is in the compost. Assesses and classified values ought to be utilized with alert. They are valuable for arranging; however, when the plant is in activity, the best technique to decide healthful and dealing with properties is to gather and examine superb delegate tests. Various characteristics of manure from different species are presented in Fig. 18.2.

18.5 Organic Manure

Natural composts have a high extent of natural matter and a modest quantity of plant supplements. They are delivered by plants or creatures and are natural in nature. By controlling the physical, synthetic, and organic nature of the dirt, they are utilized to advance soil yield. A kind of natural compost utilized in farming is excrement. By presenting natural matter and supplements, for example, nitrogen that is held in the dirt by microorganisms composts increment soil ripeness.

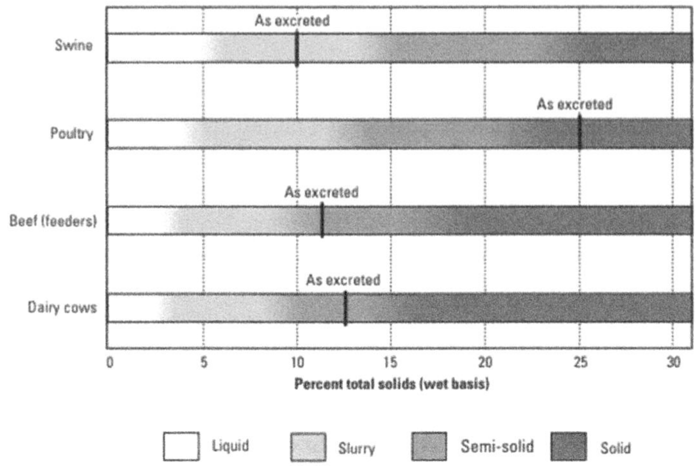

Fig. 18.2 Relative handling characteristics of different types of manure for various species

18.5.1 Bulky Organic Manures

Farmyard manure (FYM) or farming compost, farm waste, urban crop residues, dark soil, sewage, green manures, and other bulky sources of organic compounds are forms of bulky organic manures. Every one of these fertilizers has a massive appearance and offers both huge volumes of natural matter and humble degrees of plant supplements. The three most critical and frequently used cumbersome natural fertilizers are by a long shot ranch yard excrement, manure, and green compost. Various types of organic manure are shown in Fig. 18.3.

18.5.2 Effect of Bulky Organic Manures on Soil

- It fundamentally impacts the development of plants.
- Increment the dirt's substance of natural matter would increment its actual characteristics.
- Give food to soil microorganisms by supporting the natural substance in the dirt, which builds the WHC of mud soil and further develops seepage in clayey soil.
- Give soil microbes nourishment. This stimulates the activity of microorganisms, which helps transform inadequate plant nutrients into form that plants can also use.

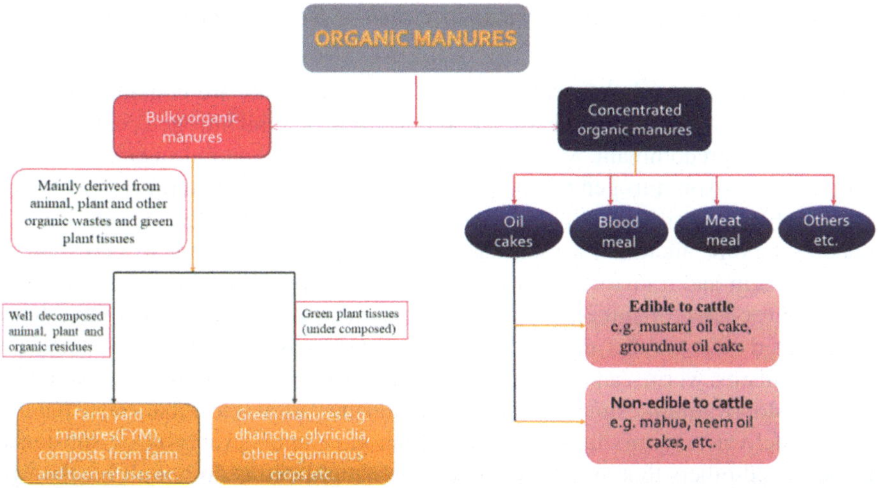

Fig. 18.3 Types of organic manure

18.5.3 Farmyard Manure (FYM)

It alludes to a decaying combination of domesticated animal's excreta and pee, as well as bedding (bedding) and scraps from roughage or feed took care of two domesticated animals. 0.5% Ns 0.2% P205, and 0.5% K_2O are found on normal in very much spoiled FYM. One of the most significant farming side effects is FYM. Tragically, around half of India's dairy cattle waste creation is currently utilized as fuel, prompting loss of rural land. Availability of different minerals have been present in Table 18.1.

- The richest among all is poultry manure.
- When contrasted with the compost part, all creatures' pee contains a higher level of N and K_2O.

Table 18.1 Average percentage of N, K_2O_5, and K_2O in the fresh excreta of farm animals

Excreta of		N (%)	P_2O_5 (%)	K_2O (%)
Cows and Bullocks	Dung	0.40	0.20	0.10
	Urine	1.00	Traces	1.35
Sheep and Goat	Dung	0.75	0.50	0.45
	Urine	1.35	0.05	2.10
Buffalo	Dung	0.26	0.18	0.17
	Urine	0.62	Traces	1.61
Poultry	–	1.46	1.17	0.62

18.5.4 Organic Manure Transformation Reactions in Soils

Plant supplements, including full scale and micronutrients, are contained in both cumbersome and concentrated natural composts, with natural nitrogen content liable to be the most predominant. Merged amino acids, proteins, amino sugars, and other natural types of soil nitrogen can be found. At the point when natural composts, for example, FYM, fertilizers, oil cakes, green excrements, and different materials are added to the dirt, microbial assault happens, bringing about complete vanishing of natural protein and transformation of the leftover nitrogen into inorganic structure through the course of mineralization. The natural contaminations that have been disintegrated into the dirt do not keep going extremely lengthy. A few fertilizer microorganisms, worms, and bugs assault them right once, especially in the event that the dirt is sloppy. Infinitesimal life forms separate the various parts that create natural waste to deliver new mixtures, some of which have unbelievably basic substance attributes and others that are fairly broad. The natural extras do not all breakdown without a moment's delay or in a similar way. While specific constituents separate quick, others do so more leisurely, nevertheless others take an extremely extensive stretch.

18.6 Various Sources of Plant Nutrients

18.6.1 Soil Sources

Because of progressing and escalated cultivating strategies, the supplement giving ability of many soils has consistently diminished. The primary drivers of low efficiency on most cultivated fields are low and disintegrating soil fruitfulness. Escalated cultivating likewise caused a deficiency in specific optional and micronutrients in the dirt. A simulation of nutrient insufficiency demonstrates that nitrogen deficit is global and that almost 49, 20, and 47% of Indian soils are similarly low in P, K, and Zn. In 125 regions, there is a lack of sulfur. In some rural creation frameworks, Fe, Mn, and B have additionally become significant limitations. Long-haul compost preliminaries in different agro-biological areas have obviously demonstrated that potassium lack will turn into the best restricting variable for crop efficiency under serious editing later on. Other manure supplements will be less successful, therefore. Moreover, it is anticipated that as much as 8 million tons of supplements are lost every year because of soil disintegration. To amplify use of accessible supplements, it is accordingly expected to restrict supplement misfortunes through fitting soil the executives' techniques, to remediate soil issues, and to utilize suitable yield types, social medicines, and editing frameworks.

18.6.2 Mineral Fertilizers

Mineral manures are fundamental for keeping horticultural result. Since it has a significant expense, it should be improved combined with the wide range of various horticultural creation factors, for example,

- Recommending manure for a whole yield developed instead of just one harvest.
- Restricting fixings like auxiliary and micronutrients are killed.
- Using the right items, such as wrapped urea and ultra granule, such as using phosphate rocks that are readily available nearby in saline soil effectively.

Along these lines, manures should be utilized in the perfect sum, with flawless timing, and from the right source.

18.6.3 Organic Sources

Natural fertilizer helps crop development and soil efficiency in different ways. Organics straightforwardly affect plant development and digestion through the take-up of humid mixtures or their breakdown items. In a roundabout way, it helps the accessibility of major and minor plant supplements by upgrading the valuable soil microorganisms and their movement.

Natural assets, which are created from creature and plant matter, are important side effects of horticulture and related callings. Natural assets incorporate components like creature squander, horticultural waste buildups, sewage, ooze, fertilizer, biofertilizers, human waste, and other modern waste. It is assessed that natural sources could deliver 9.9, 2.7, and 4.4 million tons of N, P_2O_5, and K_2O plant supplements each year, separately. 3.7, 1.1, and 1.8 million tons of nitrogen, phosphorus, and potassium, separately, are created by steers and bison compost. A large portion of it is changed over into fuel. Utilizing biogas can assist with decreasing how much supplements and fuel expected to consume compost. Creature squander water contains less nitrogen than biogas slurry, which is more gainful for plants. Night soil could likewise give around 5 mt of N, P, and K supplements whenever utilized actually. The ongoing possible accessibility of yield deposits is 400 million metric tons (mmt). A lot of buildup stays in the field in regions where mechanical reaping is utilized. Plant supplements and natural matter are lost when these deposits are signed set up. 70% of harvest buildups are rice and wheat straw. Around 33% of farming buildups are reused straightforwardly ashore, and a huge sum is reused after use as animal feed, where animal feces is utilized as FYM.

Future wellsprings of NPK supplements could incorporate fisheries, bone feast, provincial and metropolitan squanders, green manuring, agro-modern squanders, and yield buildups. In the event that those composts were painstakingly gathered, moderated, and reused, India may be fit for meeting its supplement needs and increasing its farming over the long haul. Besides there is a huge micronutrient's content of natural composts. Because of their correlative and added substance nature, the joined utilization of natural, organic, and inorganic manures in farming creation and soil efficiency expects extraordinary importance.

Soil mineral, natural, and organic wellsprings of plant supplements, notwithstanding any remaining significant sources, ought to be used really and actually in the improvement of reasonable harvests. Coordinated supplement supply additionally assists with further developing soil conditions by bringing down troublesome soil biological circumstances and raising soil creation.

18.7 Organic Cycling

18.7.1 Green Manuring

Green compost has for some time been an office utilized by ranchers. Conversely, a rancher will be unable to execute conventional green manuring in escalated horticulture by forfeiting a whole harvest season for green manuring. Green manuring, then again, assists ranchers with decreasing their compound compost necessities and is one of the best and harmless to the ecosystem strategies for natural treatment. An old practice known as "green manuring" of land for rural purposes blurred into indefinite quality by the last part of the 1960s. Since it is the least expensive wellspring of contributions for expanding soil richness and renewing plant supplements, particularly nitrogen, it requires unique consideration in the ongoing setting of a coordinated supplement supply framework.

Through the accompanying advances, green manuring assists increment with trimming creation.

- Nodulated vegetables fix barometrical nitrogen and use it to advance soil nitrogen.
- Disintegrating natural matter affects N, P, K, and micronutrients in the dirt.
- Works on the physical, substance, and organic climate of the dirt by lessening filtering and vaporous misfortunes of nitrogen and expanding the proficiency of plant supplement use.
- New cuttings of a few enduring vegetables, for example, Gliricidia filled in hedgerows and handle nooks can be utilized for take-up into the dirt as a wellspring of N on the off chance that ranchers are reluctant to commit their joined soil assets and yield contributions to green manuring. This is a run-of-the-mill practice in the southern locales of the country.

18.8 Biological Sources

The disclosure by Hellrigel and Wilfarth in 1886 that harmonious bacterial settlements could may be reap barometrical nitrogen particles to help have plants in vegetable root knobs flagged a developing acknowledgment of the significance of soil biota in richness studies. Composts with organic sources are known as biofertilizers.

The expression is currently used to depict all-natural excrement, including green fertilizer. Biofertilizers, then again, are combinations of living microorganisms that are utilized to increment soil richness and plant sustenance for more manageable farming creation. In India, biofertilizers are viewed as a reasonable option in contrast to compound manures in practical cultivating frameworks since they are economical, earth mindful, and sustainable wellsprings of mass and modest plant supplements.

18.8.1 Rhizobium Inoculants

To fix nitrogen, legumes and rhizobium collaborate. To supply up to 80% of the nitrogen needs of different pulse and legume crops, it is advised to inoculate seeds with rhizobium cultures. For the initial growth and establishment of legume seedlings, 10–20 kg of inorganic N/ha must be added to the soil. Leguminous crops have been shown to help maintain soil fertility when employed as a part of a crop rotation. Rhizobium inoculation in traditional pulses frequently increases yield by 2–3 q/ha. Rhizobium inoculation is currently a common practice for newly introduced legume crops.

18.8.2 Azotobacter Inoculants

Azotobacter is a non-symbiotic nitrogen fixer that grows in the rhizosphere and benefits plants. A number of employees have utilized azotobacter chrococcum inoculants in field research. Estimates suggest that azotobacter inoculation might save the loss of roughly 15–20 kg of N/ha and boost grain productivity by almost 10%. They also create growth hormones, which support seed germination, root growth, and plant vigor.

18.8.3 Azospirillum Inoculants

At least two species of the gram-negative nitrogen-fixing spirilla formerly known as *Spirillum lipoferum* are currently found in the genus *Azospirillum: A. brasilense* and *A. lipoferum*. In field experiments, *A. brasilense* cultures were successfully inoculated into sorghum and pearl millet, saving about 20–40 kg N/ha. Wheat responded strongly to *A. lipoferum* as well.

18.8.4 Blue-Green Algae Inoculants

An appropriate environment for algal nitrogen fixation levels ranging from 40 to 80 kg N/ha/year is a flooded rice field. Crop yield can be enhanced by 10–20% via algal implantation. There are additional claims that BGA produces chemicals that stimulate growth.

18.8.5 Azolla

Due to the heterocystous blue-green algal Anabaena azolla's presence in its ventral leaves, the water herb azolla fixes atmospheric N. India is the home of Aspergillus pinnata. Azolla has a chemical make-up of 4–6% N, 0.5–0.6% P, 2.6 percentage K, 9–10% ash, 5 percentage points crude fat, 9% crude fiber, and 20–30 percent in terms of crude protein (on a dry basis). It, therefore, can be used as green manure and is an excellent source of organic N.

18.8.6 Phosphatic Biofertilizer

Through the release of organic acids, a number of soil bacteria (Pseudomonas striata, Bacteria polymyxa), as well as fungi (Aspergillus awamori), are capable of converting insoluble phosphate into soluble forms. These acids cause the pH to decrease and dissolve phosphates that have been isolated. Furthermore, some hydroxy acids may form metal complexes with calcium and iron, efficiently emulsifying the two components and increasing the use of soil phosphorus by plants. Application of organic manure coupled with PSM can improve the P from rock phosphate's availability. On P-deficient soils, extract can significantly improve the availability of P to plants. They can also stimulate the mobility of other mineral elements like copper and zinc.

18.9 Chemical Fertilizers

Manure is any dry or fluid substance that is added to the dirt to give at least one plant supplements. Notwithstanding lime and gypsum, the expression compost alludes to economically figured out items.

18.9.1 *Nitrogen Fertilizer*

Natural and inorganic types of nitrogen both are available in the dirt. Instances of inorganic structures incorporate nitrate, nitrite, and alkali (NH^{4+}), among others (NO^{3-}). In the beginning phases of development, paddy assimilates nitrogen as NO^{3-} and NH^{4+} compounds. While NO^{3-} is lost by filtering on the grounds that it does not shape a replaceable complex with mud and natural colloids at unbiased to high pH, nitrogen as NH^{4+} structures an interchangeable complex with earth and natural colloids and is hence not lost by draining. Be that as it may, when the climate is truly acidic, it goes through a trade. In spite of the way that plants retain nitrogen as NO^{3-}, nitrate composts are not frequently utilized on the grounds that they will generally assimilate dampness. Accordingly, urea and other NH_4 composts, for example, ammonium sulfate are frequently utilized. N ought to be applied with the goal that the plant procures it all through its life cycle, as most Indian soils are nitrogen poor and yields require nitrogen all through the developing season. Nitrogen-containing composts should be regularly applied to all dirt and yields. The aggregate sum of nitrogen manures required is hence more noteworthy than the aggregate sum of different supplements.

18.9.2 *Commercial Nitrogenous Fertilizer*

Business nitrogenous manures are those that are made for an enormous scope and sold for their nitrogen focus.

18.9.2.1 Nitrogenous Fertilizers

In light of the synthetic structure in which nitrogen is joined with different components by the manure, nitrogen composts can be separated into four kinds. Different types of nitrogenous fertlizers are presented in Fig. 18.4.

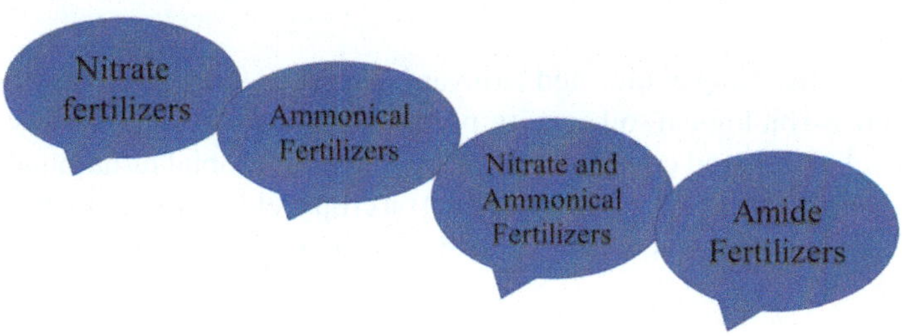

Fig. 18.4 Types of nitrogenous fertilizers

18.9.2.2 Nitrate Fertilizers

Nitrogen is combined with different components as NO^{3-}, calcium nitrate (15.5% N), sodium nitrate, or Chilean nitrate (16% N). Sodium nitrate is a business manure that is imported.

18.9.2.3 Ammonical Fertilizers

Nitrogen is blended in with different parts in ammonical (NH_4) structure in these manures:

- Ammonium sulfate (20% N)
- Ammonium chloride (24–26% N)
- Anhydrous alkali (82% N).

18.9.2.4 Nitrate and Ammonical Fertilizers

Both nitrate and ammonical nitrogen are available in these manures:

- Ammonium nitrate (33–34% N)
- Calcium ammonium nitrate (26% N)
- Nitrate of ammonium sulfate (26% N).

18.9.2.5 Amide Fertilizers

These manures have nitrogen as amide or cyanamide:

- Urea (46% nitrogen)
- Calcium cyanamide (21% nitrogen).

18.10 Conclusion

It is concluded that using composted farmyard manure as fertilizer in organic farming systems increase biological soil activity, promoting soil quality and fertility. Utilizing biodynamically prepared compost significantly alter microbial turnaround, however these changes may not always have an impact on crop yields. Organic matter improves soil, slows down land erosion, and stimulates the activity of soil bacteria, enhancing soil fertility.

References

Ali U, Rehman KU, Malik MY (2019) The influence of MHD and heat generation/absorption in a Newtonian flow field manifested with a Cattaneo-Christov heat flux model. Phys Scr 94(8):085217

Allee LL, Davis PM (1996) Effect of manure and corn hybrid on survival of western corn rootworm (Coleoptera: Chrysomelidae). Environ Entomol 25(4):801–809

Berti WR, Jacobs LW (1996) Chemistry and phytotoxicity of soil trace elements from repeated sewage sludge applications, vol 25, no 5. American Society of Agronomy, Crop Science Society of America, and Soil Science Society of America, pp 1025–1032

Durieux ME (1995) Inhibition by ketamine of muscarinic acetylcholine receptor function. Anesth Analg 81(1):57–62

Edmeades DC (2003) The long-term effects of manures and fertilisers on soil productivity and quality: a review. Nutr Cycl Agroecosyst 66(2):165–180

Giacometti C, Cavani L, Baldoni G, Ciavatta C, Marzadori C, Kandeler E (2014) Microplate-scale fluorometric soil enzyme assays as tools to assess soil quality in a long-term agricultural field experiment. Appl Soil Ecol 75:80–85

Gurevitch J, Koricheva J, Nakagawa S, Stewart G (2018) Meta-analysis and the science of research synthesis. Nature 555(7695):175–182

Hornick SB (1988) Use of organic amendments to increase the productivity of sand and gravel spoils: effect on yield and composition of sweet corn. Am J Altern Agric 3(4):156–162

Jiang G, Zhang W, Xu M, Kuzyakov Y, Zhang X, Wang J, Di J, Murphy DV (2018) Manure and mineral fertilizer effects on crop yield and soil carbon sequestration: a meta-analysis and modeling across China. Glob Biogeochem Cycles 32(11):1659–1672

Kallenbach C, Grandy AS (2011) Controls over soil microbial biomass responses to carbon amendments in agricultural systems: a meta-analysis. Agr Ecosyst Environ 144(1):241–252

Karami A, Homaee M, Afzalinia S, Ruhipour H, Basirat S (2012) Organic resource management: impacts on soil aggregate stability and other soil physico-chemical properties. Agr Ecosyst Environ 148:22–28

Latham EE (1940) Relative productivity of the A horizon of Cecil sandy loam and the B and G horizons exposed by erosion. J Am Soc Agron 32:950–954

Liu Z, Rong Q, Zhou W, Liang G (2017) Effects of inorganic and organic amendment on soil chemical properties, enzyme activities, microbial community and soil quality in yellow clayey soil. PLoS ONE 12(3):e0172767

Liu S, Pu S, Deng D, Huang H, Yan C, Ma H, Razavi BS (2020) Comparable effects of manure and its biochar on reducing soil Cr bioavailability and narrowing the rhizosphere extent of enzyme activities. Environ Int 134:105277

Luo G, Li L, Friman VP, Guo J, Guo S, Shen Q, Ling N (2018) Organic amendments increase crop yields by improving microbe-mediated soil functioning of agroecosystems: a meta-analysis. Soil Biol Biochem 124:105–115

Maillard É, Angers DA (2014) Animal manure application and soil organic carbon stocks: a meta-analysis. Glob Change Biol 20(2):666–679

Mandal S, Pu S, Wang X, Ma H, Bai Y (2020) Hierarchical porous structured polysulfide supported nZVI/biochar and efficient immobilization of selenium in the soil. Sci Total Environ 708:134831

Nowak MA, Sigmund K (1998) The dynamics of indirect reciprocity. J Theor Biol 194(4):561–574

Pu S, Yan C, Huang H, Liu S, Deng D (2019) Toxicity of nano-CuO particles to maize and microbial community largely depends on its bioavailable fractions. Environ Pollut 255:113248

Ringer MA, Andrews T, Webb MJ (2014) Global-mean radiative feedbacks and forcing in atmosphere-only and coupled atmosphere-ocean climate change experiments. Geophys Res Lett 41(11):4035–4042

Ros M, Pascual JA, Garcia C, Hernandez MT, Insam H (2006) Hydrolase activities, microbial biomass and bacterial community in a soil after long-term amendment with different composts. Soil Biol Biochem 38(12):3443–3452

Saha MR, Hasan SMR, Akter R, Hossain MM, Alam MS, Alam MA, Mazumder MEH (2008) In vitro free radical scavenging activity of methanol extract of the leaves of Mimusops elengi Linn. Bangladesh J Vet Med 6(2):197–202

Sandor JA, Eash NS (1991) Significance of ancient agricultural soils for long-term agronomic studies and sustainable agriculture research. Agron J 83(1):29–37

Tejada M (2009) Application of different organic wastes in a soil polluted by cadmium: effects on soil biological properties. Geoderma 153(1–2):254–268

Thangarajan R, Bolan NS, Tian G, Naidu R, Kunhikrishnan A (2013) Role of organic amendment application on greenhouse gas emission from soil. Sci Total Environ 465:72–96

Weidemann AG (1943) Fertilizer placement studies on Hillsdale sandy loam soil. J Am Soc Agron

Zaller JG, Köpke U (2004) Effects of traditional and biodynamic farmyard manure amendment on yields, soil chemical, biochemical and biological properties in a long-term field experiment. Biol Fertil Soils 40(4):222–229

Zhang Y, Yang S, Fu M, Cai J, Zhang Y, Wang R, Xu Z, Bai Y, Jiang Y (2015) Sheep manure application increases soil exchangeable base cations in a semi-arid steppe of Inner Mongolia. J Arid Land 7(3):361–369

Chapter 19
Current Trends and Prospects of Transforming Animal Waste into Food

Samreen Ahsan, Atif Liaqat, Adnan Khaliq, Rabia Iqbal, Muhmmad Farhan Jahangir Chughtai, Tariq Mehmood, Kanza Saeed, Shoaib Fayyaz, Aaqib Saeed, and Nimra Sameed

Abstract Animal waste production is a disquieting situation all over the world. Presently, research and development is working to utilize waste for producing value-added products (manure, animal feed, etc.), extraction of valuable compounds, renewable energy, and biodegradable packaging material production. This chapter discusses the extraction of protein (keratin, collagen, gelatin, myofibrillar, and blood protein) from animal waste and its application in food industries. Furthermore, this chapter also reviews bone meal and fish meal application and the utilization of animal protein in the development of packaging film. The review also discusses the regulatory status of waste material in packaging along with ethical and religious concerns and future prospects of animal waste in nutricosmetics, biotechnological, and dairy probiotic beverage development. The waste utilization in effective way will be an efficient tool to be a part in control of climate change.

S. Ahsan · A. Liaqat · A. Khaliq (✉) · R. Iqbal · M. F. J. Chughtai · T. Mehmood · K. Saeed · S. Fayyaz · A. Saeed · N. Sameed
Institute of Food Science and Technology, Faculty of Food, Health Science and Technology, Khwaja Fareed UEIT, Rahim Yar Khan, Pakistan
e-mail: adnan.khaliq@kfueit.edu.pk

M. Arshad (ed.), *Climate Changes Mitigation and Sustainable Bioenergy Harvest Through Animal Waste*, https://doi.org/10.1007/978-3-031-26224-1_19

Graphical Abstract

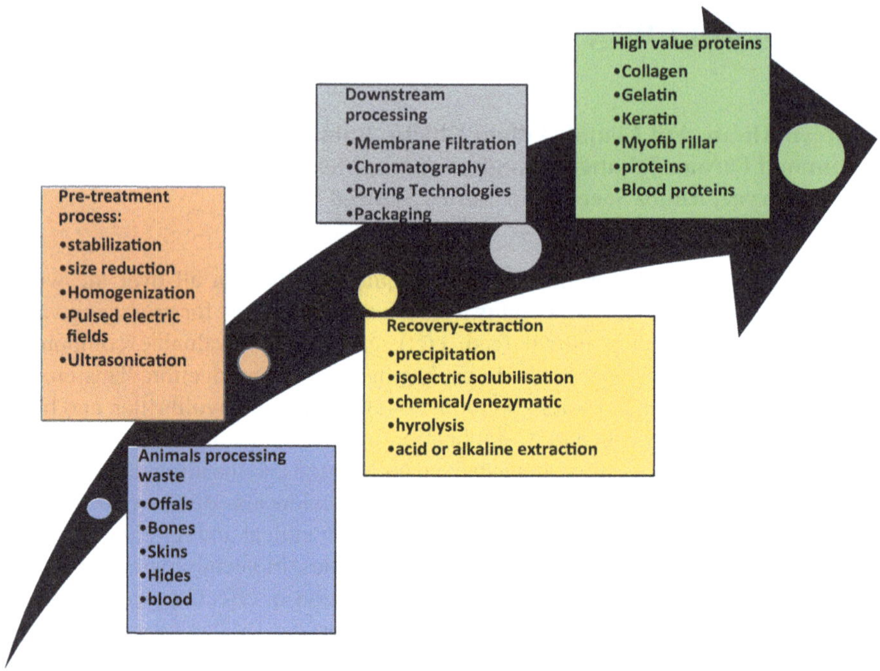

Keywords Protein · Fish bone meal · Packaging film · Climate change

19.1 Introduction

The world population is estimated to exceed 10 billion by the end of 2050 (Duque-Acevedo et al. 2020). Around the world, different countries are trying to enhance the production of food to feed ever-growing populations in the future. As a result, the industrialization of agriculture and allied sectors would further be promoted which in turn will certainly increase the production of waste. The farm mechanization preferences and progress in industrialization have increased the production of agro-industrial waste. In general, the waste produced by agricultural fields and the agri-food industry is nutrient-rich and cannot be left untreated (Sadh et al. 2018). Globally, about 50% of the total waste produced is deposited in open dumps and landfills (Millati et al. 2019).

Managing such a large amount of waste is resource-intensive and challenging work. Interestingly, it can be utilized as a raw material for the production of value-added products, extraction of compounds for use in food, production of renewable energy, and development of biodegradable packaging materials in the present times (Birania et al. 2022).

Recognizing the importance of these agro-industrial wastes, authorities propose their transformation into value-added products (Mora et al. 2019; Álvarez et al. 2010). In advanced countries, animal products are considered as main sources of protein in the diet. Globally, 45% of overall protein consumption comes from animal sources, which is predicted to rise by 135% until 2050 (Lynch et al. 2018).

Mutton, beef, poultry, and pork are the highly consumed meat types and are commonly served as processed meats, pies, burgers, and sausages. In addition, various materials as co-products can be obtained in the production of these main products and considered as secondary products. Chitin, fat, and animal protein are the chief by-products that can be obtained from poultry, livestock, aquatic species, and domestic animals and generally account for 1/3 to 1/2 of an animal's live weight (Khodaei et al. 2021). Direct consumption of meat by-products is relatively small, majorly due to cultural and traditional practices, consumer perception, and ethical and religious restrictions. In addition, public health concerns, e.g., bovine spongiform encephalopathy can be reasons of the low consumption of these by-products. Thus, worldwide efforts are being made to reduce the dependence on animal protein and to discover a model route to utilize animal waste produced during meat processing and could be effectively utilized in pharmaceuticals, animal feed, pet food, industrial applications such as plant growth stimulators, adhesives, cosmetics, textile, water treatment, and biopolymers (Lynch et al. 2018).

The utilization of by-products in any of meat industry cannot only uplift their revenue but also shows positive impact on environmental pollution with zero waste generation (Henchion et al. 2016). This waste is also thought to be a high-quality raw material with low commercial value for the production of valuable products such as chitosan, chitin, gelatin, collagen, keratin, enzymes, peptides, oils, ω-3 fatty acids, and fishmeal (Araújo et al. 2018a).

Protein remains in these products as obtained from animals are inexpensive sources. Meat by-products and wastes from slaughterhouses or processing industries are utilized as starting materials for the production of value-added products. The major steps for protein recovery from meat processing wastes include pretreatment, extraction, and downstream processing (Khodaei et al. 2021). Figure 19.1 showing processing of animal waste.

In the past three decades, one of the biggest issues discussing worldwide is "climate change". Animal waste disposal is impacting on climate by increasing surface temperature, increasing mean sea level, and variability in rainfall. There are a lot of calls published on mitigation of climate change but not on adaptation. The current review article is proposing ways by explaining that sets up waste segregation and then recycles it. Animal waste is its offal, bones, skins, hides, and blood and fortunately, all of these contain a lot of nutrients and beneficial factors that can be extracted and converted into valuable products. It would be very challenging by 2050 to fulfill the need of food for estimated 9 billion people in the challenging time period of climate change. It is the dire need to drive efforts be on track to control the rise in temperature and to achieve sustainable development goals (Batini 2019).

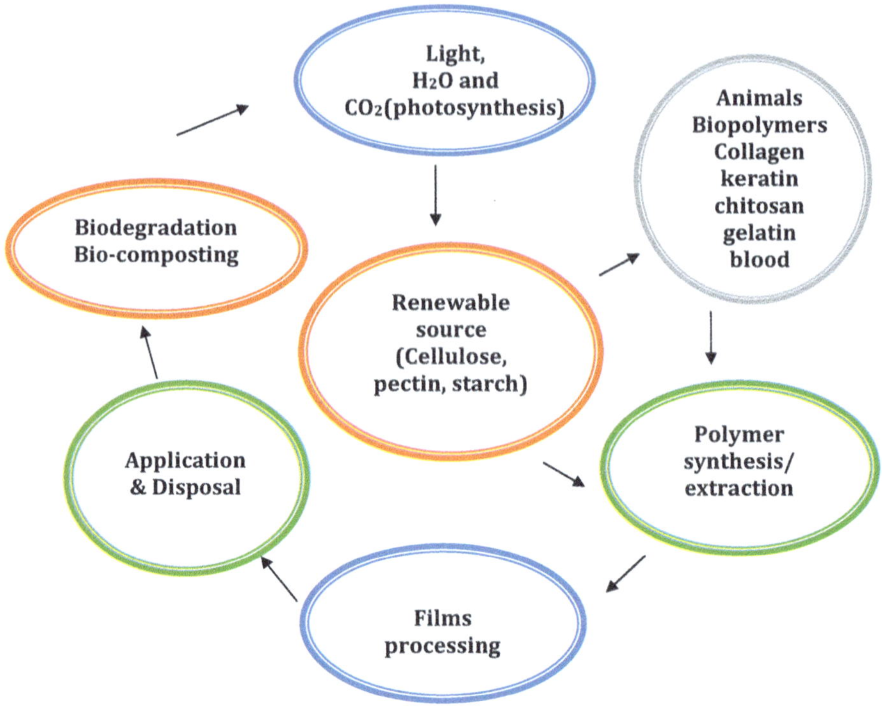

Fig. 19.1 Biodegradable polymers life cycle from natural sources (Mohanty et al. 2005)

19.2 Extraction of Protein from Animal Waste

19.2.1 Keratin Pretreatment and Extraction

Various research has been performed to isolate keratin from animal wastes in the last few years. These strategies belong to two major classes: proteolysis and protein denaturation. Based on protein denaturation, the techniques are divided into oxidative, sulfitolysis, or oxidative, whereas alkaline, enzymatic, and microbial techniques are based on protein degradation (Chilakamarry et al. 2021). Protein denaturation methods yield the highest weight keratins, but are time-consuming and use chemicals that are toxic and pollute the environment (Shavandi et al. 2017).

As keratin contains cysteine and amino acid cross linkages, its extraction using any technique either solvent or non-solvent is challenging. Methods that cause keratin denaturation include agents such as surfactants, chemicals, and hydrolytic agents that play an important part in keratin extraction by breaking sulfide bonds without affecting the keratin structure (Wang et al. 2016). First, denaturing substances and chemicals increase the ability of keratin to absorb water and swell by splitting the hydrogen and reducing the hydrophobic bonds in its structure. In the next step, sulfide bonds are broken down using chemicals that are either oxidative or reductive depending on the reaction mechanism (Kakkar et al. 2014). To improve keratin solubility in water, urea is commonly used as a denaturing agent. The effectiveness of a surfactant depends mainly on its interaction with the keratin. Anionic surfactants are reported to be more effective than cationic and neutral surfactants (Eslahi et al. 2013).

Various studies have reported sodium dodecyl sulfate (SDS) as an effective surfactant as it enhances the rate of keratin extraction through a bi-nucleophilic displacement mechanism. The process involves two reducing agents producing kerateins with thioglycolic acid and 2-mercaptoethanol as reducing agents. As the degradation of keratin takes place at pH more than 11, most of the reduction reactions are performed in alkaline conditions. For the process of reduction, mercaptoethanol is often used but it is undesirable because of its unpleasant odor and is also poisonous (Pan et al. 2016).

Peracetic acid ($C_2H_4O_3$) and performic acid (CH_2O_3) are generally used to break keratin disulfide bonds and oxidize cysteine into cysteic acid residues. Then oxidized keratin is known as keratoses that is modified chemically and can be separated at different pH and solubility into α-, β-, and γ-keratoses fractions. Another technique used to cleave sulfide bonds is sulfitolysis which uses sodium sulfite ($NaSO_3$), sodium bisulfite (NaS_2O_5), and sodium bisulfite ($NaHSO_3$) (Lee et al. 2014). To break disulfide bonds in wool, a new organ phosphonic chemical called LKS-610 was used. The chemical produces thiol groups upon interaction with disulfides and a considerable decline in cysteine content was observed (Li et al. 2019).

Alkaline extraction uses hot alkali solutions to break down amide linkages in keratin converting them into amino acids such as cysteine which is the most abundant amino acid found in keratin structure (Gupta et al. 2012). A highly concentrated basic solution can isolate hydrogen from sulfur and carboxyl groups. During this process, the decomposition of peptide chains forms alkali sulfides having an unpleasant odor. When the alkali concentration is less than 10%, there exist a direct relation between solubility of keratin and the alkali's concentration. Increasing the NaOH concentration from 15 to 38% enhances the strength of wool fibers by approximately 30%. Cysteine is among the major amino acids found in wool and is degraded in strong alkaline solutions. Process parameters must be adjusted to retain cysteine during the derivation of protein (Shah et al. 2019).

Trifluoromethanesulfonyl (1-Hydroxyethyl-3-methylimidazolium bis), as hydrophobic ionic liquid (IL), is widely used to isolate and extract keratin from chicken feathers. Investigation of the effects of IL, $NaHSO_3$ (reducing agent), temperature, and reaction time showed that keratin yield rises from 7 to 15% as the mass ratio increases (Chilakamarry et al. 2021). One of the important factors that affects keratin extraction yield is temperature. Temperature affects the yield in three different steps; (1) Increasing the temperature from 70 to 80 °C enhances the yield. (2) A rise in temperature from 80 to 90 °C had no effect on the yield. (3) An increase from 90 to 100 °C resulted in a lower yield (Ji et al. 2014).

To break down and decompose keratin into peptides, methods involving enzymes and microorganisms can also be used. The procedure used for the degradation of keratin with the help of bacteria has not been described yet. One of the microbial enzymes that breaks the bonds of keratin protein is keratinase which is produced by some microorganisms. The decomposition of keratin in the presence of microbial enzymes does not alter the structure of proteins and retains its functional properties (Chilakamarry et al. 2021; de Menezes et al. 2021).

The weight of keratin isolated by environment friendly techniques such as IL extraction, alkaline extraction, steam, and superheated water varies depending on the processing conditions. Keratin having molecular weight below 10 KDa is not suitable to be used as a structural material (Fagbemi et al. 2020). Thermal methods (superheated water, steam explosion, microwave) use steam and high pressure which destroys keratin and the yield is also less. The yield of keratin extracted using enzymes is unclear and an expensive process. More research is required to assess cost effectiveness for the process and extract keratin with similar physicochemical properties depending on the source and method (Fagbemi et al. 2020).

19.2.2 Gelatin and Collagen Extraction

The removal of non-collagen components from animal's waste is essential before collagen extraction. Various pretreatments including degreasing, swelling, and demineralization are used for this purpose. By the use of alkali treatments with sodium hydroxide (NaOH), non-collagen components are removed with the swelling of raw material. Although the concentration of NaOH and solvent temperature can be changed depending on the nature of raw material and extent of cross-linking found in collagen, the use of higher concentration for NaOH at 4–20 °C should be avoided to prevent damage in collagen structure (Zhang et al. 2013).

Demineralization of collagen is achieved with hydrochloric acid (HCl) and ethylenediaminetetraacetic acid (EDTA). EDTA, as compared to HCl at same concentration, has been reported to present better decalcification properties from animal bones and fish scales. For fat extraction, solvents such as n-hexane and ethanol are used. Pretreatment has been reported to have a significant effect on the degree of hydrolysis while collagen is being extracted from bovine animal's waste. Presently, numerous novel approaches such as high pressure processing, ultrasound, and microwave-assisted extraction are being used to facilitate pretreatment methods for isolating and extracting collagen (Chotphruethipong et al. 2019). The cavitation resulting from a high frequency ultrasound irradiation has the potential to disrupt raw material and break them into simple and small pieces thereby increasing the surface area extraction (Zou et al. 2020).

With high ultrasound frequency, the shear rate has been observed to increase with a decrease in length of bubbles showing inverse relation. Among other pretreatments, thermal techniques such as boiling and high pressure processing significantly effect the degree of hydrolysis, while collagen is being extracted from bovine animals. Boiling pretreatment generally shows better efficacy than high pressure in this regard (Hong et al. 2019).

A commonly used technique to extract collagen from animal hides is enzymatic hydrolysis. The process involves a minor chemical reaction with almost no damage and side effects to the protein's structure. The applications of pepsin as an enzymatic extraction agent have been reported in various studies. It allows breakage at the telopeptide region of the triple helix in collagen's structure producing pepsin soluble collagen (PSC). It also assists in leaching of the collagen structures increasing the extraction yields while in the solution (Jin et al. 2019).

Other enzymes being used for the extraction of collagen include trypsin, alkaline protease, lactases, and papain. Hydrolysis time, enzyme nature, concentration, and solid-to-liquid ratio affect the yield of enzymatic extraction (Grønlien et al. 2019). Following trend has been observed considering the effectiveness of enzymes for collagen extraction from bovine animals: collagenase > trypsin > pepsin (Zhang et al. 2013).

The alternative prevailing technique used for extracting collagen can be chemical electrolysis using an acid, salt, or an alkali. This one is more inexpensive and undemanding (Dhakal et al. 2018). Acid-soluble collagen (ASC) method is effective in the demineralization and extraction of collagen at low temperatures with the help of organic or inorganic acids. The extraction rates are observed to be faster with organic acids as compared to inorganic acids. Intra and intermolecular forces of the collagen helix are broken down, high molecular weight proteins are broken down into smaller units by acids such as HCl and acetic acid. The acid concentration generally varies from 0.5 to 1 M (Jafari et al. 2020). The collagen chain structure must not be affected by the acid concentration. An undesirable effect in collagen structure was seen at 1.05 M concentration of acetic acid due to side reactions. The yield of collagen extraction can also be affected by time and temperature (Blanco et al. 2019; Jafari et al. 2020). Alkaline extraction methods are less favored for collagen extraction due to poor performance, low extraction yields, and collagen denaturation (Meng et al. 2019).

Extraction methods using acids are generally an effective way for extracting collagen. But they can be influenced by various parameters such as acidity, energy consumption, and processing time thereby restricting the extraction yield resulting in an increased solubility of the final product, nutritional losses, poor function, and bitterness (Araújo et al. 2018b). The use of ultrasonic electric fields with enzyme and acidic methods can result in an increased yield, in a small amount of time. A commonly used method is combination of hydrogen bond donor and hydrogen bond acceptor called the deep eutectic solvent (DES) method (Akram and Zhang 2020; Amani et al. 2021). A supercritical fluid extraction (SFE) technique is also employed to isolate proteins using by-products. Although this method can alter the structure of the extracted protein, it is environmentally safe (Sousa et al. 2020).

Regardless of the setbacks, the acid and enzyme extraction methods are generally employed. The application of the traditional approach along with pretreatment methods improves productiveness. Although pulsed electric field (PEF), DES, SFE, and ultrasound approaches have established quite supremacy for extraction, they are more productive when combined with traditional techniques (Cao et al. 2021).

Table 19.1 is representing pretreatment and collagen extraction methods from various animal's skin.

Table 19.1 Pretreatment and collagen extraction methods from various animal's skin

Pretreatments	Collagen extraction methods and steps	References
Chicken		
Hot water bath treatment at 40 °C for 1 h	4 step extraction involving protease inhibitor solution (1 km/L NaCl and Tris-HCl 1 × 10^{-3} km L), ethylene diamine hydrochloride (24 h), acetic acid (0.5 k KM/L), and pepsin (1 g/L) followed by precipitation steps involving ammonium sulfate (25% saturation), and sodium chloride crystals	Cliche et al. (2003)
Non-collagen removal using NaOH 0.1 km/L and fat removal using butyl alcohol 10% v/v at 4 °C for 1 day	Acid hydrolysis using acetic acid 0.5 km/L at 4 °C for 42 h followed by precipitation using NaCl 2 km/L, dialysis using water and centrifugation	Du and Betti (2016)
Fat removal and de-pigmentation by centrifugation	Acid hydrolysis using acetic acid 0.5 km/L at 4 °C for 72 h followed by centrifugation and dialysis using distilled water	Jiménez Vázquez et al. (2019)
Non-collagen removal using NaOH 0.1 km/L at 4 °C for 1 day and fat removal using butyl alcohol 10% at 4 °C for 24 h	Acid hydrolysis using acetic acid 0.5 km/L for 1 day at 4 °C followed by vacuum depending on animal breed	Suurs et al. (2022)
Sheep		
Non-collagen removal using 0.1 km/L NaOH for 5–6 h at 4 °C followed by demineralization using 0.5 km/L EDTA for 48 h	Acid hydrolysis (0.5 km/L for 3 h) followed by enzyme addition (pepsin 1 g/m^3), precipitation using NaCl 2.6 km/L, centrifugation, and hydrolysis (1 km/L $NaCO_3$)	Chuaychan et al. (2015)
Washing and hair removal using deionized water	Acid enzyme treatment using acetic acid and trypsin at (0.5 km/L, 0.001 g trypsin at pH 7 for 40–60 min) followed by filtration and centrifugation	Fernandez-Hervas et al. (2007)
Goat		

(continued)

Table 19.1 (continued)

Pretreatments	Collagen extraction methods and steps	References
Non-collagen removal using NaOH 0.1 km/L for 0–48 h	Acid enzyme treatment using acetic acid 0.5 km/L and pepsin 0.1% w/v followed by precipitation using NaCl 2.6 km/L for 12 h, and centrifugation for 30 min at 4500 g	Hakim et al. (2021)
Non-collagen removal using NaOH 0.1 km/L for 1 day at 4 °C	Acidic hydrolysis using acetic acid 0.5 km/L for 24 h followed by precipitation using NaCl 2.6 km/L, centrifugation at 7000 g and dialysis using acetic acid	Wahyuningsih et al. (2018a, b)
Non-collagen removal using NaOH 0.1 km/L for 48 h at 4 °C	Acid enzyme treatment using acetic acid 0.5 km/L and 0.1% pepsin for 24 h followed by precipitation using NaCl 2.6 km/L, centrifugation at 7000 g, and dialysis using acetic acid	Wahyuningsih et al. (2018a, b)
Pig		
Degreasing through sonication using 75% sodium dodecylbenzene at 25 °C and 120 W	Acid enzyme treatment using acetic acid 0.5 km/L followed by precipitation using NaCl for 5–7 h, centrifugation at 7000 g and dialysis using acetic acid	Zhang et al. (2020)
Fat removal using ether, alkali pretreatment using NaOH 2% g/g, and sonication at 25 kHz for 40 min	Maintain alkaline pH using phosphate buffer solution followed by enzyme hydrolysis using alcalase at 1:100 w/v and centrifugation	Zhang et al. (2017)
Cattle		
Collagen was extracted from cattle skin and collagen base amalgamation for flame retardant composition was prepared and introduced into material of cellulosic textile by grafting copolymerization. Unused and raw cattle skin waste was cut into 3 to 4 mm size	Hydrolysis at alkaline pH using NaOH 3–5%, followed by filtration and neutralization using acetic acid	Nabijon et al. (2017)

(continued)

Table 19.1 (continued)

Pretreatments	Collagen extraction methods and steps	References
Salt washing using NaCl 0.5 km/L for 2 h followed by lime treatment for hair removal and non-collagen removal using NaOH 2% for 5 days	Acid enzyme treatment using acetic acid 0.5 km/L and pepsin at 1:20 w/w for 2 days followed by centrifugation at 20,000 g for 15 min, precipitation using NaCl 2 km/L for 6 h and dialysis using deionized water for 4 days	Feng and Betti (2017)
Pretreatment with hot water bath at 70 °C for 15 min	A solution of 5% calcium hydroxide and 5% acetic acid mixed with sample and hydrolyzed for 48–72 h followed by hot water bath treatment for 1 day at 70 °C and filtration	Said (2018)

19.3 Application of Animal Waste

19.3.1 Collagen as Food Additive

Substances added to food to improve the safety, texture, quality, taste, and appearance of food are known as food additives. Such substances include emulsifiers, preservatives, thickeners, and antioxidants. Collagen, a food additive, is used to enhance the rheological properties of various foods. Meat and meat products having raw materials with added collagen have better technological and rheological properties. Liverwurst or paste containing collagen has improved quality and less existence of fat caps (Hashim et al. 2015).

Heat-treated collagen fiber acts as an emulsifier in acidic products. The integrity and rheological properties of oil-in-water emulsion were evaluated. Droplet size of prepared emulsion and phase separation reduced its pH value resulting in the production of stable emulsion at acidic pH. Through high pressure homogenization, acid emulsions showed lower dispersions and six times decrease in mean surface diameter than the primary emulsions (Santana et al. 2011). Heat-treated collagen fibers are a natural alternate to man-made emulsifiers to be used in acidic foods. The heat processing in such foods decreased protein charge and solubility leading to a lowering of oil-protein interaction. Primary emulsions formed by heating collagen fiber have high creaming and emulsion rates (Aberoumand 2012).

Food-grade collagen can be used in place of lean meat in bologna. In a study, coarse bologna and bologna with fine emulsion were developed by the fibrous collagen in place of lean meat. As a result, no significant effect on shrinkage, density, stability, water content, and protein content were found. Moreover, no significant effect on color, pressure fluid, pH, or cooking loss were discovered (Kalinova et al. 2017).

Duck feet collagen was added in sardine surimi to study its effect on physiochemical properties. The added collagen resulted in improved gel hardness and strength, folding test score, and a color lightness of surimi. Furthermore, collagen proved to be an effective alternative to protein additive for improvement of surimi quality. Jelly production from the addition of chicken feet collagen also had a good response from customers (de Almeida et al. 2012; Huda et al. 2013).

19.3.2 Collagen in Beverages

Collagen-infused drinks are widely being used in the global market including various products such as soy collagen, cocoa collagen, cappuccino collagen, juice collagen, and birds nest collagen. Collagen drinks help body generate fatty tissues and stimulate the collagen making mechanism of body resulting in improved body tissues and reducing skin wrinkling and sagging (Lin et al. 2020).

Malaysian dairy industry has used collagen in their probiotic drink containing prebiotic fiber (Hashim et al. 2015). Vitagen collagen drinks facilitate the growth of healthy bacteria found in gut and also enhance skin beauty. Avon has developed "Avon Life Marine Fish Peptide Collagen Drink" which contains good quality collagen extracted from fish skin, vitamin C, and fructo-oligosaccharide. Nestle Malaysia has also launched several products including Nescafe body partner and collagen coffee containing collagen from the fish source (Binsi and Zynudheen 2019).

Collagen has a rod-like triple helix structure which is thermally unstable and plays an important role as a clarifying agent in alcoholic beverages by accumulating insoluble particles and yeast. Bovine collagen solutions could possibly be used to purify beer and yeast preparations by chemically modifying them. Collagen has a distinctive caprylic taste which can be improved by adding stevia and sucralose extract and further blending with acesulfame potassium if needed (Znamirowska et al. 2020).

19.3.3 Gelatin for Food Industry

Collagen and gelatin have wide usage as food additives and packaging materials in the food industry. Gelatin is integrated in food products to enhance their texture, color, taste, and other properties. Gelatin is majorly used as a food stabilizer and consistency enhancer as it forms stable gel and foam. It is used in water gel desserts for its melt-in-mouth property but it also forms insoluble cross-linked hydrogel that maintains its shape in swelling equilibrium (Tsykhanovska et al. 2020; Yang et al. 2020). Heat-treated collagen fiber is a natural alternative of synthetic emulsifiers. The addition of collagen in sausages improves its rheological properties and decreases fat cap of the oil-in-water emulsion. The antioxidant property of collagen hydrolysates prevents peroxidation of lipids that are dangerous for human health. This antioxidant property is linked with the radicle scavenging imidazole group of histidine (Sousa et al. 2017). To extend the shelf life of foods, collagen-based films have played a significant role. The main function of food packaging is to prevent the exchange of oxygen, moisture, and microbial activity and to maintain the sensory qualities of food material. Gelatin has been widely researched as a food packaging material because of its biodegradability, film-forming ability, and effective gas barrier quality. However, the use of gelatin has some disadvantages because of its poor strength and wide absorptive nature. It absorbs moisture in high-moisture food packaging (Fustier et al. 2015; Huang et al. 2020). A combination of gelatin with biopolymers can improve its properties by reducing mobility, improving resistance, and mechanical barrier properties. Gelatin and collagen films are used to produce sausage casings using the co-extrusion process but they are less effective due to poor mechanical properties and moisture sensitivity (Beghetto et al. 2019; Mohseni and Goli 2019). A multilayered structure can be used for this purpose containing multiple barriers to moisture and oxygen. Gelatin is the first material used as a bioactive substance carrier. Active packaging of gelatin films and coatings can be achieved by incorporating natural antioxidants and antimicrobial components. Bioactive components for active packaging could be plant extracts like rosemary, grape, lemon, and oregano. Extracts can be inactivated by heat, light, and oxygen (Benayahu et al. 2020; Regubalan et al. 2018). Researchers have tried to encapsulate solids, liquids, and gas in microcapsules to protect their functional components. Encapsulation is also a solution to control and decrease taste and smell of vegetable extracts. Gelatin encapsulation technique can be used to manufacture food formulations.

19.3.4 Keratin in Feedstock

Chemical thermal hydrolysis of chicken feathers results in feather hydrolysates that are rich in amino acids and polypeptides. They have similar composition to soybean protein and cotton seed protein used as a dietary supplement for feeding ruminants. Enzymatic-alkaline hydrolysis of feather keratin for feeding was performed (Ramakrishnan et al. 2018). Enzymatic modification through enrichment with lysine leads to an increased nutritional value. A horn meal is made by putting raw horns and hoofs under high pressure. Animal feed can be prepared by using keratin which proves as a useful protein (Brandelli et al. 2015).

19.3.5 Applications of Fish Collagen in Drug Delivery

Fish collagen has potential role in biomedical applications including wound dressing and tissue engineering techniques. However, the applications of collagen for drug delivery are limited. Recently, the researchers are gaining interest in the applications of polymeric matrices and nano and micro-particles of collagen isolated and extracted from fish for drug delivery applications (Felician et al. 2018). To study this, collagen extracted from marine sponges and jellyfish was evaluated. Fish collagen was extracted and prepared using freeze-drying technique. The extract was then incorporated with growth factors such as fibroblast growth factor in microspheres for controlled delivery. Results revealed that microspheres were dispersed uniformly into the porous structures of collagen extract and had better biocompatibility as opposed to collagen without microspheres. This allows microspheres to be used as controlled growth factors to wound site in the body (Cao et al. 2015). Table 19.2 gives protein contents and various applications of fish organs. The gels made from fish collagen extracted from skin of an eel fish were also studied for drug delivery applications. Results revealed that certain drugs including tetracycline and ampicillin were delivered effectively as the zone of inhibition appeared against different bacterial strains (Veeruraj et al. 2013). In another research, fish collagen encapsulated with negative nanoliposomes was studied for enhanced topical delivery. Results revealed that fish collagen enhanced the protein expressions of type I collagen suggesting its applications in cosmetic industries (Seo et al. 2018). Fish collagen extracted from *Lates calcarifer* scale was incorporated with *Bixa rollana* plant extract to study anticancer and antibacterial properties. Results revealed that microspheres showed consistent release of plant extract and exhibited antimicrobial properties against both, i.e., gram positive and gram negative bacteria. These studies elaborate the potential of collagen protein for the delivery of anticancer agents (Muthukumar et al. 2014).

19.3.6 Food Applications of Bone Meal

Bone meal is a combination of roughly and finely ground animal bones and waste from slaughterhouse. It is being used as a feed additive to provide monogastric animals with phosphorus (P) and calcium (Ca) in the form of hydroxyapatite. It acts as a slow organic fertilizer, and it provides plants with small amounts of calcium, nitrogen, and phosphorus.

Bone meal is utilized as a major source of phosphorus, calcium, and traces of different elements. Calcium is the basic constituent of teeth and bones. It is necessary for blood clotting, transmission of nerve impulses, muscle contraction, hormone production, and several other reasons. Stability of cell membrane is also stimulated and maintained by calcium (Coutand et al. 2008; Hendriks et al. 2002).

19.3.6.1 Dietary Supplement

Along with many other meals, bone meal, specifically meat meal, is used as a mineral/food supplement for cattle and hoofed animals. Transmissible spongiform encephalopathy, primarily known as mad cow disease in cattle, is spread by improper

Table 19.2 Protein contents and various applications of fish organs

Major organs of fish	Applications	Protein content (%)	References
Fins	Biomedical purposes, pharmaceutical, and cosmetics	17–20	Mahboob (2015)
Skin	Biomedical purposes, pharmaceutical, and cosmetics	25–28	Mahboob (2015)
Meat	Supplementation of snacks and pasta	20–22	Nawaz et al. (2019)
Egg shells	Calcium supplement and food fortification	4–8	Waheed et al. (2019)
Hairs	Antimicrobial and antioxidant	85	Pleissner et al. (2011)
Scale	Biomedical purposes, pharmaceutical, and cosmetics	22–25	Mahboob (2015)
Fish processing waste	Antimicrobial and antioxidant	11.5	Pleissner et al. (2011)
Fish processing waste	Foaming capacity, emulsifying characteristics, water retention capacity, gelling activity, and oil absorption	16–31	Ghaly et al. (2013)

use of bone and meat meal in animal nutrition. Salmonella contamination can be reduced and controlled by proper temperature control (Jiang et al. 2011).

Historically, bone meal has been used as a dietary calcium supplement for humans (Mendez and Dale 1998). Studies have shown that lead and calcium have a same atomic structure as they have in their ionic forms (Pb^{2+}, Ca^{2+}) and thus leads to the potential for lead accumulation in bones (Potera 2009). In 1970's, Allison Hayes an American actress was poisoned with a calcium supplement made from lead-rich horse bones, prompting the Environmental Protection Agency to develop stricter import regulations. In addition, in 2013, study of broth-based products of chicken bones found out that they contain more quantity of lead than present in the tap water.

19.3.6.2 Dosing Format

For bone meal, there is no Recommended Dietary Allowance (RDA). For adults aged from 19 to 50 years, the recommended daily allowance for calcium is 1000 mg/day and for women with more than 50 and men more than 70, the recommended daily dose is 1200 mg/day.

The RDA of calcium for children aged between 4 and 8 years is 1000 mg/day. The RDA of calcium for children aged between 9 and 18 years is 1300 mg/day (Atuah and Hodson 2011; Ross et al. 2011; Tangke et al. 2020).

Pregnant or breastfeeding women require some more calcium than RDA. However, it should be taken with the prime concern of healthcare provider.

19.3.6.3 Toxicity and Side Effects

In bone meal, some trace elements may be helpful. However, lead is high in bone meal. High levels of mercury may also be present there. This causes major doubts about its use as a nutritional or dietary supplement. Bone meal contains much more lead than refined calcium carbonate (Atuah and Hodson 2011).

Bovine spongiform encephalopathy (BSE) or "mad cow disease" can also be transmitted through bone meal. Bone meal and other by-products of animals that may be used as basic animal feed or supplements have been held responsible to transmit BSE. The processing procedure determines the presence of an infectious agent. There is no research or studies that claim that meal obtained from bones is suitable for human consumption (Ali et al. 2018).

There are no well-known food or drug interactions with bone meal.

19.4 Food Applications of Fish Meal

The highest quality fish meal comes from raw fish. However, to prevent oil and protein breakdown, raw fish is often handled and treated by draining, chilling (water chilling systems, mixing of ice with the fish), or chemical preservation (formaldehyde or sodium nitrite).

Fish and other seafood might be the most leading human food after cereals, providing averagely 15 percent of the global intake of protein contents. Muscles of lean fish contain 18–25% protein by weight, which is almost equivalent to poultry or beef, but contains far fewer calories. In fish, 4 to 10 cal are present in just only one gram of protein, as opposed to 10–20 cal/g of protein for lean meats. For fatty meats, caloric contents may jump up to 30 cal/g (Subhan et al. 2021).

Fish feed is the largest non-recurring operating cost in aquaculture production. Therefore, they constantly try to make food cheaper by using cheap ingredients. Therefore, they constantly try to make feed cheaper by using cheap ingredients. Fish meal is a major ingredient in several feeds and is mainly substituted due to high cost and limited global supply.

Animal feed is made from fish meal because it is rich in protein contents moreover it is also a good source of phosphorus, calcium, and other minerals. Popularity of fish meal is particularly in feed prepared for water life due to its high protein content and appropriate balance of amino acid for water life (Yilmaz and Ozmen 2020).

The raw or canned fish consists of 3 main factors: solids (dry defatted material), water, and oil. Fishmeal is made through a series of steps that include cooking, dehydration, pressing, and grinding. While cooking, at a temperature of 85–90 °C, screw press is used to press the cooked fish, which helps in the removal of liquid and "press cake" is left behind. The supernate is drained, and the remaining solid mixture is centrifuged in order to obtain "still water", from which is removed by evaporation. The pressed cake and stick water are homogenized before they enter the dryer to produce fishmeal with a moisture content of approximately 10%. During each stage of processing, different variations may arise, which will lead to obtaining fishmeal of different quality (Wood et al. 1985).

Crude protein levels, fat contents, and ash contents in good quality fish vary from 66%, 8–11%, and less than 12%, respectively. Good quality fishmeal has a crude protein level of over 66%, a fat content of about 8–12%, and ash usually less than 12%. In developing and tropical countries, sometimes "fishmeal" is processed after meat is sun-dried and milled and grind. This type of fishmeal has very high ash contents and comparatively low in protein. By-products obtained from fish that include concentration of fish protein with relatively more protein content (higher than 70%) (Watanabe et al. 2001).

In 1980s, fish silage which was acid-preserved was heavily promoted as a means of preserving raw or trash fish and for making aquaculture feed by mixing such silage with other feeds, although this practice is not extensively used (Watanabe et al. 2001).

Fish meal, roughly grinded powder from flesh of cooked fish. Although previously important as fertilizer, now fishmeal is used mainly for animal feed, particularly for poultry, pigs, pets, and farmed fish. Certain types of fatty fish, such as anchovies, menhaden, sardines, and herring are a major source of fishmeal and its byproduct like fish oil (Wood et al. 1985).

For processing into meal, sliced fish passes through long cookers of steam on a screw conveyor. Oil and water are than removed by pressing the cooked flesh of fish (that spoils rapidly during storage conditions). Hot air is used to dry the pressed fish cake, resulting in food with a high content of vitamin B12 and up to 50% of protein contents.

Roe, might be female fish eggs mass (hard roe), or sperm of male fish, i.e., mass or milk (soft roe), which are thought to be food. Eggs of different fish are consumed as edible food, after it is salted or smoked. Hard sturgeon eggs, from which caviar is made, is most valued. When preparing caviar, the egg mass of freshly caught fish is removed and passed through a very fine sieve, hence, separation of the eggs takes place. Moreover, cleaning the eggs from external pieces of fat and tissue. Then, salt is added to make them preserve and enhance the flavor. Smoked cod roe is eaten frequently in Scandinavia and Great Britain. Salted carp, cod roe, or mullet are the basis of taramasalata which is a Greek appetizer (Bledsoe et al. 2003).

Soft caviar can be fried or poached and is sometimes served as an appetizer or light first course. Other roes' fish that are particularly well eaten are mackerel, herring, sole roe, salmon, and shad.

Scrod is a young fish, particularly fish is cut up and boned for smoking or cooking. The origin is believed to come from an Old Dutch word that means "to grind". In around 1841, it seemed to be used for the first time (Shenderyuk and Bykowski 2020).

Sashimi, a specialty of Japanese main course food, where raw fresh fish is served. The fish should be absolutely fresh and is cut into very thin slices or quarter to about half an inch (0.75–1.5 cm) thick slices, cubes, or strips, depending on the nature and type of the fish. Wasabi and soy sauce are added to the sashimi. Japanese meal must contain Sashimi as a food starter, served before food, while the flavor is still clear so that its nuances can be adopted.

The commonly used fishes are harvested from oceans: yellowtail, tuna, sea bream, flounder, and mackerel. Fish harvested from freshwater includes carp and perch. They can also be eaten as raw, as are prawns, abalone, clams, and lobsters. With sashimi, sake is traditionally taken in its liquid state (Cozzo and Smart 2020).

Kippers, a most traditional and common British breakfast dish, consisting of herring aged by kippering-cut, gutted, cleaned, salted, and then smoked. It may be sautéed, fried or grilled.

The best quality kippers possess pale copper color and harvested from the Isle of Man, Scotland, and northern England. Kipper is often served with lemon and butter but sometimes with a poached egg. The dish may be sometimes served at dinner or with tea (Alfian et al. 2020).

Fish sauce, in Southeast Asian cooking, a liquid condiment made by fermenting saltwater or freshwater fish with salt in large quantities. After several months, the resulting brown liquid, which might be rich in proteins, is drained and bottled. Sometimes, it is provided with stay time to mature in glass in the presence of sunlight or bottles made of clay before its use. Worldwide, it is recognized with different names like nuoc am in Vietnam, in Thailand, it is known as nam pla, in Cambodia, it is known tuk trey, patis in the Philippines, ketjap ikan in Indonesia, and ngan-pya-ye in Myanmar. Fish sauce is widely utilized as soy sauce in regions like Vietnam and Thailand. Oyster sauce made by Chinese chefs has a similar preparation method that is being used particularly in Cantonese dishes (Choksawangkarn et al. 2018).

Fish oil, a fatty oil extracted from the fish body, that is used in the production of several different products such as cooking oil, margarine, soaps, sealants, cosmetics, paints, candles, lubricants, industrial coatings, and water repellents. Fish oil is also extensively used in leather tanning, rubber production, and the production of various chemicals used to make synthetic wax. Sardines, herring, and menhaden are the main sources of fish oil. During the production of fishmeal, oil and water are squeezed out of cooked fish and separated by a centrifuge. Before storage, the oil is additionally purified by a centrifuge. Fish oil is rich in unsaturated lipids. Oil extracted from fish liver was once considered to be a major source of vitamins A and D. These vitamins are now being synthetically produced at relatively cheaper prices (Sidhu 2003). Table 19.3 shows the protien extraction from fish waste and applications in food.

Table 19.3 Protein extraction from fish waste and its application in food

Source	Protein	Application	References
Skin of sucker catfish (*Pterygoplichthys pardalis*)	Type I collagen	Fresh cheese	Nurubhasha et al. (2019)
Grass carp collagen (GCC) films	Collagen	Pork preservation	Ameer et al. (2020)
Alaska pollock (*Gadus chalcogrammus*) and the pacific cod (*Gadus macrocephalus*), waste from fish cutting (heads, swim bladders, fins, skin, and bones)	Gelatin > 80% protein	Fish culinary products	Zarubin et al. (2021)
Several fish species	Fish protein hydrolysates (FPHs)	Alternative for synthetic antioxidants rich source of bioactive peptides	Desai et al. (2022)
Fish skin and cartilage tissue of salmon fish and fileting waste (heads, fins, vertebral bones) of cartilaginous fish species	Protein hydrolysates characterized by a high protein content (up to 80.0%)	Biologically active food additives, multicomponent food system	Zarubin et al. (2020)
Pepsin-solubilized collagens (PSC)	Type I collagen	The resulting PSC from the five tissues would all be potentially useful commercially	Liu et al. (2012)
Swim bladders of *Catla catla* (Catla)	Gelatin	Flow properties of gelatin revealed non-Newtonian and pseudoplastic behavior	Chandra and Shamasundar (2015)
Nile tilapia (*Oreochromis niloticus*)	Fish protein hydrolysate (FPH)	Diets for carnivorous and omnivorous shrimp	Silva et al. (2014)

19.5 Application of Animal Wastes in Food Packaging

In recent decades, the demand for packaging materials has also increased with the growth of the food industry. The conventional petroleum-based plastics are not biodegradable and pose serious environmental problems, such as threats to marine life including plants and animals and causes environmental pollution with deterioration of quality of air (Zhong et al. 2020). In addition, burning results in the release of toxic gases (carbon dioxide, carbon monoxide, chlorine, 1,3-butadiene, furans, amines, dioxins, etc.), reducing air quality, increasing the threat of global warming, and contributing to many health concerns. The waste may contain such substances that are non-degradable and it negatively impact on environmental and health of human being, have fueled worldwide interest in finding environment friendly alternatives (Luckachan and Pillai 2011; Shaikh et al. 2021).

In contrast, biodegradable material in waste in non-synthetic contains no toxic effect and can be easily degraded due to its environment friendly polymers that have capacity to breakdown in home and industrial composting condition. These biodegradable polymers could be the outcome of natural resources or from by-products and wastes of agricultural and animal processing. Animal processing by-products are low-value and underutilized non-meat materials that are generally produced from meat processing or slaughterhouses, such as skin, blood, and viscera. There is a long list of biopolymers produced from animal waste but the most highlighted in value addition are gelatin, collagen, keratin, myofibrillar protein, and chitosan. They have many uses in the food and pharmaceutical industries, but a significant amount is underutilized and could potentially be used to produce bio-plastics. This section summarizes research progress on the use of meat processing by-products to produce biodegradable polymers with a focus on food industry applications. Additionally, current industry status and regulations for utilization of biodegradable are also discussed. The plastic material is the major concern of environmental pollution and in ocean is about 100–200 million tons of plastic waste, and it is estimated that 8 million tons of plastic are dumped off every year (Kurtela et al. 2019).

Figure 19.1 shows life cycle for biodegradable polymers from natural source. Different food categories have separate storage and transportation conditions, e.g., fruits and vegetables preservation requires a slow rate of respiration and transpiration, which usually depends on equilibrium relative humidity, temperature, light and gas (O_2, CO_2, and ethylene). Dairy products such as milk, cheese, and cream must be protected from oxidation and microbial growth, so extrinsic factors including oxygen, light, and humidity must be carefully considered. Meat products transportation showed discoloration, which can be avoided by vacuum packaging or controlled atmosphere. Eco-friendly biopolymer packaging is widely used to ensure the safety and quality of meat products (Chen et al. 2019).

19.5.1 Biodegradable Packaging Materials

Packaging materials that are usually biodegradable are derived from natural resources that are sustainable or food or by-products of agricultural (Fig. 19.2). Studies have shown that animal proteins are highly nutritious and have techno functional properties such as gelling properties, emulsification, water, and oil holding capacity (Pérez-Andrés et al. 2019). These proteins are natural, inexpensive, and abundant, and their remarkable functional properties may make them excellent candidates for the production of biodegradable films. In addition, protein films are fully compostable and provide a source of nitrogen, thus exerting a fertilizing effect during decomposition

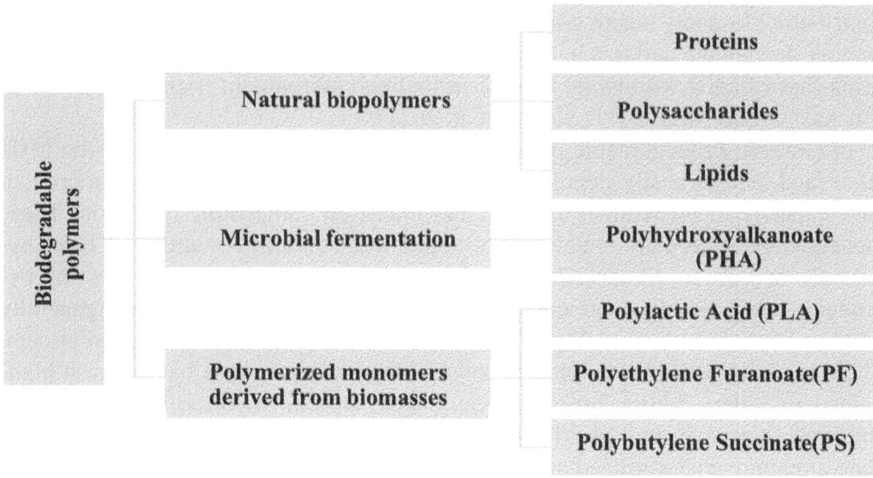

Fig. 19.2 Biodegradable polymers from different sources

in the soil. The biodegradable films are usually produced from proteins (gelatin, collagen, etc.) obtained from animal waste (da Rocha et al. 2014).

19.5.2 Development of Packaging Films from Animals' Proteins

19.5.2.1 Collagen Films

The skin, bones, ligaments, tendons and cartilage of pigs, cattle, and aquatic animals' are rich source of collagen. Their skin and peel have 30–35% protein. The collagen is fibrous protein that give structural support to different organs of body and ensure the strength and flexibility required for effective movement, tissue regeneration, and repair via processes of mechanical and chemical conversion (Sorushanova et al. 2019). Collagen is generally colorless and opaque and exhibits remarkable viscoelastic properties with high tensile strength and low elongation (Avila Rodríguez et al. 2018).

19.5.2.2 Gelatin Films

Gelatin is a water-insoluble protein produced from moderate hydrolysis of collagen (by chemical, enzymatic, or thermal methods). The type of collagen, the source, and the age of the animal are important factors that affect the properties of gelatin of particular interest are, fish skin, bones, scales, cooked filets, solid waste from surimi processing, offal from processed or semi-processed fish products, farmed alligator bones, and giant Red Sea cucumbers are examples of potential marine by-products. It can be used as a source of collagen and gelatin (Gómez-Guillén et al. 2011).

Studies have shown that gelatin films are suitable for food packaging applications due to their high mechanical strength, transparency, and barrier properties (Khodaei et al. 2020). However, the mechanical and inhibitory properties of gelatin films strongly depend on the amino acid composition and molecular weight of the polymer. Different amino acid compositions and processing conditions between species affect the molecular weight distribution of gelatin. According to the US Food and Drug Administration (FDA), EC, FAO and WHO, gelatin is a safe additive for food industry applications and can be widely used in biodegradable or especially edible packaging. Due to high hygrometry, gelatin layers in contact with food and surfaces containing high amounts of moisture show high swelling and dissolution capacity. A microbial transglutaminase, widely used to enhance the mechanical and inhibitory properties of gelatin or protein films, has also been approved as a generally recognized as safe (GRAS) substance by the FDA since 1998.

19.5.3 Keratin Films

It is a structural protein of horns, wool, hoofs, hair, nails, etc., and insoluble in nature. The keratin and creatine proteins exhibit mutual resemblance with each other to perform the functions. The creatine proteins are rich in cysteine (sulfur-containing amino acids) which link the surrounding cysteine residues through intermolecular and intramolecular interactions to build disulfide and hydrogen bridges. These bonds give keratin a crystalline structure, high stiffness, and elasticity potential (Barone and Arikan 2007). It is noticed that 65 million tons of feathers and 2.5 million tons of wool are globally produced every year. In this production, 1.5 million tons of wool is used for low/small value and in textile industry and other part is burnt and employed for landfilling (Shavandi et al. 2017). Various researchers reported that keratin films on mixing with glycerol give clearer films with pronounced mechanical strength (Yin et al. 2013). Although these films are generally very weak but can be used as humidity sensors for rough surfaces being cost effective and porous in nature (Hamouche et al. 2018).

19.5.4 Myofibrillar Proteins

These muscular proteins, on the basis of their solubility in water and location in skeletal muscle are divided into sarcoplasmic, interstitial, and myofibrillar. These form 60% of muscle proteins as the main component and responsible to perform remarkable functions of emulsification, gelling properties, and solubility (Dong et al. 2020).

The research has been conducted to study the effect of processing conditions on muscular proteins derived from fish waste. Later on, these proteins were used in film production to observe its characteristics. It was noticed that the produced films exhibit the uniform structure, low water vapor permeability, good mechanical properties, making it suitable for food packaging materials due to its 35.96% softener, drying temperature of 25.96 °C, and 1.13% protein content. Another study investigated the effects of various fatty acids (palmitic acid, caproic acid, and stearic acid), including surfactant (sodium lauryl sulfate) on the physical properties of fish filet-derived muscular (myofibrillar) proteins As a result, it was found that the addition of surfactants and fatty acids significantly improves the elongation and flexibility of the film and also increases the soluble rate of the prepared film (da Silva Pereira et al. 2019).

19.5.5 Blood Proteins

It is an excellent source of heme iron and essential amino acids and considered as one of the main protein-rich by-products of the meat industry. In present era, blood after collection from animal slaughterhouses is being used in pet food production, biogas production, biotechnology, and biofertilizers manufacturing industries with its less amounts for food applications. The physical properties of the prepared films of blood were studied by several authors at varying concentration of glycerol and plasma protein concentrations. The researcher Nuthong and their team (2009) (Nuthong et al. 2009) prepared films from plasma of pig in first attempt. During study, it was noticed that glycerol decreased the moisture resistance while increasing the transparency and elasticity of the films. It was also observed that the relatively high solubility of blood-based films limit their usage in food applications. The same authors conducted the another study by using glyoxal and caffeic acid as cross-linking agents in order to reduce the solubility of porcine bases plasma membranes in aqueous solutions. The findings were that the use of glyoxal in food is limited due to its toxic effects, while caffeic acid has a negative effect on the appearance of the films. The effects of mixing of chitosan and porcine plasma proteins at different levels were investigated by Samsalee and Sothornvit (2020) in another study. The addition of chitosan increased the mechanical properties and thermal stability while decreasing the water solubility, water vapor permeability, and transparency of the blend films. The same scientists reported that heating of porcine plasma protein solution followed by homogenization tends to improve the seal and mechanical strength of the plasma layer. Furthermore, the mechanical properties and water resistance of the resulting films could be improved by incorporation of microencapsulated turmeric oil or chitosan. These active package help to increase the storage life of packaged rice grains up to 50 days in comparison with untreated plasma-based protein films (40 days) (Samsalee and Sothornvit 2020). Development of biodegradable films from animal waste has been shown in Table 19.4.

Table 19.4 Biodegradable packaging films developed from animal wastes

S. No.	Source	Extraction method	Results	References
1	Fish skin	Acidic extraction	The thickness and elastic behavior of gelatin film were increased by adding palm oil and clove oil can enhance antimicrobial property	e Silva et al. (2021)
2	Chicken Skin	Alkali extraction with NaOH	The rice flour in films enhances water vapor pressure and thermal properties of films and decreases solubility, UV, and light transmission. Blended films with 20% rice flour gave best result	Soo and Sarbon (2018)
3	Bovine bones	Acid treatment	Addition of microbial transglutaminase enhances the molecular weight, stability of network structure, and mechanical strength, and the resultant films were insoluble in water	Ma et al. (2020)
4	Chicken feathers	Alkaline hydrolysis	Keratin from feathers of chicken in biodegradable films showed a tensile strength of 3.62 MPa and Young's Modulus of 1.52 MPa, and elongation at break was 15.8% that proved its suitability for bioplastic films	Liu et al. (2018)
5	Sheep wool keratin	Alkaline mild oxidative method	Keratin from wool of sheep showed increase in transparency, thermal stability, and UV barrier strength	Fernández-d'Arlas (2019)

(continued)

Table 19.4 (continued)

S. No.	Source	Extraction method	Results	References
6	Goat hoof	Soxhlet apparatus	Blood fibrin, keratin, gelatin, mupirocin used in biodegradable films of wound healing. The study showed cell viability, biocompatibility, cell adhesion, and proliferation of blended polymers	Sellappan et al. (2022)

19.5.6 Regulatory Status for Animal Waste Based Packaging Material

The European Union (EU) has developed regulatory policies since the 1970s on the quality control and analysis of food packaging materials to regulate both commercial reasons and consumer health. The Food and Drug Administration (FDA) defined food packaging/contact materials to ensure safety of food under the intended usage conditions. Migration testing must be prerequisite to demonstrate that the packaging material should be safe for use in agricultural product packaging and not transfer toxic substances to the product. According to the EN 17033 standard of European Union, these materials must use as edible films or coatings for food should exhibit GRAS status, with the exception of an allergen and it must be indicated on the label In addition to these, biodegradable mulch films used in horticulture and agriculture made of thermoplastic materials must degrade at least 90% for 48 months at 25 °C (Nieto 2009).Conclusively, petroleum-based plastics are the predominantly used polymers in the food packaging sector over the past decades, and this increased production has raised concerns about environmental contamination. However, innovations for increasing annual production and demand for biodegradable packaging materials in the food and agriculture sector lead to the pronounced production and processing of animal-derived products. These tools, to produce the non-petroleum biodegradable plastics from co-products or even waste, can provide more sustainable solutions to support the concepts of circular bio-economy, and also helpful to optimize the use of natural resources.

19.6 Religious and Ethical Concerns Related to Food Application of Animal Wastes

19.6.1 Religious Concerns

Despite the immense efforts to maximally utilize animal wastes in food sector, consumer has religious and ethical concerns in terms of religious prohibitions, safety of these waste products, and fraudulent usage of these products in food. Although scientific data have shown some of these concerns are meaningless and have not ground. However, concerns and anxiety expressed by the consumer are real and warrant a serious effort by government, researchers, and food manufacturers to diminish these fears and so ensure the sustainable utilization of these products.

Religion is associated with morality, deity, one's worldview of daily life. A person must strictly adhere to religion in relation to anything which he consumes. Animal waste could be a source of valuable bioactive compounds but in many regions of world it is regarded as dirty, dangerous, and disgusted. Halal is a concept that defines anything to be allowed according to the Islamic laws commonly employed with all kind of food products and food ingredients and food contact materials (Herpandi et al. 2011). Halal must cover the whole process beginning from the selection of source material, preparation, storage, handling, transportation, and distribution till consumption point. According to an estimate, there will be more than 31% of Muslims worldwide and to fulfill the consumption needs provision of halal food sources is essential (Hackett et al. 2017). Therefore, there is great need of halal food source authentication due to the strict adherence to religious obligation of consuming halal only. Gelatin and gelatin-based products are considered as mashbooh/doubtful because porcine (haram) gelatin is most abundant so traceability of gelatin is important to safeguard consumer from food frauds and to protect the consumer health. In Europe, around 80% gelatin is produced from pig (haram source) and used further in food manufacturing operations. In majority, slaughtering is not performed according to the Islamic laws which make the product or process overall haram. Halal products are also gaining popularity because of consumer awareness that these products are healthier, safe, and obtained by using humane animal treatments. Thus, due to the growing concerns among consumers, the halal industry is also on rise. Islam forbids dogs, donkeys, while Jews also have a law called kosher. Buddhist and Hinduism are destined to eat vegetables or not to eat cattle. Waste products obtained after slaughtering such as animal hides, cattle and poultry skins, bone, and fish meal are processed further to obtain secondary products such as keratin, gelatin, collagen find application in food industries in manufacturing operations. The halal status of these secondary material depends upon the nature of raw material or practices used. Muslims and Jews do not consume hides, skin, and blood of animals because of religious dictates enshrined in Halal and kosher dietary laws. Quran also explains that Muslims should eat halal or good food, therefore, halal status of food is a major concern for Muslims;

O Muhammad! Tell them; "I do not find in what has been revealed to me anything forbidden for anyone who wants to eat unless it is carrion, outpoured blood and the flesh of swine, all of which is unclean; or that which is profane having been slaughtered in a name other than that of Allah. But whosoever is constrained to it by necessity, neither desiring to disobey nor exceeding the limit of necessity - your Lord is surely All-Forgiving, All-Compassionate". (Al -An"Am:145).

19.6.2 Ethical Concerns

Animal-based secondary food products contain an abundant quantity of proteins with vitamins, minerals, and essential amino acids. The fraudulent usage of these products for economic gains is a significant concern. Partial replacement and adulteration with prohibited ingredients offer a major economic incentive to manufacturer by misleading the consumer. Inconsistency in legislations, misleading labeling leads consumers to doubt the quality of these products. Efforts to ensure the usage of these products are not abused at manufacture level as however the issue has not yet properly addressed.

Vegan community consumes only plant-based diet and avoid consuming animal-based products for ethical reasons. Therefore, it is imperative to adhere strictly with labeling laws compelling food manufacturers to clearly elaborate the usage of these products on label in layman's language to help individuals of such community to select for right food choices. Furthermore, technical labeling is not helpful and needs to be revised declaring the source of food ingredient must be deemed necessary not voluntary.

The belief that animal waste products is highly loaded with pathogenic microflora and toxic residues is another reason people are against utilization of these ingredients in food. The concern is almost genuine as these products get easily contaminated with spoilage microflora through poor processing operations or if inherent the situation can worsen. FDA in 2005, proposed the usage of cattle spinal cord and brains, from those cattle not inspected or not passed for human consumption prohibited to be used in food or feed ingredient to control the spread of Bovine Spongiform Encephalopathy (BSE) (Alao et al. 2017). Animal intestines can be used as food but these are loaded with microorganisms when separated so must be cleaned and boiled before use. In contrast to edible food portion obtained from animals, consumer does not appreciate the animal waste products in disparity. Transglutaminase may improve the texture of secondary products but may create such chemical linkages which make food product indigestible or a threat to health of consumer. Other concerns include danger of consuming untested products, disgust toward animal waste product ("Yuck factor"). The societal challenges surrounding animal waste products are mainly framed primarily in ethical and consumer acceptance, therefore, need is to address genuine concerns and barriers. Existing research data on attitude and acceptability of animal waste derived food products ranging from highly hostile to highly supportive, but limited research in this perspective shows it a source of disagreement.

19.7 Future Prospects

In past few years, global population rise exhibited trends of exponential escalation (Li et al. 2019). In order to meet the ever-increasing demands of food commodities, need of the hour is to scale up the utilization of wastes into sustainable food sources (Kumar et al. 2022). Animal waste can be of good use in this regard to owing to their nutrient-rich composition. In order to safely convert animal wastes into eatables, development novel techniques is the primary task of modern developmental science (Kumar et al. 2022). Protein wastes of animal origin are the future to reduce global malnutrition issues (Colgrave et al. 2021) like Kwashiorkor, marasmus, or marasmic-kwashiorkor (protein energy malnutrition) (Leij-Halfwerk et al. 2019). In addition to it, collagen and keratin isolates of animal waste have inexhaustible biotechnological applications (Timorshina et al. 2022). Due to fibrillar structure, these isolates can be bioengineered in to multiple formats and forms to generate target drug delivery systems, nanoparticles, biodegradable food packaging material, hydrogels, and scaffold regenerative medicine (Varghese et al. 2020).

In food processing industry, animal protein isolates like collagen can be utilized in its native form as collagen biomaterial (food packaging material) and as food additive (Tian et al. 2022). Edible bio-coatings for food commodities like sausages prepared using protein isolates can retain both moisture and oxygen thus preventing lipid oxidation, which ultimately leads to preservation of organoleptic characteristics of stored food (Titov et al. 2020). Keratin-based bioplastic is thermostables and strong moisture resistant making them a suitable material to create food packaging material which can provide an additional protective barrier to food against ultraviolet radiations (Ramakrishnan et al. 2018). Fish oil is highly nutritious but unstable under environmental extremities (Gammone et al. 2018). In order to shield this rich nutritional supplement against the deteriorative impact of ultraviolet radiation, (Tokarczyk et al. 2021) must be kept in safe conditions. Due to emulsifying properties of low molecular weight keratin, it enhances encapsulation stability of fish oil (Mishra et al. 2021).

Similarly, collagen can be isolated from other animal wastes likes eggshells which have higher biosafety with low incidence of allergic and autoimmune reactions which can be used to develop dietary supplements (Xiao et al. 2021). These supplements will support development and upholding of body tissues, skin, hair, and nail tissues particularly in old age people (Senadheera et al. 2020). Development of nutricosmetics is the future use of collagen isolates from animal bones, hides and feathers which can be used to create functional foods like high protein baked crackers (Khodaei et al, 2020; Halim 2021). Similarly, Malaysia Dairy Industry is working on development of nutritious probiotic drinks with added collagen peptides to improve digestion. These peptide collagen drinks can be a revolutionary natural high-quality peptides and fructo-oligosaccharides, thus, improving global nutritional status (Hashim et al. 2015).

19.8 Conclusion

Isolated compounds from animal wastes have been found to have crucial application in food and beverage processing industry. Among them, protein isolates like collagen and keratin are of key significance to their diverse applications as protein dietary supplements, food product coatings, meat processing carriers, edible films, and food additives. Still there one certain safety concerns associated with animal waste utilization in direct food applications, as these wastes harbor potential pathogens, antibiotic resistant bacteria, and other chemical toxins which require advance and specialized procedures for eradication prior food application. Similarly, animal waste pathogen, like swine hepatitis virus, *Cryptosporidium parvum*, and *Salmonella* spp. pose potential risk to human health.

There are certain concerns raised by various religious societies and practitioners regarding application of animal wastes, particularly blood in food intended for human consumption. Although, blood can be used as nutritional additive, pharmaceutical, emulsifier, clarifier, and stabilizer. Although, blood application can be a significant contributor in food value addition but in religion like Islam human consumption of blood is strictly prohibited due spongiform encephalopathy transmission. Animal wastes can be used in food products but require advanced levels of pretreatment to ensure safety and maximum nutritional outputs for consumers.

References

Aberoumand A (2012) Comparative study between different methods of collagen extraction from fish and its properties. World Appl Sci J 16(3):316–319

Akram AN, Zhang C (2020) Effect of ultrasonication on the yield, functional and physicochemical characteristics of collagen-II from chicken sternal cartilage. Food Chem 307:125544

Alao BO, Falowo AB, Chulayo A, Muchenje V (2017) The potential of animal by-products in food systems: production, prospects and challenges. Sustainability 9(7):1089

Alfian RL, Iskandar J, Iskandar BS, Ermandara DP, Mulyanto D, Partasasmita R (2020) Fish species, traders, and trade in traditional market: case study in Pasar Baru, Balikpapan City, East Kalimantan, Indonesia. Biodiversitas J Biol Divers 21(1)

Ali AAM, Ahmed EA, Ahmed NM (2018) Study on the safety and chemical composition of meat and bone meal. Sudan University of Science and Technology

Álvarez J, Otero L, Lema J (2010) A methodology for optimising feed composition for anaerobic co-digestion of agro-industrial wastes. Biores Technol 101(4):1153–1158

Amani H, Shahbazi M-A, D'Amico C, Fontana F, Abbaszadeh S, Santos HA (2021) Microneedles for painless transdermal immunotherapeutic applications. J Control Release 330:185–217

Ameer K, Jiang G-H, Amir RM, Eun J-B (2020) Antioxidant potential of *Stevia rebaudiana* (Bertoni). In: Pathology. Elsevier, pp 345–356

Araújo C, Rodrigues A, Joele MP, Araújo E, Lourenço L (2018a) Optimizing process parameters to obtain a bioplastic using proteins from fish byproducts through the response surface methodology. Food Packag Shelf Life 16:23–30

Araújo ÍBdS, Bezerra TKA, Nascimento ESd, Gadelha CAdA, Santi-Gadelha T, Madruga MS (2018b) Optimal conditions for obtaining collagen from chicken feet and its characterization. Food Sci Techn 38:167–173

Atuah L, Hodson ME (2011) A comparison of the relative toxicity of bone meal and other P sources used as remedial treatments to the earthworm *Eisenia fetida*. Pedobiologia 54:S181–S186
Avila Rodríguez MI, Rodríguez Barroso LG, Sánchez ML (2018) Collagen: a review on its sources and potential cosmetic applications. J Cosmet Dermatol 17(1):20–26
Barone JR, Arikan O (2007) Composting and biodegradation of thermally processed feather keratin polymer. Polym Degrad Stab 92(5):859–867
Batini N (2019) Transforming agri-food sectors to mitigate climate change: the role of green finance. Vierteljahrshefte Zur Wirtschaftsforschung 88(3):7–42
Beghetto V, Gatto V, Conca S, Bardella N, Scrivanti A (2019) Polyamidoamide dendrimers and cross-linking agents for stabilized bioenzymatic resistant metal-free bovine collagen. Molecules 24(19):3611
Benayahu D, Pomeraniec L, Shemesh S, Heller S, Rosenthal Y, Rath-Wolfson L, Benayahu Y (2020) Biocompatibility of a marine collagen-based scaffold in vitro and in vivo. Mar Drugs 18(8):420
Binsi P, Zynudheen A (2019) Functional and nutraceutical ingredients from marine resources. In: Value-added ingredients and enrichments of beverages. Elsevier, pp 101–171
Birania S, Kumar S, Kumar N, Attkan AK, Panghal A, Rohilla P, Kumar R (2022) Advances in development of biodegradable food packaging material from agricultural and agro-industry waste. J Food Process Eng 45(1):e13930
Blanco M, Vázquez JA, Pérez-Martín RI, Sotelo CG (2019) Collagen extraction optimization from the skin of the small-spotted catshark (*S. canicula*) by response surface methodology. Mar Drugs 17(1):40
Bledsoe G, Bledsoe C, Rasco B (2003) Caviars and fish roe products
Brandelli A, Sala L, Kalil SJ (2015) Microbial enzymes for bioconversion of poultry waste into added-value products. Food Res Int 73:3–12
Cao C, Xiao Z, Ge C, Wu Y (2021) Animal by-products collagen and derived peptide, as important components of innovative sustainable food systems—a comprehensive review. Crit Rev Food Sci Nutr, 1–25
Cao H, Chen M-M, Liu Y, Liu Y-Y, Huang Y-Q, Wang J-H, Chen J-D, Zhang Q-Q (2015) Fish collagen-based scaffold containing PLGA microspheres for controlled growth factor delivery in skin tissue engineering. Colloids Surf, B 136:1098–1106
Chandra M, Shamasundar B (2015) Rheological properties of gelatin prepared from the swim bladders of freshwater fish Catla catla. Food Hydrocolloids 48:47–54
Chen M, Cheng J, Zhang J, Chen Y, Zeng H, Xue L, Lei T, Pang R, Wu S, Wu H, Zhang S, Wei ZY, Ding Y, Wu Q (2019) Isolation, potential virulence, and population diversity of Listeria monocytogenes from meat and meat products in China. Front Microbiol 10:946
Chilakamarry CR, Mahmood S, Saffe SNBM, Arifin MAB, Gupta A, Sikkandar MY, Begum SS, Narasaiah B (2021) Extraction and application of keratin from natural resources: a review. 3 Biotech 11(5):1–12
Choksawangkarn W, Phiphattananukoon S, Jaresitthikunchai J, Roytrakul S (2018) Antioxidative peptides from fish sauce by-product: isolation and characterization. Agric Nat Resour 52(5):460–466
Chotphruethipong L, Aluko RE, Benjakul S (2019) Enhanced Asian sea bass skin defatting using porcine lipase with the aid of pulsed electric field pretreatment and vacuum impregnation. Process Biochem 86:58–64
Chuaychan S, Benjakul S, Kishimura H (2015) Characteristics of acid- and pepsin-soluble collagens from scale of seabass (*Lates calcarifer*). LWT Food Sci Technol 63(1):71–76. https://doi.org/10.1016/j.lwt.2015.03.002
Cliche S, Amiot J, Avezard C, Gariepy C (2003) Extraction and characterization of collagen with or without telopeptides from chicken skin. Poult Sci 82(3):503–509. https://doi.org/10.1093/ps/82.3.503

Colgrave ML, Dominik S, Tobin AB, Stockmann R, Simon C, Howitt CA, Belobrajdic DP, Paull C, Vanhercke T (2021) Perspectives on future protein production. J Agric Food Chem 69(50):15076–15083

Coutand M, Cyr M, Deydier E, Guilet R, Clastres P (2008) Characteristics of industrial and laboratory meat and bone meal ashes and their potential applications. J Hazard Mater 150(3):522–532

Cozzo D, Smart NP (2020) Sashimi: cutting up CSI-FiSh secret keys to produce an actively secure distributed signing protocol. Paper presented at the International Conference on Post-Quantum Cryptography

da Rocha M, Loiko MR, Tondo EC, Prentice C (2014) Physical, mechanical and antimicrobial properties of Argentine anchovy (*Engraulis anchoita*) protein films incorporated with organic acids. Food Hydrocolloids 37:213–220

da Silva Pereira GV, da Silva Pereira GV, de Araujo EF, Neves EMPX, Joele MRSP, Lourenço LdFH (2019) Optimized process to produce biodegradable films with myofibrillar proteins from fish byproducts. Food Packag Shelf Life 21:100364

de Almeida PF, de Araújo MGO, Santana JCC (2012) Collagen extraction from chicken feet for jelly production. Acta Scientiarum. Technol 34(3):345–351

de Menezes CLA, Santos RdC, Santos MV, Boscolo M, da Silva R, Gomes E, da Silva RR (2021) Industrial sustainability of microbial keratinases: production and potential applications. World J Microbiol Biotech 37(5):1–17

Desai AS, Brennan M, Gangan S, Brennan C (2022) Utilization of fish waste as a value-added ingredient: sources and bioactive properties of fish protein hydrolysate. In: Sustainable fish production and processing. Elsevier, pp 203–225

Dhakal D, Koomsap P, Lamichhane A, Sadiq MB, Anal AK (2018) Optimization of collagen extraction from chicken feet by papain hydrolysis and synthesis of chicken feet collagen based biopolymeric fibres. Food Biosci 23:23–30

Dong M, Xu Y, Zhang Y, Han M, Wang P, Xu X, Zhou G (2020) Physicochemical and structural properties of myofibrillar proteins isolated from pale, soft, exudative (PSE)-like chicken breast meat: Effects of pulsed electric field (PEF). Innov Food Sci Emerg Technol 59:102277

Du L, Betti M (2016) Chicken collagen hydrolysate cryoprotection of natural actomyosin: Mechanism studies during freeze-thaw cycles and simulated digestion. Food Chem 211:791–802. https://doi.org/10.1016/j.foodchem.2016.05.092

Duque-Acevedo M, Belmonte-Urena LJ, Cortés-García FJ, Camacho-Ferre F (2020) Agricultural waste: review of the evolution, approaches and perspectives on alternative uses. Glob Ecol Conserv 22:e00902

e Silva NdS, de Souza Farias F, dos Santos Freitas MM, Hernández EJGP, Dantas VV, Oliveira MEC, Joele MRSP, Lourenço LdFH (2021) Artificial intelligence application for classification and selection of fish gelatin packaging film produced with incorporation of palm oil and plant essential oils. Food Packag Shelf Life 27:100611

Eslahi N, Dadashian F, Nejad NH (2013) An investigation on keratin extraction from wool and feather waste by enzymatic hydrolysis. Prep Biochem Biotechnol 43(7):624–648

Fagbemi OD, Sithole B, Tesfaye T (2020) Optimization of keratin protein extraction from waste chicken feathers using hybrid pre-treatment techniques. Sustain Chem Pharm 17:100267

Felician FF, Xia C, Qi W, Xu H (2018) Collagen from marine biological sources and medical applications. Chem Biodivers 15(5):e1700557

Feng M, Betti M (2017) Transepithelial transport efficiency of bovine collagen hydrolysates in a human Caco-2 cell line model. Food Chem 224:242–250. https://doi.org/10.1016/j.foodchem.2016.12.044

Fernández-d'Arlas B (2019) Tough and functional cross-linked bioplastics from sheep wool keratin. Sci Rep 9(1):1–12

Fernandez-Hervas F, Celma P, Punti I, Cisa J, Cot J, Marsal A, Manich A (2007) The enzyme activity of trypsen on sheepskin trimmings in a two-step collagen extraction process. J Am Leather Chem Assoc 102(01):1–9

Fustier P, Achouri A, Taherian AR, Britten M, Pelletier M, Sabik H, Villeneuve S, Mondor M (2015) Protein–protein multilayer oil-in-water emulsions for the microencapsulation of flaxseed oil: effect of whey and fish gelatin concentration. J Agric Food Chem 63(42):9239–9250

Gammone MA, Riccioni G, Parrinello G, D'orazio N (2018) Omega-3 polyunsaturated fatty acids: benefits and endpoints in sport. Nutrients 11(1):46

Ghaly A, Ramakrishnan V, Brooks M, Budge S, Dave D (2013) Fish processing wastes as a potential source of proteins. Amino acids and oils: a critical review. Microb Biochem Technol 5(4):107–129

Gómez-Guillén M, Giménez B, López-Caballero Ma, Montero M (2011) Functional and bioactive properties of collagen and gelatin from alternative sources: a review. Food Hydrocolloids 25(8):1813–1827

Grønlien KG, Pedersen ME, Sanden KW, Høst V, Karlsen J, Tønnesen HH (2019) Collagen from Turkey (*Meleagris gallopavo*) tendon: a promising sustainable biomaterial for pharmaceutical use. Sustain Chem Pharm 13:100166

Gupta A, Kamarudin NB, Kee CYG, Yunus RBM (2012) Extraction of keratin protein from chicken feather. J Chem Chem Eng 6(8):732

Hackett C, Stonawski M, McClendon D (2017) The changing global religious landscape. J Progr Res Chem, 1–45

Hakim T, Pratiwi A, Jamhari J, Fitriyanto N, Rusman R, Abidin M, Erwanto Y (2021) Extraction of collagen from the skin of kacang goat and production of its hydrolysate as an inhibitor of angiotensin converting enzyme. Trop Anim Sci J 44(2):222–228

Halim A (2021) Fortification of rice crackers made from brown rice flour and black rice flour with protein from oyster and mussel powder. A dissertation submitted in partial fulfilment of the requirements for the Degree of Master of Science at Lincoln University. Lincoln University

Hamouche H, Makhlouf S, Chaouchi A, Laghrouche M (2018) Humidity sensor based on keratin bio polymer film. Sens Actuators, A 282:132–141

Hashim P, Ridzwan MM, Bakar J, Hashim MD (2015) Collagen in food and beverage industries. Int Food Res J 22(1):1

Henchion M, McCarthy M, O'Callaghan J (2016) Transforming beef by-products into valuable ingredients: which spell/recipe to use? Front Nutr 3:53

Hendriks W, Butts C, Thomas D, James K, Morel P, Verstegen M (2002) Nutritional quality and variation of meat and bone meal. Asian Australas J Anim Sci 15(10):1507–1516

Herpandi H, Huda N, Adzitey F (2011) Fish bone and scale as a potential source of halal gelatin. J Fish Aquat Sci 6(4):379–389

Hong H, Fan H, Chalamaiah M, Wu J (2019) Preparation of low-molecular-weight, collagen hydrolysates (peptides): current progress, challenges, and future perspectives. Food Chem 301:125222

Huang T, Tu Z, Zou Z, Shangguan X, Wang H, Bansal N (2020) Glycosylated fish gelatin emulsion: rheological, tribological properties and its application as model coffee creamers. Food Hydrocolloids 102:105552

Huda N, Seow E, Normawati M, Aisyah NN, Fazilah A, Easa A (2013) Effect of duck feet collagen addition on physicochemical properties of surimi. Int Food Res J 20(2)

Jafari H, Lista A, Siekapen MM, Ghaffari-Bohlouli P, Nie L, Alimoradi H, Shavandi A (2020) Fish collagen: extraction, characterization, and applications for biomaterials engineering. Polymers 12(10):2230

Ji Y, Chen J, Lv J, Li Z, Xing L, Ding S (2014) Extraction of keratin with ionic liquids from poultry feather. Sep Purif Technol 132:577–583

Jiang X, Yang Z, Han L, Liu X (2011) Discrimination of meat and bone meal in concentrate supplement by near-infrared microscopic imaging. Nongye Jixie Xuebao= Transa Chin Soc Agric Mach 42(7):155–159

Jiménez Vázquez J, San Martín Martínez E (2019) Collagen and elastin scaffold by electrospinning for skin tissue engineering applications. J Mater Res 34(16):2819–2827. https://doi.org/10.1557/jmr.2019.233

Jin H-X, Xu H-P, Li Y, Zhang Q-W, Xie H (2019) Preparation and evaluation of peptides with potential antioxidant activity by microwave assisted enzymatic hydrolysis of collagen from sea cucumber acaudina molpadioides obtained from Zhejiang Province in China. Mar Drugs 17(3):169

Kakkar P, Madhan B, Shanmugam G (2014) Extraction and characterization of keratin from bovine hoof: a potential material for biomedical applications. Springerplus 3(1):1–9

Kalinova G, Marinova M, Mechkarova P, Mladenova D, Grigorova E (2017) Determination and assessment of the content of collagen in minced meat and meat products offered at the Bulgarian Market. Bul J Vet Med 20(1):437–441

Khodaei D, Álvarez C, Mullen AM (2021) Biodegradable packaging materials from animal processing co-products and wastes: an overview. Polymers 13(15):2561

Khodaei D, Hamidi-Esfahani Z, Lacroix M (2020) Gelatin and low methoxyl pectin films containing probiotics: film characterization and cell viability. Food Biosci 36:100660

Kumar P, Mehta N, Abubakar AA, Verma AK, Kaka U, Sharma N, Sazili AQ, Pateiro M, Kumar M, Lorenzo JM (2022) Potential alternatives of animal proteins for sustainability in the food sector. Food Rev Int, 1–26

Kurtela A, Antolović N (2019) The problem of plastic waste and microplastics in the seas and oceans: impact on marine organisms. Croatian J Fish 77(1):51–56

Lee H, Noh K, Lee SC, Kwon I-K, Han D-W, Lee I-S, Hwang Y-S (2014) Human hair keratin and its-based biomaterials for biomedical applications. Tissue Eng Regen Med 11(4):255–265

Leij-Halfwerk S, Verwijs MH, van Houdt S, Borkent JW, Guaitoli P, Pelgrim T, Heymans MW, Power L, Visser M, Corish CA, de van der Schueren MAE (2019) Prevalence of protein-energy malnutrition risk in European older adults in community, residential and hospital settings, according to 22 malnutrition screening tools validated for use in adults $\geq$ 65 years: a systematic review and meta-analysis. Maturitas 126:80–89

Li B, Yao J, Niu J, Liu J, Wang L, Feng M, Sun Y (2019) Study on the effect of organic phosphonic compounds on disulfide bonds in wool. Text Res J 89(13):2682–2693

Lin P, Hua N, Hsu Y-C, Kan K-W, Chen J-H, Lin Y-H, Kuan C-M (2020) Oral collagen drink for antiaging: antioxidation, facilitation of the increase of collagen synthesis, and improvement of protein folding and DNA repair in human skin fibroblasts. Oxidative Med Cell Longevity 2020

Liu D, Liang L, Regenstein JM, Zhou P (2012) Extraction and characterisation of pepsin-solubilised collagen from fins, scales, skins, bones and swim bladders of bighead carp (*Hypophthalmichthys nobilis*). Food Chem 133(4):1441–1448

Liu S, Huang K, Yu H, Wu F (2018) Bioplastic based on 1, 8-octanediol-plasticized feather keratin: a material for food packaging and biomedical applications. J Appl Polym Sci 135(30):46516

Luckachan GE, Pillai C (2011) Biodegradable polymers-a review on recent trends and emerging perspectives. J Polym Environ 19(3):637–676

Lynch SA, Mullen AM, O'Neill E, Drummond L, Álvarez C (2018) Opportunities and perspectives for utilisation of co-products in the meat industry. Meat Sci 144:62–73

Ma Y, Yang R, Zhao W (2020) Innovative water-insoluble edible film based on biocatalytic crosslink of gelatin rich in glutamine. Foods 9(4):503

Mahboob S (2015) Isolation and characterization of collagen from fish waste material-skin, scales and fins of *Catla catla* and *Cirrhinus mrigala*. J Food Sci Technol 52(7):4296–4305

Mendez A, Dale N (1998) Rapid assay to estimate calcium and phosphorus in meat and bone meal. J Appl Poultry Res 7(3):309–312

Meng D, Tanaka H, Kobayashi T, Hatayama H, Zhang X, Ura K, Yunoki S, Takagi Y (2019) The effect of alkaline pretreatment on the biochemical characteristics and fibril-forming abilities of types I and II collagen extracted from bester sturgeon by-products. Int J Biol Macromol 131:572–580

Millati R, Cahyono RB, Ariyanto T, Azzahrani IN, Putri RU, Taherzadeh MJ (2019) Agricultural, industrial, municipal, and forest wastes: an overview. Sustain Resour Recovery Zero Waste Approaches, 1–22

Mishra A, Pradhan D, Biswasroy P, Kar B, Ghosh G, Rath G (2021) Recent advances in colloidal technology for the improved bioavailability of the nutraceuticals. J Drug Deliv Sci Technol 65:102693

Mohanty AK, Misra M, Drzal LT (2005) Natural fibers, biopolymers, and biocomposites. CRC Press

Mohseni F, Goli SAH (2019) Encapsulation of flaxseed oil in the tertiary conjugate of oxidized tannic acid-gelatin and flaxseed (*Linum usitatissimum*) mucilage. Int J Biol Macromol 140:959–964

Mora L, Toldrá-Reig F, Reig M, Toldrá F (2019) Possible uses of processed slaughter byproducts. In: Sustainable meat production and processing. Elsevier, pp 145–160

Muthukumar T, Sankari D, Tamil Selvi A, Sastry T (2014) Preparation, characterization, and in vitro bioactivity of Bixa Orellana extract-impregnated collagen microspheres. J Mater Sci 49(16):5730–5737

Nabijon N, Ahmed MR, Adkham R, Heng Q (2017) Extraction of collagen from cattle skin and synthesis of collagen based flame retardant composition and introduction into cellulosic textile material by graft copolymerization. Asian J Chem 29(11):2470–2474

Nawaz A, Xiong Z, Xiong H, Chen L, Wang Pk, Ahmad I, Hu C, Irshad S, Ali SW (2019) The effects of fish meat and fish bone addition on nutritional value, texture and microstructure of optimised fried snacks. Int J Food Sci Technol 54(4):1045–1053

Nieto MB (2009) Structure and function of polysaccharide gum-based edible films and coatings. In: Edible films and coatings for food applications. Springer, pp 57–112

Nurubhasha R, Sampath Kumar N, Thirumalasetti SK, Simhachalam G, Dirisala VR (2019) Extraction and characterization of collagen from the skin of *Pterygoplichthys pardalis* and its potential application in food industries. Food Sci Biotechnol 28(6):1811–1817

Nuthong P, Benjakul S, Prodpran T (2009) Effect of some factors and pretreatment on the properties of porcine plasma protein-based films. LWT Food Sci Technol 42(9):1545–1552

Pan F, Lu Z, Tucker I, Hosking S, Petkov J, Lu JR (2016) Surface active complexes formed between keratin polypeptides and ionic surfactants. J Colloid Interface Sci 484:125–134

Pérez-Andrés JM, Álvarez C, Cullen P, Tiwari BK (2019) Effect of cold plasma on the techno-functional properties of animal protein food ingredients. Innov Food Sci Emerg Technol 58:102205

Pleissner D, Wimmer R, Eriksen NT (2011) Quantification of amino acids in fermentation media by isocratic HPLC analysis of their α-hydroxy acid derivatives. Anal Chem 83(1):175–181

Potera C (2009) Extra protection for pregnant women: calcium supplement reduces blood lead. National Institute of Environmental Health Sciences

Ramakrishnan N, Sharma S, Gupta A, Alashwal BY (2018) Keratin based bioplastic film from chicken feathers and its characterization. Int J Biol Macromol 111:352–358

Regubalan B, Pandit P, Maiti S, Nadathur GT, Mallick A (2018) Potential bio-based edible films, foams, and hydrogels for food packaging. In: Bio-based materials for food packaging. Springer, pp 105–123

Ross AC, Manson JE, Abrams SA, Aloia JF, Brannon PM, Clinton SK, Durazo-Arvizu RA, Gallagher JC, Gallo RL, Jones G, Kovacs CS, Mayne ST, Rosen CJ, Shapses SA (2011) The 2011 dietary reference intakes for calcium and vitamin D: what dietetics practitioners need to know. J Am Diet Assoc 111(4):524–527

Sadh PK, Duhan S, Duhan JS (2018) Agro-industrial wastes and their utilization using solid state fermentation: a review. Bioresour Bioprocess 5(1):1–15

Said M (2018) Synthesis of collagen from Bali cattle's hide using a combination of acid and alkali on the extracting process. J Indonesian Trop Anim Agric 43(3):247–256

Samsalee N, Sothornvit R (2020) Characterization of food application and quality of porcine plasma protein-based films incorporated with chitosan or encapsulated turmeric oil. Food Bioprocess Technol 13(3):488–500

Santana R, Perrechil F, Sato A, Cunha R (2011) Emulsifying properties of collagen fibers: effect of pH, protein concentration and homogenization pressure. Food Hydrocolloids 25(4):604–612

Sellappan LK, Anandhavelu S, Doble M, Perumal G, Jeon J-H, Vikraman D, Kim H-S (2022) Biopolymer film fabrication for skin mimetic tissue regenerative wound dressing applications. Int J Polym Mater Polym Biomater 71(3):196–207

Senadheera TR, Dave D, Shahidi F (2020) Sea cucumber derived type I collagen: a comprehensive review. Mar Drugs 18(9):471

Seo J, Kim M-J, Jeon S-O, Oh D-H, Yoon K-H, Choi YW, Bashyal S, Lee S (2018) Enhanced topical delivery of fish scale collagen employing negatively surface-modified nanoliposome. J Pharm Investig 48(3):243–250

Shah A, Tyagi S, Bharagava RN, Belhaj D, Kumar A, Saxena G, Saratale GD, Mulla SI (2019) Keratin production and its applications: current and future perspective. In: Keratin as a protein biopolymer, pp 19–34

Shaikh S, Yaqoob M, Aggarwal P (2021) An overview of biodegradable packaging in food industry. Curr Res Food Sci 4:503–520

Shavandi A, Silva TH, Bekhit AA, Bekhit AE-DA (2017) Keratin: dissolution, extraction and biomedical application. Biomater Sci 5(9):1699–1735

Shenderyuk VI, Bykowski PJ (2020) Salting and marinating of fish. In: Seafood: resources, nutritional composition, and preservation. CRC Press.

Sidhu KS (2003) Health benefits and potential risks related to consumption of fish or fish oil. Regul Toxicol Pharmacol 38(3):336–344

Silva J, Ribeiro K, Silva J, Cahú T, Bezerra R (2014) Utilization of tilapia processing waste for the production of fish protein hydrolysate. Anim Feed Sci Technol 196:96–106

Soo P, Sarbon N (2018) Preparation and characterization of edible chicken skin gelatin film incorporated with rice flour. Food Packag Shelf Life 15:1–8

Sorushanova A, Delgado LM, Wu Z, Shologu N, Kshirsagar A, Raghunath R, Mullen AM, Bayon Y, Pandit A, Raghunath M, Zeugolis DI (2019) The collagen suprafamily: from biosynthesis to advanced biomaterial development. Adv Mater 31(1):1801651

Sousa RO, Martins E, Carvalho DN, Alves AL, Oliveira C, Duarte ARC, Silva TH, Reis RL (2020) Collagen from Atlantic cod (*Gadus morhua*) skins extracted using CO_2 acidified water with potential application in healthcare. J Polym Res 27(3):1–9

Sousa SC, Fragoso SP, Penna CR, Arcanjo NM, Silva FA, Ferreira VCS, Barreto MDS, Araújo IBS (2017) Quality parameters of frankfurter-type sausages with partial replacement of fat by hydrolyzed collagen. LWT Food Sci Technol 76:320–325

Subhan F, Hussain Z, Tauseef I, Shehzad A, Wahid F (2021) A review on recent advances and applications of fish collagen. Crit Rev Food Sci Nutr 61(6):1027–1037

Suurs P, van den Brand H, Daamen WF, Barbut S (2022) Properties of different poultry skins sources in relation to co-extruded sausage casings. Food Hydrocolloids 125:107434. https://doi.org/10.1016/j.foodhyd.2021.107434

Tangke U, Katiandagho B, Rochmady R (2020) Nutritional adequacy rate (RDA) and nutritional value information (ING) of Tuna Kering Kayu Fish canned with Tuna fish bone flour substitution. Agrikan: Jurnal Agribisnis Perikanan 13(2):352–357

Tian X, Zhao K, Teng A, Li Y, Wang W (2022) A rethinking of collagen as tough biomaterials in meat packaging: assembly from native to synthetic. Crit Rev Food Sci Nutr, 1–21

Timorshina S, Popova E, Osmolovskiy AJP (2022) Sustainable applications of animal waste proteins. Polymers 14(8):1601

Titov E, Sokolov AY, Litvinova E (2020) Structural and mechanical aspects of creating coatings based on biopolymers. Paper presented at the Materials Science Forum

okarczyk G, Bienkiewicz G, Biernacka P (2021) Susceptibility to oxidation of selected freshwater fish species lipids as a potential source of fish oil in dietary supplements. Int J Food Sci 2021

Tsykhanovska I, Alexandrov A, Lazarieva T (2020) Functional and technological properties of the food additive magnetofood in the production of marshmallows

Varghese SA, Rangappa SM, Siengchin S, Parameswaranpillai J (2020) Natural polymers and the hydrogels prepared from them. In: Hydrogels based on natural polymers. Elsevier, pp 17–47

Veeruraj A, Arumugam M, Balasubramanian T (2013) Isolation and characterization of thermostable collagen from the marine eel-fish (*Evenchelys macrura*). Process Biochem 48(10):1592–1602

Waheed M, Butt MS, Shehzad A, Adzahan NM, Shabbir MA, Suleria AR, Aadil RM (2019) Eggshell calcium: a cheap alternative to expensive supplements. Trends Food Sci Technol 91:219–230

Wahyuningsih R, Nurliyani R, Pertiwiningrum A, Rohman A, Fitriyanto NA, Erwanto Y (2018a) Optimization of acid soluble collagen extraction from Indonesian local "Kacang" goat skin and physico-chemical properties characterization. Chem Eng Trans 63:703–708

Wahyuningsih R, Rusman, Nurliyani, Pertiwiningrum A, Rohman A, Fitriyanto NA, Erwanto Y (2018b) Optimization of conditions for extraction of pepsin-soluble collagen from Indonesian local "Kacang" goatskin by response surface methodology. Am J Anim Vet Sci 13(2). https://doi.org/10.3844/ajavsp.2018.70.75

Wang B, Yang W, McKittrick J, Meyers MA (2016) Keratin: structure, mechanical properties, occurrence in biological organisms, and efforts at bioinspiration. Prog Mater Sci 76:229–318

Watanabe T, Aoki H, Watanabe K, Maita M, Yamagata Y, Satoh S (2001) Quality evaluation of different types of non-fish meal diets for yellowtail. Fish Sci 67(3):461–469

Wood JF, Capper BS, Nicolaides L (1985) Preparation and evaluation of diets containing fish silage, cooked fish preserved with formic acid and low-temperature-dried fish meal as protein sources for mirror carp (Cyprinus carpio). Aquaculture 44(1):27–40

Xiao N, Huang X, He W, Yao Y, Wu N, Xu M, Du H, Zhao Y, Tu Y (2021) A review on recent advances of egg byproducts: preparation, functional properties, biological activities and food applications. Food Res Int 147:110563

Yang X, Li A, Li X, Sun L, Guo Y (2020) An overview of classifications, properties of food polysaccharides and their links to applications in improving food textures. Trends Food Sci Technol 102:1–15

Yilmaz M, Ozmen SF (2020) Radiological risk assessment of fish feed and feed raw materials. Aquac Res 51(6):2190–2196

Yin X-C, Li F-Y, He Y-F, Wang Y, Wang R-M (2013) Study on effective extraction of chicken feather keratins and their films for controlling drug release. Biomater Sci 1(5):528–536

Zarubin N, Strokova N, Kharenko E, Bredikhina O (2020) Protein hydrolysate as a product of biotechnological processing of fish filleting waste. Int Multidiscip Sci GeoConf 20(6.1):295–301

Zarubin NY, Kharenko EN, Bredikhina OV, Arkhipov LO, Zolotarev KV, Mikhailov AN, Nakhod VI, Mikhailova MV (2021) Application of the Gadidae fish processing waste for food grade gelatin production. Mar Drugs 19(8):455

Zhang X, Xu S, Shen L, Li G (2020) Factors affecting thermal stability of collagen from the aspects of extraction, processing and modification. J Leather Sci Eng 2(1):19. https://doi.org/10.1186/s42825-020-00033-0

Zhang Y, Ma L, Cai L, Liu Y, Li J (2017) Effect of combined ultrasonic and alkali pretreatment on enzymatic preparation of angiotensin converting enzyme (ACE) inhibitory peptides from native collagenous materials. Ultrason Sonochem 36:88–94. https://doi.org/10.1016/j.ultsonch.2016.11.008

Zhang Y, Olsen K, Grossi A, Otte J (2013) Effect of pretreatment on enzymatic hydrolysis of bovine collagen and formation of ACE-inhibitory peptides. Food Chem 141(3):2343–2354

Zhong Y, Godwin P, Jin Y, Xiao H (2020) Biodegradable polymers and green-based antimicrobial packaging materials: a mini-review. Adv Ind Eng Polym Res 3(1):27–35

Znamirowska A, Szajnar K, Pawlos M (2020) Probiotic fermented milk with collagen. Dairy 1(2):126–134

Zou Y, Yang H, Zhang X, Xu P, Jiang D, Zhang M, Xu W, Wang D (2020) Effect of ultrasound power on extraction kinetic model, and physicochemical and structural characteristics of collagen from chicken lung. Food Prod Process and Nutr 2(1):1–12

Chapter 20
Manure-Associated Veterinary Antibiotics; Ecological Consequences and Mitigation Strategies

Muhammad Adil, Amar Nasir, Sher Zaman Safi, Muhammad Arshad, Ans Nadeem, and Aftab Hussain

Abstract The worldwide application of veterinary antibiotics in animal production and health ranges from curative to non-therapeutic uses. Typically, used veterinary antibiotics include penicillins, cephalosporins, tetracyclines, sulfonamides, macrolides and fluoroquinolones. The consumption-wise top three veterinary antibiotics are tetracyclines, sulfonamides and macrolides, respectively. Majority of the antibiotics are eliminated either unaltered or in terms of metabolites in the urine or feces of treated animals. Manure-based veterinary antibiotics are fated to be taken up by soil components, broken down by soil microbes, absorbed by the plants, moved to ground and surface waters via runoff and leaching, and eventually damage both terrestrial and aquatic ecosystems. Environmental contamination of residual veterinary antibiotics is offering prospective environmental and health hazards, including the occurrence and propagation of antibiotic-resistant pathogens, contamination of food products with antibiotic residues, and detrimental effects on non-target organisms. Prudent use of veterinary antibiotics, anaerobic fermentation, composting, bioremediation and oil-capture technique are recommended for the mitigation of manure-derived residual veterinary antibiotics.

M. Adil (✉)
Pharmacology & Toxicology Section, University of Veterinary & Animal Sciences, Lahore, Jhang Campus, Jhang 35200, Pakistan
e-mail: muhammad.adil@uvas.edu.pk

A. Nasir · A. Nadeem · A. Hussain
Department of Clinical Sciences, University of Veterinary & Animal Sciences, Lahore, Jhang Campus, Jhang 35200, Pakistan

S. Z. Safi
Faculty of Medicine, Bioscience and Nursing, MAHSA University, Selangor, Malaysia

M. Arshad
Department of Basic Sciences, University of Veterinary and Animal Sciences, Lahore, Jhang-Campus, Jhang, Pakistan

M. Arshad (ed.), *Climate Changes Mitigation and Sustainable Bioenergy Harvest Through Animal Waste*, https://doi.org/10.1007/978-3-031-26224-1_20

Graphical Abstract

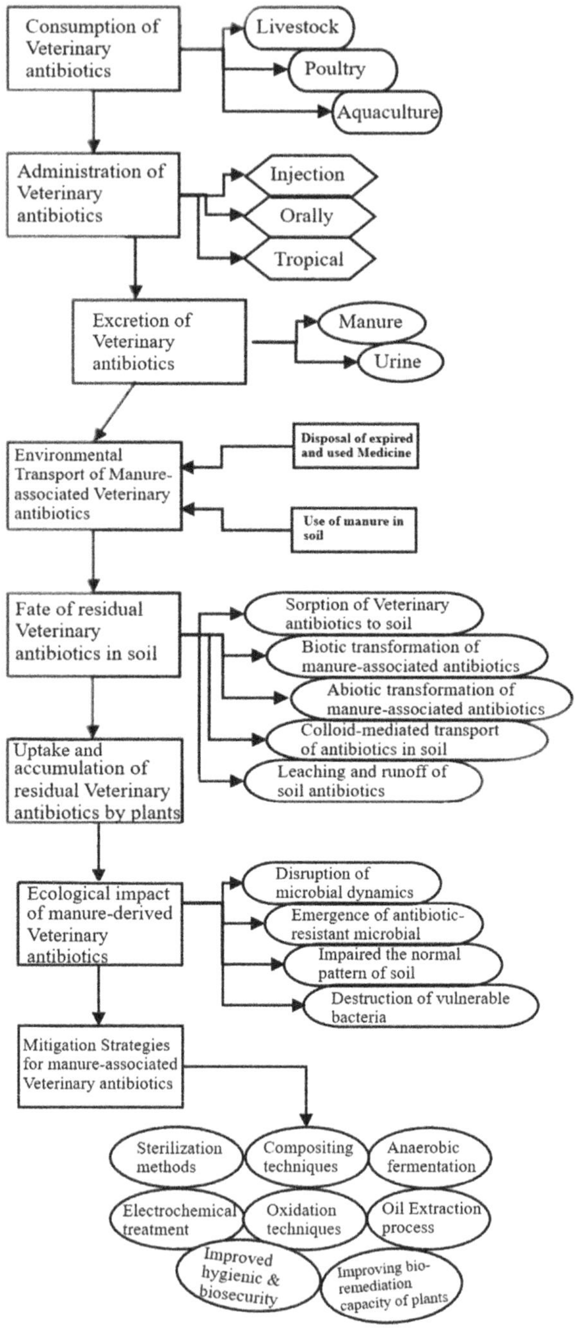

Keywords Veterinary antibiotics · Environmental contaminants · Animal manure

20.1 Introduction

The necessity of animal-derived food has immensely increased owing to rising global population coupled with a shift in dietary preferences toward meat consumption (Zilio et al. 2020). In response, livestock production has accelerated to cope with the consumers' demand and food security. It has led to the emergence of environmental issues including the residual spread of veterinary antibiotics due to their non-pragmatic use in enhancing animal growth and preventing ailments (Ye et al. 2018; Gurmessa et al. 2020). Antimicrobials have been extensively employed to boost animal growth, and treat/prevent maladies in the livestock industry (Ramaswamy et al. 2010). Veterinary antibiotics are globally manufactured in large quantities on annual basis. The worldwide use of these antimicrobials was estimated to be 93,309 tons in 2017 and is anticipated to reach 105,596 tons by 2030 (Van Boeckel et al. 2015; Tiseo et al. 2020). Generally, these antimicrobials are not completely absorbed and biotransformed in animal bodies and thus largely eliminated through feces and urine (Sarmah et al. 2006). A huge volume of animal dung is globally excreted, ultimately added to the soil and mainly utilized as manure-based fertilizer (Peidong et al. 2009). Subsequently, high levels of several antibacterial drugs have been detected in the manure of animal origin. Thus, animal manure has been identified as a substantial stock of antimicrobials and antimicrobial-resistant bacteria (Yang et al. 2014).

Despite the sporadically reported low concentrations of antibiotics in animal manure, inclusive data about the estimated levels of most veterinary antibiotics is still lacking (Qian et al. 2016). A major proportion of antibiotics remains in the soil following the application of manure-based fertilizer (Sun et al. 2017a). Antibiotic residues are capable of long-term retention in the soil with subsequent detrimental effects on agricultural crops and soil microflora (Hashmi et al. 2017). Accordingly, the residual concentrations of antibiotics may offer a potential hazard to agricultural products by disrupting seed germination and plant growth (Hu et al. 2010; Gros et al. 2019). Apart from supplying the essentially-required nutrients and enhancing soil fertility, the spreading of manure-based fertilizer also adds the antibiotic residues, their metabolic products and drug-resistant microorganisms to the agricultural land (Ding et al. 2014). Moreover, these antibiotic-resistant microbes can be transmitted to humans and animals through the consumption of contaminated fruits and vegetables, grown in manure-fertilized soil (Tien et al. 2017). Improved management and treatment of manure can prevent the occurrence, retention and dissemination of antibiotic residues and resultant ecotoxic effects (Durso and Cook 2014). Besides, the underlying mechanisms of manure-associated antimicrobial resistance and its environmental propagation require exigent and thorough elucidation (Sanderson et al. 2016). This chapter describes the manure-associated veterinary antibiotics in terms of ecological impacts and mitigation strategies.

20.2 Consumption Trend of Veterinary Antibiotics in Livestock Sector

Veterinary antibiotics are used as an essential component of livestock, poultry and aquaculture feeds. The current sequence of antibiotics consumption in animal farming is swine > poultry > cattle (Kim et al. 2011). Almost 90% of the currently used antibiotics are naturally produced by bacteria, fungi or semi-synthetically prepared from natural sources, while, others are exclusively of synthetic origin (Von Nussbaum et al. 2006). Tetracyclines, β-lactams, fluoroquinolones, lincosamides, aminoglycosides, macrolides, sulfonamides and ionophores are the principal groups of antibiotics utilized in livestock farming (Kuppusamy et al. 2018). Around 90% of the antimicrobials consumed in UK, whereas more than half of that in Denmark and Korea, are collectively contributed by macrolides, sulfonamides and tetracyclines (Kim et al. 2011). Up to 150 veterinary antibiotics have been approved to-date for use in animal agriculture and 80% of those are employed for non-therapeutic purposes (Ventola 2015). Antibiotics are administered at sub-therapeutic doses in animals to promote their growth, more frequently among swine, cattle, poultry and fish in farm settings. Furthermore, specific veterinary antibiotics are employed at certain developmental phases, while, others can be given till the animal is slaughtered (Kumar et al. 2005b). Antibiotics used as growth enhancers are known to impede the subclinical microbial infections and enhance the uptake and utility of the nutritional components. Moreover, these feed additives also minimize the bacterial viability and proliferation in the gut resulting in the better availability of feed energy that otherwise would have been lost owing to microbial fermentation (Gaskins et al. 2002).

It is predicted that the world-wide use of antimicrobials will keep on rising persistently in connection with the progressively increasing food demand of the fast growing population. The globally recorded antimicrobial usage in 2010 is projected to rise by 67% during the subsequent 20 years (Van Boeckel et al. 2015). In 2003, China manufactured about 65% and 60% of the global production of oxytetracycline and penicillin, respectively (Yang et al. 2010). Annually, about 90% of livestock farms in the United States employ 11.2 million kg of antibiotics as growth-boosters (Arikan et al. 2009). In fact, China, Russia, Brazil, South Africa and India are projected to exhibit a doubling of their antibiotic consumption by 2030. About 80% of the marketed antibiotics in United States are primarily used for growth enhancement and disease prevention in animals (Ventola 2015). Despite consuming 5,000 tons of veterinary antibiotics during 2005 (Kumar et al. 2005b), the European Union has prohibited the consumption of antimicrobials as feed additives since 2006 (Europa 2005). Table 20.1 represents the data on country/territory-wise annual consumption of veterinary antibiotics.

Table 20.1 Country/territory-wise annual consumption of veterinary antibiotics

Country/territory	Antibiotic consumption (tons)	Year	Reference
UK	897	2000	Thiele-Bruhn and Aust (2004)
China	6000	2003	Zhao et al. (2010)
European Union	5000	2005	Kumar et al. (2005b)
United States	13,067	2009	FDA (2010)
Brazil	5600	2010	Gelband et al. (2015)
Germany	1900		
India	1890		

20.3 Excretion of Veterinary Antibiotics in Animal Manure

Injectables, topical preparations and feed additives are the typical dosage forms of veterinary antibiotics meant for parenteral, cutaneous and oral routes of administration, respectively. Majority of the administered antibiotics undergo biotransformation, and virtually 30–90% of the given drug is eliminated as metabolites in feces and urine (Lin et al. 2017). Antibiotic excretion varies on the basis of animal species, development stage and dose. Antibiotics that are in animal's manure are released as conjugated metabolites, parent substances, and oxidation or hydrolysis products, depending on the chemical nature of drug and specie of target animal (Tolls 2001). The excretion rates of macrolides, tetracyclines and sulfonamides were recorded as 50–65%, 65–100% and 90–100%, respectively (Arikan et al. 2009; Halling-Sørensen et al. 2001; Winckler and Grafe 2001). More than 72% of tetracyclines and 90% of fluoroquinolones were eliminated as parent molecules in the excreta of orally-treated animals (Winckler and Grafe 2001; Sukul et al. 2009).

Even though, many antibiotics are broken down in the bodies of animals and become ineffective, the conjugated metabolites could be subjected to deconjugation reaction in urine and fecal material (Sarmah et al. 2006). So, antibiotics can be found in fresh animal feces either as metabolites or as the original molecules. The majority of antibiotics remain in fecal excreta and form stable compounds with soluble organics. When dung is spread to agricultural areas, a portion of the antimicrobials becomes portable and pollutes the ambience, including surface and groundwater. An earlier investigation revealed that pig dung had the greatest concentrations of tetracycline residues, followed by chicken and cow feces (Zhang et al. 2008). Considering the worldwide annual consumption of 63 million kilograms of antibiotics in livestock sector and 30–90% of an administered antibiotic being excreted in animal waste, the overall quantity of residual antibiotics in animal manure and manure-associated wastewater is estimated as 18.9–56.7 million kilograms per year. The bulk of dung and wastewater having manure is utilized as a source of agricultural fertilizer. Veterinary antibiotics and their metabolic products from animal waste lead to environmental contamination following the application of manure on farmland. Despite the fact that microorganisms can destroy veterinary antibiotics during animal waste storage (for instance, in storing sheds and ponds) and management (e.g., composting), their remains are regularly identified in dung compost and lagoon water containing manure.

20.4 Environmental Transfer of Manure-Associated Veterinary Antibiotics

Veterinary antibiotics and their metabolic products are passed into the environment via the disposal of surplus or expired medicines and utilization of animal excreta (i.e., dung or dung-contaminated drained water) on the soil. Antibiotics produced from animal dung can be absorbed in farming lands and degraded by means of various abiotic and biological processes (Song and Guo 2014). Antibiotics may be transmitted to adjacent watershed by leaching and/or runoff from dung-laden soils (Du and Liu 2012) or can be incorporated into the food chain following their absorption by the crop plants (Sallach et al. 2018) as shown in Fig. 20.1. Antibiotic pollutants are now commonly identified in the environment and are regarded as a rising worldwide concern (Kuppusamy et al. 2018).

Antimicrobials in the dung of grazing cattle and the water of aquaculture ponds are discharged directly into the environment. As an organic fertilizer, the heaped dung (in lagoons, slurry stores and storage sheds) and treated material (e.g., compost and sewage sludge) are frequently applied in agriculture by spreading, injection or irrigation. The environmental stability of veterinary antibiotics is highly dependent on soil characteristics, pH, temperature, ultraviolet radiation and animal manure. Antibiotics degrade at a slower pace when temperature is low. The high concentration of organic matter facilitates the binding of fluoroquinolones and sulfonamides to animal manure (Marengo et al. 1997). These antimicrobials are unaffected by increased aeration and warmth in manure, and therefore released into the environment as pharmacologically-active substances (Winckler and Grafe 2001).

Fig. 20.1 Entry of veterinary antibiotics into the food chain through animal manure

20.5 Fate of Residual Veterinary Antibiotics in Soil

20.5.1 Adsorption of Veterinary Antibiotics to Soil

Adsorption to organic matter and minerals prolongs the retention of veterinary antibiotics in the soil (Zitnick et al. 2011). Apart from physico-chemical properties, the pH, nature of organic matter and mineral concentration of soil, as well as prevailing environmental conditions, govern the sorption of veterinary antibiotics to soil particles (Batchelder 1982; Lützhøft et al. 2000; Thiele 2000; Doretto and Rath 2013). The adsorption of sulfathiazole and sulfamethazine to loamy soil and sand was significantly affected by pH (Kurwadkar et al. 2007). Moreover, the organic nature and structural diversity of veterinary antibiotics further confound their adsorption to the soil constituents. Strong adsorption to organic matter has been documented for some antibiotics such as fluoroquinolones and sulfonamides (Marengo et al. 1997; Thiele-Bruhn and Aust 2004). Likewise, efficient adsorption of tetracyclines occurs to the soil, clay and sediments, depending upon the environmental conditions (Rabølle and Spliid 2000; Allaire et al. 2006). Sorption and desorption are essential mechanisms influencing the transport, accessibility for breakdown and ecological outcome of veterinary antibiotics in soil and aquatic environments (Fig. 20.2). Most hydrophobic antibiotics are primarily adsorbed via partitioning into soil organic matter (Tolls 2001). For antibiotics that are ionizable or hydrophilic, partitioning into soil organic matter may not be the major adsorption process.

Several environmental parameters affect the retention and stability of veterinary antibiotics in the soil as well as their propagation and likely ecological effects. Environmental temperature and availability of oxygen influence the degradation of veterinary antibiotics in soil. Trimethoprim and sulfamethoxazole were slowly dissipated from mineral soils subjected to anaerobic conditions (Lin and Gan 2011). Increased soil humidity leads to conversion of chemicals into solution form, thus augmenting the likelihood of microbial degradation. Enhancement of soil humidity was linked with reduction in the half-life of sulfadimethoxine (Wang et al. 2006b). Likewise, the half-life of virginiamycin was inversely proportional to soil pH (Weerasinghe and Towner 1997).

20.5.2 Chemical Modification of Manure-Associated Antibiotics

Generally, most of the veterinary antibiotics undergo degradation in soil, having a half-life of not more than thirty days. Nevertheless, some antibiotics including virginiamycin, sarafloxacin and roxithromycin are retained in the soil for more than

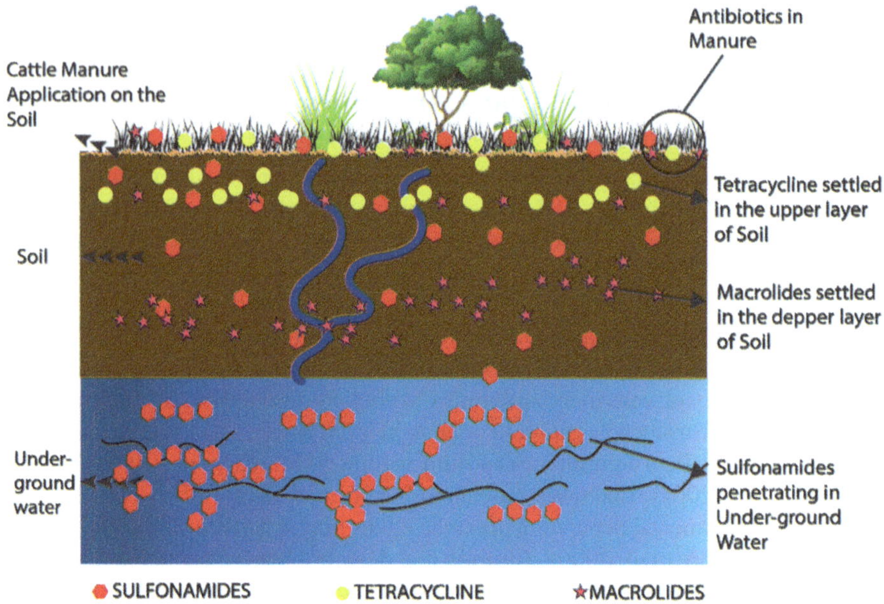

Fig. 20.2 Sorption process of veterinary antibiotics

2 months (Song and Guo 2014). Hydrolysis, oxidation, photolysis, organic transformation and microbial breakdown are various mechanisms of antibiotic decomposition occurring in the soil. In most instances, the breakdown of veterinary antibiotics may produce additional hydrophilic metabolites (Bradford et al. 2008). Certain antibiotics, for example, erythromycin, tylosin and flavomycin, are completely broken at 20–30 °C during 30 days, while only a tiny part of other antimicrobials, like ciprofloxacin and virginiamycin, pass through the breakdown process even 30–80 days, later. Cold and dark storage of many antibiotics results in their longer half-life, indicating that their retention in deeper land layers and deeper waterways for extended period of time (Hektoen et al. 1995). During the storage, transport, and soil application of solid dung or manure-holding wastewater, antibiotic residues may pass through a sequence of biotic and abiotic transformations.

20.5.2.1 Biotic Transformation of Manure-Associated Antibiotics

Biodegradation is crucial in governing the environmental fate of antibiotic pollutants. Due to the high microbial activity of animal dung, antibiotic residues may be easily degraded by manure-associated bacteria. However, many veterinary antibiotics cannot be broken down entirely in animal dung, resulting in the distribution of veterinary antibiotic residues to farm fields through the use of dung in the soil (Feng et al. 2017). Farming lands often include a variety of microorganisms known to degrade the veterinary medicines (Dantas et al. 2008). Temperature and incubation conditions are the primary environmental parameters that determine biodegradation rates of dung-associated antibiotics owing to their substantial effect on microbial activity (Wang et al. 2006a). Biodegradation under anaerobic circumstances has been recorded to be quite sluggish than that occuring under the aerobic conditions (Yang et al. 2009).

20.5.2.2 Abiotic Transformation of Manure-Associated Antibiotics

Hydrolysis, photolysis, and/or redox transformation are the major abiotic transformation processes for veterinary antibiotics. The mechanisms of abiotic transformation for the majority of antibiotics are determined by their physico-chemical characteristics (for instance, the solubility, molecular structure, hydrophobicity and speciation) and environmental conditions, i.e., temperature and pH (Jeon et al. 2014). The majority of veterinary antibiotic are hydrolytically stable under normal ecological circumstances. The presence of ample quantity of water in animal manure and soil renders the hydrolysis as the primary mechanism for environmental dissipation of antibiotics. Sulfonamides, macrolides and ß-lactam antibiotics are prone to hydrolysis (Huang et al. 2001). Antimicrobials like β-lactams and sulfonamides can be broken down via hydrolysis in some situations (Li et al. 2011). Conversely, photolysis forms the minor pathway for antibiotic degradation on account of limited exposure to sunlight (Beausse 2004). Some antibiotics possibly go through direct photolysis in water or soil under sunlight (Niu et al. 2013). Some veterinary antibiotics including oxytetracycline and chlortetracycline are susceptible to photolysis by sunlight (Conde-Cid et al. 2018). Due to the occurrence of photosensitizers in dung, photocatalytic degradation frequently plays a key role in manure-associated antibiotics, in addition to direct photolysis (Sukul et al. 2008). Besides, several veterinary antibiotics are known to undergo oxidative decarboxylation and hydroxylation reactions (Al-Ahmad et al. 1999).

20.5.3 Colloid-Mediated Transfer, Leaching and Runoff of Veterinary Antibiotics

The principal component of colloids, dissolved organic matter, may affect the adsorption of antibiotics to soil components (Dodgen and Zheng 2016). Large amounts of colloids, including dissolved organic matter, are present in dung and manure-associated wastewater, and the soil administration of animal manures may concurrently increase colloids and antibiotic residues into the settings. The transport of antibiotic in soil is increased by their interaction with effluent-borne colloids (Zou and Zheng 2013). Antibiotics leach into deeper soil or groundwater according to the absorption capacity of the topsoil (Hamscher et al. 2003). Antibiotic concentration in the overland flow of irrigated agricultural soil following pig slurry administration was analyzed for four hours. Peak concentrations of sulfachloropyridazine and oxytetracycline were recorded in runoff from the tramline plot at 0.70 mg/L and 0.072 mg/L, respectively (Kay et al. 2005). Antibiotics may penetrate aquatic habitats through urban sewage systems, effluent from animal rearing and leaching runoff from terrestrial ecosystems. Antimicrobials in aquatic environments might be transmitted to agroecosystems through irrigation and sedimentation. The movement of volatile organic compounds within soils and into ground and surface waters is caused by runoff and leaching. On manured fields, the leaching of sulfonamides into groundwater has been documented (Hamscher et al. 2005).

20.6 Uptake and Deposition of Residual Veterinary Antibiotics by Plants

Animal manure is employed to fulfill the organic matter and nutrients requirement of agricultural land. Nevertheless, this process can provoke the inadvertent release of antibiotic residues into agricultural soil and consequent uptake by plants (Kumar et al. 2005a; Chung et al. 2017). The uptake, disposition and accumulation of antibiotic residues in plants are dependent upon their species, cultivars and developmental stage, concentration and physico-chemical characteristics of drug, as well as the growth conditions and soil properties (Pan and Chu 2017). Root crops that are directly connected with soil, like radish, potato and carrot are highly contaminated with antibiotic residues. Detectable concentrations of levamisole, trimethoprim and florfenicol were recorded in carrot and lettuce, obtained from antibiotic-contaminated sandy soil (Boxall et al. 2006). Moreover, sulfonamides such as sulphapyridine, sulfathiazole and sulfamethazine were accumulated in bitter melon, lettuce, sweet potato, carrot, green pepper, white gourd and Chinese cabbage (Bao et al. 2010). Nevertheless, tylosin was not uptaken by cabbage, green onion and corn on account of its larger molecular size (Kumar et al. 2005a).

20.7 Ecological Impact of Manure-Derived Veterinary Antibiotics

Land application of animal manure not only contributes to soil fertility through the supply of essential nutrients but also leads to disruption of microbial dynamics following the introduction of microorganisms and antibiotic-resistant microbial genes (Poulsen et al. 2013; Ding et al. 2014). Manure-based antibiotics considerably alter the relative fecundity of soil microflora by selectively targeting the range of susceptible microorganisms (Ding and He 2010). The selective inhibition or destruction of vulnerable bacteria by antibiotics provides conducive environment for the growth and dominance of resistant strains. Selective destruction of soil bacteria ascribed to oxytetracycline and sulphapyridine resulted in the overgrowth of fungi (Thiele-Bruhn and Beck 2005). Besides, other antibiotics such as tylosin and apramycin also impaired the normal pattern of soil microflora (Westergaard et al. 2001; Xiaoping et al. 2004). The extent of antibiotic-induced modification of soil microflora varies with the characteristics of soil, microbial dynamics and concentration of antibiotics (Zielezny et al. 2006; Čermák et al. 2008). Even though the relative fractions of Gram-negative and Gram-positive bacteria remained unaltered following the addition of sulfadiazine, widespread resistance to tetracycline significantly diminished this ratio (Hund-Rinke et al. 2004; Hammesfahr et al. 2008).

Several antibiotics including tylosin, sulfadiazine and ciprofloxacin have been implicated in disrupting the respiration rate and biomass synthesis of soil (Demoling and Bååth 2008; Kotzerke et al. 2008; Näslund et al. 2008). The enzymatic and microbial activities of soil were affected by trimethoprim, tetracycline, sulfamethoxazole, tylosin, sulfamethazine and chlortetracycline (Liu et al. 2009). The translocation of soil-adsorbed antibiotics can contaminate both groundwater and surface water by means of leaching and erosion (Pedersen et al. 2003). Contrary to pesticides, the regulatory standards for inspecting residual concentrations of veterinary antibiotics in crops have not been established. The storage of veterinary antibiotics in edible portions has been linked with growth impairment in crops and potential health concerns following their entrance into the food chain (Zheng et al. 2014; Ahmed et al. 2015). Exposure to sulfadiazine, enrofloxacin and sulfamonomethoxine has been associated with deleterious effects in tomato, Chinese cabbage and wheat (Jin et al. 2009). Additionally, the contaminated fruits and vegetables grown on manured-amended soil and consumed in raw form can transmit the drug-resistant microorganisms to humans.

The indiscriminate usage of antibiotics may expedite the emergence of drug-resistant microbial strains and consequent unurable infections (Kumar et al. 2005a). The environmental transmission of antibiotic resistance from animal husbandry through animal manure has been well-established (Heuer et al. 2011). Animal manures, particularly those of cattle, sheep and pigs are more frequently contaminated with resistance genes based on the rate of antimicrobial consumption (Enne et al. 2008; McKinney et al. 2010). Multi-drug-resistant bacteria were isolated from the fecal material of pigs (Van den Bogaard et al. 2000). The vertical or horizontal transmission of antibiotic-resistance genes between human and veterinary pathogens may cause the emergence of untreatable forms of campylobacteriosis, colibacillosis, salmonellosis (typhoid) and other infections (Vaagland et al. 2004; Sun et al. 2017b).

20.8 Mitigation Strategies for Manure-Associated Veterinary Antibiotics

Control measures are requisite to curb the ecotoxic implications of veterinary antibiotics following the use of animal dung as fertilizer. Removal of antimicrobial residues and antibiotic-resistant bacteria is necessary before applying animal manure on agricultural land. Effective sterilization methods and composting techniques should be devised for the decontamination of animal manure. Appropriate control strategies are critically needed for precluding the leaching and subsequent dissemination of pathogens during storage, transportation and composting of manure. Composting and anaerobic fermentation can effectively decrease the level of veterinary antibiotics in animal manure. Besides, another efficient and relatively less-expensive technique of electrochemical treatment has been postulated for the decontamination of animal manure (Laridi et al. 2005). Liquid obtained from manure lagoons requires proper disinfection prior to its environmental disposal. Efficient oxidation techniques using hydrogen peroxide, ozone and ultraviolet radiation can augment the breakdown of veterinary antimicrobials (Li et al. 2008).

Manure-containing wastewater can be subjected to oil extraction process for removing lipophilic veterinary antibiotics (Dodgen et al. 2018). However, the adequate elimination of poorly hydrophobic antibiotics may necessitate several oil-capturing cycles. The phytoremediation capacity of certain woody plants has been successfully employed for the elimination of sulphonamides (Michelini et al. 2012). Besides, certain phosphate-solubilizing bacterial species like Bacillus and Pseudomonas can be used to improve the bioremediation capacity of plants. The efficacy of psychrophilic anaerobic digestion in eliminating antibiotics residues and antibiotic-resistant bacteria from animal manure necessitates further exploration (Massé et al. 2010).

Improved hygienic and biosecurity measures can reduce the consumption of veterinary antibiotics through minimizing the likelihood of microbial infections in animals (Alban et al. 2013). Another strategy for the rational application of antimicrobials in veterinary practice is the selection of exact drug and appropriate dosage regimen for the right patient/condition. Antimicrobials must be reserved for curing microbial infections in animals following the appropriate susceptibility assessment and specific prescription guidelines. Several government organizations are appraising the application of antibiotics in livestock and poultry sectors. Antibiotics that are primarily consumed for curing human infections are banned for consumption in veterinary medicine in some European countries including Denmark, Sweden and Germany (Teillant and Laxminarayan 2015). Similarly, antibiotics are no longer employed as growth promoters in animal production in New Zealand, South Korea, Mexico and Australia (Gelband et al. 2015), whereas China, Russia, Philippines, Argentina, South Africa, Japan, India, Brazil and Indonesia have not banned the consumption of antibiotics for growth enhancement in animals (Laxminarayan et al. 2015). As the chronic ecological effects caused by lower antibiotic concentrations in land, plants and water are primarily unidentified and hard to probe, the gradual exclusion of veterinary antibiotics from animal feed has been recommended. The discovery and development of green and sustainable growth promoting agents as alternatives to antibiotics should be targeted by future research.

20.9 Conclusions and Future Considerations

A wide range of antibiotics are considerably utilized in veterinary practice and animal farming for the treatment and/or prophylaxis of infectious diseases as well as for growth promotion purpose. Accordingly, the residual amounts of these antibiotics are released in animal manure and subsequently lead to environmental contamination with potentially adverse ecological consequences. Besides disrupting the health and well-being of humans, animals and plants through their direct toxicological impacts, these residual veterinary antibiotics also enhance the dissemination of drug-resistant microbial pathogens. Judicious use of veterinary antibiotics and appropriate mitigation strategies are requisite for preventing the undesirable consequences associated with the release of residual veterinary antibiotics in animal excreta. Moreover, the exploration of efficient and safe alternatives to antibiotic growth promoters and manure-based fertilizers is also suggested. Finally, the development of highly sensitive and more accurate methods for detection, quantification and inactivation of residual veterinary antibiotics in animal manure should be targeted by future research studies.

References

Ahmed MBM, Rajapaksha AU, Lim JE, Vu NT, Kim IS, Kang HM, Lee SS, Ok YS (2015) Distribution and accumulative pattern of tetracyclines and sulfonamides in edible vegetables of cucumber, tomato, and lettuce. J Agric Food Chem 63(2):398–405

Al-Ahmad A, Daschner F, Kümmerer K (1999) Biodegradability of cefotiam, ciprofloxacin, meropenem, penicillin G, and sulfamethoxazole and inhibition of waste water bacteria. Arch Environ Contam Toxicol 37(2):158–163

Alban L, Dahl J, Andreasen M, Petersen J, Sandberg M (2013) Possible impact of the "yellow card" antimicrobial scheme on meat inspection lesions in Danish finisher pigs. Prev Vet Med 108(4):334–341

Allaire S, Del Castillo J, Juneau V (2006) Sorption kinetics of chlortetracyline and tylosin on sandy loam and heavy clay soils. J Environ Qual 35(4):969–972

Arikan OA, Mulbry W, Rice C (2009) Management of antibiotic residues from agricultural sources: use of composting to reduce chlortetracycline residues in beef manure from treated animals. J Hazard Mater 164(2–3):483–489

Bao Y, Li Y, Mo C, Yao Y, Tai Y, Wu X, Zhang Y (2010) Determination of six sulfonamide antibiotics in vegetables by solid phase extraction and high performance liquid chromatography. Environ Chem 29(3):513–518

Batchelder A (1982) Chlortetracycline and oxytetracycline effects on plant growth and development in soil systems. J Environ Qual 4:675–678

Beausse J (2004) Selected drugs in solid matrices: a review of environmental determination, occurrence and properties of principal substances. TrAC, Trends Anal Chem 23(10–11):753–761
Boxall AB, Johnson P, Smith EJ, Sinclair CJ, Stutt E, Levy LS (2006) Uptake of veterinary medicines from soils into plants. J Agric Food Chem 54(6):2288–2297
Bradford SA, Segal E, Zheng W, Wang Q, Hutchins SR (2008) Reuse of concentrated animal feeding operation wastewater on agricultural lands. J Environ Qual 37(S5):S-97–S-115
Čermák L, Kopecký J, Novotná J, Omelka M, Parkhomenko N, Plháčková K, Ságová-Marečková M (2008) Bacterial communities of two contrasting soils reacted differently to lincomycin treatment. Appl Soil Ecol 40(2):348–358
Chung HS, Lee Y-J, Rahman MM, Abd El-Aty A, Lee HS, Kabir MH, Kim SW, Park B-J, Kim J-E, Hacımüftüoğlu F (2017) Uptake of the veterinary antibiotics chlortetracycline, enrofloxacin, and sulphathiazole from soil by radish. Sci Total Environ 605:322–331
Conde-Cid M, Fernández-Calviño D, Nóvoa-Muñoz J, Arias-Estévez M, Díaz-Raviña M, Fernández-Sanjurjo M, Núñez-Delgado A, Álvarez-Rodríguez E (2018) Biotic and abiotic dissipation of tetracyclines using simulated sunlight and in the dark. Sci Total Environ 635:1520–1529
Dantas G, Sommer MO, Oluwasegun RD, Church GM (2008) Bacteria subsisting on antibiotics. Science 320(5872):100–103
Demoling LA, Bååth E (2008) No long-term persistence of bacterial pollution-induced community tolerance in tylosin-polluted soil. Environ Sci Technol 42(18):6917–6921
Ding C, He J (2010) Effect of antibiotics in the environment on microbial populations. Appl Microbiol Biotechnol 87(3):925–941
Ding G-C, Radl V, Schloter-Hai B, Jechalke S, Heuer H, Smalla K, Schloter M (2014) Dynamics of soil bacterial communities in response to repeated application of manure containing sulfadiazine. PLoS ONE 9(3):e92958
Dodgen L, Zheng W (2016) Effects of reclaimed water matrix on fate of pharmaceuticals and personal care products in soil. Chemosphere 156:286–293
Dodgen LK, Wiles KN, Deluhery J, Rajagopalan N, Holm N, Zheng W (2018) Removal of estrogenic hormones from manure-containing water by vegetable oil capture. J Hazard Mater 343:125–131
Doretto KM, Rath S (2013) Sorption of sulfadiazine on Brazilian soils. Chemosphere 90(6):2027–2034
Du L, Liu W (2012) Occurrence, fate, and ecotoxicity of antibiotics in agro-ecosystems. A review. Agron Sustain Dev 32(2):309–327
Durso LM, Cook KL (2014) Impacts of antibiotic use in agriculture: what are the benefits and risks? Curr Opin Microbiol 19:37–44
Enne VI, Cassar C, Sprigings K, Woodward MJ, Bennett PM (2008) A high prevalence of antimicrobial resistant Escherichia coli isolated from pigs and a low prevalence of antimicrobial resistant E. coli from cattle and sheep in Great Britain at slaughter. FEMS Microbiol Lett 278(2):193–199
Europa (2005) Ban on antibiotics as growth promoters in animal feed enters into effect. Europa, Brussels
FDA (2010) 2009 summary report on antimicrobials sold or distributed for use in food-producing animals. Food and Drug Administration, Rockville. www.fda.gov/downloads/ForIndustry/UserFees/AnimalDrugUserFeeActADUFA/UCM231851.pdf. Accessed 25 Nov 2022
Feng L, Casas ME, Ottosen LDM, Møller HB, Bester K (2017) Removal of antibiotics during the anaerobic digestion of pig manure. Sci Total Environ 603:219–225
Gaskins H, Collier C, Anderson D (2002) Antibiotics as growth promotants: mode of action. Anim Biotechnol 13(1):29–42
Gelband H, Miller P, Molly PS, Gandra S, Levinson J, Barter D, White A, Laxminarayan R (2015) The state of the world's antibiotics 2015. Wound Healing Southern Africa 8(2):30–34
Gros M, Mas-Pla J, Boy-Roura M, Geli I, Domingo F, Petrović M (2019) Veterinary pharmaceuticals and antibiotics in manure and slurry and their fate in amended agricultural soils: findings from an experimental field site (Baix Empordà, NE Catalonia). Sci Total Environ 654:1337–1349

Gurmessa B, Pedretti EF, Cocco S, Cardelli V, Corti G (2020) Manure anaerobic digestion effects and the role of pre-and post-treatments on veterinary antibiotics and antibiotic resistance genes removal efficiency. Sci Total Environ 721:137532

Halling-Sørensen B, Jensen J, Tjørnelund J, Montforts M (2001) Worst-case estimations of predicted environmental soil concentrations (PEC) of selected veterinary antibiotics and residues used in Danish agriculture. In: Pharmaceuticals in the environment. Springer, pp 143–157

Hammesfahr U, Heuer H, Manzke B, Smalla K, Thiele-Bruhn S (2008) Impact of the antibiotic sulfadiazine and pig manure on the microbial community structure in agricultural soils. Soil Biol Biochem 40(7):1583–1591

Hamscher G, Pawelzick HT, Höper H, Nau H (2005) Different behavior of tetracyclines and sulfonamides in sandy soils after repeated fertilization with liquid manure. Environ Toxicol Chem Int J 24(4):861–868

Hamscher G, Pawelzick HT, Sczesny S, Nau H, Hartung J (2003) Antibiotics in dust originating from a pig-fattening farm: a new source of health hazard for farmers? Environ Health Perspect 111(13):1590–1594

Hashmi MZ, Mahmood A, Kattel DB, Khan S, Hasnain A, Ahmed Z (2017) Antibiotics and antibiotic resistance genes (ARGs) in soil: occurrence, fate, and effects. In: Xenobiotics in the soil environment. Springer, pp 41–54

Hektoen H, Berge JA, Hormazabal V, Yndestad M (1995) Persistence of antibacterial agents in marine sediments. Aquaculture 133(3–4):175–184

Heuer H, Schmitt H, Smalla K (2011) Antibiotic resistance gene spread due to manure application on agricultural fields. Curr Opin Microbiol 14(3):236–243

Hu X, Zhou Q, Luo Y (2010) Occurrence and source analysis of typical veterinary antibiotics in manure, soil, vegetables and groundwater from organic vegetable bases, northern China. Environ Pollut 158(9):2992–2998

Huang C-H, Renew JE, Smeby KL, Pinkston K, Sedlak DL (2001) Assessment of potential antibiotic contaminants in water and preliminary occurrence analysis. J Contemp Water Res Educ 120(1):4

Hund-Rinke K, Simon M, Lukow T (2004) Effects of tetracycline on the soil microflora: function, diversity, resistance. J Soils Sediments 4(1):11–16

Jeon DS, Oh T-K, Park M, Lee DS, Lim YJ, Shin JS, Song SG, Kim SC, Shinogi Y, Chung DY (2014) Reactions and behavior relevant to chemical and physical properties of various veterinary antibiotics in soil. J Fac Agric Kyushu Univ 59:391–397

Jin C, Chen Q, Sun R, Zhou Q, Liu J (2009) Eco-toxic effects of sulfadiazine sodium, sulfamonomethoxine sodium and enrofloxacin on wheat. Chinese cabbage and tomato. Ecotoxicology 18(7):878–885

Kay P, Blackwell PA, Boxall AB (2005) Transport of veterinary antibiotics in overland flow following the application of slurry to arable land. Chemosphere 59(7):951–959

Kim K-R, Owens G, Kwon S-I, So K-H, Lee D-B, Ok YS (2011) Occurrence and environmental fate of veterinary antibiotics in the terrestrial environment. Water Air Soil Pollut 214(1):163–174

Kotzerke A, Sharma S, Schauss K, Heuer H, Thiele-Bruhn S, Smalla K, Wilke B-M, Schloter M (2008) Alterations in soil microbial activity and N-transformation processes due to sulfadiazine loads in pig-manure. Environ Pollut 153(2):315–322

Kumar K, Gupta SC, Baidoo S, Chander Y, Rosen CJ (2005a) Antibiotic uptake by plants from soil fertilized with animal manure. J Environ Qual 34(6):2082–2085

Kumar K, Gupta SC, Chander Y, Singh AK (2005b) Antibiotic use in agriculture and its impact on the terrestrial environment. Adv Agron 87:1–54

Kuppusamy S, Kakarla D, Venkateswarlu K, Megharaj M, Yoon Y-E, Lee YB (2018) Veterinary antibiotics (VAs) contamination as a global agro-ecological issue: a critical view. Agr Ecosyst Environ 257:47–59

Kurwadkar ST, Adams CD, Meyer MT, Kolpin DW (2007) Effects of sorbate speciation on sorption of selected sulfonamides in three loamy soils. J Agric Food Chem 55(4):1370–1376

Laridi R, Drogui P, Benmoussa H, Blais J-F, Auclair JC (2005) Removal of refractory organic compounds in liquid swine manure obtained from a biofiltration process using an electrochemical treatment. J Environ Eng 131(9):1302–1310

Laxminarayan R, Van Boeckel T, Teillant A (2015) The economic costs of withdrawing antimicrobial growth promoters from the livestock sector, pp 1–42

Li D, Yang M, Hu J, Ren L, Zhang Y, Li K (2008) Determination and fate of oxytetracycline and related compounds in oxytetracycline production wastewater and the receiving river. Environ Toxicol Chem Int J 27(1):80–86

Li X, Zheng W, Machesky ML, Yates SR, Katterhenry M (2011) Degradation kinetics and mechanism of antibiotic ceftiofur in recycled water derived from a beef farm. J Agric Food Chem 59(18):10176–10181

Lin K, Gan J (2011) Sorption and degradation of wastewater-associated non-steroidal anti-inflammatory drugs and antibiotics in soils. Chemosphere 83(3):240–246

Lin W, Flarakos J, Du Y, Hu W, He H, Mangold J, Tanaka SK, Villano S (2017) Pharmacokinetics, distribution, metabolism, and excretion of omadacycline following a single intravenous or oral dose of 14C-omadacycline in rats. Antimicrob Agents Chemother 61(1):e01784-e11716

Liu F, Ying G-G, Tao R, Zhao J-L, Yang J-F, Zhao L-F (2009) Effects of six selected antibiotics on plant growth and soil microbial and enzymatic activities. Environ Pollut 157(5):1636–1642

Lützhøft H-CH, Vaes WH, Freidig AP, Halling-Sørensen B, Hermens JL (2000) 1-Octanol/water distribution coefficient of oxolinic acid: influence of pH and its relation to the interaction with dissolved organic carbon. Chemosphere 40(7):711–714

Marengo JR, Kok RA, O'Brien K, Velagaleti RR, Stamm JM (1997) Aerobic biodegradation of (14C)-sarafloxacin hydrochloride in soil. Environ Toxicol Chem Int J 16(3):462–471

Massé D, Masse L, Xia Y, Gilbert Y (2010) Potential of low-temperature anaerobic digestion to address current environmental concerns on swine production. J Animal Sci 88(suppl_13):E112–E120

McKinney CW, Loftin KA, Meyer MT, Davis JG, Pruden A (2010) Tet and sul antibiotic resistance genes in livestock lagoons of various operation type, configuration, and antibiotic occurrence. Environ Sci Technol 44(16):6102–6109

Michelini L, Reichel R, Werner W, Ghisi R, Thiele-Bruhn S (2012) Sulfadiazine uptake and effects on *Salix fragilis* L. and *Zea mays* L. plants. Water Air Soil Pollut 223(8):5243–5257

Näslund J, Hedman JE, Agestrand C (2008) Effects of the antibiotic ciprofloxacin on the bacterial community structure and degradation of pyrene in marine sediment. Aquat Toxicol 90(3):223–227

Niu J, Zhang L, Li Y, Zhao J, Lv S, Xiao K (2013) Effects of environmental factors on sulfamethoxazole photodegradation under simulated sunlight irradiation: kinetics and mechanism. J Environ Sci 25(6):1098–1106

Pan M, Chu L (2017) Transfer of antibiotics from wastewater or animal manure to soil and edible crops. Environ Pollut 231:829–836

Pedersen JA, Yeager MA, Suffet I (2003) Xenobiotic organic compounds in runoff from fields irrigated with treated wastewater. J Agric Food Chem 51(5):1360–1372

Peidong Z, Yanli Y, Yongsheng T, Xutong Y, Yongkai Z, Yonghong Z, Lisheng W (2009) Bioenergy industries development in China: dilemma and solution. Renew Sustain Energy Rev 13(9):2571–2579

Poulsen PH, Al-Soud WA, Bergmark L, Magid J, Hansen LH, Sørensen SJ (2013) Effects of fertilization with urban and agricultural organic wastes in a field trial—prokaryotic diversity investigated by pyrosequencing. Soil Biol Biochem 57:784–793

Qian M, Wu H, Wang J, Zhang H, Zhang Z, Zhang Y, Lin H, Ma J (2016) Occurrence of trace elements and antibiotics in manure-based fertilizers from the Zhejiang Province of China. Sci Total Environ 559:174–181

Rabølle M, Spliid NH (2000) Sorption and mobility of metronidazole, olaquindox, oxytetracycline and tylosin in soil. Chemosphere 40(7):715–722

Ramaswamy J, Prasher SO, Patel RM, Hussain SA, Barrington SF (2010) The effect of composting on the degradation of a veterinary pharmaceutical. Biores Technol 101(7):2294–2299

Sallach JB, Bartelt-Hunt SL, Snow DD, Li X, Hodges L (2018) Uptake of antibiotics and their toxicity to lettuce following routine irrigation with contaminated water in different soil types. Environ Eng Sci 35(8):887–896

Sanderson H, Fricker C, Brown RS, Majury A, Liss SN (2016) Antibiotic resistance genes as an emerging environmental contaminant. Environ Rev 24(2):205–218

Sarmah AK, Meyer MT, Boxall AB (2006) A global perspective on the use, sales, exposure pathways, occurrence, fate and effects of veterinary antibiotics (VAs) in the environment. Chemosphere 65(5):725–759

Song W, Guo M (2014) Residual veterinary pharmaceuticals in animal manures and their environmental behaviors in soils. In: He Z, Zhang H (eds) Applied manure and nutrient chemistry for sustainable agriculture and environment. Springer Netherlands, Dordrecht, pp 23–52. https://doi.org/10.1007/978-94-017-8807-6_2

Sukul P, Lamshöft M, Kusari S, Zühlke S, Spiteller M (2009) Metabolism and excretion kinetics of 14C-labeled and non-labeled difloxacin in pigs after oral administration, and antimicrobial activity of manure containing difloxacin and its metabolites. Environ Res 109(3):225–231. https://doi.org/10.1016/j.envres.2008.12.007

Sukul P, Lamshöft M, Zühlke S, Spiteller M (2008) Photolysis of 14C-sulfadiazine in water and manure. Chemosphere 71(4):717–725. https://doi.org/10.1016/j.chemosphere.2007.10.045

Sun J, Zeng Q, Tsang DC, Zhu L, Li X (2017a) Antibiotics in the agricultural soils from the Yangtze River Delta, China. Chemosphere 189:301–308

Sun W, Qian X, Gu J, Wang X-J, Zhang L, Guo A-Y (2017b) Mechanisms and effects of arsanilic acid on antibiotic resistance genes and microbial communities during pig manure digestion. Biores Technol 234:217–223

Teillant A, Laxminarayan R (2015) Economics of antibiotic use in US swine and poultry production. Choices 30(1):1–11

Thiele-Bruhn S, Aust M-O (2004) Effects of pig slurry on the sorption of sulfonamide antibiotics in soil. Arch Environ Contam Toxicol 47(1):31–39

Thiele-Bruhn S, Beck I-C (2005) Effects of sulfonamide and tetracycline antibiotics on soil microbial activity and microbial biomass. Chemosphere 59(4):457–465

Thiele S (2000) Adsorption of the antibiotic pharmaceutical compound sulfapyridine by a long-term differently fertilized loess Chernozem. J Plant Nutr Soil Sci 163(6):589–594

Tien Y-C, Li B, Zhang T, Scott A, Murray R, Sabourin L, Marti R, Topp E (2017) Impact of dairy manure pre-application treatment on manure composition, soil dynamics of antibiotic resistance genes, and abundance of antibiotic-resistance genes on vegetables at harvest. Sci Total Environ 581:32–39

Tiseo K, Huber L, Gilbert M, Robinson TP, Van Boeckel TP (2020) Global trends in antimicrobial use in food animals from 2017 to 2030. Antibiotics 9(12):918

Tolls J (2001) Sorption of veterinary pharmaceuticals in soils: a review. Environ Sci Technol 35(17):3397–3406

Vaagland H, Blomberg B, Krüger C, Naman N, Jureen R, Langeland N (2004) Nosocomial outbreak of neonatal *Salmonella enterica* serotype Enteritidis meningitis in a rural hospital in northern Tanzania. BMC Infect Dis 4(1):1–5

Van Boeckel TP, Brower C, Gilbert M, Grenfell BT, Levin SA, Robinson TP, Teillant A, Laxminarayan R (2015) Global trends in antimicrobial use in food animals. Proc Natl Acad Sci 112(18):5649–5654

Van den Bogaard A, London N, Stobberingh E (2000) Antimicrobial resistance in pig faecal samples from the Netherlands (five abattoirs) and Sweden. J Antimicrob Chemother 45(5):663–671

Ventola CL (2015) The antibiotic resistance crisis: part 1: Causes and threats. Pharm Therapeut 40(4):277

Von Nussbaum F, Brands M, Hinzen B, Weigand S, Häbich D (2006) Antibacterial natural products in medicinal chemistry—exodus or revival? Angew Chem Int Ed 45(31):5072–5129

Wang Q, Guo M, Yates SR (2006a) Degradation kinetics of manure-derived sulfadimethoxine in amended soil. J Agric Food Chem 54(1):157–163

Wang QQ, Bradford S, Zheng W, Yates S (2006b) Sulfadimethoxine degradation kinetics in manure as affected by initial concentration, moisture, and temperature. J Environ Qual 35(6):2162–2169
Weerasinghe CA, Towner D (1997) Aerobic biodegradation of virginiamycin in soil. Environ Toxicol Chem Int J 16(9):1873–1876
Westergaard K, Müller A, Christensen S, Bloem J, Sørensen S (2001) Effects of tylosin as a disturbance on the soil microbial community. Soil Biol Biochem 33(15):2061–2071
Winckler C, Grafe A (2001) Use of veterinary drugs in intensive animal production. J Soils Sediments 1(2):66–70
Xiaoping D, Yingjian S, Zhenjun S, Jianzhong S (2004) Effects of Apramycin on microbial activity in different types of soil. Ecol Environ 13(4):565–568
Yang J-F, Ying G-G, Zhou L-J, Liu S, Zhao J-L (2009) Dissipation of oxytetracycline in soils under different redox conditions. Environ Pollut 157(10):2704–2709
Yang Q, Ren S, Niu T, Guo Y, Qi S, Han X, Liu D, Pan F (2014) Distribution of antibiotic-resistant bacteria in chicken manure and manure-fertilized vegetables. Environ Sci Pollut Res 21(2):1231–1241
Yang Y, Chen D, Huang M (2010) The source of antibiotics in the environment and progress of its ecological impact research. Environ Sci Manag 35(1):140–143
Ye Z-L, Deng Y, Lou Y, Ye X, Chen S (2018) Occurrence of veterinary antibiotics in struvite recovery from swine wastewater by using a fluidized bed. Front Environ Sci Eng 12(3):1–7
Zhang H, Zhang M, Gu G (2008) Residues of tetracyclines in livestock and poultry manures and agricultural soils from North Zhejiang Province. J Ecol Rural Environ 24(3):69–73
Zhao L, Dong YH, Wang H (2010) Residues of veterinary antibiotics in manures from feedlot livestock in eight provinces of China. Sci Total Environ 408(5):1069–1075
Zheng W, Wiles KN, Holm N, Deppe NA, Shipley CR (2014) Uptake, translocation, and accumulation of pharmaceutical and hormone contaminants in vegetables. In: Retention, uptake, and translocation of agrochemicals in plants. ACS Publications, pp 167–181
Zielezny Y, Groeneweg J, Vereecken H, Tappe W (2006) Impact of sulfadiazine and chlorotetracycline on soil bacterial community structure and respiratory activity. Soil Biol Biochem 38(8):2372–2380
Zilio M, Orzi V, Chiodini M, Riva C, Acutis M, Boccasile G, Adani F (2020) Evaluation of ammonia and odour emissions from animal slurry and digestate storage in the Po Valley (Italy). Waste Manage 103:296–304
Zitnick K, Shappell N, Hakk H, DeSutter T, Khan E, Casey F (2011) Effects of liquid swine manure on dissipation of 17β-estradiol in soil. J Hazard Mater 186(2–3):1111–1117
Zou Y, Zheng W (2013) Modeling manure colloid-facilitated transport of the weakly hydrophobic antibiotic florfenicol in saturated soil columns. Environ Sci Technol 47(10):5185–5192

The manufacturer's authorised representative in the EU is Springer Nature Customer Service Centre GmbH, Europaplatz 3, 69115 Heidelberg, Germany. If you have any concerns regarding our products, please contact ProductSafety@springernature.com

Printed and bound by CPI Group (UK) Ltd, Croydon, CR0 4YY
30/09/2026
02225303-0017